Barry Beck
20 Wallace Ave.
Port Colborne, Ont.
834-6573

CO-AZX-656

AUTOMOTIVE SUSPENSION AND STEERING SYSTEMS

Barry Beck
20 Wallace Ave.
Port Colborne, Ont.

HARCOURT BRACE JOVANOVICH, PUBLISHERS

Technology Publications

San Diego   New York   Chicago   Austin   Washington, D.C.

London   Sydney   Tokyo   Toronto

AUTOMOTIVE SUSPENSION AND STEERING SYSTEMS

Thomas W. Birch
Yuba College

Copyright © 1987 by Harcourt Brace Jovanovich, Inc.

All rights reserved. No part of this publications may be
reproduced or transmitted in any form or by any means, electronic or mechanical,
including photocopy, recording, or any information storage and retrieval system,
without permission in writing from the publisher.

Requests for permission to make copies of any part of the work should be mailed to:
Permissions, Harcourt Brace Jovanovich, Publishers, Orlando, Florida 32887.

ISBN 0-15-504345-5

Library of Congress Catalog Card Number: 87-80185

Printed in the United States of America

Cover design: Don Fujimoto
Production supervison and interior design: WordCrafters Editorial Services, Inc.

The information contained herein has been developed from the best available sources.
Technology Publications and the author cannot assume responsibility
for the accuracy of resource data or consequences of its application.
Persons using this book are cautioned to refer to vehicle and
equipment manufacturers' manuals for specific service information.
Also be advised to observe all safety precautions herein and to
refer to manufacturers' procedures for specific vehicle and personnel safety directions.

PREFACE

Automotive Suspension and Steering Systems is for technicians who will be repairing the suspension and steering systems not only of today's complex automobiles but of the even more complex cars of tomorrow.

The evolution of automotive suspension and steering systems has included the development of ball joints, air springs, MacPherson struts, rack and pinion steering, front wheel drive, four wheel alignment, computer-controlled air suspension, computer-controlled shock absorbers, composite spring materials, and steerable rear suspensions. Computer-controlled electronic power steering and hydraulic springs are being developed now.

As automotive systems become more and more complex and precise, mechanics need to become technicians by learning repair methods that are more exacting and require far more knowledge and skill with specialized tools and equipment than did the methods of the past. Repair shops are changing also. New car dealerships play much the same roll they always have: they employ factory-trained technicians who have the specialized knowledge, tools, and equipment to repair the most complicated features of each new model. But many of the service stations and independent garages of the 1950s are gone, and in their place are shops that specialize in particular repair areas, such as air conditioning or tune-up, or in particular makes of domestic or imported cars. The repair shop of yesterday, with a mechanic who could repair any car that was driven through the door, is gone. The evolution of the automobile has made that shop and mechanic extinct.

Automotive Suspension and Steering Systems presents enough information about the systems of present-day cars to help in understanding, diagnosing, and repairing them. Some people will place much of this information in the "nice-to-know-but-not-really-necessary-for-fixing-cars" category; however, technicians should realize that a complete understanding of the why and how of a car's operation aids greatly in understanding why the car does not operate properly and how to go about repairing it. Many of our "new developments" are based on physical principles that have been understood for some time. Except for computer controls, there is not much that is really new in automotive systems.

This book describes the various styles of suspension and steering systems and their components. All The National Institute for AUTOMOTIVE SERVICE EXCELLENCE (ASE) suspension and steering tasks are explained, and review questions are presented in the style of ASE certification tests. The most common repair methods are described in a general manner.

Specialization is a definite advantage to technicians and repair shops. Rarely can any one person "know it all" about cars any more; there is simply too much to know. It is also financially prohibitive for one shop to own all the special tools and equipment needed to make every possible car repair if the equipment is not used regularly. A general repair shop, such as a dealership, might pay rent of several thousand dollars a month for a desirable location and might have a tool and equipment inventory worth over a hundred thousand dollars. That kind of shop must do a large volume of work to stay in business. It must have technicians who can use their specialized skills and knowledge to perform complex tasks and ensure that customers' cars are operating safely and efficiently.

The most thorough and reliable sources of information about a particular car are its manufacturer's service manual and training centers. However, these cover only one make, and sometimes only one model. Manuals are often followed up with specific service bulletins when a need for more information or a change in a new car part or system makes them necessary. Broader coverage is available in technicians' service manuals issued by such publishers as Chilton, Mitchell, and Motor or by aftermarket manufacturers, like Ammco, CR Industries, Dana Corp., Federal-Mogul, Hunter, Moog, TRW, or SKF. These manuals have varying amounts of information for particular cars, some of it very thorough and specific to particular cars or systems, some of it very general. *Automotive Suspension and Steering Systems* supplements the manuals by presenting the theoretical and practical knowledge needed for a complete understanding of the operation of these systems.

Technicians in different parts of the United States use different terms for the same item or task. Terms commonly used on the west coast are slightly different from those used in the east. To the greatest extent possible, the terms used in this book are those used by automobile and aftermarket manufacturers.

In describing repair procedures, the author has assumed that readers have a basic working knowledge of hand tools, fasteners, and general automotive repair procedures and of the safety precautions that should be exercised during general service and repair operations. Space does not permit including them in a book of this nature, which concentrates on a few specialized systems. Wise technicians know that improper use of a wrench or other hand tool, or of a bolt or nut, often leads to a more difficult job, and that a violation of a common-sense safety rule can cause injury to the technician, the vehicle operator, or an innocent bystander.

ACKNOWLEDGMENTS

The author is sincerely grateful to the following individuals, companies, and businesses for their assistance in the preparation of this book:

AC-Delco, General Motors Corp., Detroit, MI

Ahcon Industries, So. El Monte, CA

Alston Engineering, Sacramento, CA

American Honda Motor Company, Gardena, CA

Amermac Inc., Ellaville, GA

Ammco Tools, North Chicago, IL

Armstrong Tire Comp., New Haven, CT

Arn-Wood Co., Englewood, CO

Bear Automotive, Milwaukee, WI

Dave Becker, Marysville, CA

Bee Line Co., Bettendorf, IA

B.F. Goodrich, Akron, OH

Bilstein Corporation of America, San Diego, CA

Kerry Birch, Yuba City, CA

Brannick Industries, Fargo, ND

Bridgestone Tire Company of America, Torrance, CA

Buick Motor Div., General Motors Corp., Flint, MI

Carrera Shocks, Atlanta, GA

Chevrolet Motor Div., General Motors Corp., Warren, MI

Chrysler Corporation, Detroit, MI

CR Industries, Elgin, IL

Dorman Products, Cincinnati, OH

Everco Industries, Lincolnwood, IL

Federal Mogul Corporation, Detroit, MI

FMC Corporation, Conway, AR

Ford Motor Company, Dearborn, MI

Four-Way Shocks, San Diego, CA

Ken Gaal, Yuba City, CA

General Tire Div., GenCorp., Akron, OH

General Motors Product Service Training, Detroit, MI

General Motors Service Information, Detroit, MI

Goodyear Tire & Rubber Comp., Akron, OH

Grand Performance Tools, Toledo, OH

Guldstrand Engineering, Culver City, CA

Hickok Electrical Instrument Co., Cleveland, OH

Hunter Engineering, Bridgeton, MO

Kent-Moore Automotive Div., Sealed Power Corp., Warren, MI

Koni America, Culpeper, VA

Lisle Corporation, Clarinda, IA

Mazda Motors of America, Rancho Dominguez, CA

Mitchell Information Services, San Diego, CA

Moog Automotive, St. Louis, MO

Garland Moorehead, Colusa, CA

Neapco Inc., Pottstown, PA

John Nissen, Colusa, CA

Oldsmobile Div., General Motors Corp., Lansing, MI

OTC Div., Sealed Power Corp., Owatana, MN

Patch Rubber Co., Roanoke Rapids, NC

Perfect Circle Div., Dana Corp., Toledo, OH

Pontiac Motor Div., General Motors Corp., Pontiac, MI

Precision Switching, Spring Grove, IL

Quickor Engineering, Beaverton, OR

Radial Tire Company, West Sacramento, CA

Rubber Manufacturers Assoc., Washington, DC

Saginaw Steering Div., General Motors Corp., Saginaw, MI

Schrader Automotive Group, Nashville, TN

SKF, Automotive Products Div., St. Louis, MO

Snap-On Tools, Kenosha, WI

Specialty Products, Longmont, CO

Speedway Motors, Lincoln, NB

Sun Electric Corp., Crystal Lake, IL

Superior Pneumatic, Cleveland, OH

The Timken Co., Canton, OH

Tire Guides, Boca Raton, FL

TRW, Replacement Parts Div., Independence, OH

Uniroyal Tire Co., Troy, MI

Vette Products, St. Petersburg, FL

Volkswagen of America, Troy, MI

Western Wheel Corp., La Mirada, CA

CONTENTS

Chapter 1
INTRODUCTION TO THE CHASSIS   2
1.1   Chassis Components—Their General Purpose, 4
1.2   Suspension and Steering Systems, 4
1.3   Chassis Components, 6
1.4   Nonindependent Suspensions, 6
1.5   Independent Suspensions, 7
1.6   Swing Axle Suspension, 7
1.7   Short-Long Arm Suspension/S-L A, 9
1.8   MacPherson Strut Suspension/Strut Suspension, 10
1.9   Trailing Arm Suspension, 12
1.10 Control Arms, 12
1.11 Suspension Bushings, 13
1.12 Ball Joints, 15
1.13 King Pins, 16
1.14 Springs and Shock Absorbers, 16
1.15 Spring Types, 17
1.16 Steering System, 19
1.17 Steering Gears, 19
1.18 Steering Linkage, 21
Review Questions, 24

Chapter 2
TIRES—THEORY   26
2.1   Introduction, 28
2.2   Tire Construction, 29
    2.2.1 Ply Materials, 30
    2.2.2 Ply Arrangement, 31
    2.2.3 Belts, 32
    2.2.4 Tread, 33
2.3   Tire Sidewall Information, 34
2.4   Tire Sizing, 35
2.5   Tire Load Ratings, 37
2.6   Uniform Tire Quality Grade Labeling, 40
2.7   Speed Ratings, 41
2.8   Tubes and Tubeless Tires, 43
2.9   Replacement Tire Selection, 44

2.10 Spare Tires, 46
2.11 No-Flat, Run-Flat Tires, 46
2.12 Retreads, 47
Review Questions, 49

Chapter 3   WHEELS—THEORY 52
3.1 Introduction, 54
3.2 Construction, 54
3.3 Wheel Sizes, 56
3.4 Offset-Backspacing, 56
3.5 Aftermarket Wheels, 58
3.6 Wheel Attachment, 62
Review Questions, 65

Chapter 4   TIRE AND WHEEL SERVICE 66
4.1 Introduction to Tire Service, 68
4.2 Removing and Replacing a Tire and Wheel, 68
4.3 Lug Bolt and Stud Replacement, 71
4.4 Tire Wear Inspection, 74
4.5 Tire Rotation, 79
4.6 Removing and Replacing a Tire on a Wheel, 80
4.7 Removing and Replacing a Tubeless Tire Valve, 84
4.8 Repairing a Tire Leak, 85
4.9 Tire-Related Driving Problems, 87
4.10 Tire Pull, 88
4.11 Radial Tire Waddle, 89
4.12 Vibrations, 89
4.13 Tire and Wheel Runout, 89
    4.13.1 Radial Tire and Wheel Runout, 90
    4.13.2 Lateral Tire and Wheel Runout, 93
4.14 Tire Truing, 94
4.15 Tire Balance, 94
4.16 Wheel Weights, 96
4.17 Balancing Operations, 97
    4.17.1 Off-Car and On-Car Balancing, 98
    4.17.2 Static Balancing, Bubble Balancer, 100
    4.17.3 Static and Kinetic Balancing, Mechanical Balancer, 102
    4.17.4 Kinetic and Dynamic Balancing, Strobe Light Balancer, 104
    4.17.5 Kinetic and Dynamic Balancing, Computer Balancer, 107
Review Questions, 111

Chapter 5   WHEEL BEARINGS—THEORY 114
5.1 Introduction, 116
5.2 Bearings and Bearing Parts, 116
5.3 Bearing End Play—Preload, 119
5.4 Seals, 119

5.5   Nondrive Axle, Serviceable Bearings, 120
5.6   Nondrive Axle, Nonserviceable Wheel Bearings, 121
5.7   Drive Axle Bearing, Solid Axle, 121
5.8   Drive Axle Wheel Bearings, Independent Suspension, 124
5.9   Nonserviceable Drive Axle Wheel Bearings, 125
Review Questions, 126

Chapter 6

WHEEL BEARING SERVICE   128
6.1   Wheel Bearing Maintenance and Lubrication, 130
6.2   Wheel Bearing Problems, Diagnosis Procedure, 130
6.3   Fastener Security, 132
6.4   Repacking Serviceable Wheel Bearings, 134
        6.4.1   Disassembling Wheel Bearings, 135
        6.4.2   Cleaning and Inspecting Wheel Bearings, 137
        6.4.3   Packing and Adjusting Wheel Bearings, 141
6.5   Repairing Nonserviceable Wheel Bearings, 144
6.6   Repairing Drive Axle Bearings, Solid Axles, 145
        6.6.1   Removing a Bearing-Retained Axle, 145
        6.6.2   Removing and Replacing a Bearing on an Axle, 146
        6.6.3   Installing a Bearing-Retained Axle, 148
        6.6.4   Removing a "C" Lock Axle, 148
        6.6.6   Installing a "C" Lock Axle, 151
6.7   Repairing Serviceable FWD Front Wheel Bearings, 152
6.8   Repairing Nonserviceable FWD Front Wheel Bearings, 154
Review Questions, 156

Chapter 7

FRONT WHEEL DRIVE SHAFTS—THEORY AND SERVICE   158
7.1   Introduction, 160
7.2   Drive Shaft Construction, 160
7.3   Checking CV Joints, 164
7.4   Split CV Joint Boots, 165
7.5   Drive Shaft Removal, 165
7.6   Drive Shaft Disassembly, 168
7.7   Servicing CV Joints, 169
        7.7.1   Disassembling an Outer, Fixed-Type CV Joint, 171
        7.7.2   Assembling an Outer, Fixed-Type CV Joint, 172
        7.7.3   Disassembling an Inner CV Joint, 173
        7.7.4   Assembling an Inner CV Joint, 174
7.8   Installing CV Joint Boots, 174
7.9   Installing an FWD Drive Shaft, 176
Review Questions, 180

Chapter 8

FRONT SUSPENSION TYPES   182
8.1   Introduction, 184
8.2   Short-Long Arm, S-L A Suspensions, 184
        8.2.1   Control Arm Geometry, 186
        8.2.2   S-L A Springs and Ball Joints, 187
        8.2.3   S-L A Wear Factors, 188

8.3   Strut Suspensions, 189
    8.3.1   Modified Struts, 191
    8.3.2   Advantages of Struts, 193
    8.3.3   Strut Wear Factors, 193
8.4   Solid Axles, 193
8.5   Swing Axle, Twin I-Beam Axles, 195
8.6   Miscellaneous Suspension Types, 197
8.7   Front-Wheel Drive Axles, 197
Review Questions, 199

Chapter 9 **REAR SUSPENSION TYPES   202**
9.1   Introduction, 204
9.2   Solid Axle, RWD Suspension, 204
    9.2.1   Solid Axle, Leaf Spring Suspension, 206
    9.2.2   Solid Axle, Coil Spring Suspension, 206
9.3   Independent Rear Suspension, IRS, 208
    9.3.1   IRS, Semi-Trailing Arm Suspension, 208
    9.3.2   IRS, Trailing Arm Suspension, 209
    9.3.3   IRS, Strut Suspension, 209
9.4   Miscellaneous RWD Axle and Suspension Types, 210
9.5   FWD Rear Axles, 211
    9.5.1   FWD, Rear Solid Axle Suspension, 211
    9.5.2   FWD, Rear Trailing Arm Suspension, 211
    9.5.3   FWD, Rear Strut Suspension, 213
    9.5.4   FWD, Rear Short-Long Arm Suspension, 213
9.6   Rear-Wheel Steering, 213
Review Questions, 217

Chapter 10 **SPRINGS   220**
10.1   Introduction, 222
10.2   Sprung and Unsprung Weight, 222
10.3   Spring Rate and Frequency, 223
10.4   Wheel Rate and Frequency, 225
10.5   Spring Materials, 227
10.6   Leaf Springs, 229
10.7   Coil Springs, 232
10.8   Torsion Bars, 234
10.9   Electronically Controlled Air Suspension, 237
10.10   Aftermarket Air Suspensions, 239
10.11   Overload Springs, 240
10.12   Stabilizer, Antiroll Bar, 240
Review Questions, 244

Chapter 11 **SHOCK ABSORBERS   246**
11.1   Introduction, 248
11.2   Shock Absorber Operating Principles, 248
11.3   Shock Absorber Damping Ratios, 250
11.4   Shock Absorber Damping Force, 251

11.5   Double-Tube Shock Absorbers, 252
     11.5.1   Construction, 253
     11.5.2   Operation, 254
11.6   Single-Tube Shock Absorber Operation, 258
     11.6.1   Construction, 258
     11.6.2   Operation, 259
11.7   Strut Shock Absorbers, 259
11.8   Computer-Controlled Shock Absorbers, 260
11.9   Load-Carrying Shock Absorbers, 263
11.10   Shock Absorber Quality, 264
11.11   Shock Absorber Failure, 265
Review Questions, 266

Chapter 12

SPRING AND SHOCK ABSORBER SERVICE   208
12.1   Introduction, 270
12.2   Spring and Shock Absorber Inspection, 270
     12.2.1   Bounce Test, 271
     12.2.2   Suspension Ride Height Check, 271
     12.2.3   Visual Inspection, 274
12.3   Removing and Replacing Shock Absorbers, 275
12.4   Removing and Replacing Coil Springs, 279
     12.4.1   Removing and Replacing a Coil Spring, S-L A
             Suspension, 279
12.5   Leaf Spring Service, 285
12.6   Torsion Bar Service, 286
12.7   Strut Service, 287
     12.7.1   Special Strut Service Points, 291
     12.7.2   Strut Removal and Replacement, 292
     12.7.3   Strut Spring Removal and Replacement, 294
     12.7.4   Strut Cartridge Installation, 297
12.8   Stabilizer Bar Service, 298
12.9   Completion, 301
Review Questions, 301

Chapter 13

STEERING SYSTEMS   304
13.1   Introduction, 306
13.2   Steering Columns, 307
13.3   Standard Steering Gears, 310
13.4   Rack and Pinion Steering Gears, 314
13.5   Power Steering, 314
     13.5.1   Power Steering Pumps, 317
     13.5.2   Steering Gear Control Valve, 320
     13.5.3   Rack and Pinion Power Steering Gears, 323
     13.5.4   Standard Power Steering Gears (Integral), 324
     13.5.5   Linkage Booster Power Steering, 324
     13.5.6   Power Steering Hoses and Fluid, 326
13.6   Steering Linkage, 329
     13.6.1   Outer Tie Rod Ends, 330
     13.6.2   Tie Rods, Rack and Pinion Steering Gear, 330

13.6.3   Tie Rods, Standard Steering Gears, 332

13.6.4   Idler Arm and Centerlink, 332

13.7   Drag Links, 334

Review Questions, 335

Chapter 14   **SUSPENSION AND STEERING SYSTEMS INSPECTION   338**

14.1   Introduction, 340

14.2   Inspection Points, 340

14.3   Ball Joint Checks, 342

14.3.1   Checking a Wear Indicator Ball Joint for Excessive Clearance, 345

14.3.2   Checking a Load-Carrying Ball Joint on a Lower Control Arm for Excessive Clearance, 347

14.3.3   Checking a Load-Carrying Ball Joint on an Upper Control Arm, 350

14.3.4   Checking a Friction-Loaded/Follower Ball Joint, 351

14.3.5   Checking Ball Joint Clearance on an I-Beam or 4WD Solid Axle, 352

14.3.6   Checking King Pin Clearance, 354

14.4   Control Arm Bushing Checks, 355

14.5   Strut Rod Bushing Checks, 356

14.6   Strut Checks, 356

14.7   Strut Damper/Insulator Checks, 357

14.8   Steering Linkage Checks, 358

14.8.1   Tie Rod End Checks, 359

14.8.2   Parallelogram Steering Linkage and Idler Arm Checks, 360

14.8.3   Rack and Pinion Steering Linkage Checks, 361

14.9   Steering Gear Checks, 363

14.10   Power Steering Checks, 364

Review Questions, 368

Chapter 15   **SUSPENSION COMPONENT SERVICE   370**

15.1   Introduction, 372

15.2   Taper Breaking, 373

15.2.1   Breaking a Taper Using Shock, 374

15.2.2   Breaking a Taper Using a Separator Tool, 375

15.2.3   Breaking a Taper Using Pressure, 375

15.2.4   Removing and Replacing Ball Joint Studs Locked by Pinch-Bolts, 377

15.3   Ball Joint Replacement, 377

15.4   Removing a Control Arm, 378

15.5   Removing and Replacing a Ball Joint, 380

15.5.1   Removing and Replacing a Pressed-In Ball Joint, 381

15.5.2   Removing and Replacing a Riveted/Bolted-In Ball Joint, 383

15.5.3   Removing and Replacing a Threaded Ball Joint, 384

15.6   Control Arm Bushing Replacement, 385
    15.6.1   Removing Rubber Bushings Using an Air Hammer and Chisel, 388
    15.6.2   Removing Rubber Bushings Using a Puller Tool, 388
    15.6.3   Installing a Rubber Bushing Using a Driver, 388
    15.6.4   Installing a Rubber Bushing Using a Pressing Tool, 389
    15.6.5   Removing and Replacing Rubber Bushings Without Outer Sleeves, 391
    15.6.6   Removing and Replacing Metal Bushings, 392
15.7   Installing a Control Arm, 392
15.8   Removing and Replacing Strut Rod Bushings, 394
15.9   Removing and Replacing King Pins, 395
15.10   Rear Suspension Service, 398
15.11   Completion, 399
Review Questions, 399

Chapter 16

STEERING SYSTEM SERVICE   402
16.1   Introduction, 404
16.2   Steering Linkage Replacement, 404
    16.2.1   Removing and Replacing an Outer Tie Rod End, 405
    16.2.2   Removing and Replacing an Inner Tie Rod End, Standard Steering Gear, 407
    16.2.3   Removing and Replacing an Inner Tie Rod End, Rack and Pinion Steering, 407
    16.2.4   Removing and Replacing Inner Tie Rod End Pivot Bushings, Rack and Pinion Steering Gear with Center-Mounted Tie Rods, 414
    16.2.5   Removing and Replacing an Idler Arm, 415
    16.2.6   Removing and Replacing a Centerlink, 418
    16.2.7   Removing and Replacing a Pitman Arm, 418
    16.2.8   Removing and Replacing an Adjustable Drag Link Socket, 419
16.3   Steering Gear Service, 420
    16.3.1   Steering Gear Removal and Replacement, Standard Steering Gear, 421
    16.3.2   Steering Gear Removal and Replacement, Rack and Pinion Steering Gear, 422
    16.3.3   Steering Gear Adjustments, Standard Steering Gear, 424
    16.3.4   Steering Gear Overhaul, Standard Steering Gear, 427
    16.3.5   Steering Gear Adjustments, Rack and Pinion Steering Gear, 430
    16.3.6   Steering Gear Overhaul, Rack and Pinion Steering Gear, 431
16.4   Removing and Replacing a Steering Wheel, 433
16.5   Power Steering Service Operations, 434

16.5.1   Power Steering System Checking/Problem
          Diagnosis, 434
16.5.2   Power Steering Fluid Change and Air Bleeding, 437
16.5.3   Pressure Testing a Power Steering System, 440
16.5.4   Power Steering Pump Service, 441
16.5.5   Power Steering Gear Service, 446
16.6   Completion, 449
Review Questions, 450

Chapter 17   WHEEL ALIGNMENT—FRONT AND REAR   452
17.1   Requirements, 454
17.2   Measuring Angles, 455
17.3   Camber, 455
17.4   Camber and Scrub Radius, 456
17.5   Camber-Caused Tire Wear, 457
17.6   Camber Spread and Road Crown, 458
17.7   Camber Change, 459
17.8   Toe In, 459
17.9   Things That Affect Toe, 460
17.10   Toe-Caused Tire Wear, 461
17.11   Toe Change, 461
17.12   Caster, 462
17.13   Caster Effects, 464
17.14   Things That Affect Caster, 465
17.15   Caster and Road Crown, 466
17.16   Effects of Too Little or Too Much Caster, 466
17.17   Steering Axis Inclination (SAI), 466
17.18   Included Angle, 467
17.19   Toe Out on Turns, 469
17.20   Toe Out on Turns—Problems, 471
17.21   Set Back, 471
17.22   Rear Wheel Alignment, 471
17.23   Track/Thrust Line, 472
Review Questions, 474

Chapter 18   WHEEL ALIGNMENT—MEASURING AND ADJUSTING   476
18.1   Introduction, 478
18.2   Measuring Alignment Angles, 478
18.3   Wheel Alignment Sequence, 483
18.4   Measuring Camber, 483
18.4.1   Measuring Camber Using a Magnetic Gauge, 484
18.4.2   Measuring Camber Using an Alignment System, 486
18.5   Camber Specifications, 488
18.6   Adjusting Camber, 489
18.6.1   Adjusting Camber, Shims, 491
18.6.2   Adjusting Camber, Eccentric Cams, 494
18.6.3   Adjusting Camber, Sliding Adjustment, 495
18.6.4   Adjusting Camber, Miscellaneous Styles, 496

18.7   Measuring Caster, 501
  18.7.1   Measuring Caster Using a Magnetic Gauge, 502
  18.7.2   Measuring Caster Using an Alignment System, 504
18.8   Adjusting Caster, 505
  18.8.1   Adjusting Caster, Shims, 506
  18.8.2   Adjusting Caster, Eccentric Cams, 506
  18.8.3   Adjusting Caster, Sliding Adjustment, 507
  18.8.4   Adjusting Caster, Adjustable Strut Rod, 508
  18.8.5   Adjusting Caster, Miscellaneous Styles, 508
18.9   Measuring SAI, 510
  18.9.1   Measuring SAI Using a Magnetic Gauge, 510
18.10   Adjusting SAI, 511
18.11   Measuring Toe Out on Turns, 511
18.12   Adjusting Toe Out on Turns, 512
18.13   Measuring Toe, 513
  18.13.1   Measuring Toe Using a Trammel Bar, 514
  18.13.2   Measuring Toe Using a Light Gauge, 515
  18.13.3   Measuring Toe Using an Alignment System, 516
18.14   Toe and Steering Wheel Position, 517
18.15   Toe Specifications, 517
18.16   Adjusting Toe, 519
  18.16.1   Adjusting Toe, Standard Steering, 521
  18.16.2   Adjusting Toe, Rack and Pinion Steering, 524
  18.16.3   Adjusting Toe to Center a Steering Wheel, 524
18.17   Rear Wheel Alignment, 525
18.18   Measuring Rear Wheel Camber, 527
18.19   Measuring Rear Wheel Toe, 527
18.20   Measuring Rear Wheel Track/Thrust, 527
18.21   Adjusting Rear Wheel Camber and Toe, 528
18.22   Frame and Body Alignment, 530
18.23   Road Testing and Trouble Shooting, 531
Review Questions, 532

Chapter 19   SUSPENSION DYNAMICS   534
19.1   Introduction, 536
19.2   Cornering Force, 536
19.3   Traction/Slip Angle, 538
19.4   Oversteer/Understeer, 538
19.5   Traction Force/Circle of Traction, 539
19.6   Weight Transfer, 542
19.7   Roll Steer, 544
19.8   Roll Axis/Roll Centers, 545
19.9   Unsprung Weight, 548
19.10   Roll Resistance, 548
19.11   Wedge and Stagger, 549
19.12   Center of Gravity Location, 551
19.13   Antidive Suspension Geometry, 553
19.14   Antisquat, 553
19.15   Polar Moment of Inertia, 555

19.16   Aerodynamics, 555
19.17   Suspension Tuning/Modifications, 556
    19.17.1   Reducing Suspension Bushing Compliance/
            Deflection, 558
    19.17.2   Changing Roll Stiffness/Roll Couple Distribution,
            559
    19.17.3   Lowering the CG Height, 559
    19.17.4   Changing the Spring/Shock Absorber Rate, 561
    19.17.5   Changing Alignment Angles, 563
    19.17.6   Reducing Unsprung Weight, 564
19.18   Conclusion, 566

GLOSSARY   567

Appendix 1   ENGLISH METRIC CONVERSION   573

Appendix 2   DISTANCE AND ANGULAR EQUIVALENTS   575

Appendix 3   BOLT TORQUE TIGHTENING CHART   577

REVIEW QUESTION ANSWER KEY   579

INDEX   581

AUTOMOTIVE SUSPENSION AND STEERING SYSTEMS

Chapter 1

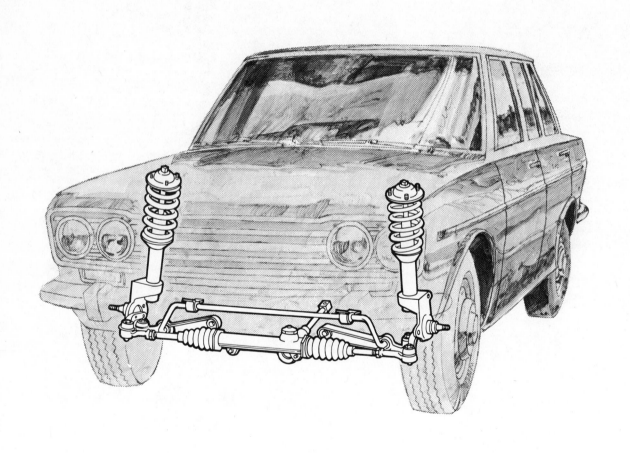

INTRODUCTION TO THE CHASSIS

After completing this chapter, you should:

- Be able to identify the major components of the suspension and steering systems.

- Have an understanding of the general purpose of these components, how they relate to each other, and the roll they play in the operation of the car.

1.1 CHASSIS COMPONENTS—THEIR GENERAL PURPOSE

We all realize that a car can move because the tires and wheels can roll along a road. Tires and wheels make a vehicle's motion possible. The car's **chassis** connects the axles of these tires and wheels to the body of the car, where the driver and passengers sit. The chassis consists of the frame or body, suspension members (springs and shock absorbers), steering gear and linkage, axles or spindles or both, and the tires and wheels. (Fig. 1–1)

If these units work well together, we might have a car that approaches the ideal. The ideal car is one that will travel smoothly on the road so the driver and passengers feel no bumps or vibrations. It will also travel in a straight line until the driver wishes to turn and will then respond quickly and easily to the turning motions of the steering wheel. The tires will also roll down the road with a minimum amount of drag, which will result in maximum tire life and fuel mileage.

1.2 SUSPENSION AND STEERING SYSTEMS

The **suspension system** includes the springs, shock absorbers, control arms, and spindle or axle. These parts should hold the tire and wheel in correct alignment with the car and the road. They also allow the tires and wheels to move up and down relative to the body over bumps and chuckholes. The tires can thus follow the road surface and maintain traction without transmitting the roughness of the road to the driver and passengers. (Fig. 1–2) Any of you who have ridden in a vehicle with a solid suspension (e.g., a farm tractor, bicycle, or forklift) have found out the value of a suspension system when you went over the first bad bump. The solid suspension probably transferred most of the bump from the road to you.

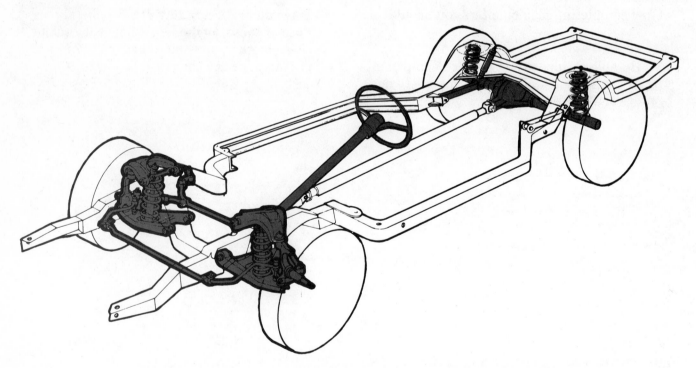

FIGURE 1–1.
A car's chassis includes the suspension and steering members, which are those parts that allow the car to roll smoothly and steer. (Courtesy of Moog Automotive)

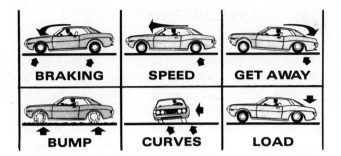

FIGURE 1-2.
The car's suspension system allows the tires to maintain contact with the road when the car is operated under various conditions. (Courtesy of Hunter Engineering)

When discussing a car's chassis, the side-to-side distance between the center lines of the tires on an axle is called the **track**. The distance between the centers of the front tires and the centers of the rear tires is called the **wheelbase**. (Fig. 1-3)

To be in correct alignment, a tire should roll on a path that is parallel to the center line of the vehicle. This is called zero/0 toe in or toe out. The tire should also be straight up and down or at a right angle (90°) to the road surface. This is called zero camber. If one or more of the tires is not positioned this way, it is out of alignment and will have to scuff or scrub sideways as it rolls down the road. This will

cause tire wear and a loss in fuel economy since a scuffing tire does not roll as freely as one that is aligned correctly.

The steering system allows the driver to control the direction the car travels. Turning the steering wheel will cause the front wheels to point in the direction the driver wants to go. The steering system consists of the steering wheel, steering gear, and tie-rods. (Figs. 1-4 & 1-5)

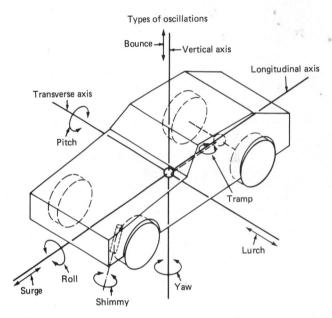

FIGURE 1-4.
The body and suspension members of a car can move in several different directions, as shown here.

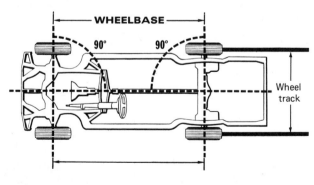

FIGURE 1-3.
The suspension system keeps the car's tires parallel to the center line of the car when the car is going straightahead. The lengthwise distance between the axles is called the wheelbase; the crosswise distance between two tires on an axle is called track. (Courtesy of Hunter Engineering)

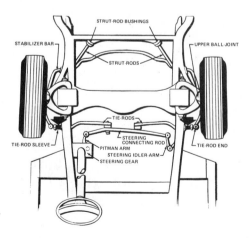

FIGURE 1-5.
The steering system allows the driver to steer the car. (Courtesy of Hunter Engineering)

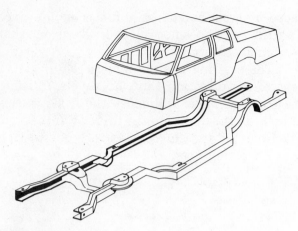

FIGURE 1-6.
On many cars, the body and the frame are two separate units that are attached to each other by bolts passing through rubber mounts.

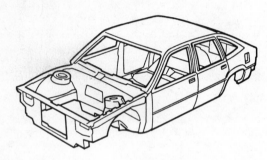

FIGURE 1-7.
A unibody car does not use a separate frame; the body parts are reinforced to provide the needed strength to keep the various systems in the correct relationship. (Courtesy of Chevrolet)

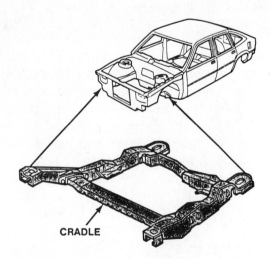

CRADLE

FIGURE 1-8.
Some unibody designs use a subframe called a cradle to support the engine, powertrain, or front suspension. (Courtesy of Chevrolet)

1.3 CHASSIS COMPONENTS

At one time the **frame** was the mounting point onto which the rest of the car was attached. The steering, suspension, engine, drive train, and body were all bolted to the frame. The frame became the skeleton and backbone of the car, and was generally made from strong steel channel or square, tube-shaped members. (Fig. 1-6)

Most new cars are now of a frameless, **unibody** type of construction. The sheet metal body of the car has reinforced sections that are strong enough for the suspension, steering, and drive train components to be connected to them. The reinforced body has made the separate frame unnecessary. (Fig. 1-7) The center and rear portions of a body are quite strong. The roof and floor pan of the body, reinforced by the door frames, make a fairly strong and rigid box. The area in front of the windshield is not nearly as strong because the only permanent body sections there are the inner fender panels. The fenders themselves are usually removable to allow easy replacement. In most cases, a unibody car will have subframe sections bolted or welded into it for the engine and front suspension. This subframe is often called an engine cradle. (Fig. 1-8)

Any severe accident that bends or twists the frame or suspension-carrying portions of a unibody car can cause the suspension mounts to change position relative to each other or the body. This movement of the suspension mounting points will change the alignment of the tires and wheels. This will usually result in increased tire wear, reduced fuel mileage, or a car that pulls to one side or wanders back and forth down the road.

1.4 NONINDEPENDENT SUSPENSIONS

The suspension arms or axles connect the wheels and tires to the frame and allow the tires and wheels to move up and down relative to the frame and body. The different styles of suspensions fall into two general classifications, independent and nonindependent. These suspension types will be briefly described in this

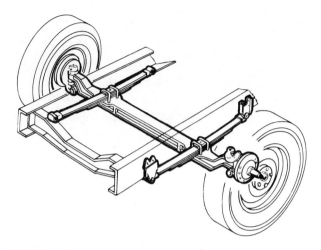

FIGURE 1-9.
A tire is mounted at each end of a solid axle; this is often called a beam axle. (Courtesy of Moog Automotive)

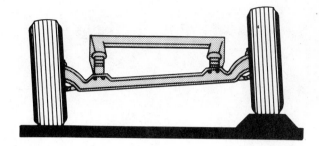

FIGURE 1-10.
When one tire of a solid axle goes over a bump, the position of the other tire is affected. (Courtesy of Hunter Engineering)

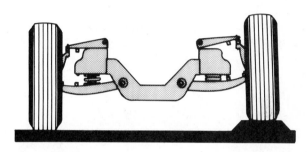

FIGURE 1-11.
An independently suspended tire will not affect any of the other tires when it encounters a bump. (Courtesy of Hunter Engineering)

chapter and again, in much more detail, in Chapters 8 and 9.

In nonindependent suspensions, the front wheels are mounted on the same axle, and the rear wheels are mounted on another, single axle. A **solid axle**, also called a **beam axle**, is used. (Fig. 1-9) In the early days of automobile development, every car had two solid axles, one in front and one in back. Many trucks and pickups still use a solid front axle. Nearly all **rear wheel drive (RWD)**, front-engine cars as well as trucks and pickups use a solid rear axle. Solid axles are strong and relatively inexpensive. They are also quite simple and trouble-free, especially on drive axles. There are several drawbacks with solid axles. First of all, they are heavy, and because of their weight, they will increase the vehicle's unsprung weight (i.e., the vehicle weight that is not on the springs). This increase in unsprung weight will require more spring and shock absorber control to keep the tires in contact with the road. Also, when one tire of a solid axle goes over a bump, the other tire on that axle will have to change position and possibly lose traction. (Fig. 1-10) A car with solid axles will usually have a harsher ride than a car with an independent suspension.

1.5 INDEPENDENT SUSPENSIONS

In independent suspension systems, each one of the front wheels and sometimes the rear

wheels are mounted on separate spindles and control arms. Mounting the wheels separately allows them to travel up or down independent of the other wheel on the same axle. (Fig. 1-11) This usually reduces the unsprung weight, which gives a softer ride with greater control of the tire and wheel position. However, an independent suspension is a more complicated system with more parts and movable bushings. There will be more parts to wear out, possibly leading to misalignment of the tires. All passenger cars, many pickups, and a few trucks have independent front suspensions. A few RWD and some **front wheel drive (FWD)** cars will have independent rear suspensions. (Fig. 1-12)

1.6 SWING AXLE SUSPENSION

The simplest type of independent suspension is the **swing axle**. It is essentially a portion of a

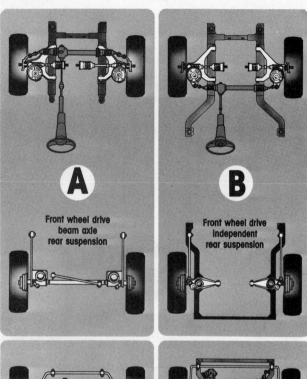

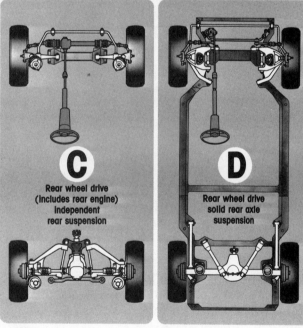

FIGURE 1–12.
The four most common combinations of front and rear suspensions are shown here. (Courtesy of Hunter Engineering)

solid axle that pivots from the frame at the inner end. The inner pivot of a nondriving, swing axle is a pivot bushing. (Fig. 1–13) The inner pivot of a driving, swing axle is usually the universal joint. The axle's outer end, which is attached to the tire and wheel, swings up and down much like a door swings on its hinge. The

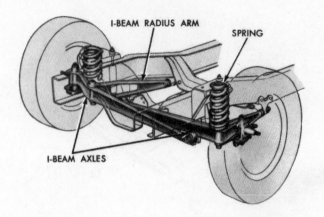

FIGURE 1–13.
A swing axle front suspension, called a twin I-beam suspension by the manufacturer is shown here. Each axle has a pivot bushing at the inner end. (Courtesy of Ford Motor Company)

only difference is that a door is vertical while the axle is horizontal. (Fig. 1–14)   A **radius rod** or **control arm** is usually used to keep the outer end from moving forward or backward. (Fig. 1–15)   Swing axles have a disadvantage in that as the tire moves up and down, it also moves in and out. The outer end travels in an arc centered at the inner pivot. This causes a track (tire-to-tire width on an axle) change and a camber (vertical tire position) change, which in turn cause tire wear or vehicle steering problems. Swing axles have been used at the rear of some

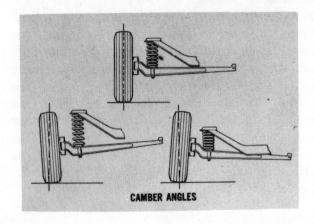

FIGURE 1–14.
As the tire moves up and down in its suspension travel, it pivots or swings from the inner pivot bushing. (Courtesy of Ford Motor Company)

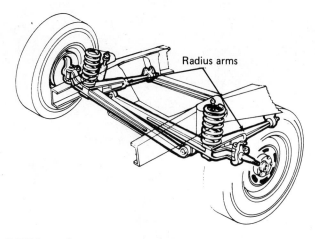

FIGURE 1–15.
A radius arm is used with twin I-beam axles to keep the axle from twisting or moving forward or rearward at the wheel end. (Courtesy of Moog Automotive)

cars and are currently being used on the front end of one make of pickups, **four wheel drives (4WD)**, and trucks.

1.7   SHORT-LONG ARM SUSPENSION/S-L A

Most RWD, domestic, American-made cars use a front suspension of the **short-long arm (S-L A)** type that is also called an **unequal arm** suspension. (Fig. 1–16)   S-L A suspensions use two control arms of unequal length, a short upper arm and a longer lower arm. The inner ends of each control arm are attached to the frame with bushings that allow the control arm to pivot. The outer ends of each control arm are attached to the steering knuckle or spindle support with either bushings or ball joints. These pivots allow steering to take place (on front suspensions) as well as vertical suspension motion.

The primary purpose of the lower control arm is to control the track width of the tires. Because of this, the lower control arm should be horizontal or nearly so. If it were mounted at an angle, vertical wheel travel would cause a severe track change. The upper control arm has the primary purpose of controlling the camber angle of the tire. This arm is seldom horizontal; it will be angled either downward or upward at the outer end. This either/or nature of the mounting angle will be discussed in Chapter 19 when roll centers are covered. (Fig. 1–17)

Using control arms of two different lengths that are mounted in a nonparallel position will cause the tire to go through a camber change as it travels up and down. This does not sound good, but the alternative is a track change, which is definitely bad. If the control arms are of equal lengths and parallel, the tire will travel in a sideways arc during vertical

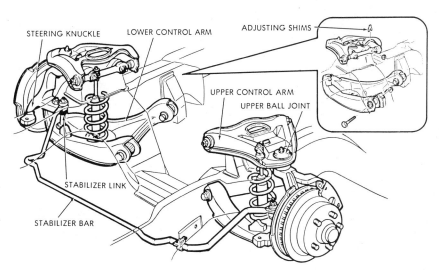

FIGURE 1–16.
An unequal-length arm or short-long arm suspension showing the relationship of the various parts. (Courtesy of Chevrolet)

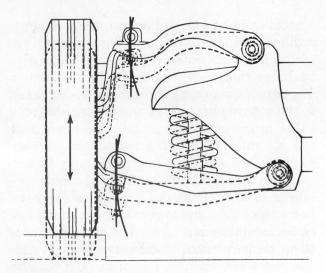

FIGURE 1-17.
During suspension travel, the outer ends of the control arms pivot on the inner bushings; there will be almost no side scrub of the tire because of the S-L A design and the horizontal position of the control arms. (Courtesy of Dana Corporation)

travel. This will cause a sideways motion and track change of the tire while the tire is going over a bump, which is called **scrub.** Scrubbing the tire sideways under this load will definitely cause tire drag and wear. (Fig. 1-18) S-L A suspensions are designed so the track changes as little as possible while the tire goes over bumps. The only way to keep the road contact track width the same while the steering knuckle moves up and down in an arc is to change the camber. The tread, at the road surface, will have negligible side scrubbing, and the camber change will not cause much tire wear during this short period of time. Improper wheel alignment can also cause tire scrub. (Fig. 1-19)

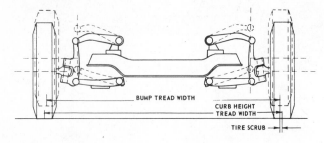

FIGURE 1-18.
If equal-length control arms are used, a track change and tire scrub would occur, which would cause tire wear. (Courtesy of Ammco)

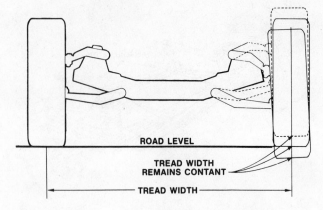

FIGURE 1-19.
The geometry of unequal length, S-L A control arms keeps tread width constant during bump travel. (Courtesy of Moog Automotive)

1.8 MACPHERSON STRUT SUSPENSION/ STRUT SUSPENSION

Another major independent suspension design is called the **MacPherson strut** or merely **strut suspension.** (Fig. 1-20) This suspension was named for the designer. Mr. MacPherson was an engineer for Ford of Britain during the introduction of strut suspensions on the 1951 Ford Consul and Zephyr. Strut suspensions are currently used on the front end of most FWD cars; on many smaller, front engine, RWD cars; and on the rear end of some RWD and FWD cars. It is sometimes called a Chapman strut when used on rear suspensions. (Fig. 1-21)

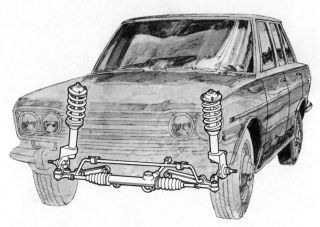

FIGURE 1-20.
A car with a MacPherson strut front suspension. (Courtesy of Moog Automotive)

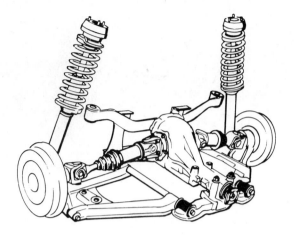

FIGURE 1-21.
A MacPherson strut rear suspension, also called a Chapman strut. (Courtesy of Moog Automotive)

The strut consists of a spindle and steering knuckle that are built onto an oversized, telescoping shock absorber. The spring is usually mounted around this shock absorber. The upper end of the strut is attached to the car using a bearing assembly that allows the steering knuckle to pivot for turns. This bearing is built into a rubber mount assembly that allows a slight angle change of the strut during suspension travel as well as a dampening of road vibrations. The lower end of the strut is attached to the car with a lower control arm, much like an S-L A suspension. Like the S-L A suspension, the lower arm controls the track width. Some cars use a modified strut suspension with the spring mounted between the lower control arm and the frame, much like most S-L A suspensions. (Figs. 1-22 & 1-23)

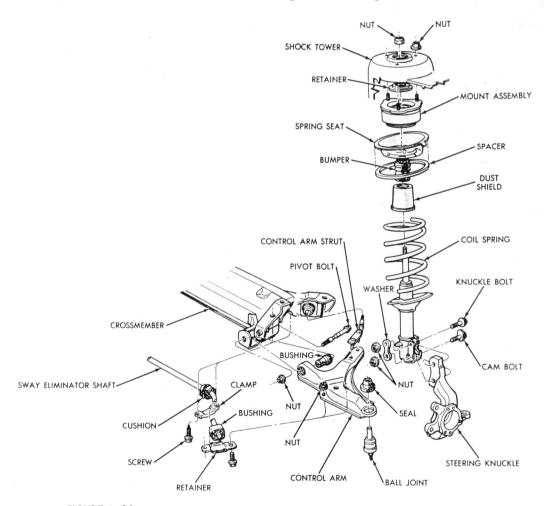

FIGURE 1-22.
An exploded strut assembly showing the various parts. This unit is for a FWD car.
(Courtesy of Chrysler Corporation)

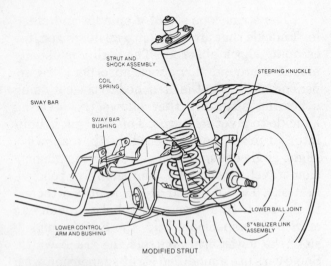

FIGURE 1-23.
A modified strut. The spring is mounted on the lower control arm, much like an S-L A suspension, but a strut is used in place of the upper control arm. (Courtesy of SKF Automotive Products)

A strut suspension with its fewer parts is simpler and usually lighter than an S-L A suspension. Because the strut takes the place of the upper control arm, more room is available in the engine compartment. This is a major reason for the popularity of strut suspensions in **transverse** (across the car) engine, FWD cars.

1.9   TRAILING ARM SUSPENSION

Another style of independent suspension involves mounting the spindle and wheel assembly onto one or a pair of **trailing arms**. A trailing arm pivots from the frame at the front (leading) end, and the spindle is at the rear (trailing) end. A leading arm is arranged just the opposite, with the pivot at the rear. This is a relatively simple suspension that has been used for front suspensions on only a few car models. It is more popular as a rear suspension. The arm(s), which pivot on metal or rubber bushings at the front to allow suspension travel, must be strong enough to withstand side loads so the tire will be kept in proper alignment. (Figs. 1-24 & 1-25)

If the trailing arm pivot is at a right angle (90°) to the car center line, there will be no camber change during suspension travel, but ve-

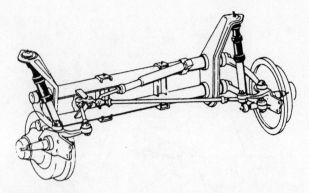

FIGURE 1-24.
A twin trailing arm front suspension and torsion bars, enclosed in the front tubes, are used for springs. (Courtesy of Moog Automotive)

hicle lean will cause an undesirable and equal change in camber angle. A more desirable camber change can be obtained by changing the position of the arm's pivot bushing. When the pivot bushings are mounted at an angle other then 90° the suspension is called a **semi-trailing arm** . (Figs. 1-26 & 1-27)

1.10 CONTROL ARMS

Control arms for S-L A and strut suspensions can use either two inner bushings on a wide, triangular-shaped, A-arm or one inner bushing

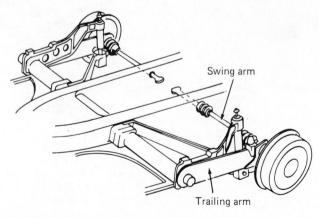

FIGURE 1-25.
A trailing arm rear suspension. This particular car uses swing arm axle shafts. (Courtesy of Moog Automotive)

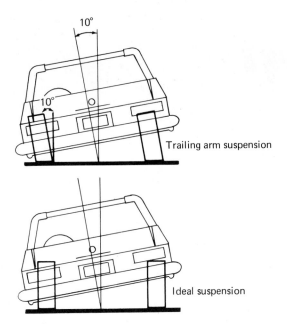

FIGURE 1–26.
If a car with trailing arm suspensions leans 10° outward on a turn, the tires will also lean 10° (top). A better situation is to allow the car to lean but keep the tires vertical.

on a simple, narrow-control arm with a **strut rod.** These arms are usually made from steel stampings. The outer end of a control arm must be able to move up and down but must not move forward or backward. Having two inner pivot points, an A-arm forms a triangle, which is one of the strongest structural shapes. The point of a triangle cannot move sideways un-

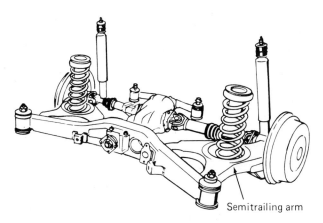

FIGURE 1–27.
This is called a semi-trailing arm rear suspension because the pivot points are at an angle. (Courtesy of Moog Automotive)

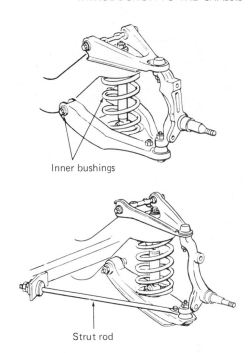

FIGURE 1–28.
The suspension at the top uses a lower A-arm; the control arm has two inner pivots. The control arm at the bottom has a single pivot plus a strut rod. (Courtesy of SKF Automotive Products)

less the sides change length or the base moves. The two inner pivots form the triangle's base on an A-arm, and the inner pivot and the strut rod bushing form the base on a control arm and strut rod triangle. All of these control arm pivots must be designed to allow the necessary movement required for steering and bump control but cannot allow free play or slop, which could cause unwanted, uncontrolled tire and wheel motion. (Fig. 1–28)

1.11 SUSPENSION BUSHINGS

Most modern cars will use a rubber, **torsilastic** bushing. (Fig. 1–29) Rubber is a favored bushing material because it requires no lubrication, will not transfer minor road vibrations, is usually silent, and offers a relatively large degree of compliance (i.e., allows for changes in position). Bushing compliance permits control arm motion without binding. Torsilastic refers to the natural, elastic nature of rubber to allow move-

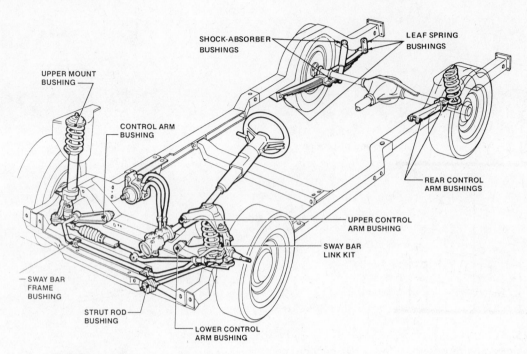

SHOCK-ABSORBER BUSHINGS

LEAF SPRING BUSHINGS

UPPER MOUNT BUSHING

CONTROL ARM BUSHING

REAR CONTROL ARM BUSHINGS

UPPER CONTROL ARM BUSHING

SWAY BAR LINK KIT

SWAY BAR FRAME BUSHING

STRUT ROD BUSHING

LOWER CONTROL ARM BUSHING

FIGURE 1–29.
Rubber bushings are used at many locations in a car's suspension system. (Courtesy of Moog Automotive)

ment of the bushing in a rotating or twisting plane. This bushing is not a bearing in the common sense. Motion is allowed by the twisting of the rubber, not by one part rotating or sliding into another. (Fig. 1–30) The inner sleeve of the bushing is usually locked to the control arm pivot shaft and is bolted solidly to the frame. The outer sleeve is pressed into the control arm, and the rubber must twist to allow the control arm to pivot. This will also cause the control arm to have a memory; that is, the rubber will want to return to the position from which it started. This action will also increase the resistance to movement of the control arms. Some sources state that 10 percent of the resistance to body roll comes from the rubber bushings. (Fig. 1–31)

Rubber bushings are also used to allow a swinging motion necessary at the end of a strut rod, stabilizer bar end link, or shock absorber. The rubber bushings provide the needed compliance and dampen road vibrations. (Figs. 1–32 & 1–33)

Metal bushings are used for some inner control arm pivots. This bushing usually resembles a bolt thread on each end of the pivot shaft and large nuts that are secured into the control arm. These nuts are threaded onto the pivot shaft threads. As the control arm pivots up and down, the nut-like bushings rotate on the threads of the shaft. Metal bushings re-

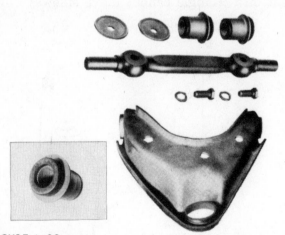

FIGURE 1–30.
Two rubber inner pivot bushings are used with this control arm. They are also called torsilastic bushings because the rubber must twist in a torsional manner. Note the notches at the end of the inner sleeve of the bushing (inset). (Courtesy of SKF Automotive Products)

FIGURE 1–31.
A complete torsilastic bushing is shown at the upper left. The bushing below it has had a section removed to show the inner sleeve; the rubber is squeezed tightly between the two metal sleeves. The bushing to the right has been pressed apart.

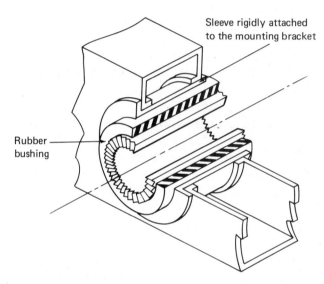

Sleeve rigidly attached to the mounting bracket

Rubber bushing

FIGURE 1–32.
The outer sleeve of a torsilastic bushing is pressed into the control arm, and the inner sleeve is locked to the mounting bracket. The rubber twists to allow control arm motion.

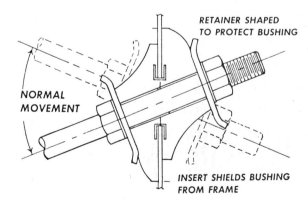

RETAINER SHAPED TO PROTECT BUSHING

NORMAL MOVEMENT

INSERT SHIELDS BUSHING FROM FRAME

FIGURE 1–33.
The rubber strut rod bushing allows enough strut rod motion for suspension travel. (Courtesy of Moog Automotive)

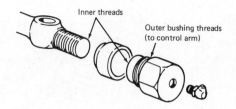

Inner threads

Outer bushing threads (to control arm)

FIGURE 1–34.
A metal bushing locks into the control arm at the outer threads; bushing motion occurs at the inner threads. (Courtesy of Moog Automotive)

quire lubrication, and they also need seals to keep the lube in and dirt and water out. They can become quite noisy if they lose their lubricant and become dry, rusty, or dirty. (Fig. 1–34)

1.12 BALL JOINTS

A ball-and-socket joint is used to provide both a swinging or pivoting motion and a rotating motion. These movements are necessary at the top and bottom of the steering knuckle of an S-L A suspension, the bottom of a strut suspension's steering knuckle, and where portions of the steering linkage connect. (Fig. 1–35) These joints must be loose enough to allow free movement but not so loose that sloppy,

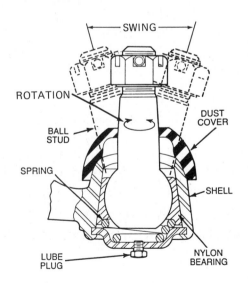

SWING

ROTATION

BALL STUD

SPRING

LUBE PLUG

DUST COVER

SHELL

NYLON BEARING

FIGURE 1–35.
A ball and socket is used for ball joints and tie rod ends. It will allow both rotary and swinging motions of the ball stud.

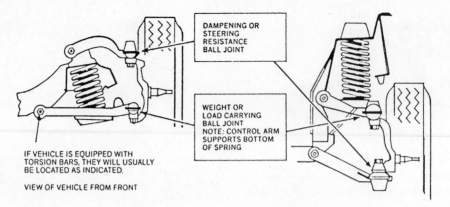

DAMPENING OR
STEERING
RESISTANCE
BALL JOINT

WEIGHT OR
LOAD CARRYING
BALL JOINT
NOTE: CONTROL ARM
SUPPORTS BOTTOM
OF SPRING

IF VEHICLE IS EQUIPPED WITH
TORSION BARS, THEY WILL USUALLY
BE LOCATED AS INDICATED.

VIEW OF VEHICLE FROM FRONT

FIGURE 1–36.
One of the ball joints used on S-L A suspensions is called a load-carrying ball joint because it helps support the car; the other one carries no weight and is called a dampening joint. (Courtesy of Dana Corporation)

uncontrolled motions result. When a ball joint carries a large amount of load, such as the vehicle loaded joint on S-L A suspensions, the vehicle load is used to keep play and clearance out of the joint. When this joint is used in a normally unloaded position such as a tie-rod end, friction inside the joint, often provided by spring tension, keeps the joint tight yet flexible. (Fig. 1–36)

1.13 KING PINS

At one time, all vehicles used a **king pin** or **king bolt** to connect the spindle to the steering knuckle or axle. The king pin provided the pivot or steering axis for steering the front wheels. Most solid axles and swing axles still use king pins. An **Elliott** style of king pin locks the steel king pin into the spindle, and the king pin rotates in a pair of bushings in the end of the axle. A **reversed Elliott** style locks the steel king pin in the end of the axle, and there are a pair of bushings in the spindle. The reversed Elliott is the most common type. (Fig. 1–37)

1.14 SPRINGS AND SHOCK ABSORBERS

The springs make the load-carrying connection between the suspension members and the frame. Springs have the ability to bend or twist

and absorb energy when they are compressed to shorter lengths. When a tire meets an obstruction, it is forced upward, and the energy of this upward motion is absorbed by the spring rather than transmitted to the frame and body of the vehicle. The spring, however, will only absorb this energy for a brief period of time; as soon as possible, it will release the energy by extending back to its original length. This will either push the tire back down on the road or lift the car if the obstacle is still under the tire.

When a spring releases its stored energy, it does so with such quickness and momentum that the end of the spring usually extends too far. The spring will go through a series of oscillations, extensions, contractions, and extensions until all of the energy in the spring is used up or released. The speed of these oscillations will depend on the natural frequency of the spring and suspension. A car with undampened

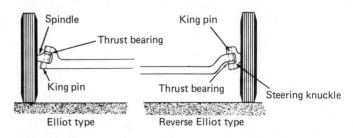

Spindle

Thrust bearing

King pin

King pin

Thrust bearing

Steering knuckle

Elliot type

Reverse Elliot type

FIGURE 1–37.
An Elliott (left) and reversed Elliott (right) king pin. The reversed Elliott is much more common. (Courtesy of Ammco)

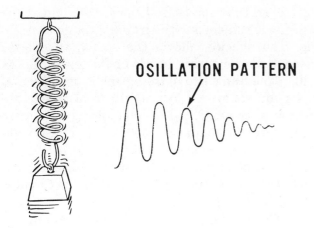

OSILLATION PATTERN

FIGURE 1–38.
When a spring is released, it bounces in a series of diminishing oscillations. A car spring tries to do the same thing unless it is dampened with a shock absorber. (Courtesy of Ford Motor Company)

springs (i.e., no shock absorbers) will tend to bounce up and down in time to these oscillations. In most cases, this bounce frequency will be disturbing to the driver or passengers. A stronger spring will oscillate at a faster frequency than a softer one. Many manufacturers will purposely mismatch the spring frequency at the front and the rear to obtain a flatter, more acceptable ride. (Fig. 1–38)

Spring oscillations are normally dampened or reduced by **shock absorbers**. Shock absorbers are very poorly named. They do not absorb shock; the springs do. The shock absorbers stop excessive spring oscillations. The shock absorber really absorbs some of the energy that was put into the spring by the bump, and it converts that energy into heat that is dissipated into the air. A shock absorber is usually mounted inside or next to each of the four springs on a car. Shock absorbers will be discussed in more detail in Chapter 11. (Fig. 1–39)

1.15 SPRING TYPES

At one time, the most commonly used spring was the **leaf spring**. It is usually a long strip of flattened spring steel that has a pivot bushing at each end. One end is attached to the frame through a bushing; the center of the spring is

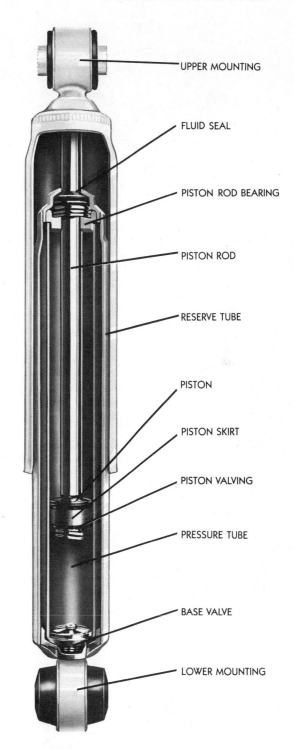

UPPER MOUNTING

FLUID SEAL

PISTON ROD BEARING

PISTON ROD

RESERVE TUBE

PISTON

PISTON SKIRT

PISTON VALVING

PRESSURE TUBE

BASE VALVE

LOWER MOUNTING

FIGURE 1–39.
A cutaway view of a shock absorber. The pressure tube is filled with oil. Suspension travel forces the piston through this oil, which generates the resistance in a shock absorber. (Courtesy of Monroe Auto Equipment)

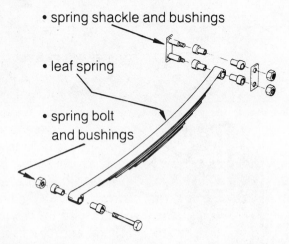

- spring shackle and bushings
- leaf spring
- spring bolt and bushings

FIGURE 1-40.
A semi-elliptic leaf spring. The axle is clamped near the center, bushings are used at each end, and a shackle is used at one end so the spring can change length. (Courtesy of Dana Corporation)

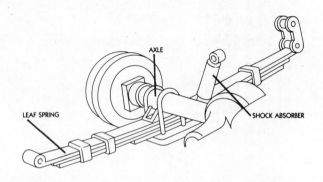

FIGURE 1-41.
A rear axle with a leaf spring. The shackle at the rear end allows the spring to change length as it flattens and bends. (Courtesy of Monroe Auto Equipment)

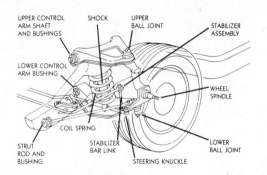

FIGURE 1-42.
An S-L A suspension with a coil spring mounted between the frame and the lower control arm. This is the most common RWD front spring, suspension arrangement. (Courtesy of Monroe Auto Equipment)

attached to the axle; and the second end of the spring is attached to the frame through a shackle. The shackle allows the spring to change length as the spring bends or flattens. (Fig. 1-40)   The frame end of the spring is usually used as a control arm; it locates the front-to-rear position of the axle. Additional spring leaves are added to the main or master leaf. These additional leaves are usually of different lengths in order to obtain a variable spring rate. Spring rates will be discussed in more detail in Chapter 10. (Fig. 1-41)

Coil springs are now the most commonly used type of spring. They require the use of control arms to locate the wheel and axle, but they give a good ride without wear or interleaf friction. Also, the coil shape often fits easily into most installations. A coil spring is simply a spring steel wire or rod that is wound into a helical, coiled shape. (Fig. 1-42)

Torsion bars are often used where coil springs would get in the way or take up too much vertical room. Some FWD cars with S-L A suspensions use torsion bars for this reason. The torsion bar is anchored to the frame at one end and attached to a lever arm (often a suspension control arm) at the other end. The control arm is free to pivot with the suspension, and it twists the torsion bar as the wheel moves up and down. An adjustment is usually built into the attachment at one end of the torsion bar to allow the bar to be twisted tighter or looser. This adjustment compensates for sag. Sag is deterioration that causes a spring to shorten and lower the car as it gets older. (Fig. 1-43)

Metal springs can be replaced by air-filled rubber bags or chambers. **Air suspension** offers a real advantage in that it allows for a change in spring rate by increasing or decreasing the air pressure inside of the air chamber. A disadvantage is that the system is rather complex. It requires an air chamber at each wheel, a height control sensor and valve at each wheel, an air compressor, and tubing to connect all of these together. (Fig. 1-44)

Modern air suspensions can be controlled and adjusted by a computer as the car moves. The computer adjusts the car's height so it stays correct regardless of weight that might

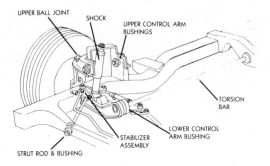

FIGURE 1–43.
This S-L A front suspension uses a torsion bar for a spring. The lower control arm connects to one end of the torsion bar to become a lever. (Courtesy of Monroe Auto Equipment)

be added or removed, such as passengers or luggage. On some cars, adjustments can also be made in the vehicle height or spring rate as the car speeds up or even as the road surface changes. Both changes improve high-speed handling, and lowering the car can improve gas mileage. (Fig. 1–45)

1.16 STEERING SYSTEM

Each front wheel is mounted onto a spindle that is attached to a steering knuckle or strut assembly. The steering knuckle pivots on ball joints, a king pin with bushings, or an upper bearing and ball joint. These pivots are called the **steering axis.** The steering knuckle has a steering arm attached to it that is connected to the tie rods and steering gear. Movement of the end of the steering arm toward or away from the center of the car causes the steering knuckle to pivot on the steering axis and the front wheels to make a right or left turn. (Figs. 1–46 & 1–47)

1.17 STEERING GEARS

Two types of steering gears are used to transmit the steering motions from the steering

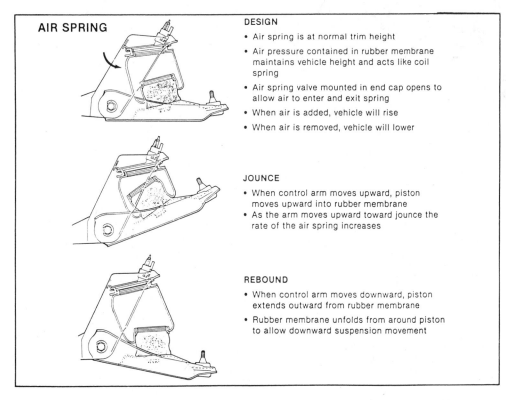

AIR SPRING

DESIGN
- Air spring is at normal trim height
- Air pressure contained in rubber membrane maintains vehicle height and acts like coil spring
- Air spring valve mounted in end cap opens to allow air to enter and exit spring
- When air is added, vehicle will rise
- When air is removed, vehicle will lower

JOUNCE
- When control arm moves upward, piston moves upward into rubber membrane
- As the arm moves upward toward jounce the rate of the air spring increases

REBOUND
- When control arm moves downward, piston extends outward from rubber membrane
- Rubber membrane unfolds from around piston to allow downward suspension movement

FIGURE 1–44.
An air spring is an air-filled, rubber container that compresses and extends during suspension travel. (Courtesy of Ford Motor Company)

SYSTEM COMPONENTS

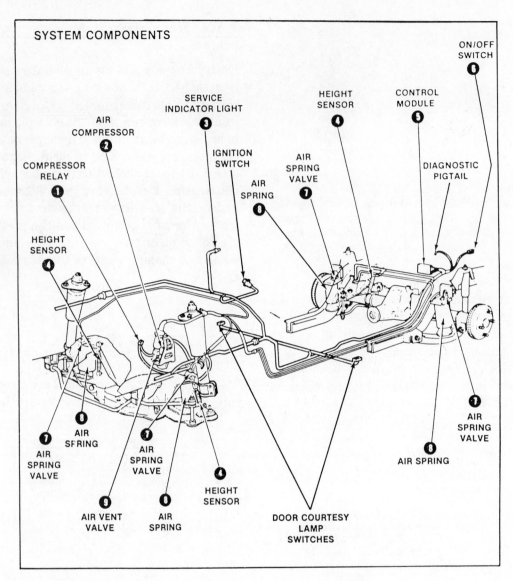

FIGURE 1–45.
Most air suspension systems use a computer control module (5) to determine if the air pressure in the springs needs to be increased or decreased. (Courtesy of Ford Motor Company)

wheel to the tie rods: the **conventional standard gear** and the **rack and pinion gear.** Both types change the rotary motion of the steering wheel into the back and forth motion of the tie rod. They also provide a gear ratio to increase the driver's effort and slow down the steering speed. The conventional steering gear has traditionally been used on larger cars; it offers the advantage of smooth, easy steering with very little road shock. Rack and pinion steering is more compact, lighter in weight, and often has a faster steering ratio. A faster steering ratio provides faster steering, but a slower ratio provides more precise steering. (Fig. 1–48)

Two main shafts are used in conventional steering gears: the **steering shaft,** which connects the worm shaft in the steering gear to the steering wheel, and the **Pitman shaft,** which with the **Pitman arm** is connected to the steering linkage. Internal gearing provides gear reduction and a 90° change in motion.

Rack and pinion gears have a steering shaft that connects the pinion gear to the steering wheel and a rack that is connected directly

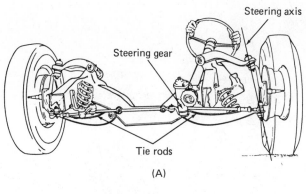

Steering axis

Steering gear

Tie rods

(A)

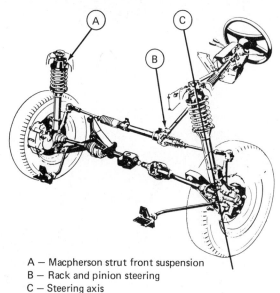

A — Macpherson strut front suspension
B — Rack and pinion steering
C — Steering axis

(B)

FIGURE 1–46.
The steering axis is the steering pivot for front tires. Note the steering axis and parts for the two major types of suspension and steering systems. (A—Courtesy of Moog Automotive; B—Courtesy of Ford Motor Company)

to the tie rods. The contact between the pinion gear mounted on the steering shaft and the rack provides gear reduction and a change in motion. Operating theory and service procedures of both styles of steering gears will be discussed in more detail in Chapters 13 and 16. (Fig. 1–49)

1.18 STEERING LINKAGE

The steering linkage consists of the tie rods and sometimes other rods or links that connect the

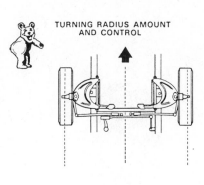

TURNING RADIUS AMOUNT AND CONTROL

THE TURNING RADIUS ARC

FIGURE 1–47.
When the steering wheel is turned, the tie rods transmit the turning force and cause the front wheels to turn. (Courtesy of Bear Automotive)

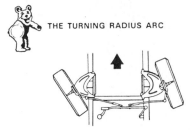

Steering shaft

Pitman shaft

FIGURE 1–48.
A standard, nonpower steering gear, the steering shaft attaches to the steering wheel, and the Pitman arm connects the Pitman shaft to the steering linkage. (Courtesy of Moog Automotive)

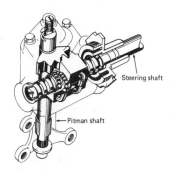
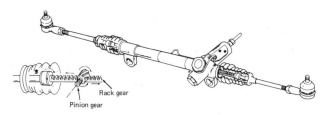

Rack gear

Pinion gear

FIGURE 1–49.
A rack and pinion steering gear with tie rods. Rotation of the pinion gear will cause the rack gear to move to the right or left (lower left). (Courtesy of Dana Corporation)

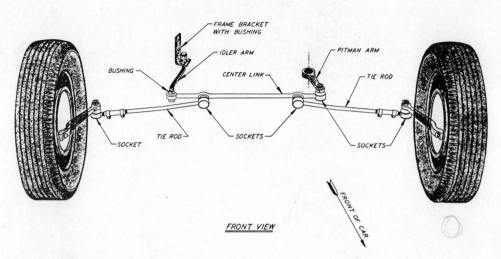

FIGURE 1-50.

A parallelogram steering linkage is commonly found on domestic RWD cars. Note the relationship of the Pitman arm, idler arm, center link, and tie rods. (© 1966, Courtesy of Replacement Parts Div., TRW, Inc.)

steering gear to the steering arms. Several types of steering linkages are used. In all cases, the steering linkage is designed to transfer steering motions to the wheels in such a way that the vertical motion of the wheels over a bump will not steer the wheels, that is called bump steer.

Most cars with conventional steering gears use a **parallelogram steering linkage**. A center link or relay rod attaches to the Pitman arm and is supported at the other end by an idler arm. The Pitman arm and idler arm are mounted in the same relative position in the car, so the center link moves back and forth, straight across the car, during steering maneuvers. Two tie rods connect the center link to each of the steering arms. Two other styles of steering linkages, Haltenberger and bell crank, are sometimes used on cars with independent suspensions. The simplest steering linkage is used with rack and pinion steering gears; a tie rod connects each end of the rack to a steering arm. (Figs. 1–50 & 1–51)

Pickups, trucks, or cars using a solid axle use a single tie rod that connects one steering arm to the other. A **drag link** connects the

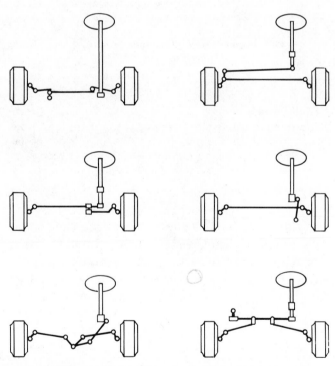

Above are shown various types of tie-rod assemblies

FIGURE 1-51.

Some of the different steering linkage arrangements. The parallelogram linkage (upper left and lower right) is the most common. (Courtesy of Hunter Engineering)

steering gear to a third steering arm which is connected to one of the steering knuckles. Sometimes, instead of using a third arm, an extension of one of the steering arms is connected to the drag link. The drag link can run either across the car to the right wheel or lengthwise to the left wheel. (Fig. 1–52)

Connections at each end of the tie rods and other links are made through a ball-and-socket joint commonly called a tie rod end. This ball-and-socket joint allows the suspension to move up and down and the linkage to swing back and forth while transmitting motion. It also eliminates free play that would cause sloppy and dangerous steering.

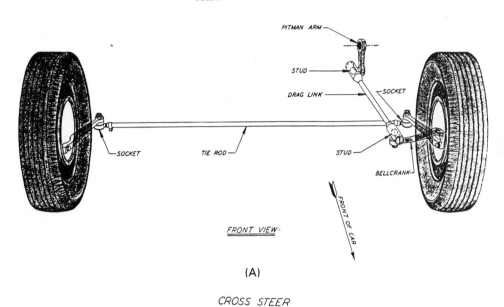

FORE-AFT AND CROSS STEER
BELLCRANK AHEAD OF AXLE.

(A)

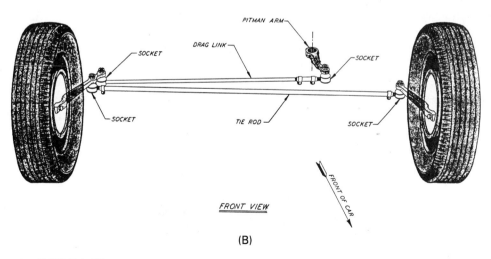

CROSS STEER

(B)

FIGURE 1–52.
Some trucks and pickups that use a solid axle also use a drag link to connect the Pitman arm to the bell crank or third arm (A). Some use a cross steer to connect the drag link to the steering arm across the car (B). (© 1965, Courtesy of Replacement Parts Div., TRW, Inc.)

REVIEW QUESTIONS

The following quiz is available for you to check the facts you have just learned. Select the best answer that completes each statement.

1. The chassis consists of the car's frame and the

   a. suspension system.
   c. car body.
   b. steering system.
   d. both a and b.

2. A front suspension system consists of the

   a. springs and shock absorbers.
   b. control arms.
   c. steering knuckle.
   d. all of the above.

3. Statement A: The distance between the tires, measured across the car, is called *track*. Statement B: *Wheelbase* refers to the distance between the front and rear tires. Which statement is correct?

   a. A only
   c. both A and B
   b. B only
   d. neither A nor B

4. Statement A: Wheels are aligned to reduce tire scrub and wear. Statement B: When a tire leans inward at the top, it is toed in. Which statement is correct?

   a. A only
   c. both A and B
   b. B only
   d. neither A nor B

5. Most modern cars are of a _____ type construction.
   A. frame plus body
   B. unibody

   a. A only
   c. both A and B
   b. B only
   d. neither A nor B

6. A vehicle with nonindependent suspension will use

   a. a swing axle.
   b. a solid axle.
   c. a 4WD axle.
   d. none of the above.

7. Which of the following is not a type of independent suspension?

   a. swing axle
   c. solid axle
   b. short-long arm
   d. trailing arm

8. Statement A: An S-L A type of suspension will cause a camber change as the wheel moves up and down. Statement B: The upper arm is shorter than the lower arm. Which statement is correct?

   a. A only
   c. both A and B
   b. B only
   d. neither A nor B

9. A strut suspension uses _____ to control the position of the tire.

   a. two equal-length control arms
   b. two unequal-length control arms
   c. a lower control arm and an oversized, telescoping shock absorber
   d. a pair of trailing arms

10. The assembly that connects the top of the strut to the fender well provides

    a. a bearing to allow steering.
    b. a flexible mounting to allow strut angle changes.
    c. a dampener to isolate road vibrations.
    d. all of the above.

11. Technician A says that an A-arm has two inner bushings. Technician B says that a strut rod is used with control arms that have a single inner bushing. Who is right?

    a. A only
    c. both A and B
    b. B only
    d. neither A nor B

12. A rubber control arm bushing will not

    a. reduce the transmission of road vibrations.
    b. operate without lubrication.
    c. cause resistance to suspension movement.
    d. allow the control arm to swing freely.

13. A ball joint or tie rod end will
    A. allow a swinging or pivoting motion.
    B. allow a rotating motion.

    a. A only   **c.** both A and B
    b. B only   **d.** neither A nor B

14. Statement A: Springs absorb energy when they are compressed. Statement B: The shock absorbers help the spring support the car's weight. Which statement is correct?

    a. A only   **c.** both A and B
    b. B only   **d.** neither A nor B

15. A car's springs can be of the _____ type.

    a. leaf spring   **c.** torsion bar
    b. coil spring   **d.** all of the above

16. A/an _____ spring system has the ability to automatically maintain a constant car height.

    a. air   **c.** coil
    b. leaf   **d.** torsion bar

17. When a car's tires meet a bump in the road,
    A. the springs and shock absorbers compress to allow upward tire travel.
    B. the springs and shock absorbers extend to allow downward tire travel.

    a. A only   **c.** either A or B
    b. B only   **d.** neither A nor B

18. The steering gear
    A. changes the rotating steering wheel into a back and forth motion.
    B. provides a gear ratio to make steering easier.

    a. A only   **c.** both A and B
    b. B only   **d.** neither A nor B

19. Statement A: A parallelogram type of steering linkage is usually used with rack and pinion steering gears. Statement B: Tie rods are used to connect the steering arms to the steering gear or steering linkage. Which statement is correct?

    a. A only   **c.** both A and B
    b. B only   **d.** neither A nor B

20. The steering linkage
    A. transmits steering motions from the steering gear to the steering knuckle.
    B. helps hold the front wheels in a straight-ahead position or turning direction.

    a. A only   **c.** both A and B
    b. B only   **d.** neither A nor B

Chapter 2

TIRES—THEORY

After completing this chapter, you should:

- Be familiar with the terms commonly used to describe tire construction, sizing, and operation.

- Have a basic understanding of how a tire is constructed.

- Understand the modern tire size designations.

- Be able to select a correct replacement tire for a car.

- Have a basic understanding of the various types of tires used for spares and the operating and service requirements of each tire type.

- Know what a recap tire is and its benefits.

2.1   INTRODUCTION

Almost everyone knows that tires are fitted on wheels, and the tires and wheels roll down the road. Most drivers also know that the tread of the tire and its grip on the road provide traction so the car can accelerate, turn corners, and stop. Some drivers have been made painfully aware that it is extremely difficult to control a car when there is not enough traction. The car can skid and cause an accident.

Most of this discussion will concentrate on passenger car tires. Much of it will also apply to truck, heavy duty off-road, farm, and industrial tires, but each tire type will differ slightly in its own way.

Traction refers to the amount of grip between the tire and the road. It can be affected by anything on the road surface including ice, water, sand, leaves, and so forth, by the depth of the tire tread, by the pattern of the tread, by the hardness or softness of the tread rubber, by the inflation pressure of the tire, by the width of the tire, by the load on the tire, by the alignment of the tire, by the temperature of the tire and the road surface, plus other things. One vehicle-handling expert who specializes in road racing cars maintains that there are 31 factors that affect tire-to-road traction. Depending on the tire tread and road surface, a tire will reach maximum traction at a slip rate of about 10 percent to 25 percent of the vehicle speed. Greater than 25 percent slippage will usually result in a skid and, with the skid, a severe loss in traction. Besides the traction loss, a skidding tire will also lose its directional control. (Fig. 2-1)   A rolling tire will usually travel in the direction that it is pointed. A skidding or sliding tire will slide to the side just about as easily as it will go in a forward direction. A tire that is spinning at a rate greater than 25 percent of the vehicle speed will lose traction and directional control in the same way as a skidding tire. Traction and its effects on handling will be discussed in Chapter 19.

The tires also serve as springs. Small bumps are absorbed as the tire sidewall flexes. We all realize that when a load is put on a tire, the tire and tread flatten at the road surface.

FIGURE 2-1.
These skid marks across the sidewalk, a fairly common sight on our roadways, show a driver that needed more traction than was available. While the tire is skidding, the driver usually loses steering ability.

This movement produces a bulge in the tire's sidewall. (Fig. 2-2)   The ability to absorb a bump will vary depending on tire construction and pressure. A nonbelted tire will usually have a softer and smoother ride than a belted tire; a fabric-belted tire will usually ride smoother than a steel-belted tire. A tire inflated to 28 psi (193 kPa) will usually have a smoother ride than one that is inflated to 34 psi (234.5 kPa). When a car is designed, differences in tire ride quality can be compensated for by changing shock absorber settings. They can also be compensated for by changing the rubber isolation bushings between the suspension and the tires or between the suspension and the frame.

A tire designer has to make numerous conflicting choices. A belted tire with a rigid tread might produce better fuel mileage, but the ride quality and traction would suffer. A softer tread rubber compound might improve traction, but tire life and possibly gas mileage would suffer. A lower, wider tire might look better and offer better dry pavement traction but vehicle ground clearance and axle revolutions per mile would be reduced and the tires would

FIGURE 2–2.
As the tread flattens for road contact, a bulge will occur in the tire's sidewall. Flexing of the sidewall in this area allows the tire to absorb small bumps.

tend to hydroplane in wet conditions. Tire design is definitely a compromise and has led to many special tires for specific purposes or conditions.

2.2   TIRE CONSTRUCTION

Today's passenger car tire begins as several layers of rubber of various mixtures or compounds, some cording, and two wire rings. A rubber compound can use from 8 to 15 different raw materials. A few of the reasons for using different rubber blends include to change the tensile strength; to increase the resistance to age, light, or different chemicals; to increase the resistance to abrasion; and to change the adhesion tendencies between the rubber and various cord materials. (Fig. 2–3)

The cording for the **plies** is formed from two or more strands of various fabrics. The cords can be twisted, braided, or woven together. The wire rings, called the **bead wires**, reinforce the bead portion. This part of the tire must stay firmly fixed in place on the wheel. The bead wires help transmit starting and stopping forces between the tire tread and the wheel as well as resist the centrifugal force that is trying to separate the tire from the wheel. The **sidewall plies**, also called **body** or **carcass plies**, reinforce the body of the tire by wrapping

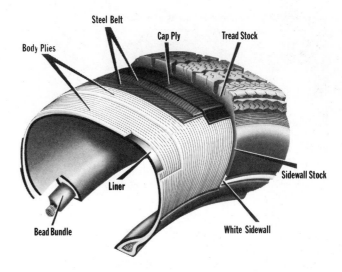

FIGURE 2–3.
A sectioned view showing a tire's internal construction. Note that the body plies wrap around the bead bundle and that the belt plies reinforce the body plies in the tread area. (Courtesy of Uniroyal)

around and attaching to the bead wires. (Fig. 2–4) These plies form a strong envelope to hold the inner liner and the air inside of the tire.

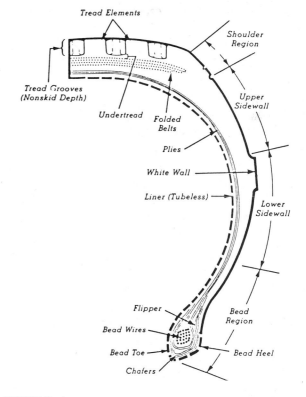

FIGURE 2–4.
This cross section of a tire shows the major internal parts. Note that the sidewall plies wrap around the bead wires. (Courtesy of B.F.Goodrich, © The B.F.Goodrich Company)

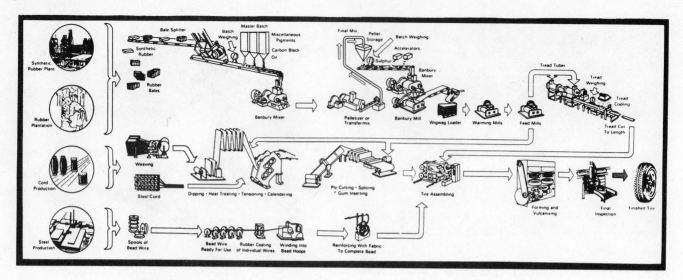

FIGURE 2-5.
The process of building a tire involves many rather complex steps. (Courtesy of Firestone)

The inner edge of the bead, where the plies wrap around the bead wires, is angled slightly so as to wedge the bead tightly onto the angled bead seat of the wheel. This ensures that the tire fits tightly and is centered on the wheel when it is inflated. (Fig. 2-5)

2.2.1 Ply Materials

Sidewall plies are commonly made from several different materials. **Cotton**, a natural fabric, was the first commonly used ply material in the early days of tire building, prior to World War II. Cotton plies were relatively weak and tended to deteriorate easily when exposed to moisture. Today the most commonly used materials for sidewall plies are **polyester, nylon,** and **rayon.** Each of these synthetic, man-made materials offers different advantages.

Rayon is fairly strong, resists fatigue, does not flat spot, and has good fabric-to-rubber adhesion characteristics. Of the three synthetic materials, rayon has the least tendency to change length because of temperature change . Nylon is a strong and very durable fabric that is relatively inexpensive. Because of its strength and its ability to dissipate heat, nylon is the most commonly used ply cord material in truck tires. When warm, nylon cords are quite flexible and tend to shrink or shorten. Nylon

cord tires tend to flat spot as they cool; that is, as the bottom portion of the cord cools, it takes the shape of the flat road surface. This tread portion tends to stay flat until it warms up again. Polyester is a thermoplastic material that flexes easily, is strong and lightweight, and resists flat spotting. Polyester is the least expensive of the three synthetic fibers. Polyester cords tend to extend or grow longer with heat; they also tend to lose strength as they get hotter. A comparison of the various body ply and belt materials is shown in Table 2-1.

The layers of sidewall plies are joined together by thin layers of rubber. A layer of rubber **laytex,** sometimes called **gum,** is coated onto the ply cords. This laytex bonds together the cords in a ply and also bonds the cord ply to the rubber layer right next to that ply. Each different type of ply cord requires a laytex mixture to suit that particular cord material. There are usually several different types of bonding laytex in a tire, one for each type of cord. Tight bonding of the materials in a tire is important to give the tire strength and to help dissipate heat. Poor ply bonding causes tire separation. (Fig. 2-6)

Another layer of rubber forms an airtight inner liner in tubeless tires. This inner liner is made from butyl rubber, which is airtight. Still another strip of rubber, called a filler, fills the

TABLE 2–1
Comparison of Different Tire Cord Materials Used in the Sidewall and Belt Plies

Cord material	Relative strength*	Where used	Comments
Cotton	2.3	Not used	Weak, rots, and mildews; poor rubber adhesion; variable quality; not used after World War II; flexible.
Steel, stranded	3.4	Belt	Heavy, strong, hard to cut; strength not affected by heat; lengthens when hot; very rigid.
Rayon	4.6	Sidewall	Medium strong, resists fatigue.
Polyester	8.8	Sidewall	Strong, light, low in cost; resists flat spots; weakens when hot; flexible.
Nylon	9.0	Sidewall	Very strong; shrinks when hot; tends to flat spot; flexible.
Fiberglass	11.0	Belt	High strength-to-weight ratio, resistant to rot; temperature stable; fairly rigid.
Aramid	17.0	Belt	Expensive; excellent physical properties; semirigid.

*Relative strength shows the approximate strength in grams per denier. Denier is a measuring system for the fineness of cord or yarn.

area where the plies wrap around the bead wires. This strip of rubber can change the rate or stiffness of sidewall flexing.

The outside layer of rubber on the sidewall protects the plies from corrosion and weather damage as well as abrasion and curb damage. Sidewall rubber must be highly resilient for good flexibility. Natural rubber is often used in radial tire sidewalls because of its excellent,

FIGURE 2–6.
A serious case of tire separation in which a portion of the sidewall and tread came loose. In this case, it was caused by a staple hole that permitted moisture to enter into the tire cords. (Courtesy of General Tire, a GenCorp Company)

natural flexibility. The sidewall rubber also provides decoration in the form of whitewalls or special lettering. The white oxide rubber used for this purpose has a tendency to crack and bond weakly to other types of rubber. This is the major reason why white rubber is used very little on high performance tires. The sidewall is also the location for any lettering that provides information on the tire.

The tread area is usually composed of two different rubber compounds. The combination of the inner and outer layers is often called **dual compounding**. The inner layer, next to the plies, is called a **cap base** or **under tread**. It is compounded for good heat dissipation. The outer rubber layer is bonded to the cap base, and the tread pattern is molded into this tough, wear-resistant layer of rubber.

2.2.2 Ply Arrangement

At one time, the sidewall plies were always used in pairs with the two plies of a pair overlapping as they crossed the tire at different angles. This angle is the reason for calling these tires **bias ply**. The cords of one ply will cross the cords of the second ply at an angle of about 30° to 40°. (Fig. 2–7) Because of the overlapping of the cord plies, bias ply construction is strong. How-

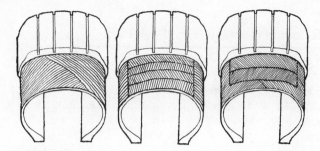

FIGURE 2-7.
Three tire sections showing three different styles of body and belt ply construction. A bias or diagonal ply is at the left; a radial ply with belt is in the center; and a bias-belted ply is at the right. Note the different angles of the body plies. (Courtesy of Goodyear Tire and Rubber Company)

ever, it also forms a tire envelope that has trouble flexing in a direction that tires need to flex. The ply laminations form a tire body that cannot change its shape very easily. The bottom area of the tire in contact with the road must flatten to provide tread contact with the road. If it does not, there will be very limited traction plus a very firm ride. (Fig. 2-8)

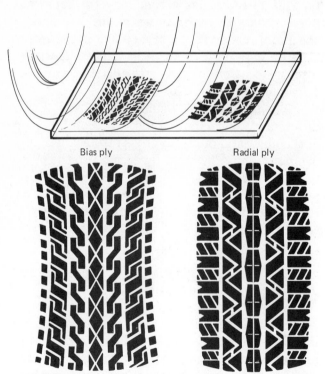

Bias ply   Radial ply

FIGURE 2-8.
As a tire rolls, the tread area must flatten into the contact pattern or tire footprint. The tread of a bias-ply tire closes up or pinches together at the center, which causes the tread to squirm across the road.

The cords in a **radial** ply run almost straight across the tire from one bead to the other. When the ply cords run in this direction, it is very easy for them to bend and allow sidewall flex. Tread to road contact is good with little to no tread squirm. Squirm causes a scuffing of the tread rubber because the tread narrows and widens as it comes in contact with the road surface. Because the sidewall can flex easier with less internal friction, a radial tire will roll easier and deliver better fuel economy. The radial sidewall ply direction, however, is vulnerable to penetration and road damage. It must be protected by a belt under the tread. The belt also helps hold the body plies in place, thus resisting the effects of centrifugal force. It also reinforces the body of the tire to provide good cornering and straightaway strength. Radial tires tend to be more expensive, partly because they are more difficult to build and partly because more expensive materials are used in them. Radial tires offer the advantages of better tire and fuel mileage, cooler and quieter operation, better dry and wet pavement handling and braking, and better resistance to sideslip caused by crosswinds. The only major drawback, besides cost, is a slightly rougher ride with more vibrations in the 20 to 30 MPH range (32 to 48 KPH).

2.2.3 Belts

The **belt**, often called **tread plies**, is made from cording that is similar to that used in sidewall plies. The usual materials are **aramid** (Kevlar, Dupont's brand name, or Flexten, Goodyear's brand name), fiberglass, rayon, and steel. Several combinations of these materials are often used for different plies on the same tire. A tire's outside diameter can be made more stable by placing a belt of a material that tends to shrink—for example, nylon—over a belt that tends to grow—for example, steel.

The belt runs around the tire in the area just under the tread. For strength, the belt has to bond tightly to both the sidewall plies and the tread rubber. The belt can be cut to width, or cut oversized with the edges folded over; the belt can also be woven or braided. (Fig. 2-

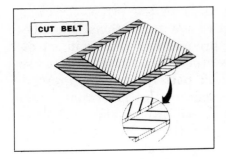

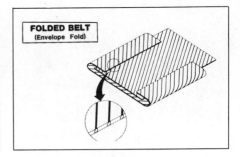

FIGURE 2–9.
Belt plies can be cut, folded, or woven. This difference, plus the way the belts are arranged in the tire, allows the tire designer a variety of ways for changing the operating characteristics of a tire. (Courtesy of the Armstrong Tire Company)

9) Also, the layers of belt plies can be the same width, called **stacked construction**, or of different widths, called **pyramid construction**. The various belt construction methods allow the tire engineer to design different handling, impact resistant, and heat build-up characteristics into the tire. Belts are always used in radial tires. When belts are used in bias tires, the tire is called **bias-belted**.

2.2.4 Tread

The tread pattern is formed in the outermost layer of rubber when the tire is vulcanized or cured in a mold. The inner surface of the mold is the same size and shape that the outside of the tire will become; that is, the tread pattern is determined by the shape of the mold surface. The "green," uncured tire is placed into the mold. Pressure from a bladder inside the tire forces the soft tread and sidewall rubber into the mold's shape while heat vulcanizes the rubber. (Fig. 2–10) Vulcanizing stiffens the rubber to develop the hardness and elasticity that we are all familiar with in tires. Before vulcanizing, the rubber is quite soft.

The rubber compound for the tread can be formulated to give the desired wear rate, ride quality, traction, weather resistance, or some other characteristic desired for a particular tire. The tread pattern is designed to provide the traction characteristics for the intended use of the tire.

The primary reason for the grooves in the tread is to let water run out from between the tire and the road, but tire stability, dry traction, wear rate, and noise are also considered when a particular tread pattern is designed. If water cannot get out from under the tread, the tire might **hydroplane**, plane, or climb up and over the layer of water much like a water skier does. A hydroplaning tire will not have any

FIGURE 2-10.
A tire mold or press has the shape and size of a tire on the inside surface. The "green," uncured tire will be forced to take the shape of the mold as it cures or vulcanizes.

traction at all because the tire actually leaves the road surface. (Fig. 2-11) Things that will increase the tire's tendency to hydroplane are small tread grooves, shallow or worn tread grooves, wider tires, lower tire pressures, and

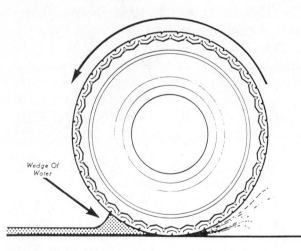

FIGURE 2-11.
Hydroplaning occurs when a wedge of water builds up in front of a tire causing the tire to ride up and over a water film. (Courtesy of B.F.Goodrich, © B.F.Goodrich Company)

higher vehicle speeds. A few street tires and many racing tires have a **directional tread** because some rain grooves work better in one direction than the other. Many tractor tires use a directional tread pattern. More aggressive tread designs or those with wider grooves and larger blocks are used for mud and snow traction. Some of these patterns become very noisy at road speeds. There is also a possibility of traction loss while cornering or braking because these rubber blocks can bend or squirm on the road surface when a traction load is put on them. (Fig. 2-12)

2.3   TIRE SIDEWALL INFORMATION

The sidewall of the tire carries most of the information the motorist needs to know about the tire. Letters are molded into the sidewall to give

- the tire size;
- the number of sidewall plies and what material is used;
- the number of tread plies and what materials are used;
- the maximum air pressure the tire can be inflated to and the maximum amount of load the tire can safely carry at that pressure;
- the tire brand name;
- the **Department of Transportation (DOT)**, number; and
- information if the tire is directional and, if so which direction it should run. (Fig. 2-13)

All passenger car tires must be approved by the Department of Transportation and carry a **DOT number** as well as the other information listed. The DOT number is coded to tell the reader what manufacturing plant made the tire (the first two letters), pertinent things about the tire construction or sizing (the middle letters), and when the tire was made (the last three numbers). The last number is the final digit of the year in which the tire was made, and the

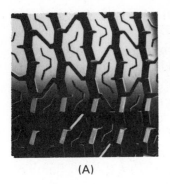
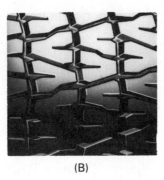
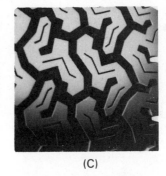
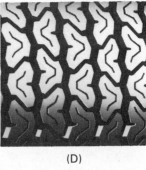

(A)   (B)   (C)   (D)

(E)   (F)

FIGURE 2–12.
A variety of tread patterns are used on tires depending on how the tire is to be used. (A) and (B) are street performance designs with good wet and dry traction characteristics. (C) and (E) are off-road designs. (D) is a street performance, mild off-road design, and (F) is a passenger car design for good tire life. (Courtesy of B.F.Goodrich, © B.F.Goodrich Company)

two numbers preceding it indicate the week it was made. (Fig. 2–14)   A DOT number ending in 306 indicates the tire was made during the 30th week of 1976 or 1986. A booklet published by Tire Guide, an independent trade publication, can be purchased in order to read the tire maker and plant location portion of the code.

2.4   TIRE SIZING

Three different tire dimensions are important to the tire purchaser: wheel size, tire width, and aspect ratio. The wheel size, in inches (sometimes millimeters), is indicated by the last two

FIGURE 2–13.
The sidewall of a tire contains a great deal of information about that tire. (Courtesy of Rubber Manufacturers' Association)

FIGURE 2–14.
A Department of Transporation (DOT) number. In this case it is DOT: JFFH MN6 063. The JF is a code for the manufacturing plant; FH and MN6 tell the manufacturer what the construction and materials are, and 063 indicates the tire was built in the sixth week of 1973 or 1983.

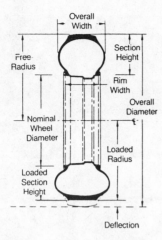

FIGURE 2–15.
Tire and wheel size designations are based on several different measurements. (Courtesy of Bridgestone Tire Company)

digits of a tire size. The most common passenger car wheel sizes are 13, 14, and 15 inches (330.2, 355.6, and 381 mm). This size is the diameter of the tire when measured between the beads across the tire and the diameter of the wheel measured between the bead flanges. A tire will only fit on a wheel of the same size and bead seat taper. Tires with a metric diameter

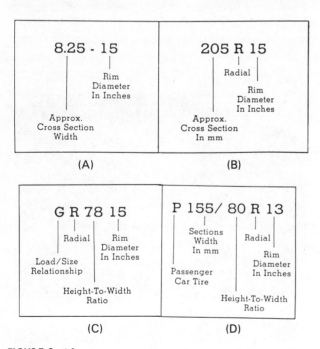

FIGURE 2–16.
Four different tire measuring systems: (A), (B), and (C) are obsolete; (D) is in current use. (Courtesy of B.F.Goodrich, © B.F.Goodrich Company)

will only fit metric-sized wheels and vice versa. (Fig. 2–15)

There have been numerous methods for measuring tire widths; the systems seem to change periodically. (Fig. 2–16)   At the present time, the **P-metric** system is being used. The "P" designates a passenger car tire; a "T" would indicate a truck tire. The letter is followed by three numbers that indicate the section width, in millimeters. (Fig. 2–17)   Some of the older measuring systems measured the section width in inches and some measure it in millimeters. The alpha-numeric system was based on the load-carrying capability of the tire.

The **aspect ratio** is the height-to-width relationship of the tire's cross section. (Fig. 2–18)   The number used will be the percentage of the height to the width. A 70 series tire will be seven-tenths or 70 percent as high as it is wide; a 50 series tire will be twice as wide as it is high. The next two numbers in the P-metric size indicate the aspect ratio. The aspect ratio will affect the tire's overall diameter, which, in turn, will affect the circumference and therefore the number of revolutions the tire must turn to cover a certain distance. (Fig. 2–19)   A P 205/ 70 R 14 tire from one particular manufacturer has an overall diameter of 25.35 inches (644 mm); this tire turns 824 revolutions to go one

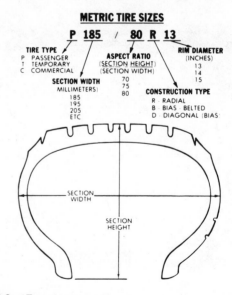

FIGURE 2–17.
The P-metric system for measuring the widths and what each portion of the code represents. (Courtesy of Oldsmobile)

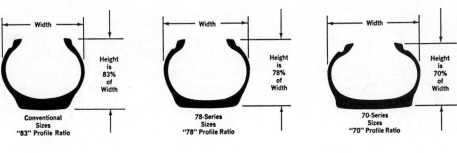

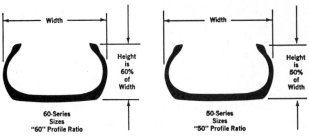

FIGURE 2–18.
The profile or aspect ratio of a tire is a comparison of the height of the tire to the width. (Courtesy of Goodyear Tire and Rubber Company)

mile (1.61 km) at 50 MPH/80.5 KPH). A P 205/60 R 14 size of the same tire has an overall diameter of 23.7 inches (602 mm); it takes 882 revolutions (58 more than the larger tire) to go the same distance. The tread width is slightly wider, 0.05 inch (1.3 mm), on the 60 series tire.

Next in the P-metric system of tire size marking is a letter to designate the type of sidewall construction. It will be one of the following: **R** for a radial ply tire, **B** for a bias-belted tire, or **D** for a diagonal/bias ply, nonbelted tire.

FIGURE 2–19.
A tire will roll down the road a distance equal to its circumference times the number of revolutions. The circumference is equal to the effective radius times two times pi ($C = 2\pi r$).

A P 205/70 R 15 tire is a P-metric passenger car tire with a section width of 205 mm (8.2 in). It has a height that is 70 percent of the section width, or about 143.5 mm (5.7 in). It is of radial construction and fits a 15 inch (38.1 mm) wheel.

2.5 TIRE LOAD RATINGS

The tire does not support the car; it contains the air that supports the car. Those of us familiar with hydraulic power transmission know that the force from a hydraulic cylinder equals the hydraulic pressure multiplied by the piston

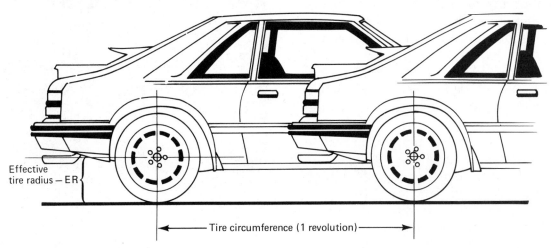

Effective tire radius — ER

Tire circumference (1 revolution)

$2 \times ER \times \pi$ = circumference
ER = effective radius

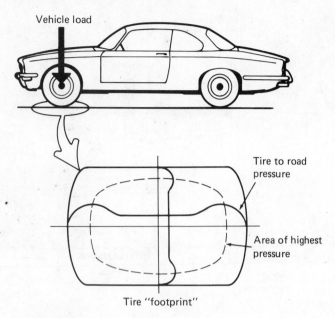

FIGURE 2-20.

The size of a tire's footprint or contact pattern is determined by the load it carries and the amount of tire pressure. The tire pressure times the contact area (for four tires) will just about equal the weight of the car. The tire-to-road pressure is highest at the outer edge because of the strength of the tread and sidewalls.

area in the cylinder. Pneumatic power transmission is essentially the same thing except the air is used to transmit the force instead of oil, and that air can be compressed into a smaller space. If pneumatics are applied to the automobile tire, it can be determined that the weight of the car is supported by four pneumatic areas, or the road contact points of the tire. If we multiply these areas, in square inches (square millimeters), by the tire pressure, in pounds per square inch (kilograms per square millimeter), we would get a result that would be very close to the weight of the car in pounds (kilograms). We all know that a tire that is going flat will spread out its tread contact area as it loses pressure. Raising the tire pressure will make the tire more round, which will reduce the tread contact area. (Fig. 2-20)

If a tire is not strong enough to contain this air pressure, it will blow out if the failure occurs quickly or simply go flat if the leak is slow. In addition to the normal, stationary vehicle load, the tire has to withstand shock loads that can be many times greater. Each time a tire hits a bump or obstruction, the load on the

tire increases substantially. (Fig. 2-21) Imagine a car traveling 55 MPH (88.5 KPH); it will be traveling 80.6 feet (24.6 meters) every second. In our minds, let's put a bump in the road that is 1 inch (25.4 mm) high. This bump will force the tire tread to rise as it passes over, but the car's weight will try to keep the wheel from lifting. So, for a brief period of time, the sidewall of the tire will flex as the tire compresses, and the pressure in the tire will increase substantially, depending on the amount of compression. Tires are built with extra strength to provide a safety margin for cases much more severe than the example just given. Imagine the pressure in the tires of a car that has just landed on the ground after flying through the air for a while in a movie chase scene. Again, if a tire is not strong enough, it will blow out. The strength of the tire is indicated by the tire's load rating, which is molded into the sidewall. In this country, it is illegal to drive a vehicle that is carrying a load greater than the load rating of its tires.

P-metric sized tires are available in the following two different load ratings: standard load tires, which can contain a maximum of 35 psi (241 kPa), and extra load tires, which can be inflated to 41 psi (283 kPa). This system determines the maximum inflation pressure for the tire. The actual load a tire can carry is deter-

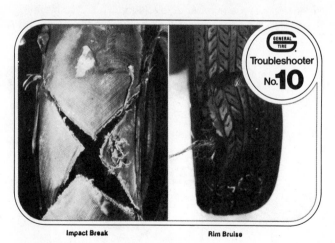

FIGURE 2-21.

These two tires show the result of the tires striking a chuckhole or obstruction in the road. The air loss that occurred could have been instantaneous; in that case it would be called a blowout. (Courtesy of General Tire, a GenCorp Company)

TABLE 2-2
Tire Load/Inflation Tables for Popular Automobile Tire Sizes

13" Rim Diameter (cont.)

Alpha* Sizes All Const.	European Metrics	P-Metric Sizes							Air pressure in PSI maximum loads in pounds											
		80 Series	75 Series	70 Series	65 Series	60 Series	55 Series	50 Series	18	20	22	23	24	26	28	29	30	32	35	36
A-13									770	810	860	880	900	940	980	1000	1020	1060		
		P165/80R13							783	816	860	871	893	926	959	981	992	1025	1069	
						P195/60R13			783	827	860	882	904	937	970	981	1003	1036	1080	
	185/70R13								915	940	965	977	990	1015	1040	1052	1065	1090	1127	1140
								P215/50R13	794	838	871	893	904	948	981	992	1014	1047	1091	
			P175/75R13						805	849	882	904	926	959	992	1014	1025	1058	1113	
	175R13								840	890	930	955	980	1030	1070	1090	1110	1150	1210	1230
B-13									840	890	930	955	980	1030	1070	1090	1110	1150		
				P185/70R13					827	871	915	926	948	992	1025	1036	1058	1091	1135	
					P195/65R13				849	882	926	948	970	1003	1047	1058	1080	1113	1157	

*This section of a tire load/inflation chart illustrates how the load-carrying ability of a tire is affected by the internal pressure. (Courtesy of Tire Guide, 1101-6 S. Rogers Circle, Boca Raton, Fla. 33432)

TABLE 2–3
Comparison of Maximum Loads

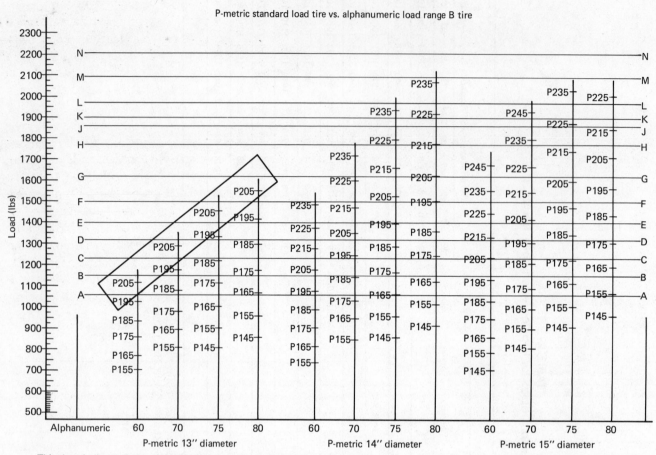

This chart indicates the maximum load that can be carried by passenger car tires of various sizes. (Courtesy of General Tire, a GenCorp Company)

mined by the tire size and the load rating. Since we ride on air, a larger tire can carry more load than a small tire. Charts, like the one in Table 2–2, are available to help determine how much actual load a particular tire can carry.

At one time, a **load range** system was used to designate a tire's load-carrying ability. Load range was related to ply ratings. Load range A was essentially the same as a two-ply sidewall; load range B equaled a four-ply; C equaled a six-ply; and so on. Because plies can be made using different sized cords, ply numbers and load ratings can vary. The load rating of the tire is still related to the strength of the sidewall plies. Tire pressure is related to load range, as shown in Table 2–3. The maximum pressure rating for load range A is 28 psi (196 kPa); for load range B it is 32 psi (221 kPa); for load range C it is 36 psi (248 kPa), and so on.

2.6   UNIFORM TIRE QUALITY GRADE LABELING

In the late 1970s, the **National Highway Safety Administration (NHSA)** established the standards for the **Uniform Tire Quality Grading Labeling (UTQGL)** of passenger car tires. This act established standards in three areas of tire performance: treadwear, traction, and temperature resistance. Every new tire is tested and graded against these standards, and the score or grade is molded into the sidewall of the tire. (Fig. 2–22)

The treadwear standard is in a state of flux with some degree of conflict between the **Rubber Manufacturer's Association (RMA)**, the **National Highway Traffic Safety Administration (NHTSA)**, some tire manufacturers, and

FIGURE 2–22.
This tire has a Uniform Tire Quality Grade Label of Tread-wear 220, Traction B, and Temperature B.

some consumers' groups. The base grade for treadwear is 100. If a tire will give twice the tread life as the base tire, it is graded 200 (the maximum grade). If it will go 50 percent farther than the base, the grade will be 150; if it goes 50 percent less than the base, the grade will be 50. There is some question as to the validity of this particular standard. Some tire manufacturers have been able to revise their scores and use this modified score in their advertising. Also, in the real world of automobile operation, treadwear is dependent on several other factors including inflation pressures, alignment, tire pressure, vehicle type, driving habits, road type, climate, and so forth. However, the treadwear grade does give an indication of probable tread life differences when comparing two different tires. (Fig. 2–23)

Traction grades are A (highest), B (middle), and C (lowest). Traction tests are used to test the tire's ability to stop on wet pavement. A-graded tires have a wet pavement coefficient of friction greater than 0.35 on concrete and 0.5 on asphalt; B-graded tires will have a coefficient of friction greater than 0.26 on concrete and 0.38 on asphalt; and C-graded tires will have a coefficient of friction less than this. Coefficient of friction is a term that refers to the relative friction between two surfaces. A tire marked C will probably have poor stopping ability on wet pavement. Traction tests give a poor indication of the tire's ability to corner;

this ability is often quite different from the tire's ability to accelerate or stop. Also, a tire with good wet pavement traction does not necessarily have good dry pavement traction.

Temperature grades are A (highest), B (middle), and C (lowest). They indicate the tire's ability to resist heat generation, its ability to get rid of heat, or both. Heat is one of the tire's worst enemies. It can reduce the bonding strength of the rubber materials, which can cause separation of the cords and plies from the rubber layers or cause separation of the rubber layers from each other. Heat will also cause the rubber in the tire to become harder and more brittle over time as the added heat causes the rubber to continue vulcanizing. A C grade corresponds to the minimum level of performance a tire must show to pass a DOT tire safety test. A and B scores indicate a tire quality somewhat better than the minimum, with A being better than B.

2.7   SPEED RATINGS

Another tire rating related to speed is commonly used in Europe. In the United States, the 55 MPH (88.55 KPH) speed limit makes this rating unnecessary. A letter (S, V, or H) is added to the tire size just after the aspect ratio; for

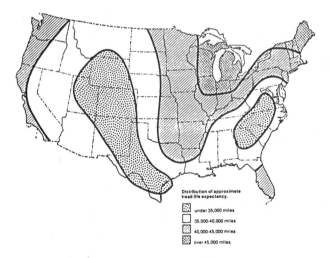

FIGURE 2–23.
Because of different road construction materials and methods and different climatic conditions, motorists in different parts of our country can expect to get different tread life from their tires. (Courtesy of Uniroyal)

example, a 195/70 H R 14 has an H speed rating. A speed rating of S indicates a tire that is speed rated to 112 MPH (180 KPH), H indicates a speed rating to 130 MPH (210 KPH), and V indicates a tire that can sustain speeds of over 130 MPH (210 KPH). These are the most common speed ratings; there are at least nineteen different ratings. In Europe, even tractor tires have a speed rating. The ones that will probably be increasingly used on passenger cars are the following: P to 93 MPH (150 KPH), Q to 99 MPH (160 KPH), R to 106 MPH (170 KPH), T to 118 MPH (190 KPH), and U to 124 MPH (200 KPH), in addition to the S, H, and

V ratings already discussed. (Table 2–4) The speed rating indicates at what speed a tire can be safely operated without danger of failure. In Europe, it is illegal to drive faster than the speed rating of the tires.

The speed itself does not damage the tire, but the tire can fail because of heat generated by a tire phenomenon called the **standing wave**. As mentioned earlier, the road contact portion of the tread and the sidewalls next to it are deformed by the weight of the vehicle. The natural elastic properties of the tire cause the tread and sidewalls to return to their normal shape as they roll away from the road surface. At high speeds, these parts of the spinning tire cannot return to their normal shape fast enough and can still be deformed when they return to the road surface. This will cause a noticeable dis-

TABLE 2–4
Speed Symbol and Load Index*

Example: 195/60R15 86H		
Speed Symbol	Speed (km/h)	Speed (mph)
A1	5	3
A2	10	6
A3	15	9
A4	20	12
A5	25	16
A6	30	19
A7	35	22
A8	40	25
B	50	31
C	60	37
D	65	40
E	70	43
F	80	50
G	90	56
J	100	62
K	110	68
L	120	75
M	130	81
N	140	87
P	150	93
Q	160	99
R	170	106
S	180	112
T	190	118
U	200	124
→ H →	210 →	130 →
V	210 Plus	130 Plus

*Note: All calculation for km/h to MPH are rounded off. Your calculation may differ. If in doubt, use RMA's recommendation. The speed symbol indicates the maximum speed at which a tire can safely carry its rated load. (Courtesy of Bridgestone Tire Company)

FIGURE 2–24.
This high speed photograph was taken of a cut steel-belted tire running on a test roll at 120 MPH (193 KPH). Note the standing wave distortion in the sidewall and tread areas. (Courtesy of Armstrong Tire Company)

tortion to develop in the tire. The standing wave creates internal friction that will cause the tire to become hotter, wear faster, and lose handling and traction capabilities. It can quickly destroy a tire. (Fig. 2–24)

2.8 TUBES AND TUBELESS TIRES

Originally, all tires used tubes to seal the air inside of the tire. A tube is much like its name, a tube of rubber. It has no end, much like a doughnut. (Fig. 2–25) A tube does have an opening at the valve stem to make it possible to put air in it or let air out of it. The valve core is threaded into the stem to provide the valve action and hold the air pressure; it is commonly called a Schrader valve. This valve is opened when a projection on the tire inflator presses inward on the stem of the valve. A cap is usually installed over the valve to protect it from dirt and other debris and to provide a secondary seal for holding the air in the tube. (Fig. 2–26)

There are two general types of tubes used with passenger car tires, radial and nonradial. A nonradial tube should only be used in a bias or a bias-belted tire. A radial tube can be used in any tire type, but it is more expensive. Ra-

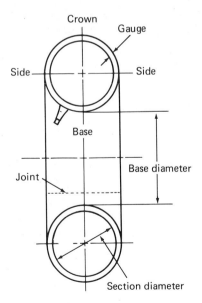

FIGURE 2–25.
An inner tube, essentially a rubber tube, is available in different sizes.

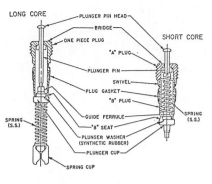

Valve core components.

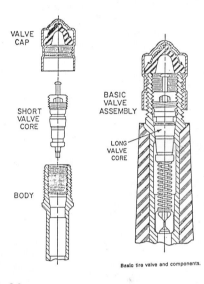

Basic tire valve and components.

FIGURE 2–26.
Tire valve cores and stems. Note the valve core sealing surfaces and valve core-to-stem sealing area. (Courtesy of Schrader Automotive Inc.)

dial tubes have improved splices where the tube is joined during manufacture. In many cases, they are also more elastic to contend with the increased sidewall flex of the radial tire.

Today, most passenger car tires are **tubeless**; that is, they do not use or need a tube. Tubeless tires have a definite safety advantage. A tube tends to deflate very rapidly or blow out when pierced. This can be compared to the action of a balloon that is stuck by a pin; the stretched out rubber will pull away from the pin hole, which gets larger, very rapidly. When pierced, the rubber in a tubeless tire will tend to move toward the hole. This causes a much slower deflation and much greater car control while the tire is losing air pressure. Tubeless tires also run cooler without the increased rubber thickness and friction from the tube.

FIGURE 2–27.
A tubeless tire's bead makes an airtight seal at the wheel's bead seat and flange, and the tire's body is airtight. The valve stem is also sealed to the wheel and allows the tire to be inflated.

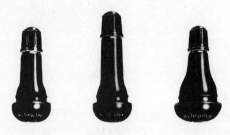

Snap-In Tubeless Tire Valves

**Metal Clamp-In
Passenger Car Valves**

FIGURE 2–28.
Snap-in and metal clamp-in tubeless tire valves are available in different diameters and lengths. (Courtesy of Schrader Automotive Inc.)

The wheel rim for a tubeless tire must be airtight, and the tire bead must make an airtight seal with the rim. The inside of the tire body is lined with an airtight layer of rubber called a butyl liner. The **valve stem** and **Schrader valve** are installed in the wheel rim. (Fig. 2–27) There are two general styles of tubeless tire valve stems used with passenger cars. The rubber, snap-in type is held in place by the compression of rubber, and the metal, clamp-in type is locked in place by tightening a nut. (Fig. 2–28) Valve stems are available for the two common wheel hole sizes of 0.453 inch (11.5 mm) and 0.625 inch (15.87 mm). Each size is available in various lengths. A valve stem should be just long enough to extend through the wheel cover to allow for easy tire inflation. Wheels smaller than 14 inches usually use the smaller 0.453-inch stem. Wheel sizes of 15 inches and larger use the 0.625-inch stem.

A tubeless tire is normally used without a tube; use of a tube will make the tire run hotter and probably shorten its life. The increased heat is caused by tube-to-sidewall chafing and the increased rubber thickness of the sidewall. Use of a tube is sometimes necessary with a porous alloy or mag wheel since porous rims will let the air leak through minute holes in the metal. It is usually better to paint the inside of the rim or treat it with sealant rather than use a tube. Spoke wheels will require a tube if the spokes extend into the wheel rim, which they usually do. A rubber or tape rim strip should also be used to prevent the tube from chafing against the spoke ends.

2.9   REPLACEMENT TIRE SELECTION

Most car owners are quite satisfied with the **O.E.M. (Original Equipment Manufacturer)** tires that they have on their car, and their major concern is how to get the most tire life from their investment. They will usually purchase a tire of the same type and size as the original for replacement. When choosing a slightly different replacement tire for these cars, it is wise to follow these recommendations:

- A replacement tire should be of the same size or slightly larger than the original. Make sure that the replacement tire has a load carrying capacity that is equal to or greater than the load to be carried.

- A tire that is one size larger (cross section) will usually fit on the wheel and in the fender well with no interference problems. It will cost slightly more but, because of its larger size, will deliver more miles before wearing out. During the life of this tire, the cost per mile will usually be slightly lower.

- Replacement tires should have tire pressure characteristics that match those on the tire inflation sticker on the car.

- When the replacement tire is a different size than the original, the rim width should be checked to insure its suitability. The fender clearance should also be checked after the tire and wheel are mounted to ensure that the tire can operate in all possible suspension travel and steering positions without interference.

Many people are not completely satisfied with their present tires and want to replace them in order to improved their car's handling or wet weather traction, increase their fuel mileage, or improve the looks of the car. When selecting replacement tires that are different from the O.E.M. tires, it is wise to follow these recommendations:

- Ideally, all four tires on a car will be of the same size and construction; it is very important that the two tires on an axle be of the same size, tread pattern, and construction. Mismatching tires on an axle can cause a pulling or self-steering to one side as the car goes down the road. (Fig. 2–29)

- If the rear tires are to be different from the front tires, never put radial-ply tires on the front with bias-ply tires on the rear. Also, never put a lower profile tire on the front. If the most responsive tires (i.e., those with better handling characteristics) are put on the front of the car, a potentially dangerous oversteer condition can develop. This will be discussed in more detail in Chapter 19.

| Construction | | FRONT TIRES (Read DOWN for Rears) | | | | | | | | | | | |
| | | DIAGONAL (BIAS) | | | | BELTED BIAS | | | | RADIAL | | | |
		Numeric & 80 Series	78/75 Series	70 Series	60/50 Series	Numeric & 80 Series	78/75 Series	70 Series	60/50 Series	Numeric & 80 Series	78/75 Series	70 Series	60/50 Series
DIAGONAL (BIAS)	Numeric & 80 Series	P	A	NO	NO	A	A	NO	NO	NO	NO	NO	NO
	78/75 Series	A	P	A	NO	A	A	NO	NO	NO	NO	NO	NO
	70 Series	A	A	P	NO	A	A	A	NO	NO	NO	NO	NO
	60/50 Series	A	A	A	P	A	A	A	A	NO	NO	NO	NO
BELTED BIAS	Numeric & 80 Series	A	A	NO	NO	P	A	A	NO	NO	NO	NO	NO
	78/75 Series	A	A	A	NO	A	P	A	NO	NO	NO	NO	NO
	70 Series	A	A	A	NO	A	A	P	NO	NO	NO	NO	NO
	60/50 Series	A	A	A	A	A	A	A	P	NO	NO	NO	NO
RADIAL	Numeric & 80 Series	A	A	A	NO	A	A	A	NO	P	A	A	NO
	78/75 Series	A	A	A	NO	A	A	A	NO	A	P	A	NO
	70 Series	A	A	A	NO	A	A	A	NO	A	A	P	NO
	60/50 Series	A	A	A	A	A	A	A	A	A	A	A	P

REAR TIRES (Read ACROSS for Fronts)

NOTES:
P: Preferred applications. For best all-around car handling performance, tires of the same size and construction should be used on all wheel positions.
A: Acceptable, but not preferred applications. Consult the car owner's manual and do not apply if vehicle manufacturer recommends against this application.
NO: NOT RECOMMENDED

FIGURE 2–29.
Ideally, all of the tires on a passenger car will be of the same size, aspect ratio, and construction type. With care, some mixing of tire type and aspect ratio can be done and safe handling retained. (Courtesy of Bridgestone Tire Company)

2.10 SPARE TIRES

From the beginning, motorists have worried about flat tires and the effect they have on vehicle operation. A **spare tire** that can be temporarily used to replace the flat tire is very common. At one time, the spare was a fifth tire and wheel of the same size and type as the four standard tires. On older cars, the spare is often the best tire from the last, worn-out set. Today, a new car might come with one of three types of temporary-use spares: **stowaway, compact,** or **temporary**. All three of these are of bias-ply construction and are useable for about 3,000 miles (4,830 Km) at speeds up to 50 MPH (80.5 KPH), unless a slower speed is required. They are all marked "temporary" on the sidewall. They offer the advantages of being lighter, which makes them easier to handle and helps improve fuel mileage, and also smaller, which allows more storage space in the car. (Fig. 2-30)

In all cases where the spare is different from the other tires on the car, driving speed while using the spare should be reduced because of the change in traction and handling. Also, if the spare is being used on the drive axle and that axle is equipped with a limited-slip differential, the spare must have the same effective diameter as the other tire on that axle.

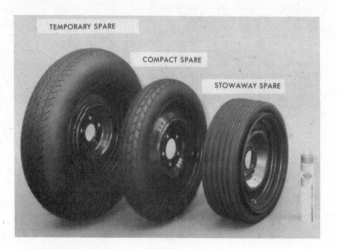

FIGURE 2-30.
Three different types of temporary use spare tires. The stowaway spare is stored flat and inflated when it is used. (Courtesy of General Motors)

A folding, stowaway spare is not inflated until it is needed and is then inflated either with a pressurized canister that is carried in the car or with a standard air hose. During inflation, the tire should be either mounted on the car axle with the lugs slightly tightened or on a tire machine to prevent personal injury. After use, this tire can be deflated, refolded, and returned to its storage space. This tire is usually not serviceable; that is, the tire and wheel are replaced as an assembly if the tire becomes worn out or damaged.

A compact spare is a narrow, lightweight, high pressure tire. It is mounted on a wheel that is usually narrower in width and larger in diameter than the standard wheels on the car. A compact spare is fully serviceable, just like the standard tires.

A temporary spare is simply a full-size, lightweight tire of the same effective diameter as the standard tires. It is usually used on cars with limited-slip differentials. This tire is also serviceable using standard equipment.

2.11   NO-FLAT, RUN-FLAT TIRES

The ideal spare tire is none at all. Then it would cost nothing, not add any weight to the car, and not take up any storage space. But, in order to operate without a spare, we need tires that either will not go flat or that can be operated with no air pressure in them.

If a pneumatic tire is operated very long without pressure, the excessive sidewall flexing will generate so much heat that the tire will start to disintegrate or even catch fire or the bead will move into the dropwell at the center of the wheel. If the latter happens, the bead can work its way off of the wheel.

No-flat or **self-sealing tires** are built with a soft, pliable inner lining that is gummy. If a puncturing object penetrates the tire, this sealant will seal to the puncturing object, or, if the object is withdrawn, the sealant will seal the hole. Self-sealing tires tend to be heavier and more expensive. They will also generate slightly more heat then a standard tire. (Fig. 2-31)

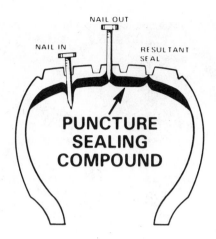

FIGURE 2–31.
A self-sealing tire has a puncture sealing compound on the inside of the tread area. This compound will seal to a projection or to the hole if the projection is removed and will prevent a flat. (Courtesy of Oldsmobile)

Run-flat tires must be strong enough to support the car, ensuring vehicle stability and mobility after being deflated. They also must have enough internal strength to keep the beads seated or be used with a wheel design that will do the same. To support the car, a run-flat tire will use either reinforced sidewalls or a raised section or inner tire inside of it. If an inner device is used, a lubricating system must also be built into the tire and wheel to reduce inner friction while the tire runs flat. Special tire bead, wheel bead flange, and drop well designs can be used to reduce the possibility of bead unseating. Run-flat tires tend to be heavier and more expensive and to have driving characteristics slightly different from standard tires.

2.12 RETREADS

When the tread of the tire becomes worn down to a depth of 2/32 inch (1.587 mm) from the bottom of the groove, a tire is considered to be worn out. (For a long time 1/32 inch was the measurement standard for the tire industry.) Many states will not only cite a motorist with worn out tires but will also remove the car from the road. Many tires will still have strong, useable bodies at this time, and it is a waste of en-

ergy and materials to discard these tires. **Retreading**, also called **recapping**, is the process of applying new tread rubber and a new tread pattern to the old body. Two slightly different processes, **hot-cap** and **cold-cap**, are used to secure the new tread. Cold-caps are now commonly called **precured** recaps.

Both processes begin with a sound, undamaged tire body or casing. The casing is carefully inspected to ensure that there are no internal defects or structural flaws. Next, the remaining old tread is ground or rasped off of the casing. This provides a clean, properly roughened rubber surface for the new tread rubber. (Fig. 2–32)

New tread rubber is then wrapped around the casing. In the hot-cap process, uncured rubber in one of two forms is used. In one form, a strip about 3/8 inch (9.5 mm) thick, as wide as the tread, and as long as the tread circumference is wrapped once around the tire. In the other form, a much longer strip of the same thickness but about 1 inch (25.4 mm) wide is wrapped around the tire the number of revolutions that is required to make up the tread width. The single wrap process can leave either too much or too little rubber at the splice junction if the tread length is not exactly right for the tire circumference. This can cause tire run-out or balance problems. Some recappers use

FIGURE 2–32.
These truck tires have had all of the remaining tread buffed off of them. They are ready for new tread rubber to be applied.

FIGURE 2-33.
New, uncured tread rubber has been applied to these tires; they are now ready to be placed into the recapping molds.

FIGURE 2-34.
A group of recapping molds used to form the new tread pattern, cure the new tread rubber, and bond it to the carcass. A section of the actual tread mold is shown at the lower left.

the splice junction to their advantage. The tire balance is checked after the body is buffed, and as the tread rubber is wrapped, the splice is placed at the light portion of the body. The overlap is then adjusted so it is the correct weight to balance the tire. The continuous wrap method does not have this junction problem. (Fig. 2-33)

After the tread has been applied, the tire is put into a mold to form the tread pattern, a process similar to the original tire molding process. (Fig. 2-34) A thick, heavy duty curing tube is placed in the tire and inflated to a pressure over 100 psi (689.5 kPa) while in the mold. This internal pressure is used to force the tread rubber into the mold to ensure complete bonding and tread design formation. The mold is heated to about 300° F (149° C) to form the tread pattern, cure the tread rubber, and bond it to the body. The secrets of getting a good recapped tire are preparing and conditioning the bonding surface on the tire body and using pure bonding cement, good bonding rubber, the correct air pressure, the correct mold temperature, and the correct curing time. After molding, the tire is ready for cleanup and to be returned to service. (Fig. 2-35)

Precured recapping uses cured tread rubber that is premolded with the tread pattern

and is the right width for the tire. This process was developed by the Bandag Corporation. An adhesive and an intermediate layer of bonding rubber, called a cushion strip, are applied to the body and the new tread is wrapped around it. Pressure is applied to squeeze the tread onto the casing, and it is then heated to about 212° F (100° C) to cure the adhesive. Precured recaps are commonly used on truck tires. Because the tread is precured and does not include the cord materials of the body, it can be molded under a much higher pressure and temperature,

FIGURE 2-35.
These tires are cured hot caps; they still need to be trimmed.

which results in a more dense, longer running tread. Precured recaps will often deliver more mileage than the original tread. Also, the lower temperature used in this process does not shorten the life of the tire body as much as the hot process. (Fig. 2–36)

Although recaps are becoming less and less common in passenger cars, they remain quite popular in the trucking industry where the price of a new tire is about $250 to $300. A worn carcass, worth about $75, is too expensive to throw away. Also, since many trucks average about 200,000 miles (321,800 km) a year and have 18 or more tires, tire costs can be significant. It is possible to save about one-fourth of the tire costs by using recaps. This can amount to a few thousand dollars a year in savings.

Retreads are less expensive than standard tires. Depending on the skill and care of the recapper and the quality of the tread rubber, they can be as strong and as durable as the original tire. There is often a problem of the recapper getting a supply of sound carcasses for a particular size and type of tire. Some recappers will recap a set of tires and return them to the owner, but this usually takes a certain amount of time. This is referred to as custom capping.

FIGURE 2–36.
A precured strip of tread rubber is being applied to this tire over a strip of cushion gum. It will need to be stitched firmly in place and then heat-bonded to the carcass.

REVIEW QUESTIONS

1. Statement A: The amount of grip between the tire and the road is called traction. Statement B: A tire has its best grip when it is slipping slightly across the road surface. Which statement is correct?

   a. A only   **c.** both A and B
   b. B only   **d.** neither A nor B

2. The bead wires of the tire
   A. clamp onto the wheel rim.
   B. help transmit starting and stopping torque from the wheel to the tire.

   a. A only   **c.** both A and B
   b. B only   **d.** neither A nor B

3. Which of the following is not a common sidewall ply material?

   a. cotton   **c.** nylon
   b. rayon   **d.** polyester

4. The cords in a tire's sidewall plies
   A. cross each other at about a 35° angle in a bias-ply tire.
   B. are used with a belt in a radial-ply tire.

   a. A only   **c.** both A and B
   b. B only   **d.** neither A nor B

5. Statement A: The belt helps reinforce the tread area of the tire. Statement B: Each

type of belt material requires a special rubber blend to bond the belt to the material next to it. Which statement is correct?

a. A only
b. B only
c. both A and B
d. neither A nor B

6. Which of the following materials are used in tire belts?

a. rayon
b. steel
c. glass fiber
d. all of the above

7. The tire mold is used to
A. form the tread pattern and sidewall markings.
B. vulcanize the rubber.

a. A only
b. B only
c. both A and B
d. neither A nor B

8. Statement A: The name of the manufacturer and the tire size must be molded into the tire sidewall. Statement B: The maximum air pressure a tire can safely hold and whether the tire has a directional tread must be molded into the sidewall. Which statement is correct?

a. A only
b. B only
c. both A and B
d. neither A nor B

9. The DOT number of a tire is coded to tell

a. what day of the year a tire was made.
b. the name and location of the plant that made the tire.
c. what the ply construction of the tire is.
d. all of the above.

10. The DOT number of a tire that ends in 506 indicates the tire was made

a. in December of 1976 or 1986.
b. on the 50th day of 1976.
c. on the 50th day of 1986.
d. any of the above.

11. A tire size of 185/70 R 14 indicates
A. the tire has a tread width of 185 mm.

B. the distance between the beads across the tire diameter is 14 cm.

a. A only
b. B only
c. both A and B
d. neither A nor B

12. A tire size of P 185/70 R 14 indicates that
A. the width of the tire's cross section is 70 percent of the tread width.
B. the tire is of a radial construction.

a. A only
b. B only
c. both A and B
d. neither A nor B

13. A P 185/70 R 14 tire is designed for use on
A. passenger cars.
B. light trucks.

a. A only
b. B only
c. both A and B
d. neither A nor B

14. Statement A: The tire does not support the car; air does. Statement B: A tire with more plies generally has a lower load rating. Which statement is correct?

a. A only
b. B only
c. both A and B
d. neither A nor B

15. Statement A: A tire with a treadwear grade of 75 should last a relatively long time and give good tire mileage. Statement B: A tire with an A traction grade should give good wet condition stopping. Which statement is correct?

a. A only
b. B only
c. both A and B
d. neither A nor B

16. Statement A: Generally, the higher the letter in a tire's speed rating, the faster the tire can be operated. Statement B: Tire speed ratings are only enforced in Europe. Which statement is correct?

a. A only
b. B only
c. both A and B
d. neither A nor B

17. A tubeless tire must

    a. be used on an airtight wheel.
    b. make an airtight connection at the wheel flange.
    c. have an airtight inner liner.
    d. all of the above.

18. Statement A: Replacement tires must be the same size or slightly larger than the O.E.M. tires. Statement B: Replacement tires can have a lower load rating than the O.E.M. tires. Which statement is correct?

    a. A only   **c.** both A and B
    b. B only   **d.** neither A nor B

19. Statement A: A temporary spare tire can be used just like the other tires on the car.

Statement B: A stowaway spare cannot be patched or repaired. Which statement is correct?

    a. A only   **c.** both A and B
    b. B only   **d.** neither A nor B

20. A retread tire has
    A. a new tire tread cut into the original tire.
    B. new tread rubber with a new tread formed into it.

    a. A only   **c.** both A and B
    b. B only   **d.** neither A nor B

Chapter 3

WHEELS—THEORY

After completing this chapter, you should:

- Be familiar with the terms commonly used to describe wheel construction, sizing, and operation.

- Have a basic understanding of how wheels are constructed.

- Understand the modern wheel size designations.

- Be able to select a correct replacement wheel for a car.

3.1 INTRODUCTION

A wheel has a fairly easy job; all it has to do is secure the tire to the hub. But, in doing this, the tire must be held exactly centered in two directions, radially and laterally. It also must be held centered under some rather severe loads. These loads are the vertical load of the vehicle's weight; the shock loads caused by bumps; the twisting, torsional loads of braking and acceleration; and the sideways load that occurs during cornering. (Fig. 3–1)

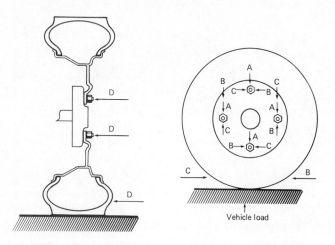

FIGURE 3–1.
A wheel is subject to forces generated by the weight of the vehicle (A), braking (B), acceleration (C), and cornering (D).

3.2 CONSTRUCTION

A wheel is made up of two main parts—the **center section**, also called a **spider** or **disc**, which attaches to the hub, and the **rim** where the tire is attached.

The **rim** is usually made from rolled steel, but it can also be made from stamped, cast, or spun aluminum or magnesium alloy. The rim has two **bead flanges** that the tire beads push against. These flanges must be strong enough to keep the tire beads in place. Internal tire pressure exerts a constant, outward pressure that forces the tire beads against the wheel flanges. Just inside of the flanges are the **bead**

seats, which are at a slight angle of about 15° for passenger cars. When you break a tire bead loose from the bead seats, you soon realize how tightly the tire beads wedge onto the tapered seat of the wheel. The flanges control the tire position in a lateral, sideways direction while the tire beads wedging onto the bead seats position the tire in a radial, vertical direction. Friction between the tire bead and the wheel bead seats and flanges transmit braking and acceleration forces from the tire to the wheel or vice versa.

Flanges are made with different heights and curvatures to match the tire bead configuration. Flange type is designated by a classification system using one or two letters. The most commonly used flange heights are: J = 0.68 inch (17.3 mm), JJ = 0.69 inch (17.5 mm), JK = 0.71 inch (18 mm), K = 0.77 inch (19.5 mm), and L = 0.85 inch (21.6 mm). Passenger car rims have a raised bead just inside of the bead seat called a **safety bead**. This safety bead helps keep the beads of a flat tire from moving into the well if the tire deflates. If a flat occured without this safety bead, the bead of the flat tire might work its way off the wheel before the driver could stop the car. The safety bead is required for DOT approval on tubeless tire wheels. (Fig. 3–2)

All passenger cars use **drop center** rims. The rim is shaped with a lowered well section near the center, and this drop center or well is

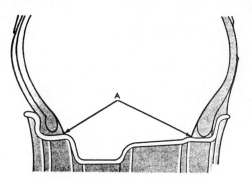

FIGURE 3–2.
Passenger car wheels have a raised area just inside of the tire bead at A; this wheel is often called a safety rim wheel. Note the dropped section of the wheel rim and that it is off-center, toward the left. (Courtesy of Chrysler Corporation)

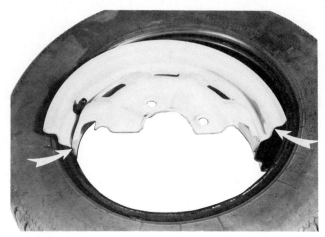

FIGURE 3-3.
The dropped section of the wheel rim allows us to remove and replace a tire on a wheel; note that when the tire bead is in the well, the other side of the bead is outside of the wheel flange (arrows).

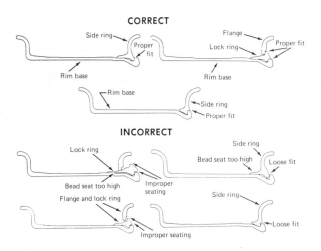

FIGURE 3-4.
Three styles of two- and three-piece rims. Internal tire pressure pushing sideways on the beads holds the side and lock rings in place. Removal of the side ring or flange and lock ring allows the tire to be slid off the side of the rim. If the side ring or lock ring is not properly seated, it can fly off the wheel with a great deal of force. (Courtesy of Goodyear)

necessary so the tire can be installed or removed. As we learned in the last chapter, the tire bead is reinforced with steel wire; it will not stretch. The wheel well is deep enough so that the bead will pass over the flange on one edge of the rim when the other side of the bead is in the well. (Fig. 3-3) Most truck and heavy equipment wheel rims are of a two- or three-piece configuration, commonly called **split rims.** One of the bead flanges is removable for tire removal or installation. Service work on this tire and wheel combination requires special skill and equipment and extreme care. (Fig. 3-4) Split rims have separated during tire inflation and even while running on the road. The flying rim flanges have killed or severely injured service personnel and bystanders. Space does not permit covering split rim service operations in this text. Instructions are published by various wheel manufacturers and the Rubber Manufacturers Association.

The center section of a passenger car wheel, like the rim, is usually made from stamped steel; it can also be made from cast, stamped, or spun aluminum or magnesium alloy. It is usually welded to the rim, but it can be riveted, or bolted, or connected using wires or spokes.

A center or spindle hole is centered in this wheel part and surrounded by three to six lug bolt holes. The size and number of lug bolts is determined by the vehicle load and expected

driving conditions. Heavy duty vehicles will commonly use six or eight lug bolts; most passenger cars use four or five. **Lug bolt circles,** also called **bolt patterns,** are of different diameters depending on the design strength or the available space. Bolt pattern sizes are usually printed as 5-4½ or 4-100 mm; this would mean there are five holes that are evenly spaced around a 4½ inch diameter circle or four holes evenly spaced around a 100 millimeter diameter circle. It is easy to measure the bolt circle diameter on even-numbered, four, six, or eight lug circles; just measure from the center of one lug bolt to the center of the lug bolt across the wheel. A three or five bolt circle diameter is difficult to measure. Most shops will use charts that specify the pattern for a particular car or a template to check a particular wheel or hub. (Fig. 3-5)

Composite wheels made from plastics will probably be commonly used in the near future. Composite wheels are made from various plastic compounds and are usually reinforced with fiberglass or carbon fibers to produce adequate strength. It is expected that composite wheels will be a stronger, lighter, less expensive, and better looking wheel that will be free from corrosion.

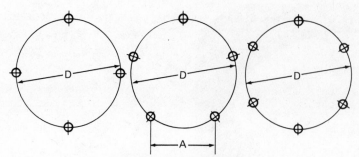

A x 1.7013 = Circle diameter for 5 hole circle

FIGURE 3–5.
The wheel's bolt circle diameter is measured as shown here. For five hole wheels, the bolt circle diameter can be determined by muliplying the center-to-center distance between two wheel studs (A) by 1.7013.

3.3 WHEEL SIZES

The two most commonly used wheel dimensions are diameter and width. These dimensions match the size of the tire. Wheel diameter is measured from bead seat to bead seat across the diameter of the wheel and must be exactly the same as the tire diameter. The flange height is not included in the wheel diameter. (Fig. 3–6)

Wheel width, measured across the rim at the inside of the bead flanges, is usually smaller than the tire width. Some tire manufacturers will specify a particular wheel "design" width for their tires. Tables are published that show recommended rim widths for each tire size. (Fig. 3–7)   If no other information is available, a rule of thumb is that the rim width should be about 75 to 80 percent of the tire cross-sectional width. A wider rim tends to reduce sidewall flexing, which results in increased steering response, stiffer ride, and increased tread shoulder wear. A narrower rim will tend to increase sidewall flexing, sidewall damage, and center tread wear, and produce a softer ride. (Fig 3–8)

3.4 OFFSET-BACKSPACING

Another important wheel dimension to consider when replacing a wheel is **offset**. This specifies where the wheel mounting face is rel-

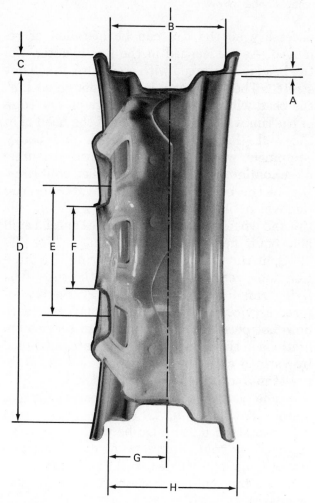

FIGURE 3–6.
The different wheel dimensions that are illustrated here are (A) bead seat angle, (B) rim width, (C) flange height, (D) rim diameter, (E) bolt circle diameter, (F) spindle hole diameter, (G) offset, and (H) back spacing.

Approved rim widths (inches)

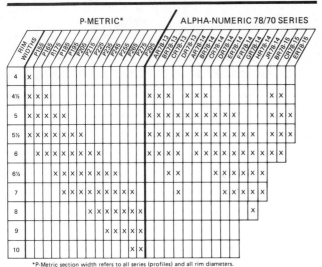

RIM WIDTHS	P155	P165	P175	P185	P195	P205	P215	P225	P235	P255	P265	P275	P295	AR78-13	BR78-13	CR78-13	DR78-13	AR78-14	BR78-14	CR78-14	DR78-14	ER78-14	FR78-14	GR78-14	HR78-14	JR78-14	BR78-15	CR78-15	ER78-15
4	x																												
4½	x	x	x											x	x	x		x	x	x							x	x	
5	x	x	x	x	x									x	x	x	x	x	x	x	x	x					x	x	x
5½	x	x	x	x	x	x	x							x	x	x	x	x	x	x	x	x	x	x			x	x	x
6		x	x	x	x	x	x	x	x					x	x	x		x	x	x	x	x	x	x	x	x		x	x
6½			x	x	x	x	x	x	x						x	x						x	x	x	x	x			
7				x	x	x	x	x	x	x							x						x	x	x	x			
8							x	x	x	x	x	x														x			
9									x	x	x	x	x																
10																												x	x

*P-Metric section width refers to all series (profiles) and all rim diameters.

FIGURE 3-7.
Charts are available to show the approved rim widths for different size tires; for example, 5-1/2, 5, 6, 6-1/2, and 7 inch wide rims are correct for a P205 tire. (Courtesy of Rubber Manufacturers' Association)

ative to the rim and determines where the tire center line will be relative to the hub. If the wheel mounting face is centered and directly in line with the wheel center line, there is zero offset. In such a case, there will be an equal amount of rim width on each side of the wheel mounting surface at the hub.

FIGURE 3-8.
Many wheels have their size stamped into the rim. This wheel is 4-1/2 inches wide, has a "J" flange, and is 13 inches in diameter; the other numbers are the manufacturer's part number.

If the wheel mounting face and tire center line are not centered or in line, the wheel has a certain amount of positive (+) or negative (−) offset, and the rim and tire will be offset from the hub. There is some controversy regarding which direction positive or negative offset refers to. Better sources agree that negative offset moves the tire and rim outboard from the mounting flange and increases the track width. The mounting flange will be inboard from the wheel center line. Positive offset (used on many FWD cars) does just the opposite. The tire is moved inward relative to the mounting flange. (Fig. 3-9)

Determining offset requires measuring the wheel width between the flanges and measuring the distance from the mounting flange to the back side of the wheel, often called **back spacing, back side spacing,** or **rear spacing.** Next, the rim width is divided by two and the wheel flange width is subtracted from the back spacing. Now, subtract the result of one-half of the rim width from the corrected back spacing and you have the amount of offset. If the back spacing dimension is less than one-half of the

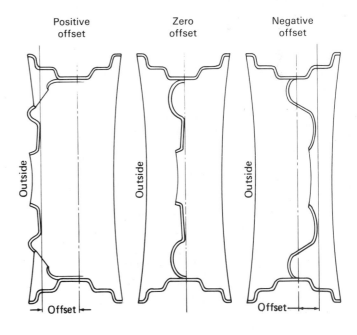

FIGURE 3-9.
If the mounting flange is centered in the wheel, the wheel has zero offset (center). If it is outward from the wheel center, the wheel has positive offset; if it is inward, the wheel has negative offset.

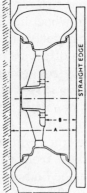

You can measure offset with or without the tire mounted on the wheel. Place wheel with the outboard side down and take dimensions "A" and "B". Offset is calculated with the formula shown. If "B" is larger than "A/2" the offset is positive. If tire is not on wheel, take measurements to edge of rim rather than tire.

$$OFFSET = \frac{A}{2} - B$$

FIGURE 3–10.
Wheel offset can be measured using the procedure shown here. (Courtesy of Bridgestone Tire Company)

rim width, the wheel has negative offset. If the back spacing dimension is greater, the offset is positive. (Fig. 3–10)

Offset is a little difficult to handle by some people and a little difficult to measure by all. Many wheel manufacturers simply use the back spacing dimension instead of offset. It is much easier to measure, and it is almost the same as using offset dimensions. (Fig. 3–11) But, you should remember that a 6-inch (15.24 cm) wheel with a 2-inch (5.08 cm) back spacing will have a different offset, −1 inch (2.54 cm) than a 5-inch (12.7 cm) wheel with the same back spacing, −1/2 inch (1.27 cm). The tire center line will

FIGURE 3–11.
Wheel back spacing is measured by placing a straight edge across the wheel at the bead flanges and measuring the distance between the mounting flange and the straight edge.

move 1/2 inch (1.27 cm), and the track will change 1 inch (2.54 cm) if one pair of these wheels is replaced by the other.

Wheel offset has several effects on vehicle operation. Negative offset (i.e., the tire center line is moved outward) will increase the track width, which will reduce the lateral weight transfer (Chapter 19). This is a benefit, but it also reduces the tire-to-fender clearance, moves the wheel bearing load outward and increases the load on the smaller wheel bearing, increases the wheel leverage on the springs and suspension, and changes the **scrub radius**. The scrub radius, also called the **king-pin offset** or **steering axis offset**, is the distance from the center of the tire to the steering axis measured at the road surface. Most RWD cars position the tire center line slightly outboard of the steering axis while most FWD cars place the tire center line inward. This can be easily seen by comparing the wheel offset of the two styles of cars. Changing the scrub radius can have drastic effects on vehicle handling, especially when a front tire meets a bump or pothole. It is highly recommended that replacement wheels have the same or very close to the same offset as the O.E.M. wheels in order to maintain the correct scrub radius. (Figs. 3–12 & 3–13)

3.5 AFTERMARKET WHEELS

Building and selling aftermarket wheels has become a rather strong industry. The major reason for replacing the O.E.M. wheels is probably cosmetic, but other common reasons are a desire to change the wheel width to accommodate wider profile tires or to change the wheel diameter to allow plus 1, 2, or 3 changes. (Fig. 3–14)

An increasingly popular concept with car enthusiasts is the **plus 1, plus 2,** or **plus 3** conversion. If you desire a wider tire with a wider/shorter aspect ratio you usually have to give up tire diameter. The plus 1 concept is to use a wheel that is one inch (2.54 mm) larger in diameter and a tire that is one step greater in aspect ratio—80 to 70, 70 to 60, 60 to 50, or even 50 to 40. Plus 2 uses a wheel that is two inches

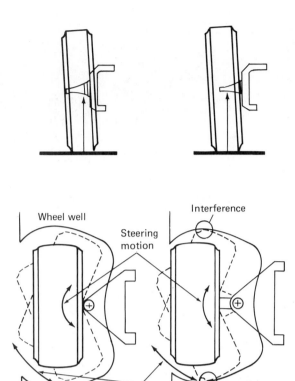

FIGURE 3–12.
If an O.E.M. wheel (top left) is replaced with a wheel having more negative offset (top right), the load on the spindle will move outward and place more load on the outer wheel bearing. The tire will also move outward in the fender well (lower right) and often cause interference during turning manuevers.

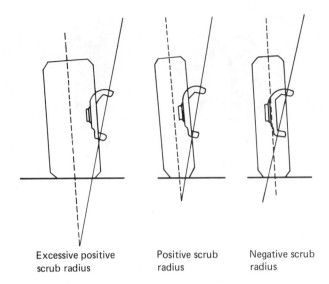

Excessive positive   Positive scrub   Negative scrub
scrub radius   radius   radius

FIGURE 3–13.
The distance between the center of the tire and the steering axis at ground level is called the scrub radius. A FWD car will normally have a negative scrub radius with the steering axis offset toward the outside of the tire (right). A RWD car will have a positive scrub radius with the steering axis offset toward the inside. Changing to a wider wheel with the same back spacing will increase the offset or scrub radius. (Courtesy of Moog Automotive)

larger in diameter with a tire that is two steps larger, and Plus 3 (not very common) uses a wheel that is three inches larger in diameter with a tire that is three steps larger. Some manufacturers are offering this type of option on their sport and performance cars. Plus 1, 2, 3 allows the motorist to end up with a wider, sportier tire that fits the car as well as the O.E.M. tire recommendations. Whenever the tire width is increased, remember to check to ensure that there is adequate side clearance for normal suspension and steering motions of the tire. (Fig. 3–15)

Many people feel that aftermarket wheels look better than the stock O.E.M. wheels, and that a change in wheel style allows them to improve the appearance of their car as well as customize it to their personal tastes. Many different styles are available to suit almost every preference.

When a wider replacement tire is desired, a change in wheel width is often mandatory. A stock wheel will often accommodate a tire one or two sizes larger than standard width, but a greater change than this will probably exceed the correct tire-to-rim width range. When wider wheels are desired, it is often possible to use an O.E.M. wheel that was available as optional equipment. Many car manufacturers offer larger or sport or rally type tires and wheels as an option. These wheels are usually wider than the standard wheel and offer the benefit of having the correct offset for a particular car. You are also assured that this wheel will have no interference problems with the brake assemblies and brake drum balance weights.

A change in wheel diameter usually cannot be done without a change in the tire's aspect ratio. Otherwise, fender clearance and gear ratios will increase with larger tires and wheels, ground clearance and gear ratio losses will increase with smaller tires and wheels, and speedometer error problems will occur.

FIGURE 3–14.
Aftermarket wheel manufacturers offer a large selection of wheels in many styles and sizes. (Courtesy of Western Wheel)

FIGURE 3-14 (Continued).

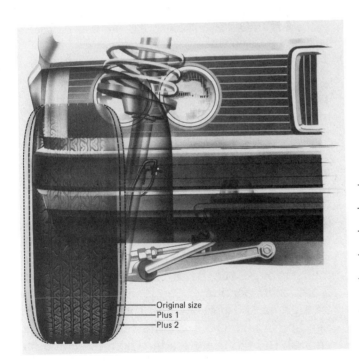

Original size
Plus 1
Plus 2

FIGURE 3-15.
The plus 1 and plus 2 concept replaces the original tire and wheel with a wider tire on a larger diameter wheel. The end result is a wider tire with close to the original overall diameter. (Courtesy of B.F.Goodrich, © The B.F.Goodrich Company)

PLUS SIZING COMPARISON

	ORIGINAL SIZE BFGoodrich Radial T A 185 70R13	PLUS 1 BFGoodrich Euro Radial T A 195 60HR14	PLUS 2 BFGoodrich Comp T A 205 50VR15
	13"	14"	15"
TIRE DIAMETER	23.40"	23.23"	23.12"
TIRE WIDTH	5.50"	5.76"	6.84"
TIRE LOAD CAPACITY	1040 lbs.	1055 lbs.	1060 lbs.
INFLATION FRONT/REAR	28/28 psi.	28/28 psi	32/32 psi
SECTION WIDTH	185	195	205

Engine RPM, gear ratio, tire diameter, and car speed are all directly related. A tire will roll down the road a distance that is equal to its circumference on each revolution. You can use the formula $C = 2 \pi R$ to compute the circumference, but be sure to use the distance from the center of the wheel to the ground as R, the tire's effective radius. This is important because car load causes the tire to bulge, which shortens this distance. Some manufacturers publish revolutions per mile specifications for their tires for easy comparison. You can use the following formula to determine engine RPM or car speed:

$$\text{Engine RPM} = \frac{\text{MPH} \times \text{Gear Ratio} \times 336}{\text{TIRE DIAMETER}}$$

$$\text{MPH} = \frac{\text{RPM} \times \text{Tire Diameter}}{\text{Gear Ratio} \times 336}$$

When using these formulas, the gear ratio equals the final drive ratio (rear or front axle ratio) times the transmission ratio. Also, the tire diameter equals the effective tire radius times two.

The quality of aftermarket wheels is ensured by the **Specialty Equipment Manufacturers Association (SEMA)**, through the SFI, SEMA Foundation, Inc. SEMA and SFI are nongovernmental agencies that were formed by concerned aftermarket manufacturers. A wheel that carries SFI certification has been manufactured to rather stringent standards, which helps ensure its safe operation.

3.6 WHEEL ATTACHMENT

Most of us are familiar with the normal **wheel lug bolt and nut**. This bolt has a serrated shank that is pressed through the hub to prevent bolt rotation, and it is usually upset or riveted at the outer side to retain it securely in place. When the wheel is installed, the tapered, conical face of the lug nut enters the tapered holes in the wheel's nut bosses to center the wheel to the hub. This allows the use of a larger hole in the wheel lug holes and bosses for easy wheel installation and removal, and the tapers ensure

FIGURE 3–16.
The wheel is centered on the hub when the tapered section of the lug nut enters the tapered boss in the wheel. Note the raised wheel section around the lug boss; it deflects slightly to help lug nut retention.

wheel centering on the hub. Some wheels have a center hole that closely fits the hub flange. These wheels are centered as the wheel's center hole fits over the hub. (Fig. 3–16)

The lug nuts must hold the wheel tightly against the hub. This tight fit reduces the bending loads on the bolts as well as the shearing loads during braking and acceleration. A properly tightened wheel will transmit about 90 percent of its load through the friction between the wheel and the hub and about 10 percent of the load through the bolts. If the lug bolts get loose, their portion of the load will increase to as high as 100 percent and they will probably bend or break. The lug bolts selected are large enough to hold these loads. The most common sizes are 7/16–20 (7/16 in. thread diameter and 20 threads per inch), 1/2–20, a few 3/8–24, M12×1.5 (12 mm thread diameter and 1.5 mm thread pitch, i.e., distance from one thread to the next), M12×1.25, M14×1.5, and a few M10×1.5. (Fig. 3–17)

A lug nut stays in place on the lug bolt because of the friction between the threads of the nut and bolt and between the face of the nut and the wheel. If the nut is tightened against a solid object, the total amount of these fric-

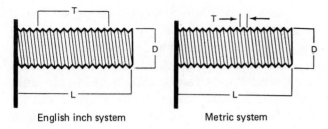

FIGURE 3–17.
A wheel stud, like other bolts, has two different measuring systems. The inch measuring system measures the length of the stud (L) in inches, the number of threads in one inch (T), and the diameter of the threads (D) in fractions of an inch. The metric measuring system measures the length of the stud (L), the pitch or distance from one thread to the next (T), and the diameter of the threads (D) in centimeters.

tions will increase rapidly as soon as the nut meets the object. They will also decrease rapidly if the nut backs off slightly. When a bolt and nut are tightened, a stretching load is placed on the bolt as soon as the clearance is

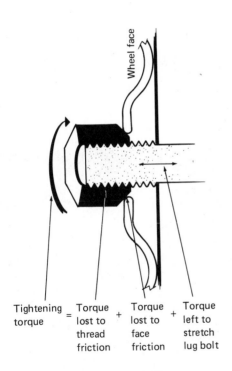

$$\text{Tightening torque} = \text{Torque lost to thread friction} + \text{Torque lost to face friction} + \text{Torque left to stretch lug bolt}$$

100% = 15-20% + 60-80% + 4-8%

FIGURE 3–18.
When a lug nut is tightened, most of the tightening torque goes to the friction at the threads and the interface between the nut and the wheel. Part of the torque is used to stretch the bolt and compress the nut boss in the wheel.

taken up. This stretching action maintains interthread and interface friction to ensure nut retention. If the bolt is overtightened, the bolt can be stretched to the fatigue or yield point, and it might break, especially if stress or road shock places a greater load on it. An overtightened lug nut can also cause distortion loads on the brake rotors, drums, or both. If the lug nut is undertightened, it will soon come loose and fall off because there will not be enough interthread friction to hold the nut in place. (Fig. 3–18)

The area immediately around the lug nut boss of most stamped wheels is usually raised slightly so this portion of the wheel center section will deflect as the lug nuts are tightened. This deflection has the effect of reducing bolt stretch as the lug nuts are tightened and also maintaining an outward pressure to maintain nut friction.

Cast wheels usually have thick center sections and normally use straight-shanked lug nuts plus a washer. The washer helps reduce galling of the wheel as the nuts are tightened and loosened. Wheel centering is accomplished by a close fit between the nut shank and the wheel holes. Nuts are available with different shank lengths, and the shank length should be slightly shorter than the thickness of the wheel boss. Some cast center section wheels will use tapered nut faces much like stamped wheels. On special installations, you should make sure there is sufficient nut-to-bolt thread contact. Sometimes, with thick wheel bosses, it is necessary to replace the lug bolts with ones having a longer thread length. (Fig. 3–19)

With either style of lug nut, the nut shank or the bolt length must be long enough to ensure adequate thread contact or else thread stripping might result. The thread contact length should equal to 0.8 times the bolt diameter for tapered seat nuts and 0.85 times the bolt diameter for straight-shank nuts—1 to 1.5 times the bolt diameter is preferred. For a 1/2-20 lug bolt (12.7 mm diameter with 20 threads per 25.4 mm), there should be a minimum of 0.8 × D or 0.8 × .5, which equals 0.4 in. (10.16 mm) of thread contact. This would mean a contact of at least eight (0.4 × 20) threads. Most people prefer a thread length with a few more threads

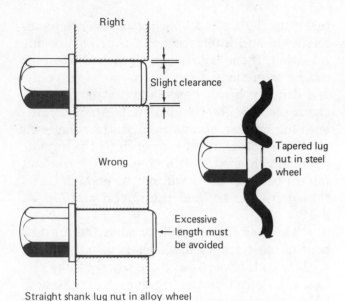

Right

Slight clearance

Wrong

Tapered lug nut in steel wheel

Excessive length must be avoided

Straight shank lug nut in alloy wheel

FIGURE 3-19.
An alloy wheel often has a thick wheel section that requires a straight-shank nut; this nut should be slightly shorter than the thickness of the wheel and have a diameter slightly smaller than the holes. A steel wheel usually uses a tapered lug nut.

than this to provide a margin of safety and a hedge against bolt and nut wear or stripping during wheel installation and removal.

Knock-off or **quick change wheels** use a large, single nut to secure the wheel to the hub. Drive and braking torque loads are transmitted by splines in the wheel center circle and hub or by a group of drive pins on the hub that fit into holes in the wheel center section. Knock-off wheels were commonly used on sports cars and also are currently in use on many racing cars. (Fig. 3-20)

(A)

(B)

FIGURE 3-20.
This race car uses pin drive, quick change wheels. Note the drive pins on the hub (A), which allow driving and stopping forces to be transmitted between the wheel and the hub. Also note the large retaining nut (B) that holds the wheel onto the hub.

REVIEW QUESTIONS

1. Statement A: The center portion of the wheel is sometimes called the disc. Statement B: The tire fits onto the rim of the wheel. Which statement is correct?

   a. A only   **c.** both A and B
   b. B only   **d.** neither A nor B

2. The tire bead seats of a wheel are
   A. formed at a slight angle.
   B. about 0.060 inch smaller than the bead of the tire.

   a. A only   **c.** both A and B
   b. B only   **d.** neither A nor B

3. The center portion of the rim is dropped downward to allow
   A. a stronger attachment to the wheel center section.
   B. the tire to be removed and replaced.

   a. A only   **c.** both A and B
   b. B only   **d.** neither A nor B

4. A wheel with a bolt pattern of 4–100 mm has
   A. four mounting holes on a 4 inch diameter circle.
   B. mounting holes in a square shape, 100 mm long, on each side of the square.

   a. A only   **c.** either A and B
   b. B only   **d.** neither A nor B

5. Statement A: The width of the wheel should equal the width of the tire tread. Statement B: The diameter of the wheel should equal the inside diameter of the tire. Which statement is correct?

   a. A only   **c.** both A and B
   b. B only   **d.** neither A nor B

6. Statement A: The scrub radius will change if the wheel is replaced with one having a different offset. Statement B: Changing the wheel offset will change the loading on the wheel bearings. Which statement is correct?

   a. A only   **c.** both A and B
   b. B only   **d.** neither A nor B

7. The back spacing on a 6-inch-wide wheel is 2 inches, so the wheel has an offset of

   a. positive 2 inches.
   b. positive 1 inch.
   c. negative 2 inches.
   d. negative 1 inch.

8. A plus 1 tire changeover means that the tires and wheels are replaced with a
   A. tire that has the next wider profile.
   B. wheel that is 1 inch larger.

   a. A only   **c.** both A and B
   b. B only   **d.** neither A nor B

9. Statement A: Most of the braking and accelerating torque between the wheel and the hub is transmitted by the lug bolts. Statement B: Improperly tightened lug bolts can cause brake rotor and drum distortion as well as lug bolt breakage. Which statement is correct?

   a. A only   **c.** both A and B
   b. B only   **d.** neither A nor B

10. The wheel is centered to the hub as the
    A. tapered lug nuts enter the tapered wheel bosses.
    B. center of the wheel is placed over the hub.

    a. A only   **c.** either A or B
    b. B only   **d.** neither A nor B

Chapter **4**

TIRE AND WHEEL SERVICE

After completing this chapter, you should:

- Be able to correctly adjust the tire pressure for a particular car.

- Be able to determine the cause of an unusual tire wear pattern and recommend the needed repair to correct it.

- Be able to determine the cause of a tire-related vibration and recommend the needed repair.

- Be able to rotate a set of tires using any of the recommended rotation patterns.

- Be able to measure tire, wheel, and hub runout in a radial and lateral direction.

- Be able to determine if a steering pull is caused by a tire-related problem and if so, correct this problem.

- Be able to static and dynamic balance a tire and wheel either on or off the car.

4.1 INTRODUCTION TO TIRE SERVICE

When a tire problem occurs and the motorist brings the car into a shop for repairs, that shop and the tire service technicians concerned take on an obligation to determine the cause of the problem and to repair it. The repair should be such that the car can be operated in a normal and safe manner for a reasonable length of time. The repairs should be made following methods that are approved by the car manufacturer, the Rubber Manufacturers' Association (RMA), and the repair industry. Also, the repairs should be made in such a way that the cost to the motorist will be as low as possible yet ensure a reasonable wage for the tire service mechanic and profit for the shop.

The service procedures given in this text will be typical for the industry. Exact methods might vary depending on the particular make or model of car that is being worked on or the particular type of equipment that is available for use in the repair. In all cases, a repair operation should begin with a base of knowledge on the part of the technician, be followed by a diagnostic procedure to determine the cause of the problem and the type of repair needed, and then end with a planned, careful, and thorough repair procedure. Good workmanship will ensure lasting repairs.

When diagnosing a problem, the technician uses knowledge and experience to arrive at the cause of the problem. The first step is to understand what the problem is. As much as possible or necessary, the motorist should be asked to describe the problem, how he or she knows that it is there, when it began, and under what speeds and driving conditions it occurred. If possible and if necessary, the technician should verify that the problem exists by road testing the car. The causes of many problems often become very obvious when they are actually seen, heard, or felt by a good technician. For example, nearly every vibration problem is caused by something that moves, spins, rotates, or rolls. If a noise or problem is checked under different driving conditions, the technician can come closer to the cause of a vibration problem. The cause of some problems can be

TABLE 4–1

This chart lists the cause and correction procedures to cure many of the problems faced by front-end mechanics; note that many of them are tire problems. (Courtesy of Chrysler Corporation)

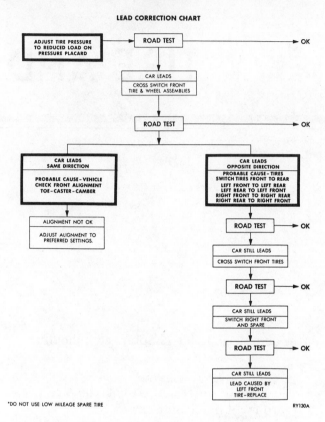

more difficult to determine, but problem diagnosis flow charts such as those shown in Tables 4-1 and 4-2 are valuable guides.

4.2 REMOVING AND REPLACING A TIRE AND WHEEL

Removing and replacing a tire and wheel on a car are common service operations. Most tire service technicians perform it so often that they appear very casual and nonchalant while doing it. A good technician has developed work habits through experience that are carefully followed, even though he or she appears casual while doing them. Hopefully, none of you will have to have a wheel come loose from a car that you have worked on to show you the effect of sloppy workmanship.

TABLE 4–2
This chart lists the procedure to follow to determine if a lead or pull is caused by a tire or wheel alignment and, if it is caused by a tire, to determine which tire is causing the problem. (Courtesy of Chrysler Corporation)

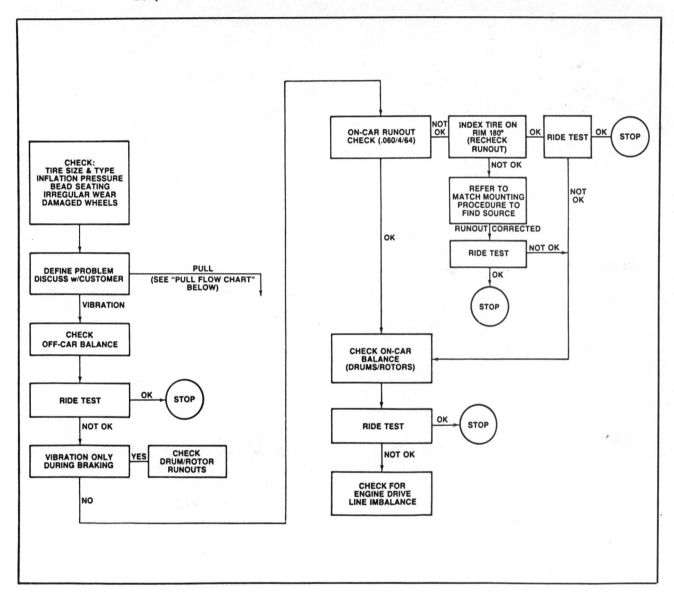

To remove a tire and wheel from a car, you should:

1. Remove any wheel covers, which are often locked.

CAUTION: Set the parking brake. If the car is on an unlevel surface, block the wheels so the car will not roll and fall off the jack or stand. (Fig. 4–1)

2. Chalk mark the end of the lug bolt or the wheel stud or bolt that is closest to the valve stem. This is done so the wheel assembly can be replaced in the same location to ensure that no change in the balance or the runout of the tire and wheel will occur. If the valve stem is positioned between two studs, mark both of them or draw a line on both the wheel and the hub. This step is unnecessary during tire replacement or rotation. (Fig. 4–2)

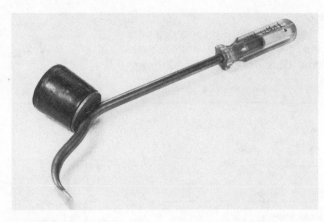

FIGURE 4-1.
The lower end of this tool is used to pry off hub caps and wheel covers, and the rubber hammer portion is used to help replace them.

FIGURE 4-2.
Placement of a chalk mark on the end of the wheel stud closest to the valve stem will help in replacing the wheel in the same position it was before removal.

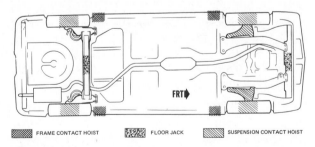

FRAME CONTACT HOIST   FLOOR JACK   SUSPENSION CONTACT HOIST

FIGURE 4-3.
The jacking and lifting points are usually shown in the manufacturer's or mechanic's service manual. These points are the places that are strong enough to support the car. (Courtesy of Oldsmobile)

3. Loosen the lug nuts using a six point lug nut socket or lug wrench.

4. Place a jack under some part of the car that is strong enough to support the weight. (Fig. 4-3) Some cars have a jacking pad at the bumper or side of the car. Other places that are commonly used are: the frame, a frame cross-member, a reinforced unibody box section, an axle, or most suspension arms. Lift the tire and wheel off the ground, and place a jack stand under some strong portion of the car. (Fig. 4-4)

NOTE: The rear axle of some FWD cars will bend if the jack is placed in the center of the axle. The axle should be lifted, one side at a time with the jack placed near the spring.

NOTE: If you are unsure where to lift the car, consult the Lifting and Jacking section of the car manufacturer's service manual or a technician's service manual.

5. Finish removing the lug nuts. If you are using an air-impact wrench, excessive speed can cause lug bolt or nut galling. Adjust the wrench

FIGURE 4-4.
Two different types of jack stands that can be placed under the car for safe support. Never work under a car that is supported by only a jack.

speed or lubricate the threads to ensure that galling does not occur.

6. Remove the tire and wheel assembly.

To install a tire and wheel assembly, you should:

1. Check the wheel nut bosses for worn or elongated holes. Damaged wheels should be replaced. Place the wheel over the lug bolts with the valve core next to the previously marked stud or studs.

2. Snug down the lug nuts making sure the tapered portion of the lug nut enters the tapered opening of the wheel nut bosses. While the tire and wheel are in the air, it is difficult to apply enough torque to completely tighten the nuts because the wheel will spin. However, they can and should be tightened enough to hold the wheel in the correct position.

3. *CAUTION: Lower the car onto the ground, and immediately complete Step 4. Don't allow yourself to forget it.*

4. Tighten the lug nuts to the correct torque using a tightening pattern that moves back and forth across, not around, the wheel. (Fig. 4–5) Lug nut torque will vary depending on the size of the lug bolt and the wheel center section material. Aluminum wheels will sometimes use a slightly higher torque because of the solid center section. (Fig. 4–6) However, a slightly lower torque is sometimes recommended to reduce nut-to-wheel galling and compression of the wheel nut bosses. If the manufacturer's torque specifications are not

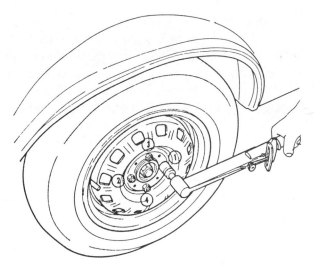

FIGURE 4–6.
A torque wrench is used to ensure that the bolts are tightened the correct amount. (Courtesy of Chrysler Corporation)

available, use the following torques based on the lug bolt size:

3/8–24 — 35—45 ft.lb. (48—61 N·M)
7/16–20 — 55—65 ft.lb. (75—88 N·M)
1/2–20 — 75—85 ft.lb. (102—115 N·M)
M10×1.5 — 40—45 ft.lb. (54—61 N·M)
M12×1.5 — 70—80 ft.lb. (95—108 N·M)
M14×1.5 — 85—95 ft.lb. (115—129 N·M)

SAFETY NOTE

It is a good practice to retighten the lug nuts after driving 10 to 20 miles (16 to 30 km). With alloy center wheels, it is a good idea to check lug nut tightness again after another 100 miles (161 km).

---

5. Replace the wheel cover or hub cap using a rubber hammer or the hammer portion of the hub cap tool.

4.3 LUG BOLT AND STUD REPLACEMENT

If a few of the threads on a lug bolt are damaged, the threads can be **chased** to repair them,

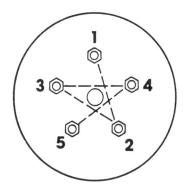

FIGURE 4–5.
Lug bolts should be tightened in a star-shaped or crisscross pattern. (Courtesy of Moog Automotive)

but if the threads are too badly damaged or the bolt is broken, the bolt should be replaced. A thread chaser, also called a chasing tool or die nut, must be of the same thread size and pitch as the lug bolt. One style of thread chaser resembles a nut and is threaded onto and then off of the lug bolt, cutting away any metal that is in the wrong place. Another style of thread chaser is split into two pieces, and the two halves are placed onto the bolt (the threads next to the rotor or drum are usually in the best shape) and then threaded off to make the thread repair. (Fig. 4–7)

The procedure used to remove a lug bolt depends on where the bolt is used and how tightly it is retained. In those cases where the rotor or drum is normally removed with the hub, the lug bolt is often upset or riveted to secure the rotor or drum to the hub and the lug bolt in place. The term **upset** refers to the metal that is deformed or moved when parts are riveted together. This type requires a little work to replace the lug bolt. In other cases, where the rotor or drum can be easily removed separate from the hub, the lug bolt can usually be tapped loose with a hammer and punch. In

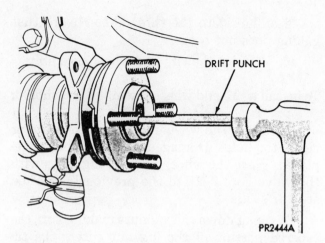

FIGURE 4–8.
A damaged lug bolt can often be removed using a hammer and punch if it is not swaged in place. (Courtesy of Chrysler Corporation)

either case, if force is needed to get the bolt loose, care should be taken to prevent bending of the hub flange, brake rotor, or drum.

To remove a lug bolt from a hub or axle flange that is separate from the rotor or drum, you should:

1. Tap on the outer end of the lug bolt using a punch and hammer. If the bolt loosens, proceed to Step 3. If the bolt does not come loose under a reasonable hammer blow, proceed to Step 2. (Fig. 4–8)

2. Position a pressing tool such as a universal joint press or a suspension bushing press across the hub or axle flange and lug bolt. Tighten the press to push the bolt inward until the bolt brakes loose and can be moved inward and out of the hub flange. (Fig. 4–9)

3. If necessary, turn the hub or axle flange to a point where there is sufficient room, and slide the broken lug bolt out of the flange.

To remove a lug bolt from a rotor and hub or a drum and hub combination, you should:

1. If the bolts are upset and you have the right size cutting tool, cut the upset metal from the lug bolt. (Fig. 4–10)

2. Place a support, often called an anvil, under the hub so the hub is well-supported yet

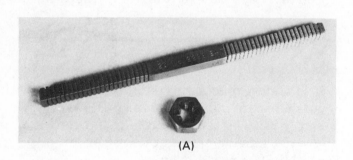

(A)

(B)

FIGURE 4–7.
A thread file, and a thread chaser, or die nut (A) can be used to repair the threads on a wheel stud, but they are rather slow. A wheel stud thread chaser (B) is faster and often does a better job; it is placed over the stud and then unscrewed to dress up the threads.

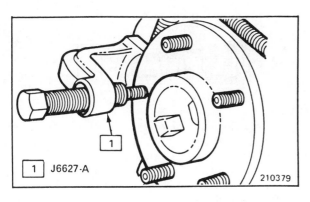

FIGURE 4-9.
A damaged wheel stud is removed using a pressing tool (J6627-A). (Courtesy of Pontiac Division, GMC)

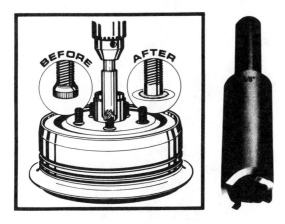

FIGURE 4-10.
If the wheel stud is upset or swaged on the outer side, the swaged shoulder can be cut off with a special cutting tool. (Courtesy of Century Wheel, Division of Precision Switching)

FIGURE 4-11.
A brake drum or rotor can be placed over this anvil that supports the hub as the wheel stud is being driven or pressed out.

there is room for the bolt to pass by. (Fig. 4-11)

OR Position a lug bolt press over the hub and rotor or drum and the damaged lug bolt. (Fig. 4-12)

3. Position either set-up in the bed of a press, and press the old lug bolt out of the hub. (Fig. 4-13)

OR Strike the old lug bolt or the arbor with a hammer to drive the old bolt out of the hub.

A replacement lug bolt should be selected that has the same diameter and length of thread and the same diameter and length of serrations as the one removed. (Fig. 4-14)   In most cases, the replacement bolt can be installed by sliding the new bolt into position, placing several flat washers over the bolt threads to help protect the nut and to help ensure a straight installation, and then tightening the lug nut onto the bolt. (Fig. 4-15)   The nut should be positioned so its flat face is against the washers while doing this. The bolt should then pull into place. If necessary, the lug bolt can be pressed into the hub and rotor or drum assembly with either a press and hub support tool or a lug bolt press using a procedure just the opposite of the removal procedure. (Fig. 4-16) Whatever method is used, do not place a pressing load across the brake rotor or drum because they can crack or become distorted.

FIGURE 4-12.
A tool set for removing and replacing wheel studs. (Courtesy of OTC)

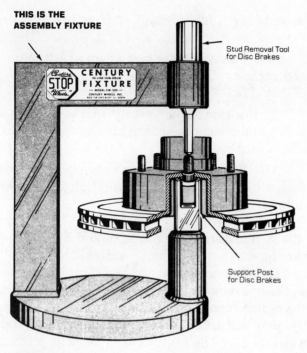

**THIS IS THE
ASSEMBLY FIXTURE**

Stud Removal Tool
for Disc Brakes

Support Post
for Disc Brakes

FIGURE 4-13.
A fixture that is set up for removal of a wheel stud. Note how the hub is supported; there is no pressure on the brake rotor. (Courtesy of Century Wheel, Division of Precision Switching)

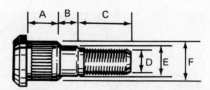

FIGURE 4-14.
The important dimensions when selecting a replacement wheel stud are (A) serration length, (B) shoulder length, (C) thread length, (D) thread diameter, (E) shoulder diameter, and (F) serration diameter plus the thread pitch and direction. Replacement wheel studs should match the original ones. (Courtesy of Dorman)

FIGURE 4-16.
Pressing a wheel stud into a hub and rotor; a press is being used to provide the pressure. (Courtesy of OTC)

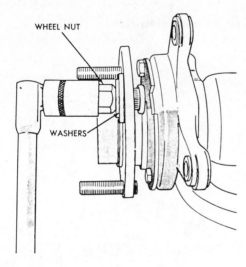

WHEEL NUT

WASHERS

FIGURE 4-15.
A new wheel stud can usually be pulled into position using a few flat washers and the lug nut. (Courtesy of Chrysler Corporation)

4.4 TIRE WEAR INSPECTION

When a tire wears out, the wear should occur evenly across the tread. When a tire is worn to a point where there is less than 2/32 inch (1/16 inch, 1.587 mm) in two adjacent tread grooves, the tire is legally worn out in many states. Because of the loss in the ability for water to run out through the tread, a worn-out tire will give poor driving control and stopping ability if operated on wet pavement. Besides being unsafe,

(A)

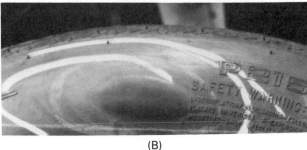

(B)

FIGURE 4-17.
A tire crayon was rubbed on tire A while it was spinning on a balancer; the darker areas show the cupping and other unusual wear patterns. The bulge in tire B is a sign of separation of the internal parts. (Tire B is courtesy of B.F.Goodrich, © The B.F.Goodrich Company)

the probability of flats and blowouts also increases as the tread wears thinner. A thinner tread becomes more vulnerable to cuts and penetration from road objects as the tread rubber wears off. Tire wear is checked visually by looking at the tread, by feeling the tread, and by

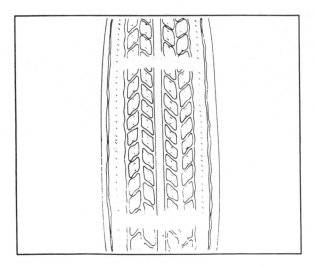

FIGURE 4-18.
Wear bars are raised strips across the bottom of the tread grooves. When they appear, the tire is worn out. (Courtesy of Oldsmobile)

using a tread depth gauge. Some tire shops rub a tire crayon on a spinning tire to make any defects show up better. (Fig. 4-17)

Tread depth can be quickly checked by observing the wear bars, also called treadwear indicators. These are raised areas, 1/16 inch (1.6 mm) high, running across the bottom of the tread grooves at six places around the tire. (Fig. 4-18)   Another easy way to check tread depth is to place a U.S. penny in the tread groove, as shown in Figure 4-19. The distance from the

FIGURE 4-19.
A penny placed upside down in the tread groove is a common check for tread wear. If you can see the top of Lincoln's head, the tire is nearly worn out.

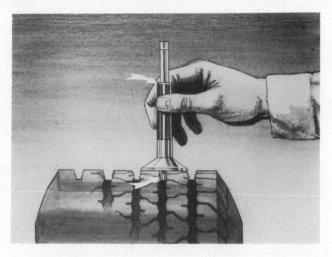

FIGURE 4-20.
A tread depth gauge. The measuring stem (lower arrow) should touch the bottom of the tread groove; the depth of the tread groove is read on the scale at the top of the main body (upper arrow). (Courtesy of B.F.Goodrich, © The B.F.Goodrich Company)

edge of the penny to the top of Abraham Lincoln's head is a little greater than 1/16 inch. A more accurate method of checking treadwear is to use a tread depth gauge. (Fig. 4-20) This gauge allows a careful check of each tread groove and will show normal or abnormal wear much better than an unaided visual inspection. A regular check with a tread depth gauge can give an early warning if inflation pressures or alignment angles are not correct.

Normal tire wear occurs evenly across the tread. If a tire wears faster on one side or location than another, something is wrong. Depending on the location of the wear, a competent technician can tell if the tire has been inflated incorrectly, driven severely, aligned wrong, or mounted on a suspension with worn parts. (Fig. 4-21)

Incorrect tire pressures will cause a tire tread to wear abnormally fast at either the cen-

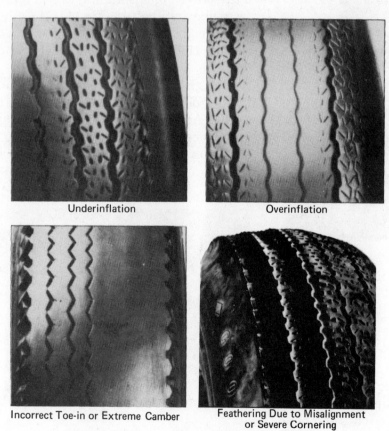

Underinflation

Overinflation

Cupping — underinflation and/or mechanical irregularities such as out-of-balance condition of wheel and/or tire, and bent or damaged wheel. Possible loose or worn steering tie-rod or steering idler arm. Possible loose, damaged or worn front suspension parts.

Incorrect Toe-in or Extreme Camber

Feathering Due to Misalignment or Severe Cornering

FIGURE 4-21.
The more commonly encountered abnormal tire wear patterns and their probable causes. (Courtesy of Ford Motor Company)

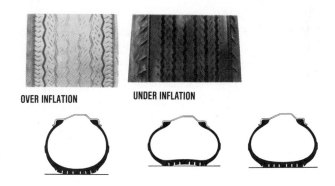

FIGURE 4-22.
Underinflation will cause the tire to wear fastest at the sides of the tread (left); overinflation will cause excessive wear at the center of the tread (middle); and correct inflation will give an even tread to road contact and wear. (Courtesy of Moog Automotive)

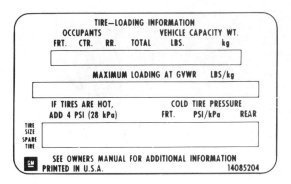

FIGURE 4-23.
All cars have a decal or placard like this to provide the correct tire inflation pressure for that particular car. (Courtesy of Oldsmobile)

ter, if the pressure is too high, or at the edges, if it is too low. (Fig. 4-22) A properly inflated tire will have a relatively even road contact pressure across the tread. Excessive pressure tends to push the center of the tire harder onto the road while too little pressure does not push hard enough. Correct tire pressure will depend on the size of the tire and the weight of the car. The recommended pressure is usually printed in the Owner's Manual and on a decal that is attached to the car. (Fig. 4-23) A belted tire will often tend to hide or reduce the effects of incorrect pressure.

Checking and maintaining the correct pressure is the most important thing a motorist can do to obtain maximum tire life and safety. When checking tire pressure, it is important to use an accurate gauge and to observe the following recommendations:

1. Check tire inflation once a month and before taking a major trip.

2. Check inflation when a change in weather temperature occurs. A temperature change of 10° F (6° C) will cause a tire pressure change of about 1 psi.

3. Check the pressure when the tires are cold, after driving only a short distance, 1 mile (1.6 km) or less.

4. Never reduce the pressure from a hot tire. The pressure of a hot tire should be greater than normal.

5. Make sure that all tire valves have a valve cap to keep dirt and moisture out of the valve.

6. Most tire manufacturers and tire service technicians recommend running a tire at the maximum inflation pressure printed on the sidewall for the best tire service and performance.

A belted-bias tire will often have rapid wear on the tread rib that is next to the outside on each side of the tire. This unusual treadwear is normal for this type of tire. It occurs because the centrifugal force acting on the edges of the belt tends to increase the road pressure on these two tread bars.

Many narrower, steel-belted radial tires mounted on front wheels will often wear the shoulder areas and leave high tread in the center. This will appear to be underinflation wear, but it can occur even if the tires are operated at maximum inflation pressures. This wear pattern is possibly caused by a fault in the tire's internal construction (pyramid belts) or by the tire rolling under during turns.

A tire that is cambered excessively will tend to have faster treadwear on the side of the tire that it is leaning outward. The treadwear will be greatest at the outer groove with less wear at the groove next to it and still less wear at each additional groove across the tread. Also, there will be a fairly sharp corner between the worn tread and the sidewall. (Fig. 4-24) This

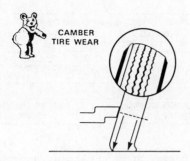

FIGURE 4–24.
Incorrect camber angles will cause a tapered tire wear that begins at one shoulder of the tire tread. (Courtesy of Bear Automotive)

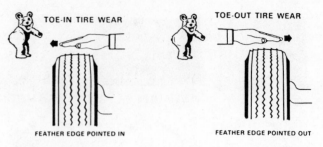

FIGURE 4–25.
Incorrect toe will cause a feather-edge wear pattern across the tire tread that can be felt as you slide your hand across the tread. (Courtesy of Bear Automotive)

indicates an alignment problem that will be covered more thoroughly in Chapters 17 and 18.

A tire that is toed-in or toed-out excessively will be forced to scuff sideways as it rolls down the road. It will tend to wear in a feather-edge or saw-tooth pattern that is easier felt than seen. (Fig. 4–25) Run your hand across the tread from the outside to the inside and then back to the outside. Toe-in wear is most noticeable as you pass your hand toward the outside; toe-out wear will cause a feather edge

FIGURE 4–26.
This tire was ruined by abnormal, heavy wear from the outer tread bar downward onto the shoulder; this wear was probably caused by driving excessively fast on corners (hard cornering) or incorrect toe-out on turns.

in the opposite direction with the sharp edge toward the outside. This is an indication of an alignment problem that will also be discussed in Chapters 17 and 18.

A wear pattern similar to positive camber wear, or wear on the outside edge, can be caused by hard cornering. This wear will tend to cause a rounded shoulder while camber wear causes a sharp corner at the edge of the tread. When a tire is driven hard into a corner, the tread will tend to roll, causing a heavy load on the shoulder and sometimes even on the sidewall. This wear pattern can also be caused by an incorrect turning radius or toe-out on turns, which will be discussed in Chapters 17 and 18. (Fig. 4–26)

Spotty wear that occurs in one or more spots around the tire is usually caused by improper balancing or worn suspension parts or shock absorbers. (Fig. 4–27) This wear pattern, often called **cupping**, is easier felt than seen, at least in the early stages. Pass your hand around the tire and feel for spotty, worn areas. Incorrect balance will cause excessive tread pressure in one or more locations because the heavy spots of the tire spin with the tire and push the tread harder into the road. Worn suspension parts will let the tire change camber or toe angle as the tire spins. These intermittent changes in alignment will cause intermittent camber or toe wear in spots of the tread. Occasionally, this type of tire wear can also be seen on properly balanced tires mounted on good suspensions. In these cases, it is possibly caused by internal tire construction in the belt area. It is also occasionally seen on the rear of FWD cars. In these cases, it is possibly started

FIGURE 4–27.
These two tires show spotty tire wear that was caused by loose suspension parts in combination with wrong alignment settings; the loose parts allowed the tires to move or dance around, producing a scrubbing action. (Courtesy of Moog Automotive)

by severe brake application causing skidding on the lightly loaded tread. One worn spot on a tire can cause the tire to hop, which, in turn, will cause additional flat spots. (Fig. 4–28)

4.5 TIRE ROTATION

Tires should be rotated around the car periodically to even out the wear through a set of tires. Turning motions, hard acceleration, or hard braking tend to increase tire wear. Tire wear normally occurs faster at some tire positions. For example, a high horsepower RWD car with an impatient driver will usually wear out the right rear tire first. An FWD car will tend to wear out the front tires first, especially if the driver is aggressive. For convenience and more stable driving, most motorists will replace the tires as a set of four with the replacement time determined by the first tire to wear out. The ideal situation is for all four tires to wear out completely at the same time. Tires are rotated so that each one spends part of its life in the high wearing position and the rest of its life in the low wearing positions.

Tires are normally rotated when an uneven treadwear is found, or about every 10,000 miles (16,100 km). It is a good practice to rotate the tires more frequently under severe driving conditions. The recommended rotation patterns are shown in Figure 4–29. At one time, it was recommended that a radial tire should never be cross-rotated so it would run in the opposite direction. This recommendation has now been dropped by the RMA (Rubber Manufacturers' Association) and several major tire manufacturers. You can use the pattern that best suits your needs.

CONDITION	RAPID WEAR AT SHOULDERS	RAPID WEAR AT CENTER	CRACKED TREADS	WEAR ON ONE SIDE	FEATHERED EDGE	BALD SPOTS	SCALLOPED WEAR
EFFECT							
CAUSE	UNDER-INFLATION OR LACK OF ROTATION	OVER-INFLATION OR LACK OF ROTATION	UNDER-INFLATION OR EXCESSIVE SPEED*	EXCESSIVE CAMBER	INCORRECT TOE	UNBALANCED WHEEL OR TIRE DEFECT*	LACK OF ROTATION OF TIRES OR WORN OR OUT-OF-ALIGNMENT SUSPENSION.
CORRECTION	ADJUST PRESSURE TO SPECIFICATIONS WHEN TIRES ARE COOL ROTATE TIRES			ADJUST CAMBER TO SPECIFICATIONS	ADJUST TOE-IN TO SPECIFICATIONS	DYNAMIC OR STATIC BALANCE WHEELS	ROTATE TIRES AND INSPECT SUSPENSION SEE GROUP 2

*HAVE TIRE INSPECTED FOR FURTHER USE.

FIGURE 4–28.
Most of the commonly encountered types of abnormal tire wear. (Courtesy of Chrysler Corporation)

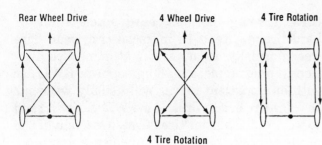

Rear Wheel Drive   4 Wheel Drive   4 Tire Rotation

4 Tire Rotation

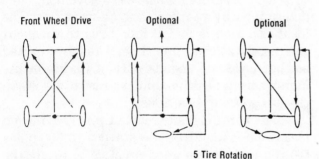

Front Wheel Drive   Optional   Optional

5 Tire Rotation

FIGURE 4-29.
Depending on the type of vehicle, here are six different tire rotation patterns. The same pattern should be used through the life of a tire set. Many people prefer using the patterns shown at the top right or center bottom for radial tires to keep the tire always rolling in the same direction. (Courtesy of Bridgestone Tire Company)

Some of the rotation patterns are four tire patterns. A standard type of spare tire can be included. If so, the spare is often placed on the right rear, and the tire that would have gone to the right rear is put in the trunk to become the new spare. Temporary use spares should not be included in any rotation pattern.

4.6 REMOVING AND REPLACING A TIRE ON A WHEEL

Tire replacement is normally done at a tire shop or a store that specializes in tire sales and service. Although the operation can be done with hand tools, a tire changing machine is usually used to make the operation of removing and replacing a tire on a wheel quick and easy. There are several makes and styles of tire changers, and space does not permit covering the opera-

tion of the individual machines. The operation of most tire changers is similar.

***CAUTION:** Tire changers operate very quickly and powerfully. They present several possibilities for injury. It is important to follow the operating directions for each particular machine to avoid the possibility of injury to yourself, the tire, or the wheel.*

To remove a tire from a wheel, you should:

1. Locate the side of the wheel with the shortest distance between the bead flange and the drop well, and clamp the wheel into the tire changer with this side up. (Fig. 4-30)

2. Remove the valve from the valve stem to ensure complete tire deflation. Also, remove all balance weights from the upper bead flange. If this tire is going to return to the same wheel after repairing, mark the location of the weights and valve stem on the tire sidewall.

3. Position the bead breaker next to the bead flange, and operate the bead breaker. (Fig. 4-31) Most tire changers will break loose both the upper and lower beads from the bead flange at the same time. The beads need to be loosened so they can be moved into the drop well during tire removal. It is sometimes necessary to ro-

FIGURE 4-30.
After positioning the tire onto the changer (narrow wheel ledge up), the hold down is installed to fasten it securely in place. (Courtesy of Rubber Manufacturers' Association)

FIGURE 4–31.
The bead breaker is positioned at the rim, right next to it on many changers, and operated to break the bead loose from the rim. Both the upper and lower beads need to be loosened. (Courtesy of Rubber Manufacturers' Association)

FIGURE 4–32.
An approved tire lubricant should be placed on the tire bead and the inside of the bead flange. (Courtesy of Rubber Manufacturers' Association)

tate the wheel assembly, and break the bead at several locations in order to get it to come loose.

NOTE: Sometimes a tight bead on a safety wheel and a tire with a limber sidewall will allow the sidewall to fold inward and the bead breaker will slide past without breaking the bead loose from the bead seat. When this happens, replace the tire valve; inflate the tire to about 5 psi (34 kPa), and try breaking the bead again.

NOTE: Care should be exercised when breaking the bead on alloy wheels. A strong bead breaker can crack or break the wheel flange if it catches it.

4. Lubricate the upper tire bead and wheel flange with tire or rubber lube. This is a lubricant that is especially formulated for this purpose; it is often a mixture of vegetable soap and water. (Fig. 4-32)

5. Insert the removal end of the tire tool between the upper tire bead and the wheel flange, and pry the bead up and over the flange. It is usually necessary to push the bead down into the well on the opposite side of the tire as the bead is being pryed upward. (Fig. 4-33)

6. Operate the tire changer to rotate the removal tool around the wheel, removing the top bead. If the tire has a tube in it, remove the tube at this time. (Fig. 4-34)

FIGURE 4–33.
The tire removal tool is used to pry the tire bead over the wheel flange; at the same time the bead at the opposite side of the tire must go into the wheel well. (Courtesy of Rubber Manufacturers' Association)

FIGURE 4–34.
The changer is operated to move the tire removal tool around the wheel, lifting the upper bead past the flange. (Courtesy of Rubber Manufacturers' Association)

FIGURE 4–35.
After lubricating the lower bead and upper bead flange, the steps shown in Figures 4–32 and 4–33 are repeated to remove the lower tire bead over the upper wheel flange. (Courtesy of Rubber Manufacturers' Association)

7. Repeat Steps 4, 5, and 6 to remove the lower bead and the tire over the upper bead flange. (Fig. 4–35)

Installing a tire onto a wheel usually follows a procedure just the reverse of removal.

To replace a tire, you should:

1. Clamp the wheel into the tire changer with the narrowest bead seat upward. (Fig. 4–36)

2. Lubricate the wheel flanges and bead seats and the tire beads with tire lube. (Fig. 4–37)

3. Slide the lower tire bead partially over the wheel flange. If a tube is to be installed, place the tube in the tire and inflate it with just enough pressure to hold its shape. (Fig. 4–38) Install the valve cap to hold the pressure in the tube.

4. Position the tire installing tool so it hooks between the wheel flange and the tire bead. (Fig. 4–39)

5. Operate the tire changer to rotate the installing tool around the wheel, sliding the tire bead over the wheel flange. When about one-half to two-thirds of the bead enters the wheel, make sure that the bead is also entering the drop well. If a tube is being used, remove the valve cap, work the valve stem through the valve hole, and install a valve stem fishing tool onto the stem to prevent it from being lost.

CAUTION: Be careful not to catch your fingers or hands between the tire and rim or between the bead installing tool and the rim while mounting the tire.

6. Repeat Steps 3, 4, and 5 to install the upper bead.

7. Inflate the tire to seat the beads using a maximum pressure of 40 psi (276 kPa). Tubeless, belted tires usually require an inflation device to rapidly move a volume of air between the bead and bead flange to get enough air into the tire quickly enough to seat the beads. A

FIGURE 4–36.
The wheel that a tire is to be mounted on should be cleaned and placed on the changer with the narrow bead ledge upward. Both bead seats should be lubricated. (Courtesy of Rubber Manufacturers' Association)

FIGURE 4–37.
The sides and base of both tire beads should be lubricated before placing the tire over the wheel. (Courtesy of Rubber Manufacturers' Association)

FIGURE 4–38.
The bottom tire bead should be pushed into the center well and worked over the wheel flange as far as possible. The installing end of the tire tool can be used to work the bottom bead completely on if necessary. (Courtesy of Rubber Manufacturers' Association)

FIGURE 4–39.
Push the top bead into the center well and rotate the tire tool to roll the rest of the bead over the wheel flange. (Courtesy of Rubber Manufacturers' Association)

bead expander can be used on nonbelted tires to force the beads up to the bead flanges for easy inflation. In either case, watch the tire beads during inflation; they should creep upward and snap into position as the tire is inflated. If a tube is used, make sure the tube is not caught between the bead and wheel flange during inflation.

CAUTION: Do not allow yourself to be over or in front of a tire that is being inflated; tires and wheels have separated violently during inflation causing severe injury. (Fig. 4-40)

FIGURE 4–40.
Do not stand over the tire while inflating it; use an extension to allow you to stay clear of the tire in case it should come apart. Inflate the tire slowly until the tire bead slides or pops up to the wheel flange. **Warning:** *If more than 40 psi (275 kPa) are required to seat the bead, deflate the tire and relubricate the bead.* (Courtesy of Rubber Manufacturers' Association)

CAUTION: If the beads do not seat with 40 psi (276 kPa) or less, break the bead down, and clean and relubricate the beads and bead seats. (Figs. 4-41 & 4-42)

8. After inflation, check the positioning tire-to-wheel flange to ensure proper bead seating. (Fig. 4-43)

FIGURE 4–41.
A portable bead seater. Air is blown from the base ring into the tire through the gap between the bead and wheel flange. A similar unit is built into most modern tire changers. (Courtesy of Ammco Tools)

FIGURE 4-42.
A bead expander. Adding air to this device will cause it to pull inward on the center of the tread and will force the beads outward against the wheel bead seats. It cannot be used on belted tires.

FIGURE 4-43.
After the tire bead is seated, check the concentric rings (arrow) next to the bead or wheel flange to ensure that they are completely seated; if they are not even, the bead is not fully seated.

4.7   REMOVING AND REPLACING A TUBELESS TIRE VALVE

A weathered, brittle rubber, snap-in type tubeless tire valve should be replaced when a tire is replaced to ensure that it will last the life of the new tire. With the tire off the wheel, valve stem replacement is an easy matter. An old valve is usually removed by either cutting the inner portion off with a knife or pulling it apart using a valve installer tool. Thread the installing tool onto the old valve, and try pulling it out of the rim. It will either break apart or pull out. Some tire shops have a special tool that will allow installing a rubber valve stem from the outside of the wheel, which makes removal of the tire from the wheel unnecessary. A metal valve is removed by removing the retaining nut and pushing the old valve stem into the wheel well.

To install a new rubber, snap-in valve, you should:

1. Position the new valve inside of the wheel, pass the valve installer through the wheel boss, and thread the tool onto the valve. (Fig. 4-44)

2. Lubricate the valve and the wheel hole with tire lube.

3. Pull the new valve into position using a prying motion of the tool. Rubber from the valve should creep outward on both sides of the rim.

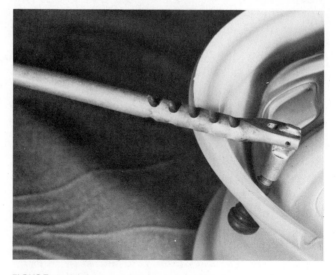

FIGURE 4-44.
After the new valve core is lubricated with tire lube, it is pulled into place from the inside of the tire.

To install a metal, clamp-in type valve core, merely lubricate the sealing area with tire lube, place the new valve stem in position, and tighten the retaining nut.

4.8 REPAIRING A TIRE LEAK

Although a punctured tire can be repaired using a plug from the outside, it is not a good practice. A plug installed from the outside is considered a temporary repair because it is not always safe or permanent. Inside tire repair offers two real advantages: (1) the use of a better patching method and (2) the ability to inspect the inside of the tire to make sure that it is safe. Failure to seal the tire's inner liner might allow air, under pressure, to enter the cord body; the air might work around the cords and cause a separation between the ply layers or between the plies and the rubber layers.

SAFETY NOTE

All good tire service technicians feel responsible for ensuring the safe operation of every repair that they make; the only way to ensure that a tire to be repaired is sound is to check the inside of it. The person who repairs an unsafe tire and sends it back on the road can be held liable for damages if that tire fails. (Fig. 4–45)

The first step in repairing a leaky tire is to locate the leak. This is usually accomplished by visually inspecting the tread area, looking for the puncturing object. If nothing is found, check for a bead leak by laying the tire down, running water around the tire bead and looking for the tell-tale bubbles.

If the leak still has not been located, immerse the tire in a test tank. Again, you will be looking for the tell-tale stream of bubbles coming from the leak. Tire repair shops will immerse the tire as the first step in looking for a leak; it is important to locate all of the leaks

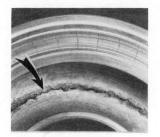

FIGURE 4–45.
This tire damage was not visible from the outside of the tire and was caused by driving a short distance while severely underinflated. It is a dangerous, nonrepairable condition. (Courtesy of Rubber Manufacturers' Association)

before any repairs are made. Also, bubbles at the wheel flange or at a weld indicate a possible crack. Cracked wheels are unsafe and should be replaced. Once the leak has been located, its location can be marked using a tire crayon, thus allowing you to find the leak after the tire has been broken down. (Fig. 4–46)

If the leak is at the valve or valve core, the defective valve stem or core will need to be replaced. If the bead is leaking, the tire bead will need to be broken loose from the bead seat so the wheel bead seat and flange and the tire bead can be cleaned or checked for cracks or breaks.

FIGURE 4–46.
A tire should be immersed in water before dismounting to show all of the leaks in the tire and any cracks or breaks in the wheel. (Courtesy of Rubber Manufacturers' Association)

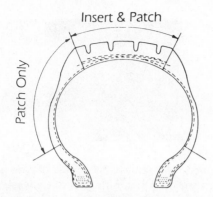

FIGURE 4-47.
The repairable areas of a radial tire; injuries to the sidewall must be no bigger than a pinhole. (Courtesy of Patch Rubber Company)

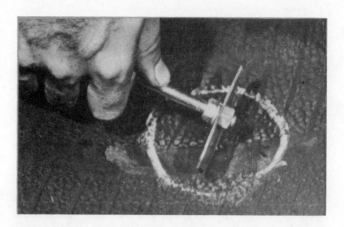

FIGURE 4-48.
The first step in repairing a tire is to clean the repair area with a prebuff cleaner and scraper. (Courtesy of Patch Rubber Company)

If the leak is in the tread portion, it can be patched. Sidewall damage to a radial tire is normally not considered to be repairable. This is because the puncturing object might have cut enough cords in the body plies to severely weaken the tire and the flexing of the sidewall might work the patch loose. However, a tire shop with the proper vulcanizing equipment can often repair sidewalls and shoulder areas if the damage is not greater than 1/4 inch (0.63 cm) in size. (Fig. 4-47)

Two methods are used to repair a hole in the tread. If the hole is big enough to cause a void, it is recommended that both should be used. A plug can be inserted through the hole from the inside or outside. This plug not only seals the air leak, but it also keeps water and dirt from entering into the belt plies. A patch on the inside of the tire is the best method of sealing a leak. Many shops prefer patches because the plug installing tool tends to enlarge the hole or damage the ply materials.

To plug and patch a leak, you should:

1. Spray an area several inches in diameter around the puncture on the inside of the tire with a prebuffing cleaner. Let it soak in for about 15 seconds, and scrape or wipe off the cleaner plus any silicons, waxes, or other contaminants from the inner liner. (Fig. 4-48)

2. Probe the hole with a probe or rasp to clean out the hole. (Fig. 4-49)

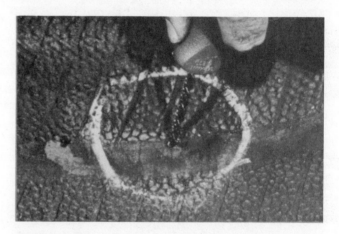

FIGURE 4-49.
Injuries larger than pinholes should be cleaned with a reamer or tire probe, following the angle of the injury. (Courtesy of Patch Rubber Company)

3. Coat the inside of the hole and the plug, also called a filler or insert, with vulcanizing cement and push or pull the plug into the hole. (Fig. 4-50)

4. Trim off the inner and outer ends of the plug with a knife.

5. Buff the inner end of the plug and the area around it using a powered buffer. Properly buffed rubber should have a dull, velvety appearance. (Fig. 4-51)

6. Coat the buffed area with vulcanizing cement. (Fig. 4-52)

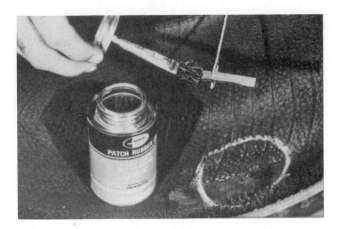

FIGURE 4–50.
Larger injuries should be plugged before patching; different styles of plugs are used for different types of tires. The injury and the plug should be coated with vulcanizing cement before inserting the plug. (Courtesy of Patch Rubber Company)

7. Allow the vulcanizing cement to dry thoroughly, about 10 to 20 minutes depending on temperature and humidity. Remove the protective backing from the patch, center the patch over the hole, and stitch the patch securely in place. Stitching is done with a special, narrow roller. You should begin at the center of the patch and work to the sides. Stitching should

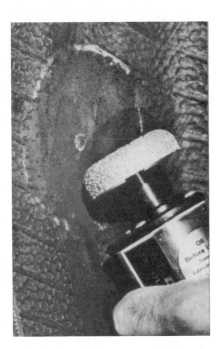

FIGURE 4–51.
After the plug is inserted, it is trimmed to length, and the area for the patch is buffed. (Courtesy of Patch Rubber Company)

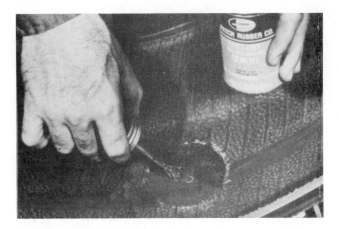

FIGURE 4–52.
A generous coating of vulcanizing cement should be painted onto the buffed area. (Courtesy of Patch Rubber Company)

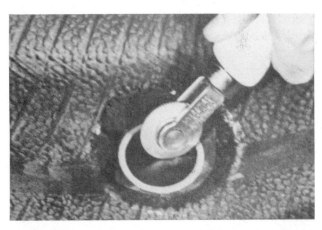

FIGURE 4–53.
The last step in patching a tire is to remove the protective backing from the patch and stitch it in place. Stitching should be from the center outward to remove any and all air bubbles. (Courtesy of Patch Rubber Company)

work any air bubbles out and press the patch firmly into contact with the tire. (Fig. 4-53)

Other methods have been developed to repair tubes and heavier truck and equipment tires. Instructions for these methods are available from manufacturers of tire repair materials.

4.9 TIRE RELATED DRIVING PROBLEMS

Faulty tires or wheels can cause several different driving problems that usually fall into the catagories of pull and vibration. Car **pull** is a tendency of the car to try to turn right or left by itself. Pull, sometimes called **lead**, is normally caused by a fault in either or both front tires or a misalignment of one or both front tires.

Car vibrations are caused by one or more things that spin or turn rapidly. Annoying vibrations can be seen, heard, or felt. Vibrations occur when the spinning objects have runout or are unbalanced. Runout occurs when an object spins off-center or wobbles. The intensity of the vibration usually depends on the amount of unbalance and the speed of the spinning object, but, in some cases, it depends on the harmonic frequency of the things affected by it. A harmonic vibration occurs when the frequency of the vibration force of an unbalanced or runout condition matches the frequency of the suspension. The harmonic frequency of the car's suspension is usually most affected by tire unbalance or runout in wheel speed frequencies equivalent to 40 to 60 MPH (64.4 to 96.6 KPH).

4.10   TIRE PULL

Tire **conicity** is a term that describes a tire's tendency to pull. Conicity is caused by a tire being built with a belt that is off-center, or by a belt that is slightly shorter on one side than the other. (Fig. 4–54)   It can also be caused by

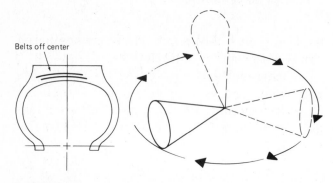

Belts off center

FIGURE 4–54.
Most tire conicity problems are probably caused by a tire being built with the belt off-center, as little as 0.025 inch (6 mm). Conicity will cause the tire to try to roll in a circle, just like a cone.

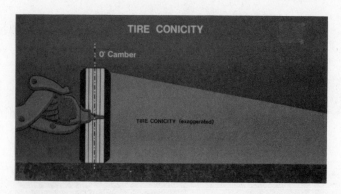

FIGURE 4–55.
Tire conicity will cause the tire to want to turn in the direction of the cone's apex and to pull or turn the steering mechanism in that direction. (Courtesy of Hunter Engineering)

the beads not being on the same plane. Those of you who have tried to roll a cone know that it will not roll straight. It rolls in a circle with the center of the circle being the point of the cone. A tire with a conicity problem will also try to roll in a circular fashion, just like the cone that it is a part of. Tire conicity problems will usually show up when a new tire is first mounted on a car, but, occasionally, a tire can develop conicity as it wears. (Fig. 4–55)

A tire pull problem usually shows up at the same time that the tires are changed or rotated. It is fairly easy to determine if a pull is a tire-caused problem. Merely switch the two front tires; if the pull direction changes, the pull is caused by one or both of the tires. Determining which of the two tires is the culprit is almost as easy. Switch one of the front tires with a rear tire, or one that you know is good. Again, if the pull changes, you have found a problem tire. If it is new, a tire with a conicity problem can be returned to the dealer, or it can be run on the rear with no problems. It is possible to run this tire on the front with an adjustment in the amount of caster spread (see Chapters 17 and 18) if the car has adjustable caster angles, but another caster adjustment will be required when that tire is changed or moved. (Table 4–2)

Torque steer is an uncommon tire problem that can also cause a pull but only during acceleration or deceleration; there will be no pull when the car is driven at even speeds. Torque steer is caused by one of the two tires on the drive axle. Some tire types are more prone to torque steer if the tire pressures are unequal. The term torque steer is usually used to describe a phenomenon encountered with FWD axles, which will be described in Chapter 7.

4.11   RADIAL TIRE WADDLE

If the belt is placed in a crooked position as the tire is built, **radial tire waddle** will occur. The belt tends to go straight down the road, and this can force the tire and wheel to waddle back and forth. Waddle usually occurs at slow speeds, about 5 to 30 MPH (8 to 50 KPH). It is most noticeable at the rear; it feels like someone is slowly shaking the back of the car back and forth in time with the tire rotation. If this tire is at the front, it can cause the front end to appear like it is moving back and forth. A tire that causes waddle will often also cause rough operation at 50 to 70 MPH (80 to 110 KPH). Roughness or harshness is a problem that can be felt or heard or both, much like vibration except that the roughness is irregular.

A tire with a tread or ply separation can cause problems similar to radial waddle. At low speeds, it can feel the same as radial tire waddle but becomes a vibration at higher speeds. A separated tire can be located by careful inspection, looking for bulges or soft spots.

4.12   VIBRATIONS

A **vibration** is a regular motion that can be felt, seen, or heard. In most cases, tire-caused vibrations are speed sensitive; that is, they come and go and change intensity depending on the speed of the car. Knowing the speed at which the vibration occurs will often give the technician a clue as to the nature of the vibration. The possible tire, wheel, and suspension related vibrations and the probable speeds at which they occur are

Tire wear (spotty or uneven)	felt at 30 to 70 MPH
Radial tire runout	felt at 20 to 70 MPH
Lateral tire runout	felt at 60 to 70 MPH
Tire balance	heard and felt at 30 to 60 MPH
Wheel bearing	heard at 0 to 70 MPH felt at 50 to 60 MPH

The procedure for eliminating a tire-caused vibration is usually to check and cure, if necessary, tire radial and lateral runout, check and cure, if necessary, faulty wheel bearings (Chapter 6), and then rebalance the tire.

Most tire shops will use a procedure similar to this to cure a tire vibration:

1. Make a preliminary check for tire defects, and mount the tire and wheel on an off-car balancer. With the tire mounted, make a more thorough check for flat spots, separation, and other defects.

2. A quick check for lateral and radial runout is made by watching the edges of the tire tread while the tire is spinning on the balancer. If it appears excessive, a more accurate check will be made. These checks will be described in Sections 4.13.1 and 4.13.2.

3. The balance of the tire and wheel is checked and corrected if necessary.

4. The tires are replaced on the car, and a test drive is made to check the results.

5. If a problem still exists, the tire and wheel are spun balanced on the car. While doing this, the operator should watch for bad wheel bearings, a bent hub or axle flange, faulty motor mounts or driveline, as well as any other defect that might cause a vibration. (Table 4–3)

4.13   TIRE AND WHEEL RUNOUT

Runout is a commonly used term to describe something that does not spin evenly/true. A tire with **lateral runout** will wobble from side to

TABLE 4-3
Most vehicle vibrations will occur at certain speed and load conditions, which can give the mechanic a valuable clue as to what is causing them. (Courtesy of Dana Corporation)

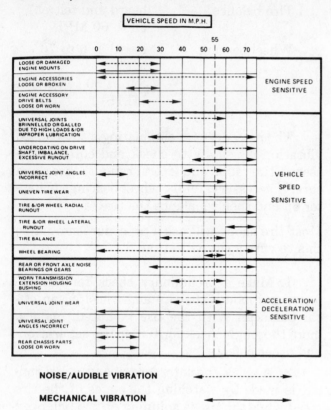

NOISE/AUDIBLE VIBRATION ← - - - - - - - →

MECHANICAL VIBRATION ← - - - - - - →

side; a tire with **radial runout** will hop or wobble up and down. A tire with runout cannot run true; it will try to move the suspension up and down or in and out, which will result in a vibration. Runout is easily seen by lifting the tire off the ground and spinning it. Watch the position of the tread relative to some stationary object. If the tread appears to move up and down, it has radial runout; if it appears to move sideways, it has lateral runout. The amount of runout is determined by measuring it with a runout gauge or a dial indicator. If a standard dial indicator is used, it should have a wide tip to ensure that the tip will not get caught in the tread grooves. (Fig. 4–56)

4.13.1 Radial Tire and Wheel Runout

Radial runout of the tire tread can be caused by an out-of-round or nonconcentric tire, an out-of-round or nonconcentric wheel, a nonconcentric hub bolt pattern, or a tire with a variation in sidewall stiffness. Tire flat spotting will affect radial runout. If a tire has been sitting for a while with the car's weight on it, the car should be driven to warm up the tires and round out any flat spots before checking runout. (Fig. 4–57)

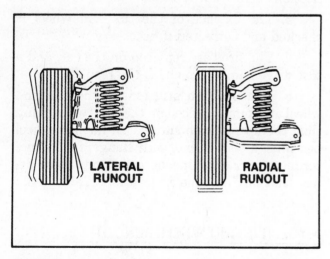

FIGURE 4–56.
Lateral tire run-out can cause a sideways tire shake while radial run-out can cause a hop. (Courtesy of Ford Motor Company)

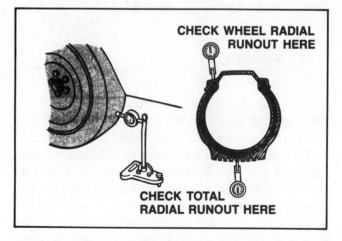

FIGURE 4–57.
Radial tire and wheel run-out are checked by positioning a dial indicator next to the tire or wheel and rotating them while watching for movement of the indicator. (Courtesy of Ford Motor Company)

Tire stiffness variation or a tire that has strong or weak portions of the sidewall or belt, can only be checked with the tire and wheel still mounted on the car and carrying the weight of the car. The device used to make this check, called loaded radial runout, is a tire problem detector or TPD. A TPD is a motor-driven roll that will rotate the tire and measure any up and down motions of the car that occur because of the tire's rotation. This device is no longer on the market in a simple form; an on-car tire truer uses a TPD sensor. A tire with a stiffness variation problem, like one with radial runout, can sometimes be saved by mounting the stiff or high spot of the tire on the low spot of the wheel. Also, it can usually be saved by truing the tire using an on-car tire truer (see Section 4.14). (Fig. 4-58)

Unloaded runout is commonly checked in garages and tire shops. The tire is lifted off the ground, an indicator is placed at the center of the tread, and the tire is rotated while watching the indicator movement. The amount of radial runout that a car will tolerate will vary depending on the size and type of tire, the weight of the car, and the tolerance or sensitivity of the driver. Radial runout tolerances are usually listed in the manufacturer's manual. A rule of thumb is that a very sensitive driver might notice and complain about 0.030 inch (0.75 mm). Most drivers will notice and find objectionable

FIGURE 4-59.
A run-out gauge is used to measure radial run-out on a tire and a wheel. (Courtesy of Hickok)

0.060 inch (1.5 mm), and everyone will probably complain about runout greater than 0.090 inch (2.3 mm). (Fig. 4-59)

If excessive runout is found at the tire tread, the high spot should be marked, and then the wheel should also be checked for runout. To measure the wheel's radial runout, the indicator is mounted at the inner side of the wheel's bead seat, and the wheel is rotated while watching for indicator movement. A more accurate method of measuring wheel runout is to remove the tire and check for runout at the tire side of the bead seats. The manufacturer's manual should be checked to determine the allowable runout. The normally accepted limit for wheel radial runout is 0.030 to 0.040 inch (0.75 to 1.00 mm), which will vary slightly depending on the type of wheel. A faulty tire is indicated if the wheel runout is acceptable and the tire runout is not. (Fig. 4-60)

Occasionally tire runout can be reduced to acceptable standards by tire matching, that is, matching the tire to the wheel. This is done on many O.E.M. tire and wheel mountings. Tire stiffness is measured, and the high (or low) spot

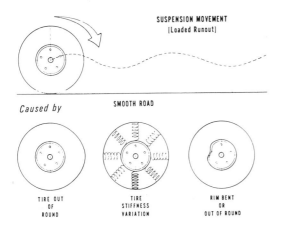

FIGURE 4-58.
The most common causes of loaded radial run-out, which can in turn cause a vertical suspension movement. (Courtesy of Oldsmobile)

MEASURING WHEEL RUNOUT

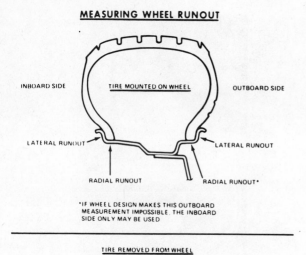

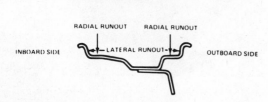

FIGURE 4-60.
The areas where wheel radial and lateral run-out are checked; the most accurate checks are with the tire off. (Courtesy of Oldsmobile)

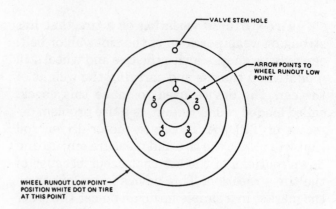

FIGURE 4-61.
Shown is one of the systems used to mark wheel run-out so the tire's high spot can be indexed to the wheel that produces a smoother running tire and wheel assembly. This is called tire matching. (Courtesy of Chevrolet)

is marked on the tire. Wheel runout is measured, and the wheel is marked. The tire is then placed on the wheel with the marks in the proper relationship so the high or stiff spot of the tire is located over the low spot of the wheel. Tire and wheel marking systems vary among manufacturers; at this time, there is no universal system. (Fig. 4-61)

Excessive wheel radial runout can be caused by lug bolt circle runout.

To measure lug bolt runout, you should:

1. Remove the wheel, and mount a dial indicator so the indicator stem is on one of the lug bolts. (Fig. 4-62)

CAUTION: Be sure the car is properly supported on a hoist or jack stands.

2. Rotate the hub slightly to get the highest possible reading, and adjust the indicator to read zero.

3. Carefully pull back the indicator stem. Do not let the indicator body move, and rotate

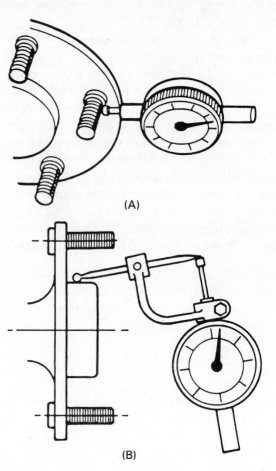

FIGURE 4-62.
Radial tire and wheel run-out can be caused by a run-out of the wheel stud circle. Wheel stud run-out can be checked using a dial indicator (A). Axle or hub, drum and wheel pilot run-out can also cause wheel radial run-out (B). It is also checked using a dial indicator. (Courtesy of Ford Motor Company)

the hub to bring the next lug bolt into position with the indicator stem.

4. Rotate the hub slightly to get the highest possible reading. Compare this reading to zero; it is the amount of runout between these two lug bolts.

5. Repeat Steps 3 and 4 on the remaining lug bolts. The lug bolt circle runout should not exceed 0.015 inch (38 mm).

NOTE: Some wheels are centered using the pilot of the hub or axle; in these cases it is important that the runout of the pilot be checked.

Lug bolt circle runout is corrected by replacing the hub or axle. The effect of lug bolt circle runout can sometimes be reduced by positioning the low spot of the wheel to coincide with the high lug bolt. Usually the quickest way to do this is to measure the tire runout, relocate the wheel one lug bolt different, and remeasure the runout. Continue doing this in the other two or three wheel positions until an acceptable position is found.

When straight shank lug nuts are used on alloy wheels, wheel runout can result from the clearance between the wheel holes and the shanks of the nuts. Some of this runout can be reduced by snugging the wheel onto the hub or axle mounting flange using two standard, tapered lug nuts to center the wheel to the lug bolts. Then two straight shanked nuts should be installed and partially tightened and the tapered nuts replaced with straight-shanked nuts. The wheel should be then tightened in place.

4.13.2 Lateral Tire and Wheel Runout

Lateral tire runout can be caused by a faulty tire, a bent or poorly machined wheel, or a bent hub or axle flange. Lateral tire runout is measured by placing an indicator at the side of the tire on a smooth section of rubber that is close to the shoulder of the tread. Next, the tire is rotated while watching the indicator movement. Limits for lateral runout are published by various manufacturers. If they are not available, use the same rule of thumb as for radial runout—0.030 inch (0.75 mm), 0.060 inch (1.5

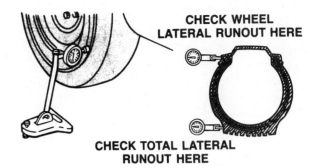

FIGURE 4-63.
Lateral run-out can be checked by placing a dial indicator at the side of the tire or wheel and rotating it while watching for movement of the indicator. (Courtesy of Ford Motor Company)

mm), and 0.090 inch (2.3 mm). If excessive lateral runout is found at the tire, mark the tire and check the wheel lateral runout. (Fig. 4-63)

Lateral wheel runout is measured by placing the indicator on the side of the bead flange and rotating the wheel while watching the indicator movement. Acceptable wheel runout is less than allowed at the tire, depending on how much closer the checking point is to the center of the hub; 0.010 inch of wobble, 5 inches out from the hub, would become 0.020 inch of wobble if it could be measured 10 inches out from the hub. A faulty tire is indicated if the wheel runout is acceptable while the tire runout is not. (Fig. 4-64)

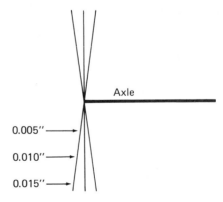

FIGURE 4-64.
The same amount of lateral run-out will be measured as different amounts depending on how far from the center of the wheel the measurements are taken: 0.005 inch of run-out at the mounting flange will cause about 0.010 inch of run-out at the rim and 0.015 or 0.020 inch of run-out at the tire.

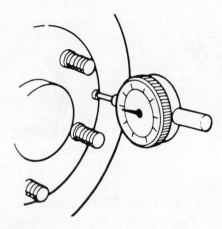

FIGURE 4-65.
Lateral tire run-out can be caused by a bent wheel mounting flange. Flange face run-out can be checked using a dial indicator. (Courtesy of Ford Motor Company)

If wheel runout is excessive, the hub or axle flange should be checked for runout. Mount the indicator on the side of the hub flange, and rotate the hub while watching for indicator movement. About 0.010 inch (0.25 mm) or less runout is acceptable. Excessive flange runout is corrected by replacing the hub or axle. The effect of excessive flange runout can sometimes be reduced by relocating the wheel on the flange using the same procedure that was described for lug bolt runout. Remount the wheel in the different positions, being careful to use the correct amount of torque and the correct pattern to tighten the lug bolts. Measure the wheel runout at each of the possible positions until an acceptable position is found. If an acceptable position cannot be found, the hub or axle will have to be replaced. (Fig. 4-65)

4.14   TIRE TRUING

Tire truing is a process of cutting rubber off of a tire in order to make it round. Tires can be trued either on or off the car, depending on the type of equipment used.

Off-car truing involves mounting the tire and wheel in a machine that rotates the tire while cutting off the high portions of the tread. A motor driven blade or a rasp-type cutter is used to remove the excess rubber, although

FIGURE 4-66.
An off-car tire truer. The motor at the upper right turns a cutter which trims the high spots off of the tire as it is turned. (Courtesy of Amermac)

there is seldom any excess rubber on a tire. This type of tire truing is not recommended by some because it reduces tire life and does not always make a permanent repair. (Fig. 4-66)

On-car tire truing involves using a machine that measures tire-loaded radial runout and cuts rubber from the edges of the tread in relationship to the runout. This machine uses a pair of rasp-type cutters, one on each side of the tread. This machine is fairly expensive, but it has a reputation of being an effective tire repair method. (Fig. 4-67) These machines should be operated following the procedure recommended by the manufacturer.

4.15   TIRE BALANCE

Tire and wheel balancing is the process of adding weights to the wheel rim in order to counterbalance heavy spots in the tire and wheel assembly and, sometimes, the brake rotor or drum. This is usually called wheel balancing. If not counterbalanced, these heavy spots are acted on by centrifugal force when the tire is

FIGURE 4–67.
An on-car tire truer. This machine uses a pair of rotating rasps (inset) to remove tread material from the outer tread ribs wherever there is a tire force variation. Force variation is measured by the transducer attached to the fender well. (Courtesy of Hunter Engineering)

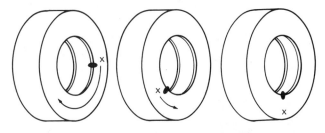

FIGURE 4–68.
A tire that is not in static balance will rotate back and forth until the heavy spot works into a position at the bottom.

spinning. Centrifugal force can cause the tire to hop up and down (or at least try to) or wiggle sideways while it is rolling. Balance problems are placed into two general classifications, **static** and **dynamic**.

Static is a term that normally refers to a stationary object or thing. As far as the tire is concerned, static balance refers to the distribution of weight around the wheel. Any heavy spots on the tire should be counterbalanced by an equally heavy weight on the other side of the wheel. (Fig. 4–68)   Static balance probably got its name because the original balancers checked balance with the tire stationary on a bubble

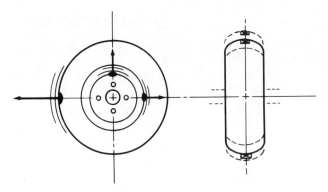

FIGURE 4–69.
When a tire is spun, differing amounts of centrifugal force can cause a tire to go out of balance if there are heavy portions on different radiuses such as the tire tread, wheel rim, or brake drum. This is called a kinetic unbalance.

balancer or turning slowly, by gravity, on a free-turning arbor. A statically balanced tire will probably spin true about 75 percent of the time, but, in some cases, it will still try to hop or wiggle. There are several styles of spinning-static balancers to correct these problems. These are called **kinetic** or **single-plane** (radial) balancers. Kinetic balancing is recommended over static balancing because a certain weight on the rim can counterbalance a certain weight on the tread while the tire is static, but when the wheel is spinning, the different radius of each point will generate differing amounts of centrifugal force, which would unbalance the tire. (Fig. 4–69)

A tire that is unbalanced dynamically will have a heavy spot on the inside or the outside. Centrifugal forces cause the heavy spot to try to move to the tire center line when the tire spins. If the heavy spot moves toward the center line when it is in a forward and then also in a rearward position, this will cause the tire to wiggle back and forth or shimmy. Dynamic unbalance is noticeable on front, steerable wheels because they are the only ones that will allow this sideways movement. Dynamic balance can only be checked by spinning the tire and wheel. A dynamic balancer is sometimes called a **two-plane** balancer because it can balance a tire in the radial plane as well as a lateral plane. Most spin balancers can be dynamic or kinetic or both dynamic and kinetic balancers. (Fig. 4–70)

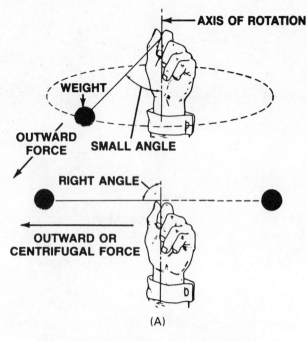

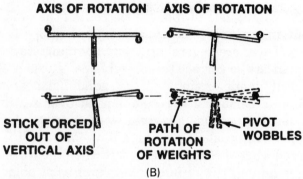

FIGURE 4-70.
A spinning weight will try to position itself at a right angle to the axis of rotation; dynamic unbalance problems result from a heavy spot at the side of a tire trying to move to the tire's center line. (Courtesy of Ford Motor Company)

4.16 WHEEL WEIGHTS

Wheel balance is achieved by attaching weight onto the wheel rim. Weights are available in various sizes, generally in one-quarter ounce increments. There are several styles of clip-on weights for steel and aluminum rims. Weights are also available in a stick-on style that are often used on alloy rims. These weights come in premolded sizes or in sticks that can be cut to the weight needed. Caution should be exer-

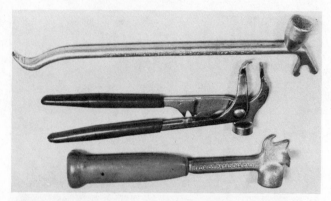

FIGURE 4-71.
Three different wheel weight tools. They each provide a hammer to drive weights onto a wheel rim and a way to pry weights off; the center tool also can trim weights smaller and tighten the spring clip.

cised when working with alloy wheels. They are easily scratched and some weight materials can quickly damage the wheel. For example, a weight containing zinc will cause rapid corrosion on a magnesium wheel. (Figs. 4-71 & 4-72)

FIGURE 4-72.
Some of the various types of wheel weights. Standard clip-on weights of two different sizes are at the center, to the left is a weight with an extra-long clip to fit under some wheel covers and two glue-on or stick-on weights, the longer one is to be cut to the correct weight. At right center is a pickup or truck weight with an overly wide clip. At the right is a weight with an alloy clip for use on alloy wheels.

Clip-on weights are installed and removed using a wheel weight tool or hammer. This tool is designed to remove weights quickly and easily plus drive them in place. Some weight tools are also made to trim the size of the weight.

4.17 BALANCING OPERATIONS

Balancing a tire usually requires the placement of a number of weights, occasionally only one but usually two or four, on the wheel rim. Two weights or two pairs of weights are generally used to static balance a tire and wheel, one on the inside and one on the outside. The amount of weight that is added should equal the amount of unbalance. If the unbalance is very slight, only one weight can be used, but its placement on only one side of the rim could upset dynamic balance. A balancer is used to tell the operator where to place the weight and give some indication of how much weight to use. (Fig. 4–73)

Two equal-sized weights are normally used to correct dynamic unbalance. These two weights must generate a force equal to the amount of unbalance, and these two weights must be placed 180 degrees apart on the rim, with one weight on the inside of the wheel and the other on the outside. (Fig. 4–74)

It is common and often less confusing to use four weights, two for static and two for dynamic balance. Some balance machines will help an experienced operator reshuffle these weights

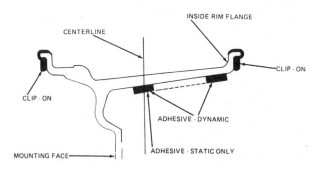

FIGURE 4–73.
Clip-on weights are secured to the rim flange. Adhesive weights can be placed at the rim center line or on any flat location inside of the rim; the closer they are to the side of the rim, the more effect they will have on dynamic balance. (Courtesy of Chevrolet)

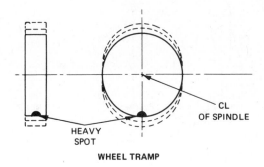

WHEEL TRAMP

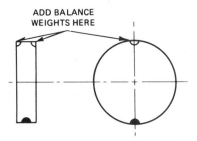

CORRECTIVE WEIGHTS

FIGURE 4–74.
Weights to correct static or kinetic unbalance are placed on the wheel, directly opposite the heavy spot. If the amount of weight to be added is more than an ounce or so, it is usually split in half and added to each side of the rim. (Courtesy of Oldsmobile)

SAFETY NOTE
Always use caution when spinning tires and wheels for balancing purposes (or for other reasons) since there are several possibilities for injury. Getting clothing or various body parts caught in the spinning tire and wheel or balancer or getting hit by a flying weight or debris from the tire and wheel are probably the most common causes of injury. Avoid them.

so placement of the static weights will reduce or eliminate the need for dynamic weights. An experienced tire and wheel balance technician will never let two weights end up 180 degrees apart on the same side of the wheel; in this position, they are balancing out each other. A computer balancer, either off-car or on-car, nor-

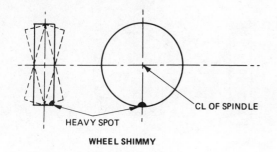

WHEEL SHIMMY

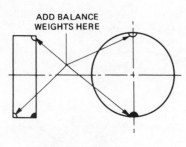

CORRECTIVE WEIGHTS

FIGURE 4–75.
Weights to correct dynamic unbalance are always placed in two locations at the inside and outside of the rim; these two weights are usually equally heavy. (Courtesy of Oldsmobile)

mally uses only one or two weights; one on the inside and one on the outside of the wheel. Occasionally, only a single weight will be needed. The computer has the ability to measure the various forces and quickly make the necessary calculations so the placement of two weights will balance both radial and lateral forces. (Fig. 4–75)

4.17.1 Off-car and On-car Balancing

Wheel balancing can be done with the tire and wheel off the car and mounted on a balancer or with the tire and wheel mounted in its normal position and the balancer attached to the car. Off-car balancing is ideal when a new tire is mounted. The wheel is already off the car, saving the removal step. Off-car balancing machines are often simpler and less expensive because the machine is only concerned with the tire and wheel, and the tire and wheel is in a definite position relative to the machine sensors. All modern, spin-type, off-car balancers are two-plane, kinetic and dynamic balancers.

A possible disadvantage with off-car balancers is that the wheel commonly mounts to the machine using the wheel's center hole. A runout and balance error can occur if the wheel's center hole and the lug bolt bosses are not concentric.

On-car balancing, often called **spin-balancing**, will probably produce the best balancing results because the tires and wheels are mounted in their running position and any slight runout errors are compensated for. Also, the brake rotor or drum is balanced with the tire so another possible unbalance is compensated for. Possible disadvantages of on-car balancing are that this balance is disrupted if the tire and wheel are moved to another location on the car or rotated on the hub or axle. Also, the balancer is more complex and often more expensive because it has to be portable and adaptable to different cars. On-car balancing can be either one- or two-plane, either kinetic or kinetic and dynamic. Nonsteerable wheels cannot be dynamically balanced on the car. A balancer cannot sense dynamic unbalance if the tire is not free to move in that direction.

On-car balancing requires something to turn or spin the wheels. Spinner motors are used to spin nonpowered wheels, the front wheels of an RWD car, or the rear wheels of an FWD car. A spinner uses an electric motor that drives the tire through a machined metal drum that is pressed against the tread or shoulder of the tire. Spinners are available in different horsepower ratings. A spinner that is too small will not be powerful enough to spin large and heavy wheels at full speed without overheating. Spinners are also affected by brake drag. Sometimes it is necessary to loosen up the shoes to allow sufficient speed and coast time. A brake is provided on spinners to allow stopping the tire quickly when desired. Spinners are often used to spin a tire and wheel up to speed to see if that particular tire and wheel are causing a vibration or to check a wheel bearing. (Fig. 4–76)

A drive wheel, the front wheels of an FWD car or rear wheels of an RWD car, are normally spun by the car, after starting the engine and putting the transmission into gear. In most cases, just the wheel being checked is lifted off

FIGURE 4–76.
A spinner motor is positioned to check tire balance. Spinner motors are used to spin nondrive wheels fast enough to cause an unbalance to show up. (Courtesy of Hunter Engineering)

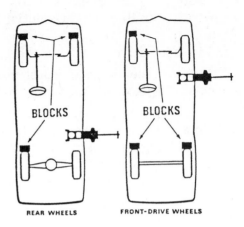

FIGURE 4–77.
Drive wheels are spun using the car's engine; a jack is used to lift the tire to be checked off the ground while blocks are placed in front of the remaining tires. Both drive tires should be lifted if the car has a limited-slip differential. (Courtesy of Hunter Engineering)

the ground. On cars with limited-slip differentials, both drive wheels must be off the ground. Use of the car's engine to drive the wheels will ensure that the differential gears are properly lubricated during the balancing operation. Severe differential wear has occurred on some cars through the use of a spinner motor on one tire of a drive axle.

SAFETY NOTE

The following are some cautions to observe when using the engine to spin the drive wheels:

- Blocks should be placed in front of one or more of the tires remaining on the ground. (Fig. 4–77)
- The power should be applied gradually to make sure the car does not move forward.
- No one should be permitted in front of a car during this operation.
- When only one wheel is off the ground, the differential gears will cause a 2:1 speed increase of the spinning wheel; its speed will

be double that read on the speedometer. Excessive tire speed can cause the tire to explode; it is recommended to limit the speed to 40 MPH (65 KPH). (Fig. 4–78)

- The wheels on some FWD cars will hang down far enough to place severe operating angles on the axle shafts and universal joints. They should either be operated for only short periods of time at these angles or blocked partially up to limit the severity of these angles.

FIGURE 4–78.
This tire exploded from spinning too fast; it ruined the tire and damaged the fender well. When spin balancing a tire or trying to free a stuck car, do not spin the tire to extremely high speeds. (Courtesy of General Tire)

The procedure to spin balance the drive wheels on a car with a limited-slip differential is:

1. Lift both drive wheels off the ground.

2. Remove the tire and wheel not being balanced, and replace two lug nuts to retain the brake drum.

3. Spin the first tire and wheel to be balanced by starting the engine and placing the transmission in gear; in this case, the speedometer will be reading tire speed. Balance the first tire and wheel.

4. Replace the unbalanced tire and wheel, and balance it. There is no need to remove the previously balanced tire and wheel because it is balanced.

4.17.2 Static Balancing, Bubble Balancer

A bubble balancer is an off-car, single plane, static balancer. At one time, bubble balancers were the most common type of balancers. They are simple, reliable, inexpensive, fairly fast, and fairly accurate.

To balance a tire and wheel using a bubble balancer, you should:

1. Clean any heavy deposits of dirt from the wheel, and remove all of the old weights.

2. Adjust the balancer feet so the balancer is level; the control lever must be in the Off position while leveling. The balancer head will not pivot in the Off position. (Fig. 4–79)

3. Place the tire and wheel on the balancer. The outside of the wheel should be up, the wheel's mounting flange should be contacting the balancer flange completely, and the tire's valve stem should be in line with the triangle or arrow mark on the center level.

4. Move the control lever to the On position. This will allow the balancer head and the tire to pivot, the heavier portion of the tire and wheel will lower, and the bubble will move off center, toward the lighter side. The distance that the bubble travels from the center gives an indication of the amount of unbalance. (Fig. 4–80)

FIGURE 4–79.
A bubble balancer. Note the three adjustable feet (A) that are used to level the device and the control handle (B) that is moved downward to allow the balancer head to pivot.

FIGURE 4–80.
Note the center bubble (arrow). This wheel is out of balance. Weight should be added to the rim directly opposite to the bubble. Any weight left on the wheel should be removed.

5. Rotate the tire, wheel, and balancer head until the lighter side is closest to you; the bubble should be between you and the center of the balancer.

6. Place a group of four equally sized weights on the wheel rim directly in front of you. When the proper weights have been selected, the bubble will move past center toward the other side. If the correct starting point has

FIGURE 4–81.
Four equal-sized weights have been placed near the wheel rim in line with the bubble; these weights were heavy enough to cause the bubble to move past the center.

FIGURE 4–82.
The four weights were divided into two pairs, and they were moved apart enough to cause the bubble to move to the center. A mark was placed on the tire tread in line with each of the weights.

been selected, the bubble will pass through the center on its way to the other side. (Fig. 4–81)

7. With the bubble past the center, move two of the weights slightly to one side and the other two the same distance to the other side. Continue moving the two pairs of weights, making sure that each pair stays the same distance from the starting point, until the bubble is centered or until there is a 90° angle between the two pairs of weight. If you need to move the weights to a 90° angle or beyond, select two

pairs of lighter weights, and repeat this step. (Fig. 4–82)

8. After you have selected the right-size weights and located their correct position, move the control lever to the Off position, and place a chalk mark on the tire tread in line with each pair of weights.

9. Remove the tire and wheel from the balancer, and secure a weight on the inside bead flange one-half inch inward, so the weights are closer together, from each of the chalk marks. (Fig. 4–83)

FIGURE 4–83.
The wheel has been removed from the balancer, and one of the weights has been installed in line with each chalk mark.

FIGURE 4-84.
The wheel has been placed back on the balancer, and the two upper weights have been adjusted to improve the balance and are being installed on the rim.

10. Replace the tire and wheel on the balancer (remember to align the valve stem with the reference mark). Move the control lever to the On position, and place the two remaining weights on the rim in line with the chalk marks. If necessary, adjust their position to center the bubble.

FIGURE 4-85.
A mechanical balancer head has two internal weights. The location of the weights is moved by gripping one of the green knobs; the amount of weight is changed by gripping one of the red knobs. (Courtesy of Hunter Engineering)

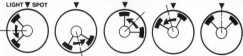

11. Move the control lever to the Off position, and secure the last two weights in position. (Fig. 4-84)

4.17.3 Static and Kinetic Balancing, Mechanical Balancer

This system uses a single-plane, on-car, spin-balancer that attaches to the wheel. Internal weights can be adjusted while the tire and wheel and the balancer unit are spinning. After these weights have been adjusted so that the assembly spins true with no vibration, the assembly is stopped. The amount and location of weight to add will be indicated by the balancer unit. (Fig. 4-85)

To spin balance a tire and wheel using a mechanical balancer, you should:

1. Lift the tire and wheel off the ground, and support the car in a safe manner so the tire to be balanced is several inches off the ground.

2. Remove the wheel cover, trim, or beauty ring to gain access to the inside of the outer bead seat. Also remove any heavy dirt deposits and all of the old weights.

3. Select the correct size wheel adapter, and move the cam levers to the unlocked position. The red side of the flags should appear. (Fig. 4-86)

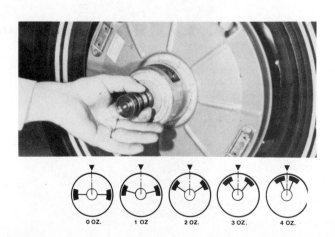

FIGURE 4–86.
A mechanical balancer wheel adapter. Swinging the two cam levers will allow inserting the adapter into the wheel rim. Be sure to clean the rim before installing the adapter. (Courtesy of Hunter Engineering)

FIGURE 4–87.
After the wheel adapter is inserted and locked into the wheel rim, its security must be checked by pulling firmly outward. If it comes out, it needs to be reattached in a better way. (Courtesy of Hunter Engineering)

4. Place the adapter ring into the rim with the adapter ring notch in line with the valve stem, and push the adapter into the rim until the adapter flange contacts the rim. If the adapter ring will not fit completely into the rim, reduce its size by turning the two thumb screws equally.

5. Swing the cam levers, both at the same time, to a locked position with the black flags showing. If it takes a lot of force to swing the cam levers, reduce the size of the adapter ring.

6. *CAUTION: Check for proper mounting by grasping the adapter ring with both hands and pulling outward with a firm pull. This is important since you or others around you could be injured if the ring comes loose while spinning at high speeds. If the adapter ring can be pulled loose, adjust it to a larger size, and repeat Steps 3, 4, 5, and 6. (Fig. 4–87)*

7. Open the four locks on the balancer unit, and place the unit over the four mounting studs on the wheel adapter ring. (Fig. 4–88)

8. *CAUTION: Close all four latches to lock the balancer onto the adapter. Make sure that all of the latches are closed, and gently tug on the balancer to ensure that it is securely attached.*

9. Turn either or both red knobs until the pointer reads about 1 1/2 ounces. Note that the two red knobs turn in reverse of each other and change the amount of weight and that the two green knobs also turn in reverse of each other and cause the inner unit to turn relative to the mounting body.

10. Spin the tire and wheel up to speed, and watch for vibrations. These are most noticeable by watching the bumper, radio an-

FIGURE 4–88.
Four cam locks are used to secure the balancer head to the wheel adapter. (Courtesy of Hunter Engineering)

FIGURE 4–89.
The balancer head is tuned by gripping one of the four knobs while watching some part of the car, often the bumper, for vibration. When the vibration disappears, the amount of unbalance will be read on the balancer head. Note that the operator's thumb is pointing in the direction of wheel rotation. (Courtesy of Hunter Engineering)

FIGURE 4–90.
The amount of weight to be added is indicated in the balancer window (inset), and the weight should be added to the rim in line with the "add weight here" arrow. (Courtesy of Hunter Engineering)

tenna, an open door or hood, or a crumpled shop cloth placed on top of the fender.

CAUTION: Be careful around this spinning balancer unit.

11. If a definite vibration is noticeable, place your hand over the green knobs with your thumb pointing in the direction of wheel rotation and lightly grip or pinch one of the green knobs. If the vibration gets worse, shift to the other knob. If the vibration gets better, continue pinching the knob until the vibration reduces to a minimum and starts getting worse. Then shift to the other knob, and return to the best point. This is much like tuning a radio. The most accurate tuning is done with the tire coasting down through the speed where the vibration is most noticeable. If the tire gets too slow, speed it back up, and then resume tuning. (Fig. 4–89)

12. Shift your fingers to one of the red knobs, and pinch it. If the vibration gets worse, shift to the other red knob. If the vibration gets better, continue pinching until the vibration reduces to a minimum and then gets slightly worse. Then, shift to the other red knob, and tune back to the best point. Fine tune using the green knobs and then go back to the red knobs.

13. Stop the assembly, turn the tire so the "add weight here" arrow is pointing upward and add the amount of weight indicated in the balancer window to the top of the wheel. Larger amounts of weight, over three ounces, should be split so half the weight is added to the outside and half to the inside of the rim. (Fig. 4–90)

4.17.4 Kinetic and Dynamic Balancing, Strobe Light Balancer

A **strobe light balancer**, sometimes called an **electronic balancer**, is an on-car, two-plane, spin-balancer. These units sense tire and wheel unbalance by means of a pick up, also called a transducer unit, that attaches to the suspension using a magnet. Vibrations of the suspension cause the strobe light to flash in time with the vibration; the flashes of the strobe light can cause the spinning tire and wheel to appear stopped. By noting the position of the wheel, the operator can tell where to add weights. By noting the reading on an indicator dial, the operator can tell the approximate amount of weight to add. The pick up unit is placed under the suspension in a vertical position while doing a kinetic balance. It is placed alongside of the suspension in a horizontal position while dynamic balancing.

To balance a tire and wheel using a strobe light balancer, you should:

1. Raise and support the car by the frame or body subframe in a safe manner so the tire to be balanced hangs a few inches off the ground.

2. Remove all of the old wheel weights and any heavy dirt deposits.

3. Position the pick up unit under a suspension member, usually the lower control arm or axle. The probe should be vertical and placed so it makes solid contact with the suspension member as close to the tire as possible. Slide the probe magnet up to contact the suspension member, and tighten the lock screw. (Fig. 4-91)

4. Place the strobe light about a foot away from the tire with the light facing the tire. Tap the top of the tire; the light should flash in response. If it does not, check the pick up probe or the light unit. (Fig. 4-92)

5. If the strobe light unit has a "sensitive-normal" switch, switch it to normal. If the unit has a "front-rear" switch, switch it to whichever axle you are working on.

FIGURE 4-92.
A pickup probe is placed under a front suspension in a kinetic balance position. Note that the strobe light is placed so it will flash on the tire. (Courtesy of Hunter Engineering)

6. Spin the tire and wheel; the strobe light will usually begin flashing and a reading will appear on the meter. Increase the tire speed until the meter increases to its highest point and then begins to drop back and move to a lower reading. The highest reading is called a peak and the dropping back is called a break. Increase the speed past a peak and break point, and let the tire coast down in speed. Use the strobe light to note the position of the tire at the break point on the indicator; it should appear stopped because of the strobe effect. When noting the tire position, use the valve stem, hub cap, or tire markings as a reference. In brightly lit shops, it might be necessary to pick up the strobe light and move it around the tire to locate a reference mark. Some operators place a chalk mark or a piece of tape on the sidewall for a more visible reference point. (Fig. 4-93)

CAUTION: Always be careful while working around the spinning tire and wheel assembly.

7. Stop the tire, and return it to the position where it appeared to be at the break point. With the pickup unit vertical and under the tire, the heavy portion of the tire is now downward, at the 6 o'clock position. Weights need to be added at the top of the wheel, or 12 o'clock position.

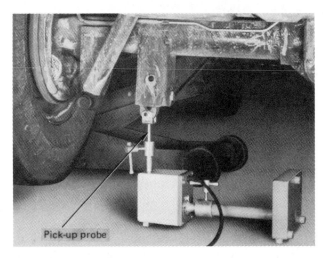

FIGURE 4-91.
A pick-up probe or transducer is in position to sense a kinetic unbalance condition of a rear tire. Note the jack placement causes the tire to hang a few inches off the ground. (Courtesy of Hunter Engineering)

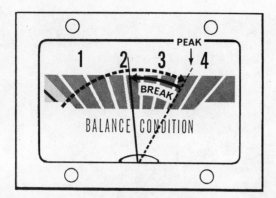

FIGURE 4-93.
When the tire is spun, the balancer will reach its highest reading or peak point and then at the break point, move back downward. Strobe light balancing is done at the peak and break points. These points are where the amount of unbalance is indicated on the meter and where the heavy location is found using the strobe light. (Courtesy of Hunter Engineering)

8. Add weight to the rim, using the number where the indicator needle was at the break point as a guide to how much weight to add (in ounces). (Fig. 4-94)

9. Repeat Steps 6, 7, and 8, and note the position of the weight that was added. Also note the indicator needle. If it does not move or if it stays in the good or green zone, the tire and wheel are kinetically balanced. If you are balancing a front tire, go to Step 11. If you are

FIGURE 4-94.
In kinetic balancing, the heavy spot is at the 6 o'clock position and weight is added at 12 o'clock. As this tire is spinning, the chalk mark on the tire is located at 9 o'clock with the strobe light. (Courtesy of Hunter Engineering)

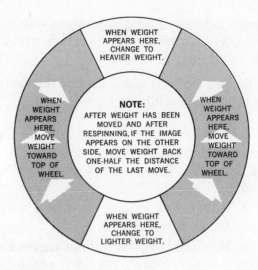

FIGURE 4-95.
After weights are added, the tire should be respun. If an unbalance still exists, use this chart to relocate or adjust the weight. (Courtesy of Hunter Engineering)

balancing a rear tire, you are done. If the needle moves past the good zone, the balance weights need adjusting. If the weight appears between 5 and 7 o'clock, it is too heavy; too much weight was added. If the weight appears between 11 and 1 o'clock, it is too light; add more weight. If the weight appears between 1 and 5 o'clock or 7 and 11 o'clock, it is in the wrong position; shift the weight an inch or so toward the top of the wheel, toward 12 o'clock. (Fig. 4-95)

10. Adjust the weight amount or position and repeat Steps 6, 7, 8, and 9 until the meter indicates a balanced tire and wheel or you begin moving a weight back and forth with no apparent progress. Sometimes a tire and wheel that are dynamically unbalanced will cause difficulties while kinetically balancing.

11. Turn the steering wheel so the front of the tire to be balanced is outward, and reposition the pick up unit with the probe horizontal. The magnet can be attached to a brake backing plate, rotor splash shield, or a steering arm. (Fig. 4-96)

12. Repeat Steps 6 and 7. If the indicator needle stays in the good zone, the tire and wheel are dynamically balanced. If not, the tire should be returned to the position it appeared to be in at the break point. Because the probe has been repositioned, the heavy portion of the tire is

FIGURE 4–96.
The pickup probe should be located in a horizontal position when dynamically balancing the tire. (Courtesy of Hunter Engineering)

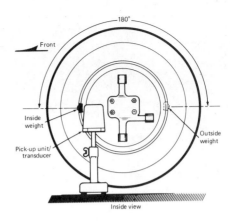

FIGURE 4–97.
Dynamic weights are added in pairs; one right next to the pickup at the inside of the wheel and one to the outside, one-half turn away.

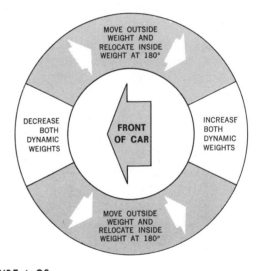

FIGURE 4–98.
After adding dynamic weights, the tire should be respun; if an unbalance still exists, use this chart to adjust the weights. (Courtesy of Hunter Engineering)

now in line with the probe and on the outside of the tire at 3 o'clock or on the inside of the tire at 9 o'clock. (This is true for a right-side tire with the pick up probe toward the front of the tire).

13. Add a two ounce weight to the inside of the wheel at 3 o'clock, in line with and next to the pick up probe. Add another two ounce weight to the outside at 9 o'clock, exactly one-half turn from the inner weight. This is for a right-side tire; a left-side tire would have the weights added at the inside at 9 o'clock and the outside at 3 o'clock. (Fig. 4–97)

14. Repeat Steps 6, 7, and 12. If the indicator reading stayed in the good zone, the tire is kinetically and dynamically balanced. If the needle showed a higher reading, note the position of the outside dynamic weight. If it is in line with the pick up probe, both weights are too heavy; reduce both of them, equally. If the outer weight is in line but one-half turn (180°) from the pickup probe, the two weights are too light, increase both of them. If the outer weight was between 10 and 2 o'clock or 4 and 8 o'clock, rotate it toward the rear of the car. Rotate the inside weight the same amount so it stays 180° from the outer weight. (Fig. 4–98)

15. Repeat Steps 6, 7, 12, and 14 until the tire and wheel are dynamically balanced.

4.17.5 Kinetic and Dynamic Balancing, Computer Balancer

Most **computer balancers** are off-car balancers. They are two-plane balancers, balancing in both planes at the same time. They have the reputation of being fast, accurate, and easy to use. Most computer balancers can be forced to give a kinetic only balance by entering an extremely narrow wheel width; this is rarely recommended. After the operator mounts the wheel,

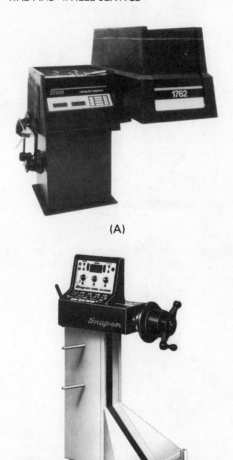

(A)

(B)

(C)

FIGURE 4–99.
Off-car computer wheel balancers. (A is courtesy of Sun Electric Corporation; B is courtesy of Snap-On; C is courtesy of Hunter Engineering)

enters information about the tire and wheel, and starts the procedure, the balancer goes through a short spin cycle, stops the rotation, and then gives a readout of how much and where to add weight on each side of the wheel. (Fig. 4–99)

To balance a tire using a computer balancer, you should:

1. Place the wheel on the mounting cone, and hand tighten the hub nut to secure it in place. Remove any heavy dirt deposits and all the old weights from the wheel. (Figs. 4–100 & 4–101)

2. Measure the machine-to-rim distance, and enter it into the machine. (Figs. 4–102 & 4–103)

3. Using the special calipers, measure the rim width, and enter it into the machine. (Fig. 4–104)

4. Note the wheel diameter, and enter it into the machine.

5. Lower the protective cover or safety hood, and press the start button. The machine will now spin the tire for a few seconds, stop spinning after a preset time, and display the amount of unbalance for each side of the tire.

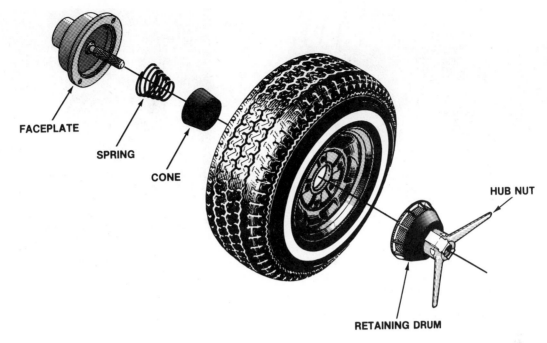

FIGURE 4–100.
Most computer balancers use a quick-mount method of mounting the wheel onto the balancer. Most balancers center the wheel using the center hole (shown). (Courtesy of Sun Electric Corporation)

FIGURE 4–101.
This adapter allows a wheel to be centered onto the balancer by the lug bolt holes rather than the wheel center; some wheels are not made very accurately, which can cause run-out when balanced from the center hole. (Courtesy of Ahcon Industries, Inc.)

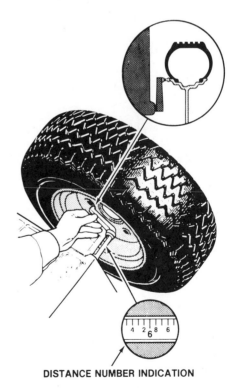

FIGURE 4–102.
Gauging the wheel-to-balancer location. This is one of three dimensions that need to be programmed into the computer. (Courtesy of Sun Electric Corporation)

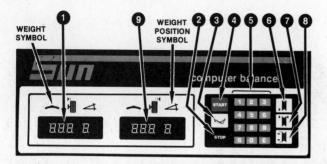

FIGURE 4-103.

The control panels for two different computer balancers. On control panel A, 1 is the weight readout for the inside of the wheel, 2 is the stop button, 3 is to select a special mode for styled wheels, 4 is the start button, 5 is keyboard to input dimensions, 6 selects rim distance, 7 selects rim width, 8 selects rim diameter, and 9 is the readout for the outside weight. On control panel B, the amount of unbalance will be readout at the upper and lower planes where "rdy" appears. (A is courtesy of Sun Electric Corporation; B is courtesy of Hunter Engineering)

6. Raise the protective cover, and rotate the tire slowly by hand until the inner or upper (depending on the machine type) indicator light comes on. Attach the displayed amount of weight to the inside or upper edge of the wheel rim in line with the weight location mark on the machine. Continue rotating the tire until the second indicator light comes on, and attach an amount of weight equal to the display to the outside or lower side of the wheel rim. (Fig. 4-105)

7. Verify the balance job by repeating Step 5. The weight displays should show zero or ready for weight amounts. If not, adjust the amount or location of the weights as indicated.

An on-car, computer balancer combines much of the operating simplicity of a computer balancer (i.e., digital readout of weight amount and location) with the ability to correct for wheel mounting runout errors and brake rotor or drum unbalance. A major problem in developing an on-car balancer of this type is in the construction of portable pick up units or an unbalance sensor that can give the computer an accurate signal. The car is placed on special

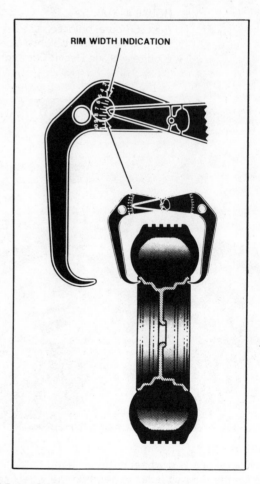

FIGURE 4-104.

Gauging the rim width using a rim caliper. (Courtesy of Sun Electric Corporation)

support stands that have the sensors mounted in them. Wheel location is determined by the machine in reference to strips of reflective tape that are placed on the tire. This tape reflects an infrared light beam that is sent from the balancer or remote stand.

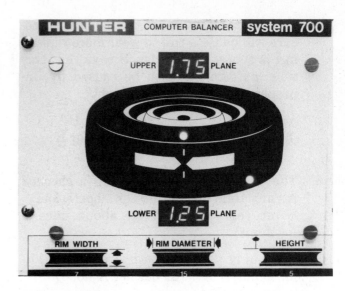

FIGURE 4–105.
This control panel indicates that 1.75 ounces of weight should be added to the upper edge of the rim with the wheel in this position. The wheel should be turned to the left until the center light comes on, and the 1.25 ounces of weight should be added to the lower rim. (Courtesy of Hunter Engineering)

REVIEW QUESTIONS

1. Technician A says that it is a good idea to mark a lug bolt before removing a wheel so the wheel can be replaced in the same position. Technician B says that relocating a wheel on a hub can change runout and balance conditions. Who is right?

   a. A only   **c.** both A and B
   b. B only   **d.** neither A nor B

2. When a wheel is replaced, the lug bolts should be torque tightened to ensure that the
   A. wheel does not come loose.
   B. brake rotor does not become distorted.

   a. A only   **c.** both A and B
   b. B only   **d.** neither A nor B

3. Technician A says that all replacement lug bolts are the same except for the thread size. Technician B says that a lug bolt can often be installed using a few flat washers and the lug nut. Who is right?

   a. A only   **c.** both A and B
   b. B only   **d.** neither A nor B

4. A tire is worn heavily on the inside of the tread with a fairly sharp corner at the sidewall; this wear was probably caused by too much

   a. negative caster.
   b. negative camber.
   c. toe in.
   d. positive camber.

5. Technician A says that too much air pressure will cause a tire to wear more rapidly in the center. Technician B says that this wear pattern can also be caused by improper alignment. Who is right?

    a. A only   **c.** both A and B
    b. B only   **d.** neither A nor B

6. The air pressure in a tire should be checked
    A. any time during the car's operation.
    B. at regular intervals, about once a month.

    a. A only   **c.** both A and B
    b. B only   **d.** neither A nor B

7. A tire is removed from a wheel by
    A. pushing the bead into the drop center at one side of the wheel.
    B. prying the bead over the wheel flange.

    a. A only   **c.** A and then B
    b. B only   **d.** neither A nor B

8. When removing and replacing a tire on a wheel, the tire bead and the bead seats should be
    A. clean.
    B. lubricated with motor oil.

    a. A only   **c.** both A and B
    b. B only   **d.** neither A nor B

9. Technician A says that a leaky tire can be safely sealed with a plug inserted through the tread. Technician B says it is a good practice to always inspect the inside of the tire before repairing it. Who is right?

    a. A only   **c.** both A and B
    b. B only   **d.** neither A nor B

10. Faulty construction of a radial tire can cause

    a. a side pull.
    b. a back and forth wobble.
    c. a hop up and down.
    d. any of these.

11. Technician A says that a side-to-side steering wheel shake is probably caused by radial tire runout. Technician B says that this condition can be cured with a static balance of the tire. Who is right?

    a. A only   **c.** both A and B
    b. B only   **d.** neither A nor B

12. Radial tire runout will cause the tread of the tire to move _____ while the tire is rotated.
    A. in and out
    B. sideways

    a. A only   **c.** both A and B
    b. B only   **d.** neither A nor B

13. Technician A says that lateral tire runout can sometimes be improved by remounting the wheel in a different position on the hub. Technician B says that it can usually be improved by remounting the tire on the wheel. Who is right?

    a. A only   **c.** both A and B
    b. B only   **d.** neither A nor B

14. On-car tire truing can often cure
    A. radial tire pull.
    B. loaded radial runout.

    a. A only   **c.** both A and B
    b. B only   **d.** neither A nor B

15. Technician A says that all of the weights can be placed on the inner side of the wheel when statically balancing a tire. Technician B says that some clip-on weights can damage aluminum alloy wheels. Who is right?

    a. A only   **c.** both A and B
    b. B only   **d.** neither A nor B

16. Technician A says that all on-car balancing systems dynamically balance the tire and wheel. Technician B says that a dynamic balance should cure a steering wheel side shake condition. Who is right?

a. A only   **c.** both A and B
b. B only   **d.** neither A nor B

17. When spin balancing the drive wheels and using the car's engine to spin the tires, use caution because
    A. spinning the tire too fast can cause it to explode.
    B. the car can move forward as power is applied to the drive wheels.

    a. A only   **c.** both A and B
    b. B only   **d.** neither A nor B

18. When kinetically balancing a tire using a strobe light balancer, the
    A. pick up probe is placed in a vertical position.
    B. weights are added at the 12 o'clock position on the tire.

    a. A only   **c.** both A and B
    b. B only   **d.** neither A nor B

19. A tire is being dynamically balanced using a strobe light balancer, and an unbalance condition equal to six ounces is located with the pickup probe at the 3 o'clock position. Where should the weight be added to correct this problem?
    A. Three ounces on the inside at 3 o'clock and three ounces on the outside at 3 o'clock.
    B. Three ounces on the inside at 3 o'clock and three ounces on the outside at 9 o'clock.

    a. A only   **c.** both A and B
    b. B only   **d.** neither A nor B

20. The advantage with computer balancers is that they are
    A. accurate and sensitive.
    B. fast.

    a. A only   **c.** both A and B
    b. B only   **d.** neither A nor B

Chapter 5

WHEEL BEARINGS—THEORY

After completing this chapter, you should:

- Be familiar with the terms commonly used with wheel bearings.

- Understand the various types of wheel and axle bearing arrangements.

5.1   INTRODUCTION

The wheel bearing's job is to allow the hub, wheel, and tire to rotate freely while it holds them in alignment with the car's steering knuckle or axle. In doing this, the wheel bearing has to absorb the vertical load of the car plus any side loads that result from cornering maneuvers as well as other forces. In addition, on drive axles, the bearing allows the drive axle to rotate while transmitting driving torque to the wheel and tire.

A mechanic has five basic styles of wheel bearings to learn that are based on the type and function of the bearings. As mentioned earlier, the term **wheel bearing** is normally used for the bearings on all front wheels and nondrive axle rear wheels. The term **axle bearings** is commonly used for RWD drive axles. The five various bearing types are:

1. nondrive axle, serviceable bearings—front wheels on RWD and rear wheels on FWD cars

2. nondrive axle, nonserviceable bearings—rear wheels on some FWD cars

3. drive axle bearings on a solid axle—rear wheels on most RWD cars

4. drive axle bearings on independent suspension—rear wheels on independent rear suspension (IRS) and front wheels on FWD cars

5. drive axle, nonserviceable bearings—front wheels on some FWD cars

The nonserviceable styles are very similar to each other. They both support the weight of the car, with the drive axle style capable of transmitting torque from the axle to the wheel. Like their name implies, nonserviceable bearings are not serviced. There is no service required or possible for the bearing unit. If the bearing is faulty, the bearing assembly is removed and replaced with a new one; **R & R** is the common term for this.

There is also a strong similarity between the nondrive axle serviceable bearings and the independent rear suspension drive axle bearings in that they are usually a pair of tapered roller bearings. They both normally require occasional cleaning, repacking with grease, and a clearance or preload adjustment.

5.2   BEARINGS AND BEARING PARTS

The term **bearing**, when used by most technicians and manufacturers, usually refers to either ball, roller, or tapered roller bearings. These bearing types are often referred to as frictionless bearings. They are called **frictionless** because the balls or rollers, fitted between the two races, roll easily whenever either of the two races rotates. A **bushing**, commonly used for crankshaft bearings, has a sliding action. Bushing friction is usually reduced by grease or a constant flow of oil. (Fig. 5-1)

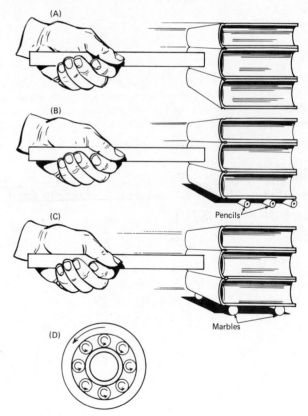

FIGURE 5-1.
Try pushing a stack of books using a stick or ruler as shown in (A). You will note a certain amount of drag. Place a few pencils (B) and then some marbles (C) under the books and try pushing them again. You should notice a definite reduction in the amount of drag and friction. The rolling action of the pencils and marbles is similar to the action in a bearing (D).

Frictionless bearings are usually made up of three major parts: (1) a cone or inner race, (2) a cup or outer race, and (3) the balls or rollers. (Fig. 5-2) All of these parts are made from hardened steel alloys and are precision ground to very close size and finish tolerances. The bearing balls or rollers are usually placed in a cage or separator so they will not rub against each other. The cage also ensures that the balls or rollers stay spread out in the correct spacing for proper load distribution. Sealed bearings position a seal between the inner and outer races on one or both sides of the bearing. The seal is used to keep lubrication in the bearing and dirt out. A frictionless bearing must be lubricated to reduce friction and prevent wear. Lubricants also carry heat away from the bearing and protect the metal surfaces from corrosion. Dirt and other abrasive materials must be kept out to prevent bearing damage or wear.

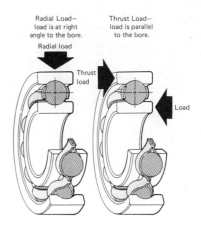

FIGURE 5-3.
The two basic loads that a bearing is subject to are radial loads from a right angle to the bore and thrust loads parallel to the bore.

Ball bearings run in concave grooves that are ground into each of the races. A ball bearing usually has the ability to control radial movement and loads as well as thrust movement. (Fig. 5-3) For example, it can support a shaft and allow the shaft to rotate while keeping the shaft from moving sideways. Because the balls only contact the races in a tiny spot, the amount of side or radial load and end or thrust load that a ball bearing can support is rather limited. Note that only a few of the balls are carrying the load at one time. The balls on the nonloaded side of the bearing are not really doing much. Bearing failure will be discussed more thoroughly in Chapter 6 when wheel bearing service is covered. (Fig. 5-4)

Roller bearings can carry a greater side load than ball bearings because the rollers, being longer, have a much greater load-carrying surface area. A roller bearing cannot control end thrust. When mounted where side thrust control is important, thrust bearings are used along with the roller bearings. Very thin roller bearings are called **needle bearings**. Needle bearings can carry still more load because there are more of the thin needles and, therefore, more surface area to transfer loads. (Fig. 5-5)

Tapered roller bearings are used when side thrust as well as radial loads need to be controlled. Front wheel bearings are the perfect example of this. The wheel bearings carry the ra-

FIGURE 5-2.
A disassembled tapered roller bearing. The cup is also called the outer race and the cone the inner race. (Courtesy of The Timken Company)

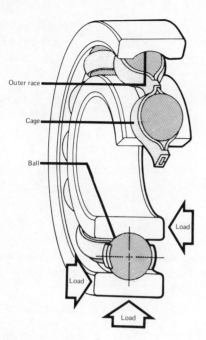

FIGURE 5–4.
A cutaway view of a ball bearing.

dial load of the car's weight as well as the side load that is trying to take the hub off of the spindle. Tapered roller bearings are used in pairs with the tapers of each bearing facing in opposite directions to each other. It is these tapers that provide sideways control. (Fig. 5–6)

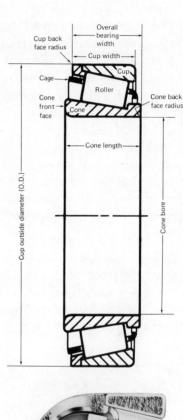

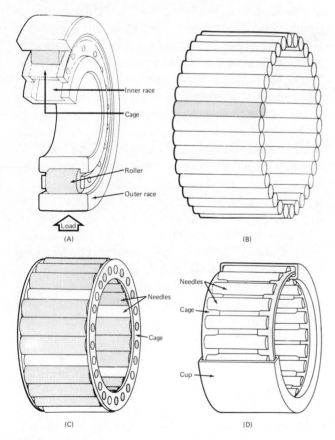

FIGURE 5–5.
A cutaway view of a roller bearing (A), a group of free needle bearings (B), a caged needle bearing (C), and a needle bearing with an outer shell or cup.

FIGURE 5–6.
A section view of a tapered roller bearing showing the normally used dimensions. (Courtesy of The Timken Company)

5.3 BEARING END PLAY—PRELOAD

An important step when adjusting bearings is to set the correct bearing **end play**. Too much end play gives a loose, sloppy fit, which would let the wheel wobble and change alignment angles in addition to reducing the cone-to-bearing contact. As far as the bearing is concerned, it will get maximum life if it has a free running clearance (i.e., no preload) with no appreciable end play. Free play also provides room for expansion as the bearings, shafts, and housings heat up during operation. A bearing will have end play if there is a slight clearance between the balls or rollers and the races. A bearing will have preload when there is a pressure between the balls or rollers and the races. (Fig. 5–7)

Preload is used on bearings when the shaft must not change position. The pinion gear and shaft in a rear axle transmit a large amount of torque, and, because of gear pressure, the gears try to move lengthwise as well as sideways. The pinion bearings must be preloaded to keep the gear from changing position under these heavy loads. Bearing preload increases the friction of the bearing and consumes power. If possible, bearings are adjusted to a slight end play for freer running. The exact amount of end play for a particular bearing set is usually specified by the vehicle manufacturer. A rule of thumb for the clearance of wheel bearings is 0.001 to 0.005 inch (0.03 to 0.13 mm) of end play.

5.4 SEALS

Frictionless bearings are always used with some sort of seal at each side of the bearing. The seal is used to keep the lubricant in the bearing and dirt and other foreign material out of the bearing. Many seals are designed to do both operations at the same time. A lip seal is the most common type of wheel or axle bearing seal. (Fig. 5–8)

The sealing lip of the seal, sometimes called a wiping lip, is made from a flexible ma-

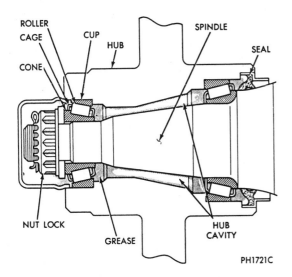

FIGURE 5–7.
A hub, bearing, and spindle assembly as used on many nondrive wheels. Bearing end-play or preload is adjusted with the nut just inside of the nut lock. (Courtesy of Chrysler Corporation)

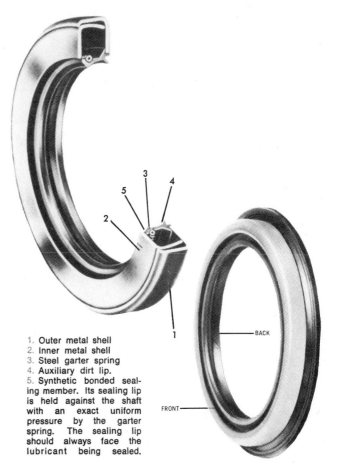

1. Outer metal shell
2. Inner metal shell
3. Steel garter spring
4. Auxiliary dirt lip.
5. Synthetic bonded sealing member. Its sealing lip is held against the shaft with an exact uniform pressure by the garter spring. The sealing lip should always face the lubricant being sealed.

FIGURE 5–8.
A typical seal showing its various parts. (Courtesy of CR Industries)

terial, usually neoprene rubber. Different materials can be used depending on the speed of the seal, the type of lubricant, and the operating conditions where the seal is used. The flexible lip is usually molded into the seal's outer case. The seal's housing forms a **static (stationary)** seal to the hub or axle housing when it is pressed into position in its bore. The seal lip forms a **dynamic (moving)** seal by wiping oil or grease off of the shaft. The open side of the seal lip always faces toward the lubricant (inside) so that any internal pressure will increase, not decrease, the wiping pressure between the seal lip and the shaft. A garter spring is often placed around the seal lip to increase this wiping pressure. Most seal failures occur when this wiping lip fails. Some seals will have a second lip that faces outward. This lip stops dirt and grit from working its way under the inner sealing lip.

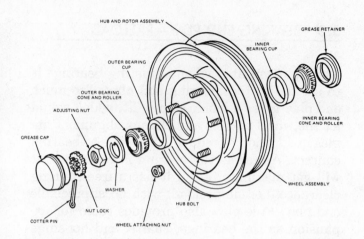

FIGURE 5-9.
An exploded view of the hub and wheel bearing from a nondrive wheel. Note the use of a cotter pin and nut lock to hold the adjusting nut. (Courtesy of Ford Motor Company)

5.5 NONDRIVE AXLE, SERVICEABLE BEARINGS

Nondrive axle, serviceable bearings are very common and are used on the front wheels of most RWD cars and the rear wheels of many FWD cars. A pair of tapered roller bearings are placed over the spindle with the cups pressed into the hub. The bearings are placed so the smaller diameters of the bearing tapers are towards each other to give the bearing a large amount of control over end play. (Fig. 5-9)

The larger of the two bearings normally carries most of the radial load. This bearing is positioned close to the center of the tire where most of the vehicle load is. It is also located over the larger and stronger portion of the spindle. The smaller bearing serves primarily to keep the hub from wobbling. With the large bearing, it controls end play.

End play of most front wheel bearings is adjusted by the tightness of the spindle nut. Turning the nut inward will decrease the end play and, if turned far enough, will preload the bearings. After the bearing end play is adjusted, the nut is locked in place to hold this

adjustment. The usual lock is a cotter pin placed through a hole in the end of the spindle and a slot in the castellated nut or castellated style of nut lock. Other methods of locking the nut in place include

- using a second nut tightened against the adjusting nut, called a **lock** or **jam nut**;
- bending or staking a portion of the adjusting nut into a groove in the spindle;
- tightening a clamp portion of the adjusting nut onto the spindle;
- bending a sheet metal washer over the adjusting nut and also over the jam nut; this washer is often kept from rotating by a tang of the washer that fits a groove in the spindle (truck and some 4WD); and
- tightening set screws in the lock nut against the adjusting nut; these set screws pass through slots in the washer that is tanged to the spindle (truck and 4WD). (Fig. 5-10)

Occasionally, this bearing is adjusted by changing spacer shims or washers that are positioned between the two bearings. Thicker shims will give more end play and less preload. (Fig. 5-11)

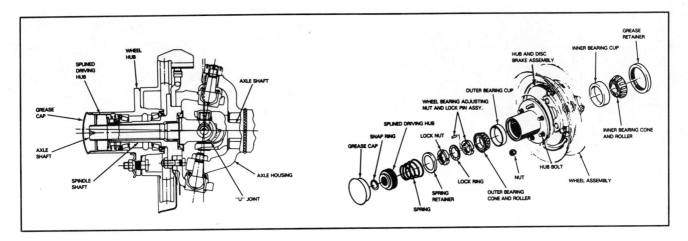

FIGURE 5-10.
A front hub from a 4WD vehicle. Note the different style of adjusting nut lock.
(Courtesy of Ford Motor Company)

5.6   NONDRIVE AXLE, NONSERVICEABLE WHEEL BEARINGS

Nondrive axle, nonserviceable wheel bearings are used on the rear wheels of some FWD cars. This bearing assembly is permanently lubricated, adjusted, and sealed during manufacture and requires no further service. In fact, it is not possible to service it internally. This bearing assembly bolts to the car's suspension, and the tire and wheel and the brake drum or rotor bolt to it. If this bearing becomes noisy, rough or loose, the whole assembly is removed and replaced with a new unit. (Fig. 5-12)

5.7   DRIVE AXLE BEARING, SOLID AXLE

A drive axle bearing is used on the rear end of most RWD cars. The bearing, commonly called an axle bearing, is at the outer end of the axle

① ②

WITH WHEEL ROTATING,
TIGHTEN ADJUSTING NUT,
TO 23-34 N·m (17-25 LB-FT)

BACK ADJUSTING
NUT OFF 1/2 TURN

③ ④

TIGHTEN ADJUSTING
NUT TO 1.1-1.7 N·m
(10-15 LB-INS)

INSTALL THE LOCK
AND A NEW COTTER PIN

FIGURE 5-11.
One method of adjusting the wheel bearings on a nondrive wheel. (Courtesy of Ford Motor Company)

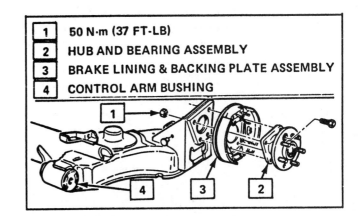

1	50 N·m (37 FT-LB)
2	HUB AND BEARING ASSEMBLY
3	BRAKE LINING & BACKING PLATE ASSEMBLY
4	CONTROL ARM BUSHING

FIGURE 5-12.
A rear suspension with brake and hub and bearing assemblies from an FWD car. This hub and bearing cannot be disassembled or serviced. It is replaced when there is a problem. (Courtesy of Pontiac Division, GMC)

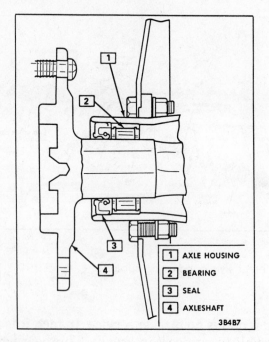

FIGURE 5-13.
The outer end of a drive axle showing one style of axle bearing. This style of axle is called a semi-floating axle because the axle's inner end is supported by a gear. The outer end of this axle supports some of the car's weight. (Courtesy of Buick)

housing, and the axle, with the tire and wheel bolted to it, turns in this bearing.

Most passenger cars use a style of axle and bearing that is called a **semi-floating axle.** (Fig. 5-13) The inner end of the axle **floats** in the axle gear inside the differential. The term "floats" indicates that the axle is not directly supported by a bearing; the gear into which the axle is splined supports the axle. The axle bearing supports the vertical load on the axle but not the side loads. Larger pickups and trucks use **full-floating axles,** and the bearing does not touch the axle at either end. The rear hubs of a full-floating axle use a large pair of tapered roller bearings that are very similar to common front-wheel bearings, except much larger. A full-floating axle carries no vehicle load, with only torque going to the rear tires. (Fig. 5-14)

Two major styles of bearings are used on semi-floating axles. These styles somewhat determine how the axle is held in the axle housing. In some axles, the bearing's inner race is pressed onto the axle, and a secondary retainer is pressed onto the shaft right next to the bear-

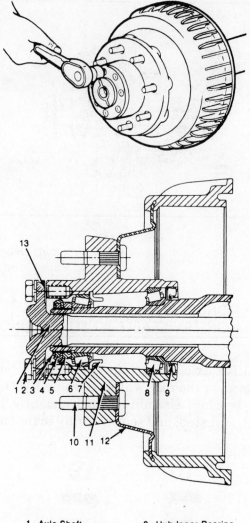

1. Axle Shaft	8. Hub Inner Bearing
2. Shaft-to-Hub Bolt	9. Oil Seal
3. Retainer	10. Wheel Bolt
4. Key	11. Hub Assembly
5. Adjusting Nut	12. Drum Assembly
6. Hub Outer Bearing	13. R.T.V.
7. Snap Ring	

FIGURE 5-14.
The outer end of a full-floating style of drive axle as used on many trucks and large pickups. Note that the drive axle (1) does not carry any vehicle loads; they are carried by the bearings (6) and (8). The bearings are adjusted and held onto the hollow spindle by the adjusting nut (5), key (4), and retainer (3). (Courtesy of Chevrolet)

ing. The retainer helps ensure that the axle does not slide out of the bearing and the axle housing. The outer race of the axle bearing fits snugly into the axle housing and is held in place by a bearing retainer that is bolted to the axle housing. The brake backing plate is usually held

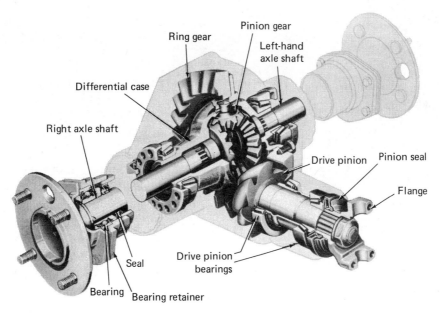

FIGURE 5-15.
A rear axle assembly that uses a bearing retained axle. The bearing is pressed on the axle and secured in the housing by the outer retainer. (Courtesy of Ford Motor Company)

in place by the same bolts. This arrangement is often called a **bearing-retained axle.** (Fig. 5-15)

The other style of axle uses a "C" clip to keep the axle in the housing; it is usually called a **"C" clip axle.** The outer end of this axle shaft, just inboard of the wheel mounting flange, is hardened and ground smooth to serve as the inner race of a roller bearing. (Fig. 5-16) The outer bearing race fits snugly into the axle housing with the axle passing through it. The "C" clip is placed into a groove in the inner end of the axle. This "C" clip also fits into a recess in the axle gear and is locked in place when the differential pinion shaft is installed. The axle is prevented from sliding outward by the "C" clip and inward by the differential pinion shaft. (Fig. 5-17) If this axle breaks outboard of the "C" clip, the outer portion of the axle, with the tire and wheel and brake drum attached, can slide out of the housing, at least until the tire runs into the fender.

Both of these types of bearings are normally lubricated by either a mist of gear oil from the axle gears or by grease that was packed and sealed into the bearing during man-ufacture. A lip seal is often placed just inboard of the axle bearing to keep excess gear oil from passing by the bearing, out the end of the axle housing, and onto the brake shoes.

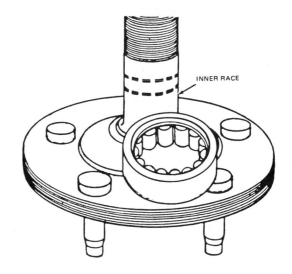

FIGURE 5-16.
The inner bearing race of a "C" lock retained axle is a smooth portion of the axle. (Courtesy of Ford Motor Company)

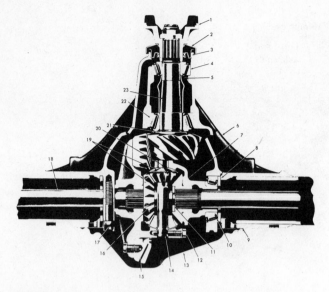

FIGURE 5-17.

A cutaway view of a rear axle assembly showing how the "C" lock (11) fits into the differential side gear (16) and is kept from coming out by the differential pinion shaft (14). The axle cannot slide outward because of the "C" lock or inward because of the pinion shaft. (Courtesy of Chevrolet)

5.8 DRIVE AXLE WHEEL BEARING, INDEPENDENT SUSPENSION

The drive axle wheel bearing is found at the rear drive axle on a car with independent rear suspension or on the front end of many FWD cars. (Fig. 5-18) Cars with independently suspended driving wheels usually transmit power to the tire through a short **stub axle**. This axle is connected to the differential by a short drive shaft or **half shaft**. The ends of these drive shafts are connected to universal joints on most RWD cars or to **CV joints (constant velocity universal joints)** on all FWD cars and some RWD cars. The CV joints are needed on FWD cars to allow sufficient turning angles and to reduce drive train vibrations while turning. (Fig. 5-19)

Independent rear suspension RWD cars usually use a pair of tapered roller bearings to support the stub axle. Like other paired, tapered roller bearings, these bearings allow the

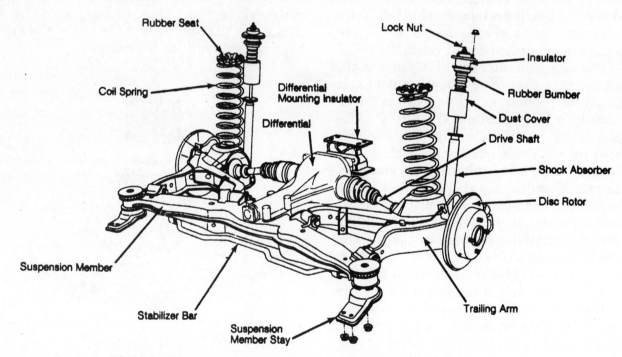

FIGURE 5-18.

An independent rear suspension drive axle, each drive shaft or axle has two universal joints, one next to the differential and one next to the trailing arm. (Courtesy of Mitchell Information Services, Inc.)

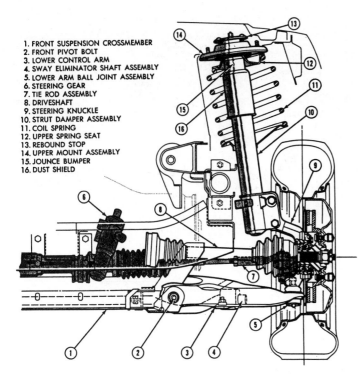

1. FRONT SUSPENSION CROSSMEMBER
2. FRONT PIVOT BOLT
3. LOWER CONTROL ARM
4. SWAY ELIMINATOR SHAFT ASSEMBLY
5. LOWER ARM BALL JOINT ASSEMBLY
6. STEERING GEAR
7. TIE ROD ASSEMBLY
8. DRIVESHAFT
9. STEERING KNUCKLE
10. STRUT DAMPER ASSEMBLY
11. COIL SPRING
12. UPPER SPRING SEAT
13. REBOUND STOP
14. UPPER MOUNT ASSEMBLY
15. JOUNCE BUMPER
16. DUST SHIELD

FIGURE 5–19.
The front suspension of an FWD car showing the drive axle or driveshaft, (8). (Courtesy of Chrysler Corporation)

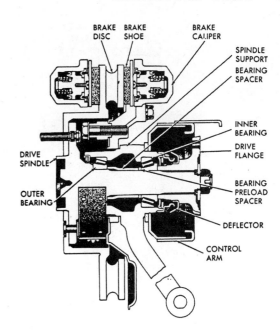

FIGURE 5–20.
The rear-wheel spindle and bearing assembly from a RWD, IRS car. The wheel bearing clearance is adjusted by changing the bearing preload spacer. (Courtesy of Mitchell Information Services, Inc.)

shaft to rotate while eliminating any side motions. The bearings are often mounted a few inches apart in the bearing support to give them better leverage in controlling shaft position. The spindle support is connected to the suspension members in such a way that it can control the alignment of the tire and wheel as the tire moves over the road surface. Lubrication of these bearings is by periodic disassembly, cleaning, and repacking with grease. Some manufacturers use permanently packed and sealed bearings. The end play or preload of this bearing type is usually controlled by a spacer between the bearings. (Fig. 5–20)

Most FWD front wheel bearings must be compact to fit in the small space provided for them. The stub axle is often the splined extension of the outer portion or housing of the CV joint. The splines of the CV joint housing pass into the splines in the front hub, and these two are held together by a nut at the outer end. The hub is supported by a pair of ball or tapered

roller bearings that are mounted in the steering knuckle. These bearings can be packed with lubricant and sealed during manufacture or require periodic lubrication. The vehicle manufacturer's guidelines should be checked to determine the maintenance requirements for a specific car. Some manufacturers recommend that these bearings be replaced whenever the front hub is removed from the spindle. Bearing end play or preload is usually controlled by the size of the parts. As they are assembled and tightened into place, the adjustment is automatically made. (Fig. 5–21)

5.9 NONSERVICEABLE DRIVE AXLE WHEEL BEARINGS

Nonserviceable drive axle wheel bearings are found at the front end of some FWD cars. They are very similar to the nondrive axle, nonserviceable wheel bearings. The only real difference is that the drive axle bearing is hollow with a splined hole in the hub so the CV joint can

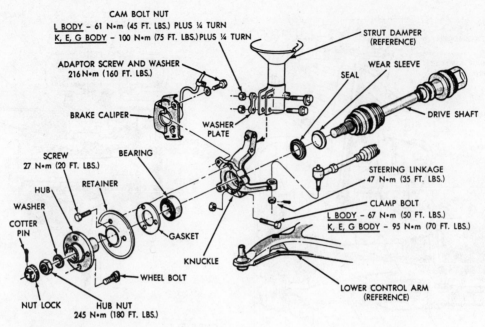

FIGURE 5-21.
An exploded view of an FWD car front knuckle. This bearing is not serviced; it is replaced when there is a problem. (Courtesy of Chrysler Corporation)

attach to it. Like other FWD wheel bearings, this assembly attaches to the steering knuckle, and like other nonserviceable bearing assemblies, this unit requires no maintenance and is serviced by replacement. (Fig. 5-22)

FIGURE 5-22.
This FWD car uses a nonserviceable front-wheel bearing assembly. The bearing is replaced with the hub if there is a problem. (Courtesy of Pontiac Division, GMC)

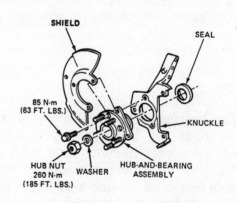

REVIEW QUESTIONS

1. Statement A: Most of the rear wheel bearings, hubs, and spindles of FWD cars are similar to those on the front wheels of RWD cars. Statement B: Some wheel bearings can not be adjusted. Which statement is correct?

   a. A only
   b. B only
   c. both A and B
   d. neither A nor B

2. A frictionless bearing uses _____ as the rolling element of the bearing.

   a. balls
   b. rollers
   c. needles
   d. all of these

3. The purpose of the cage in a ball bearing is to
   A. keep the balls separated and properly spaced.
   B. hold the balls between the races.

   a. A only
   b. B only
   c. both A and B
   d. neither A nor B

4. Statement A: A roller bearing can usually carry more side and thrust load than a ball bearing. Statement B: Needle bearings can carry greater thrust loads than ball bearings. Which statement is correct?

   a. A only   **c.** both A and B
   b. B only   **d.** neither A nor B

5. Technician A says that free movement of a hub in and out is called bearing end play. Technician B says that preload causes a drag on the bearings. Who is right?

   a. A only   **c.** both A and B
   b. B only   **d.** neither A nor B

6. The inner race of a tapered roller bearing is called
   A. a cup.
   B. a cone.

   a. A only   **c.** both A and B
   b. B only   **d.** neither A nor B

7. Technician A says that the inner lip of a seal should always point toward the outside to keep dirt from entering under the seal lip. Technician B says that the sealing pressure of a lip seal is often increased by a garter spring. Who is right?

   a. A only   **c.** both A and B
   b. B only   **d.** neither A nor B

8. Bearing preload ensures that the shaft or hub will not
   A. move in or out.
   B. rock sideways.

   a. A only   **c.** both A and B
   b. B only   **d.** neither A nor B

9. The lip seal used in the front hub of an RWD car will make a
   A. dynamic seal with the spindle.
   B. dynamic seal with the hub.

   a. A only   **c.** both A and B
   b. B only   **d.** neither A nor B

10. Technician A says that the bearings used for the front wheel bearings for most RWD cars are of the tapered roller type. Technician B says that the spindle nut is used to adjust the bearing end play. Who is right?

    a. A only   **c.** both A and B
    b. B only   **d.** neither A nor B

11. Nonserviceable wheel bearing assemblies can be found on the
    A. rear wheels of some RWD cars.
    B. front and rear wheels of some FWD cars.

    a. A only   **c.** both A and B
    b. B only   **d.** neither A nor B

12. Most RWD cars use a rear axle of the _____ floating type.

    a. one-quarter   **c.** semi
    b. three-quarter   **d.** full

13. The rear axle shaft of most RWD cars is held in place by
    A. the axle bearing and retainer.
    B. a "C" shaped lock.

    a. A only   **c.** both A and B
    b. B only   **d.** neither A nor B

14. The front wheel bearings of an FWD car fit between the

    a. spindle and the hub.
    b. steering knuckle and the hub.
    c. steering knuckle and the CV joint housing.
    d. spindle and the CV joint housing.

15. The front wheel bearings of an FWD car are
    A. adjusted by the tightness of the hub nut.
    B. repacked in a manner similar to the front wheel bearings on an RWD car.

    a. A only   **c.** both A and B
    b. B only   **d.** neither A nor B

Chapter 6

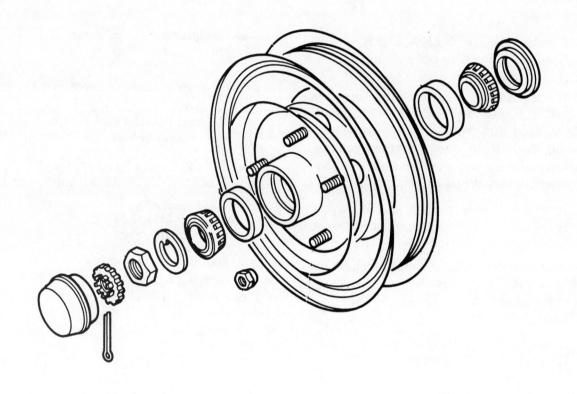

WHEEL BEARING SERVICE

After completing this chapter, you should:

- Be able to determine the cause of wheel or axle bearing problems.

- Be able to clean, inspect, repack, reassemble, and adjust serviceable wheel bearings.

- Be able to remove and replace nonserviceable wheel bearings.

- Be able to remove and replace axle bearings and seals.

6.1 WHEEL BEARING MAINTENANCE AND LUBRICATION

Periodic maintenance is normally required on the serviceable types of wheel bearings used on nondrive axles. The other styles of bearings are usually serviced on an "as needed" basis. If the bearing is noisy, leaking grease, or loose, it should be serviced. The term **service** means different things, depending on the type of bearing. It can mean removing and replacing the entire bearing assembly for nonserviceable units. On the rear axle of an RWD car, it usually means removing an axle shaft to replace a bearing or seal. On nondrive axle bearings, it usually involves disassembling the hub and bearings, cleaning and repacking the bearings, and adjusting the bearing end play or preload during reassembly. The rear axle of an independent rear suspension, RWD car usually has similar service requirements. In cases where you are not sure of the lubrication or service requirements for a particular bearing set, it is wise to check the car manufacturer's or technician's service manual.

6.2 WHEEL BEARING PROBLEMS, DIAGNOSIS PROCEDURE

Faulty or excessively loose wheel or axle bearings can appear to the driver of the car as noise, road wander, wheel shake, play in the steering, cuppy tire wear, or a low brake pedal on disc brakes. If some or all of these problems are encountered, a systematic procedure should be followed to determine if loose or faulty wheel or axle bearings are at fault and which particular one(s) are faulty.

The first step in diagnosing faults such as these is to check the tires for proper inflation and wear. After adjusting the tire pressure to normal, if necessary, and ensuring that the tires are in a safe condition, road test the car. When road testing, first drive at varying speeds until you find the conditions that make the problem show up. In cases where noise is the complaint, try to determine which corner of the car the

noise is coming from. Then, if traffic and road conditions permit, make slow-to-moderate left and right turns. Faulty wheel or axle bearings will usually change noise level as the vehicle load on them changes because of the weight transfer. During the road test, apply the service and parking brakes lightly. A reduction in noise level as the brake shoes apply pressure on the drums and rotors indicates a faulty wheel bearing. Use caution while doing this. In some cars the parking brake can lock up the rear wheels and cause a skid. Apply the brakes gently and be ready to release them quickly if necessary. (Fig. 6-1)

If the road test confirms a problem, raise the car on a hoist and check for loose bearings. Try pushing the tire and wheel straight up and down, pushing straight in and out, and rocking it sideways. Depending on the bearing type, some of these motions are not permitted and some of them are. A vertical motion is not permitted in any wheel or axle bearing type. More than barely perceptible in and out motion is permitted only on "C" lock axle, RWD rear axle bearings. On these axles, end play up to 0.010 inch (0.26 mm) is permitted on most axles; some axles are allowed up to 0.030 inch (7.9 mm). Tapered roller wheel and axle bearing sets (i.e., a pair of bearings) should have about 0.001 to 0.005 inch (0.03 to 0.13 mm) of end play. Nonserviceable wheel and axle bearings are allowed up to 0.005 inch (0.13 mm) of end play.

When rocking a tire and wheel to check for bearing looseness, grip the tire at the top and bottom. Push inward with one hand while pulling outward with the other. Then reverse these motions. A perceptible motion is acceptable on tapered roller wheel bearings and axle bearings on independent suspensions. More than barely perceptible rocking motion is not permitted on ball type wheel or rear drive axle bearings. (Fig. 6-2)

Next, spin a nondrive tire and wheel by hand or by a wheel balancer spinner motor. Spin a drive axle tire and wheel with the engine. As the tire spins, listen for a harsh, grating sound. You can often confirm the sound origin and the bearing roughness by gently placing your finger tips on the steering knuckle or axle housing as close to the bearing as possible. A rough

SEALED WHEEL BEARING DIAGNOSIS

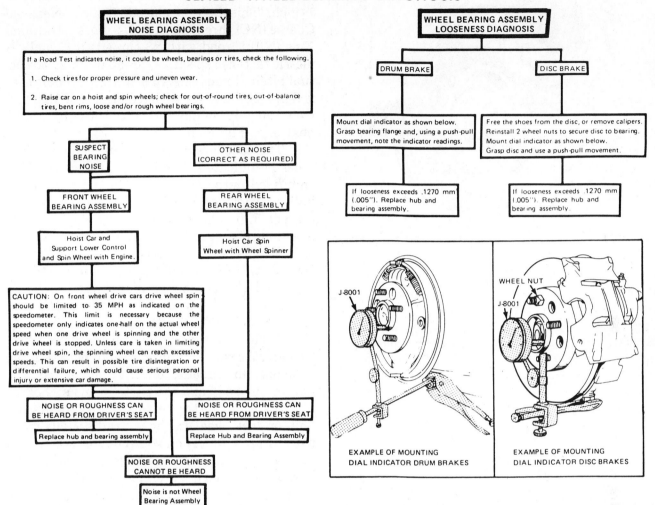

FIGURE 6-1.
This chart lists the procedure for diagnosing problems with sealed, nonservice-able wheel bearings. Except for the repair steps, the same procedure is used for serviceable bearings. (Courtesy of Pontiac Division, GMC)

wheel or axle bearing will cause a rough, irregular feel.

The proper repair method can be decided on after the faulty bearing has been located and the type of bearing has been determined. Loose bearings on some types can be adjusted; others will require replacement. Rough bearings of some types require replacement of the whole assembly; on others, only the faulty bearing part needs replacement.

SAFETY NOTE:

Always use caution when moving or working around a spinning tire and wheel. When spinning a tire with the engine, limit the tire speed to about 55 to 70 MPH (88 to 113 KPH). Don't forget that the differential gears can cause one of the tires to run at twice the speed shown on the speedometer and excessive speed can cause a tire explosion. Higher speeds are not necessary.

FIGURE 6-2.
When checking for loose wheel bearings, push in at the top of the tire while pulling outward at the bottom; then reverse these pressures while feeling for excessive play. A small amount of play is normal. (Courtesy of Hunter Engineering)

6.3 FASTENER SECURITY

Fasteners (commonly called bolts, nuts, or screws) used on steering and suspension parts must be strong enough to hold the parts securely in place and provide a high degree of safety. The size and strength of the fastener are designed to provide this security. They are also designed with some method of keeping them in place.

The size of a bolt or screw is determined by measuring the diameter of the threads. Traditionally, in the United States this diameter was sized by an inch or a fraction of an inch; for example, 1/4, 5/16, 3/8, and so forth. Today, most bolt diameters are sized by the metric system and are a certain number of millimeters (mm) in diameter. Some of the common sizes used in cars are 6, 6.3, 8, 10, 12, 14, 16, and 20 mm. When a bolt or nut is replaced, the thread pitch is also important. Thread pitch is the distance between the threads. One system of bolt sizing specifies thread pitch by the number of threads in one inch. A 1/4-20 would be a bolt that is 1/4 inch in diameter with 20 threads in 1 inch. This bolt is classified as a **National Coarse (NC)** thread bolt. A 1/4-28 is a **National Fine (NF)** thread with 28 threads in 1 inch. A metric fastener specifies thread pitch by the actual pitch dimension. An 8 × 1.25 is a bolt that is 8 mm in diameter with a distance of 1.25 mm from the point of one thread to the point on the next one. (Fig. 6-3)

The tensile strength of an inch-sized fastener is indicated by a group of lines that are embossed onto the head of the bolt. No lines indicate a grade 2 (weakest), three lines indicate a grade 5, five lines indicate a grade 7, and six lines indicate a grade 8 (strongest). The number of lines plus two is the grade of the bolt. A metric bolt will use a number embossed onto the bolt head; this number corresponds to the bolt's strength, with a larger number indicating a stronger bolt. (Fig. 6-4)

A nut is normally held in place on a bolt by interthread friction, friction between the bolt and nut threads, and friction between the bolt or nut head and the part. To ensure bolt or nut retention, cotter pins through the nut and bolt or lock washers have traditionally been used. (Fig. 6-5) The metal cutting action of a lock washer increases the resistance to turning of a fastener in a counterclockwise direction. A class of nuts and bolts referred to as **prevailing torque** nuts or bolts is becoming very common. These are commonly called **self-locking** nuts or bolts. (Fig. 6-6) This group of fasteners uses a distortion of the bolt or nut, a nylon insert in

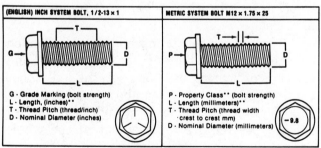

(ENGLISH) INCH SYSTEM BOLT, 1/2-13 × 1	METRIC SYSTEM BOLT M12 × 1.75 × 25
G - Grade Marking (bolt strength) L - Length, (inches)** T - Thread Pitch (thread/inch) D - Nominal Diameter (inches)	P - Property Class** (bolt strength) L - Length (millimeters)** T - Thread Pitch (thread width crest to crest mm) D - Nominal Diameter (millimeters)

*The property class is an Arabic numeral distinguishable from the slash SAE English grade system.
**The length of all bolts is measured from the underside of the head to the end.

FIGURE 6-3.
The standard dimensions used to determine bolt sizes vary slightly between the English and the metric systems. (Courtesy of Ford Motor Company)

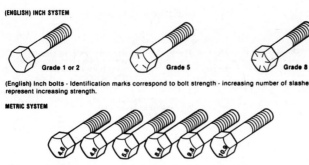

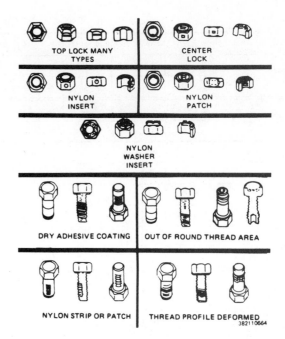

(ENGLISH) INCH SYSTEM

Grade 1 or 2   Grade 5   Grade 8

(English) Inch bolts - Identification marks correspond to bolt strength - increasing number of slashes represent increasing strength.

METRIC SYSTEM

Metric bolts - Identification class numbers correspond to bolt strength - increasing numbers represent increasing strength.

FIGURE 6–4.
The strength or grade of a bolt is indicated by the markings on the bolt head. How much the bolt can hold before breaking and how tight it is tightened is determined partially by the grade. (Courtesy of Ford Motor Company)

FIGURE 6–6.
Today there is a large variety of prevailing torque nuts and bolts to lock the nut onto the bolt. These are commonly called self-locking nuts and bolts. (Courtesy of Oldsmobile)

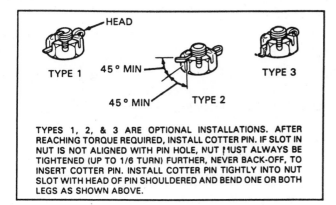

HEAD

TYPE 1   45° MIN   TYPE 3

45° MIN   TYPE 2

TYPES 1, 2, & 3 ARE OPTIONAL INSTALLATIONS. AFTER REACHING TORQUE REQUIRED, INSTALL COTTER PIN. IF SLOT IN NUT IS NOT ALIGNED WITH PIN HOLE, NUT MUST ALWAYS BE TIGHTENED (UP TO 1/6 TURN) FURTHER, NEVER BACK-OFF, TO INSERT COTTER PIN. INSTALL COTTER PIN TIGHTLY INTO NUT SLOT WITH HEAD OF PIN SHOULDERED AND BEND ONE OR BOTH LEGS AS SHOWN ABOVE.

FIGURE 6–5.
Cotter pins have been used for many years to lock the nut on the bolt in critical areas. (Courtesy of Chevrolet)

the threads of the bolt or nut, or a dry or liquid adhesive coating on the bolt or nut. Several brands of liquid lock washer are available to the technician for this same purpose. These are usually anaerobic materials; that is they harden or set when oxygen is removed.

The use of prevailing torque fasteners by manufacturers has placed some added responsibilities on the front-end technician. If these fasteners wear, they might lose their ability to stay put, and they might work loose. A suspension or steering part must not come loose after the car has been worked on. Reuse of a worn prevailing torque nut does not always provide the same degree of security as the reuse of a nut locked by a cotter pin or lock washer.

SAFETY NOTE
The following are some rules to observe for the safe reuse of prevailing torque fasteners:

1. The nuts and bolts must be clean.
2. The nuts and bolts must be inspected for damage. Discard any that show signs of abuse, overtightening, cracks, elongation, or wear.
3. Start the nuts and bolts by hand to observe the feel; also observe the feel (amount of turning resistance) before they seat. The nut or bolt should develop as much resistance as indicated in Figure 6–7. Any prevailing torque fasteners that turn too easily should be replaced.
4. Replace all damaged or rusty fasteners with new parts of equal or greater strength.

METRIC SIZES

		6 & 6.3	8	10	12	14	16	20
NUTS AND ALL METAL BOLTS	N•m	0.4	0.8	1.4	2.2	3.0	4.2	7.0
	In. Lbs.	4.0	7.0	12	18	25	35	57
ADHESIVE OR NYLON COATED BOLTS	N•m	0.4	0.6	1.2	1.6	2.4	3.4	5.6
	In. Lbs.	4.0	5.0	10	14	20	28	46

INCH SIZES

		2.50	.312	.375	.437	.500	.562	.625	.750
NUTS AND ALL METAL BOLTS	N•m	0.4	0.6	1.4	1.8	2.4	3.2	4.2	6.2
	In. Lbs.	4.0	5.0	12	15	20	27	35	51
ADHESIVE OR NYLON COATED BOLTS 382110665	N•m	0.4	0.6	1.0	1.4	1.8	2.6	3.4	5.2
	In. Lbs.	4.0	5.0	9.0	12	15	22	28	43

FIGURE 6–7.
A prevailing torque nut or bolt is worn out if it turns too easily. If it takes less torque than indicated in this chart to turn it before seating, the nut or bolt should be replaced. (Courtesy of Oldsmobile)

When prevailing torque nuts are used on ball joints or tie rod ends, it is difficult to thread the nut onto the stud. The tapered stud is kept from rotating by the locking of the tapered stud into the tapered hole. The turning resistance of the prevailing torque nut is great enough to cause the stud to rotate. Turning the nut merely rotates the stud. This job is made possible by first tightening the stud into the boss to set the taper using an ordinary nut of the correct size or a special nut-like tool. After the taper is locked, the nut or special tool is removed, and the prevailing torque nut is installed to the correct torque. A special wedge can also be used to hold the stud while the nut is tightened. (Fig. 6–8)

6.4 REPACKING SERVICEABLE WHEEL BEARINGS

Most RWD front wheel bearings and FWD rear wheel bearings should be repacked at regular intervals. This operation includes disassembly, cleaning, packing with grease, reassembling, and adjusting. The interval, which varies with different car manufacturers, is about 15,000 to 20,000 miles or greater. In some cases, bearing service is required only during a brake reline.

METRIC BALL STUD INSTALLATION

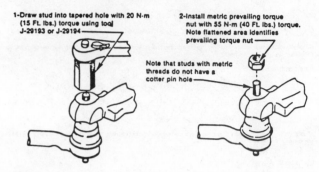

1-Draw stud into tapered hole with 20 N·m (15 Ft. lbs.) torque using tool J-29193 or J-29194

2-Install metric prevailing torque nut with 55 N·m (40 Ft. lbs.) torque. Note flattened area identifies prevailing torque nut

Note that studs with metric threads do not have a cotter pin hole

FIGURE 6–8.
It is often difficult to install a prevailing torque nut onto a ball joint or tie rod end stud because the stud rotates easier than the nut does. The stud taper can be locked using a special tool or plain nut of the correct size before installing the prevailing torque nut. (Courtesy of Chevrolet)

Whenever the wheel or axle bearings are serviced, the brake shoes and the drums or rotors should be checked for wear. The average driver has only a vague idea of what the brake shoes are or where they are located. The average driver doesn't realize the damage that will be done to the drum or rotor shortly after the brake shoes wear out or the poor and unsafe stop that will result. This same person usually appreciates any information that you can provide that will make his or her car perform better, safer, or at a lower cost. Most technicians feel a strong responsibility to inform the car owner about things that are or might become unsafe.

SAFETY NOTE
The following good service practices should be observed while servicing wheel or axle bearings:

1. Always lift the car at the correct lifting points and support it with carefully placed jack stands.

2. Never just add grease to a bearing; always clean, inspect, and repack it with new grease.

3. Never let grease or solvent get on the braking friction surfaces; always clean up any spilled grease, solvent, or greasy fingerprints.

4. Never let brake calipers or drive shafts hang by their hoses or universal joints; always support them and protect the rubber hoses and CV joint boots.

5. Don't let a tire and wheel fall and bounce during removal; keep these heavy parts under control.

6. Keep a clean, neat work area; slide the tire and wheel under the car and out of the way during service operations that require tire and wheel removal.

7. Always replace the seal with a new one; reusing an old seal is taking a chance on ruining the brakes or losing a bearing because of possible grease leakage.

8. Always replace lock washers and cotter pins; use all recommended locking devices and tighten nuts and bolts to the correct torque.

9. Remember that the service operations you are performing should last for a few years and thousands of trouble- free miles.

6.4.1 Disassembling Wheel Bearings

Disassembly is the first step when repacking serviceable wheel bearings.

To disassemble wheel bearings, you should:

1. Raise the car so the tires are off the ground.

CAUTION: Position a jack stand under a secure portion of the car and lower the car onto the jack stand or raise and support the car on a hoist.

2. If disc brakes are used on this wheel, remove the tire and wheel as described in Section 4.6. If drum brakes are used, the tire and wheel

with the brake drum and hub can usually be removed as a unit.

3. On cars equipped with disc brakes, remove the caliper as described in a technician's service manual and suspend it from the steering knuckle or other convenient point using mechanic's wire.

SAFETY NOTE:

Do not let the caliper hang from the hose; if this hose is bent too sharply, it can fail and cause a brake loss.

---

4. Remove the grease or dust cap from the wheel hub; a dust cap remover will do this quickly and easily. Slip joint pliers can also be used for this. (Figs. 6–9 & 6–10)

5. Locate and remove the locking device for the spindle nut. The most common type is a cotter pin. Straighten the bent end using a pair of dikes (diagonal cutting or side cutting pliers). Grip the head of the cotter pin with the dikes, and pry it out of the spindle and spindle nut. Another good tool for this is a cotter pin puller. If a staked spindle nut is used, some manufacturers recommend merely unscrewing the nut. The staked portion will bend out of the way during removal. Other manufacturers rec-

FIGURE 6–9.
The wheel bearing dust cover can be removed using a pair of special dust cover pliers. (Courtesy of CR Industries)

FIGURE 6-10.
A pair of water pump or slip joint pliers can be used to remove the dust cover. (Courtesy of Federal Mogul Corporation)

FIGURE 6-12.
After the nut and washer have been removed, rock the hub to work the outer bearing out of the hub. (Courtesy of CR Industries)

ommend that you bend the staked portion of the nut upward using a small, sharp chisel or using a drill and drill bit to drill through the staked portion of the nut. Be careful not to drill the spindle. In either case, the staked nut should be replaced. (Fig. 6-11)

6. Remove the spindle nut. It can usually be removed with your fingers; if not, use slip joint pliers or a wrench to unscrew it.

7. Remove the washer and outer bearing and bearing cone. Rocking the tire and wheel

or hub will usually work the bearing out to where you can grasp it. (Fig. 6-12)

8. Slide the hub off of the spindle, and pry the seal out of the back of the hub using a seal puller. (Fig. 6-13) Alternate methods of removing the seal are (1) to tap the inner bearing and bearing cone and seal out using a large wooden dowel and hammer, (Fig. 6-14) or (2) to replace the nut back on the spindle and, while

FIGURE 6-11.
The cotter pin can usually be removed easily by gripping the head of the cotter pin (arrow) in the jaws of a pair of diagonal or side cutting pliers (dikes) and prying the pin out.

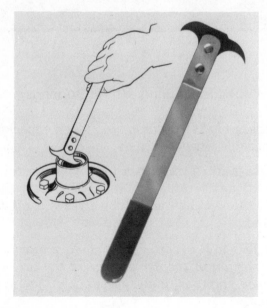

FIGURE 6-13.
After the seal puller has been hooked into the seal, a prying action (in this case, toward the left) will lift the seal out of the hub. (Courtesy of Specialized Products)

FIGURE 6–14.
A wooden dowel or punch can be placed on the inner face of the inner bearing and tapped to remove the seal and inner bearing. (Courtesy of CR Industries)

FIGURE 6–15.
The inner bearing and seal can be removed with the spindle nut. Thread the nut a few turns onto the spindle and pull the hub outward in a quick motion so the bearing catches on the nut.

sliding the hub off of the spindle, catch the bearing on the spindle nut so that the bearing and seal are jerked out of the hub. (Fig. 6–15) *Neither of these alternate methods is strongly recommended as they might bend the bearing cage.*

6.4.2 Cleaning and Inspecting Wheel Bearings

After disassembling the spindle, the bearings, and hub should be cleaned to get rid of all of the old grease and the dirt or metal fragments it might contain and to allow a thorough inspection. (Fig. 6–16)

To clean and inspect wheel bearings, you should:

1. Wipe off the spindle and inspect it for damage. The bearing cones must have a slip fit on the spindle. This fit should leave a marking on the spindle to show that the cone has been creeping, slowly rotating about one revolution per mile. Bearing creep keeps changing the loaded part of the cone and produces longer cone and bearing life. (Fig. 6–17)

2. Also check the spindle in the area where the seal runs; it should be clean and smooth.

3. Wipe all of the old grease out of the hub,

and wipe the bearing cups clean. Inspect the cups for pitting, spalling, or other signs of failure. If the cup is damaged, the cup and the bearing and cone should be replaced. The bearing might appear good, but it is possible to miss seeing some damage because of the cage.

NOTE: Many technicians use the part number from the old bearing or seal to ensure getting the correct replacement.

FIGURE 6–16.
After the spindle has been cleaned and inspected, coat it lightly with grease. Note the marks (arrows) where the cones of the bearings have been creeping on the spindle. This is good. (Courtesy of Federal Mogul Corporation)

Damage to the race

These dents result from the rollers "hammering" against the race. It's called brinelling.

Dents like this usually come from mishandling. The bearing should be discarded.

Under load, a hairline crack like this will lead to serious problems. Discard the bearing.

Always check for faint grooves in the race. This bearing should not be reused.

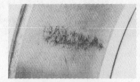

Regular patterns of etching in the race are from corrosion. This bearing should be replaced.

Light pitting comes from contaminants being pressed into the race. Discard the bearing.

In this more advanced case of pitting, you can see how the race has been damaged.

Pitting eventually leads to "spalling," a condition where the metal falls away in large chunks.

If corrosion stains haven't etched into the surface yet, try removing them with emery cloth.

Line etching looks like cracks. If the etching can be removed, the bearing can be reused.

This condition results from an improperly grounded arc welder. Replace the bearing.

Discoloration is a result of overheating. Even a lightly burned bearing should be replaced.

Damage to the rollers

This is a normally worn bearing. If it doesn't have too much play, it can be reused.

This bearing is worn unevenly. Notice the stripes. It shouldn't be reused.

When just the end of a roller is scored, it's from excessive preload. Discard the bearing.

Grooves like this are often matched by grooves in the race (above). Discard the bearing.

When corrosion etches into the surface of a roller or race, the bearing should be discarded.

Any damage that causes low spots in the metal, renders the bearing useless.

This is a more advanced case of pitting. Under load, it will rapidly lead to "spalling."

In this "spalled" roller, the metal has actually begun to flake away from the surface.

If light corrosion stains can be removed with emery cloth, the bearing can be reused.

This is "line etching" from corrosion, not a crack. If you can remove it, reuse the bearing.

When "fluting" shows up on the race, it'll appear on the rollers too. Discard the bearing.

Discoloration comes from overheating. When you see it, play it safe and discard the bearing.

FIGURE 6–17.
After the bearings and bearing cups have been cleaned, they should be carefully inspected for reuse. These are the commonly found causes of failure. (Courtesy of CR Industries)

D80L-927-A

T77F-1102-A

(A)

(B)

FIGURE 6-18.
A faulty bearing cup can be removed by pulling it out with a puller (A) or driving it out with a punch (arrow) (B). (A is courtesy of Ford Motor Company)

4. Check the cups to ensure that they are tight in the hub. A loose cup will usually require replacement of the hub.

5. If a cup is damaged, it can be pulled out of the hub using a puller or driven out using a punch and hammer. (Fig. 6-18) The new cup is installed using a cup driving tool and hammer. (Fig. 6-19) Be sure the new cup is fully seated. The hammer blows will make a different, more solid sound when the cup becomes seated.

6. Wash the bearings and cones in clean solvent, and blow them dry using shop air. The direction of the air blast should always be parallel to the rollers. An air blast across the rollers doesn't really clean the bearing, and it will tend to spin the bearing too fast. (Fig. 6-20)

CAUTION: DON'T *spin a bearing at high speeds. A spinning bearing might generate*

(A)

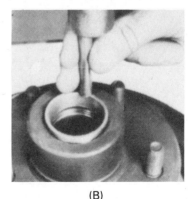

(B)

FIGURE 6-19.
A new bearing cup can be installed quickly and easily with a bearing cup driver (A). Make sure that you hear the bearing seat completely. If a cup driver is not available, a cup can be driven in place using a soft metal punch. Be careful to drive the cup in straight, and make sure it is installed completely. (A is courtesy of CR Industries; B is courtesy of The Timken Company)

(A)

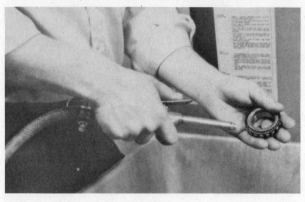

(B)

FIGURE 6–20.
The bearing should be washed in clean solvent and then dried with clean, compressed air. Never spin the bearing with the air pressure while drying it. (Courtesy of CR Industries)

enough centrifugal force to explode. Also, running the bearing at high speeds without lubrication could damage the bearing. Spinning a bearing with shop air might sound neat, but it does not do any good, only possible harm.

7. Rewash and dry the bearings until they are thoroughly clean.

8. Inspect the bearings for roller or cage damage. The commonly encountered causes of bearing failure are shown in Figure 6–17. All damaged or questionable wheel bearings should be replaced.

9. If the bearings, cones, cups, or spindle show signs of misalignment, check for a bent spindle. A spindle runout gauge can be placed over the spindle and adjusted to a sliding fit on the inner bearing (see Figure 6–21). Next the

dial indicator is adjusted to zero and the gauge unit is rotated around the spindle. A bent spindle is indicated if the dial indicator reading changes. Another check for a bent spindle is to measure the distance from a straight edge to the location of the outer bearing cone on each side and top of the spindle. The measurements should be the same.

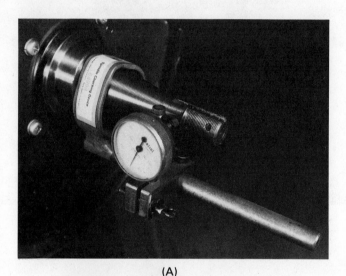

(A)

(B)

FIGURE 6–21.
If a bent spindle is suspected, it can be checked using a spindle run-out gauge (A) or a straight edge and caliper (B). The run-out gauge is set and then rotated around the spindle while watching the indicator dial. The straight edge is placed against the inner cone surface while measurements are made to the outer cone surface on both sides of the spindle. If the measurements are different, the spindle is bent. (B is courtesy of Volkswagen of America)

6.4.3 Packing and Adjusting Wheel Bearings

Wheel bearings should be packed with suitable grease and adjusted to the correct end play or preload during assembly. A good grade of chassis grease can be used to pack wheel bearings if one of the uses, printed on the container, states that it is suitable for wheel bearings. Many technicians prefer to use a grease formulated especially for wheel bearings. It is recommended to use **high temperature** or **heavy duty** grease if the car is to be driven hard or if the brakes will get a lot of use. Packing a bearing fills the area between the cage and the cone, alongside of the rollers or balls, with grease.

A properly adjusted tapered roller wheel bearing should have 0.001 to 0.005 inch (0.03 to 0.13 mm) of end play. With this clearance, you should notice a perceptible play if you rock the tire and wheel. Ball type wheel bearings should have zero end play; they are normally adjusted to a slight preload. Ball bearings should not have a perceptible play when the tire and wheel are rocked.

FIGURE 6-22.
The quickest way to pack a bearing is to use a bearing packer. Pushing downward on the packer forces grease into the area inside of the cage. (Courtesy of Kent Moore)

To pack and adjust wheel bearings, you should:

1. Pack the large, inner bearing first. The outer, small bearing should be packed when you are ready to install it. Some technicians prefer to pack both bearings at the same time and place the small bearing on a clean shop towel until it is needed. Packing is normally done with a bearing packer but can be done by hand. When using a bearing packer, which is much faster, follow the procedure for that particular packer. (Fig. 6-22) When packing by hand, place a tablespoon of grease in the palm of your hand, and push the bearing, open side of the cage downward, into the grease and against the palm of your hand. Repeat this step until you see grease oozing through the upper part of the cage. Work all the way around the bearing so you fill the entire cage. (Fig. 6-23)

2. After packing the bearing, smear a liberal coating of grease around the outside of the cage and rollers.

3. Smear a coating of grease around the inside of both bearing cups. It is also recommended to place a ring of grease in the hub, just inside of the cups. This ring of grease will act like a dam to prevent the grease from running out of the bearing and into the hub cavity when the grease gets hot. Smear a thin film of grease on the spindle, also being sure to include the area where the seal lip rubs. (Fig. 6-24)

4. Place the packed inner bearing into its cup.

5. Position the seal so the seal lip faces inward, toward the grease. Using a suitable driver, drive the seal into the hub so it is even with the end of the hub. The back side of a bearing cup installer is often convenient and of the proper size. On seals without a lip, the seal is usually positioned so the part numbers face outward. (Fig. 6-25)

6. Wipe away any grease that might be left on the outside of the seal or on the end of the bearing cone. Any grease left on the ends of the cones or on the side of the retaining washer might cause a bearing to change ad-

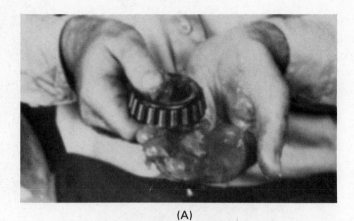

(A)

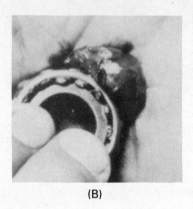

(B)

FIGURE 6-23.
A bearing can be repacked by placing grease in the palm of your hand (A) and then repeatedly forcing the bearing into the grease until grease comes up between the rollers and the cage (B). (A is courtesy of CR Industries; B is courtesy of the Timken Company)

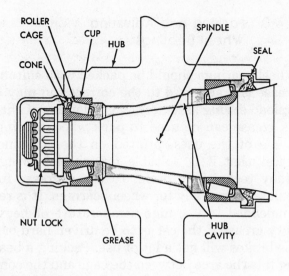

FIGURE 6-24.
Grease should fill the shaded areas between the hub and spindle to ensure that the bearings stay lubricated. (Courtesy of Chrysler Corporation)

a. While rotating the hub by hand, tighten the spindle nut to about 15 to 20 foot pounds (20—27 N•M) of torque. This ensures complete seating of the bearings. This is about as tight as you can tighten the nut using slip joint pliers and normal hand pressure.

FIGURE 6-25.
After the cup has been greased and the inner bearing installed in the cup, the area between the cone and the cup should be filled with grease and the top edge of the cone (arrow) should be wiped clean of grease.

justment. Grease that stays in these areas during adjustment can be squeezed out by side motions of the vehicle. It is said that each layer of grease might increase bearing clearance by 0.001 inch (0.025 mm). (Fig. 6-26)

7. Slide the hub partially onto the spindle.

8. Repeat Steps 1, 2, 3, and 4 to pack the outer bearing, and slide it onto the spindle and into the hub.

9. Install the washer and spindle nut.

10. Adjust the wheel bearing clearance as recommended by the car manufacturer. If the specifications are not available, the following is a standard method of adjusting tapered roller bearings:

FIGURE 6-26.
With the seal lip pointing inward, the new seal should be installed using a flat, properly sized seal driver. (Courtesy of The Timken Company)

b. Back off the spindle nut one-quarter to one-half turn.

c. Retighten the spindle nut using your thumb and forefinger. This is equivalent to about 5 inch pounds (0.56 N·M) of torque. (Fig. 6-27)

11. Replace the nut lock if used, and position a new cotter pin through the nut or nut lock and the spindle. Using the dikes, bend the long leg of the cotter pin outward, upward, and across the end of the spindle. The other leg can be cut off. Occasionally, if there is a static suppression spring in the dust cap, it is necessary to bend the two legs of the cotter pin sideways and around the spindle nut. (Fig. 6-28)

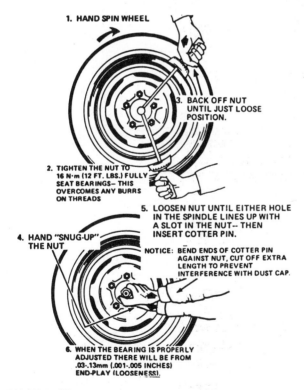

FIGURE 6-27.
Wheel bearings are first tightened to seat the bearings and then adjusted to the correct running clearance. This is a commonly used procedure. Be sure always to rotate the hub while seating the bearings. (Courtesy of Chevrolet)

(A)

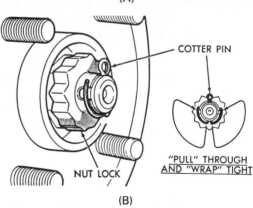

(B)

FIGURE 6-28.
After the bearing is adjusted, the cotter pin is installed, trimmed if necessary, and bent across the end of the spindle (A) or wrapped around the spindle (B). Method B is required for clearance in some dust covers. (A is courtesy of the Federal Mogul Corporation; B is courtesy of Chrysler Corporation)

FIGURE 6–29.
Wheel bearing adjustment can be checked using a dial indicator. A correctly adjusted tapered roller bearing set will have about 0.001 to 0.005 inch of end play. (Courtesy of The Timken Company)

12. If you have doubts about the adjustment, measure the bearing or hub end play by mounting a dial indicator on the hub with the indicator stylus on the end of the spindle. Pull outward on the hub, adjust the dial indicator to zero, push inward on the hub, and read the amount of travel on the dial. (Fig. 6–29)

13. Replace the dust cap. Some technicians like to wipe a film of grease around the inside of the cap or around the lip of the cap to serve as a water seal.

14. Replace the caliper, being sure to follow the recommended procedure and tightening specifications.

15. Replace the wheel, being sure to tighten the lug bolts correctly.

CAUTION: Make sure that all nuts and bolts are properly tightened and secure and the brake pedal has a normal feel before moving the car.

6.5 REPAIRING NONSERVICEABLE WHEEL BEARINGS

As stated earlier, the only repair for a faulty nonserviceable wheel bearing is to remove and replace the hub and bearing assembly. This fairly simple procedure will vary slightly on different makes and models of cars. It is wise to follow the procedure described in the manufacturer's service manual.

To remove a nonserviceable hub and bearing assembly, you should:

1. Raise and support the car securely on a hoist or jack stands as previously described.

2. Remove the wheel.

3. On drum brake cars, remove the drum. On disc brake cars, remove the brake caliper and then remove the rotor.

CAUTION: Be sure to support the caliper to prevent damage to the brake hose.

4. Remove the bolts securing the hub and bearing assembly to the control arm or axle. Note that a hole is provided in the mounting flange so that a socket and extension bar can be used for removal and installation. (Fig. 6–30)

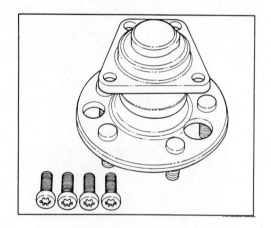

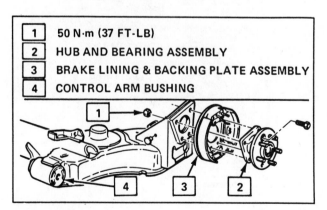

1	50 N·m (37 FT-LB)
2	HUB AND BEARING ASSEMBLY
3	BRAKE LINING & BACKING PLATE ASSEMBLY
4	CONTROL ARM BUSHING

FIGURE 6–30.
This permanently sealed and adjusted hub and bearing assembly is removed and replaced with a new one if it becomes faulty. (Courtesy of Pontiac Division, GMC)

To install a hub and bearing, you should:

1. Make sure the recess in the control arm or axle in which the hub and bearing assembly fits is clean.

2. Install the hub and bearing assembly into position, and tighten the bolts to the correct torque.

3. Install the brake drum or rotor. On disc brake cars, install the caliper and tighten the caliper mounting bolts to the correct torque.

4. Install the wheel, being sure to tighten the lug bolts correctly.

6.6 REPAIRING DRIVE AXLE BEARINGS, SOLID AXLES

Traditionally, drive axle work has not been done by the front-end technician. It has been included in this text for use by the technician who works in a general repair shop and because the evolution of the automobile is blurring or changing the traditional areas of repair. A front end on an FWD car includes the drive axle and axle shafts. The rear end of an FWD car has wheel bearings much like the front end of RWD cars.

If a drive axle bearing becomes loose, rough, or noisy, the bearing must be replaced. If a grease leak develops from the end of the axle housing, the axle seal or bearing, depending on the axle design, must be replaced. To replace either the bearing or the seal, the axle must be removed from the axle housing. As previously mentioned, there are two ways of securing the axle in the housing. The inner end of the axle in both designs is supported by the axle gear in the differential unit inside of the axle housing.

A **bearing-retained axle** uses a bearing that is pressed onto the axle shaft and held in the axle housing by a retainer. A "C" **lock axle** holds the axle in place using a "C"-shaped piece of metal that fits in a groove in the inner end of the axle. This axle slides through the axle bearing that is pressed into the end of the axle housing.

All "C" lock designs use a seal that is separate from the bearing. Some bearing-retained designs use a seal incorporated in the bearing. Some designs use a separate seal or both.

Axle bearing and seal repair procedures will vary on different cars. It is recommended that you always follow the procedure listed in a technician's service manual.

6.6.1 Removing a Bearing-retained Axle

To remove a bearing-retained axle, you should:

1. Raise and support the car securely on a hoist or jack stands.

2. Remove the wheel.

3. Mark the brake drum or rotor next to the previously marked stud to ensure replacement in the same location. On disc brake cars, remove the caliper and secure it to the axle using wire.

4. Remove the brake drum or rotor.

5. Remove the nuts and bolts securing the bearing retainer and, usually, also securing the backing plate to the end of the axle housing. Note the hole that is provided in the axle flange to allow the use of a socket and extension during removal and installation of these nuts or bolts. (Fig. 6–31)

FIGURE 6–31.
When removing an axle that is retained by the bearing, the nuts holding the bearing retainer and the brake backing plate onto the axle housing must be removed. (Courtesy of CR Industries)

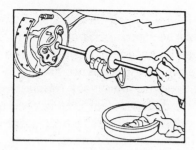

FIGURE 6-32.
This special adapter allows a slide hammer to be attached to the end of the axle to help pull the axle bearing out of the axle housing. (Courtesy of OTC)

6. Attach a slide hammer to the axle flange. Using whatever force that is required, remove the axle from the housing. As you slide the axle and bearing out of the housing, support the axle so it does not drag over the seal. Also make sure the bearing passes through the backing plate bore and the backing plate stays at the end of the rear axle housing. (Figs. 6-32 & 6-33)

6.6.2   Removing and Replacing a Bearing on an Axle

This operation requires an axle bearing removal and installer set and a lot of caution. A large amount of pressure is usually required to push an axle out of a bearing, and there is only

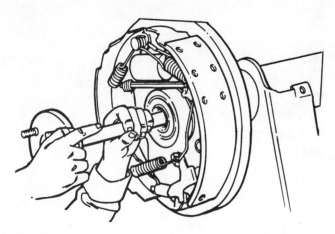

FIGURE 6-33.
The axle should be supported as it is removed or installed in the housing to keep it from dragging across and damaging the seal. (Courtesy of Ford Motor Company) .

enough room to support the bearing by the outer race.

SAFETY NOTE
Force must be put on the axle and inner bearing race across the balls to the outer race. This force causes an outward pressure on the outer bearing race, and this outward force can cause the race to fracture and explode violently. Newer axle bearing removal tools enclose the bearing in an adapter and provide some degree of protection. If the bearing is exposed during removal operations, most technicians place a shield around the bearing during this operation to contain the explosion if it should occur. (Fig. 6-34)

FIGURE 6-34.
Axle bearings can explode violently as they are being pressed off of the axle. An old generator or starter case can be used as a scatter shield when using press adapters that do not enclose the bearing.

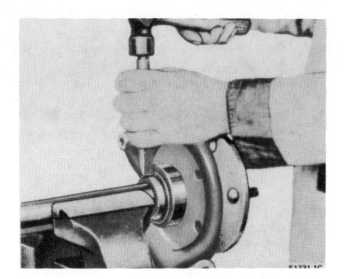

FIGURE 6-35.
Before pressing an axle bearing off, make about five deep chisel marks in the bearing retainer to expand it and loosen it from the shaft. Be careful not to mark or damage the seal area on the shaft next to the retainer on some axles. (Courtesy of Ford Motor Company)

To remove an axle bearing, you should:

1. Place the axle shaft over a large vise or anvil so it is supported under the bearing retainer ring. Using a large chisel and hammer, slice straight into the retainer in four to six places. Usually one good blow at each location is sufficient. This action should stretch the retainer so it becomes loose on the shaft. Note that the area next to the bearing is used for a seal surface on some axles; it must not be damaged. A short length of exhaust tubing or pipe can be slid over the axle to provide a shield. (Fig. 6-35)

2. Slide the bearing retainer plate toward the axle flange, and attach the bearing removal adapter onto the bearing. (Fig. 6-36)

3. Place the axle and bearing remover into a press, and press the axle out of the bearing.

CAUTION: After the axle is pushed a few inches, it will fall free. Be ready to catch it to prevent injury or any damage to the lug bolts. (Fig. 6-37)

4. Remove the bearing retainer plate, and clean the axle and retainer plate in solvent.

To install an axle bearing, you should:

1. Place the bearing retainer plate over the axle next to the axle flange.

2. Place the new bearing over the axle. Note that some bearings have an inner and outer side; they should be positioned correctly.

3. Support the bearing by the inner race, and press the axle completely into the bearing. (Fig. 6-38)

4. Place the new bearing retainer ring onto the axle, support the retainer ring on a press plate, and press the axle into the ring so the ring is solidly against the bearing.

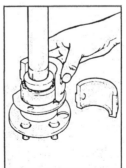

Positive Grip pulling collars grip bearing firmly.

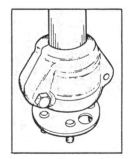

Pulling collets completely encompass bearing during removal.

FIGURE 6-36.
Some axle bearing puller or installer sets use a collet to enclose the bearing and attach it to the puller tube. (Courtesy of OTC)

FIGURE 6-37.
After attachment to the axle, the axle and puller assembly are placed in a press to push the axle out of the bearing. (Courtesy of OTC)

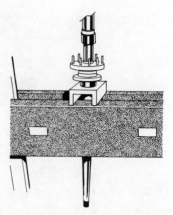

FIGURE 6-38.
As the new bearing is pressed onto the axle, make sure that the adapters support the inner race of the bearing. A new retainer is pressed on after the bearing is in place. (Courtesy of OTC)

6.6.3 Installing a Bearing-retained Axle

Axle shaft installation is essentially the reverse of the removal process. If seal replacement is necessary, on those axles that use a seal separate from the bearing, the seal should be removed and replaced while the axle is removed. Use a procedure like that described in Section 6.6.5 to replace the seal.

To install a bearing-retained axle, you should:

1. Slide the axle into the housing, being careful not to damage the axle seal, if used.

2. When the axle stops a few inches from being fully installed, grip the axle by the flange so the inner end is lifted enough to enter the axle gear in the differential. It might be necessary to rotate the axle to align the splines.

Slide the axle inward so the bearing enters the housing.

3. Install the bearing retaining flange nuts and bolts, and tighten them to the correct torque.

4. Replace the brake drum or rotor and caliper.

5. Replace the wheel.

6. Check the axle lubricant level, and add lubricant if needed.

6.6.4 Removing a "C" Lock Axle

To remove a "C" lock-retained axle, you should:

1. Raise and support the car securely on a hoist or jack stands.

2. Remove the wheel.

3. Mark the brake drum or rotor next to the stud that was previously marked. On disc brake cars, remove the caliper and secure it to the axle with wire.

4. Remove the drum or rotor.

5. Remove any dirt that is around the rear axle cover. Position a drain pan under the cover, and remove the cover retaining bolts. Note that most of the rear axle grease will drain out as you loosen the cover.

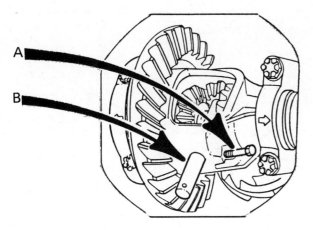

FIGURE 6-39.

When removing an axle that is retained by a "C" lock, the differential pinion shaft locking bolt (A) is removed and then the differential shaft (B) is removed. (Courtesy of Ford Motor Company)

6. Remove the differential pinion shaft lock bolt and the pinion shaft. Note that the pinion shaft will not slide inward far enough for removal; it has to slide outward. If necessary, drive the axle shaft inward from the end opposite the lock pin hole. When the lock pin hole is past the differential case, insert a punch into the hole so the shaft can be pulled out. (Fig. 6-39)

7. Slide the axle inward until the "C" lock can be removed from the axle, and remove the "C" lock. (Fig. 6-40)

8. Slide the axle out of the housing. Be sure to support it to prevent damage to the seal.

6.6.5 Removing and Replacing a Bearing and Seal for a "C" Lock Axle

The outer race of this bearing is fitted tightly into the outer end of the axle housing; the inner race is a hardened and ground portion of the axle. When an axle is removed, this portion of the axle should be inspected for damage; a rough or worn surface will require axle replacement. One after-market supplier supplies a bearing that is positioned slightly differently to permit reuse of a worn axle. The axle seal is positioned just outboard of the bearing with the seal lip running against the axle shaft (Fig. 6-41)

To remove a bearing or seal, you should:

1. Pry out the axle seal using a seal puller or pry bar. (Fig. 6-42)

2. Attach a bearing puller or hook to a slide hammer; position the puller inside the bearing and pull the bearing from the housing. (Fig. 6-43)

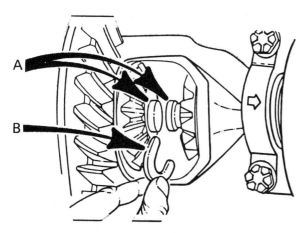

FIGURE 6-40.

With the differential pinion shaft removed, the axles (A) can be slid inward far enough to remove the "C" locks (B). (Courtesy of Ford Motor Company)

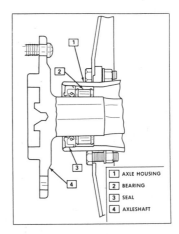

1	AXLE HOUSING
2	BEARING
3	SEAL
4	AXLESHAFT

FIGURE 6-41.

The axle serves as the inner bearing race on "C" lock axles. The axle will easily slide out of the bearing. If the bearing or seal surface is damaged, the axle will usually need to be replaced. (Courtesy of Buick)

FIGURE 6-42.
The seal on "C" lock axles can be easily pried out of the housing. Some bearing-retained axle housings use a seal farther inside of the housing that requires the use of a puller. (Courtesy of CR Industries)

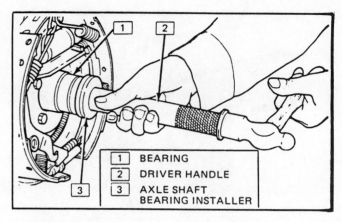

1	BEARING
2	DRIVER HANDLE
3	AXLE SHAFT BEARING INSTALLER

FIGURE 6-44.
A new bearing should be driven in using a flat, correctly sized driving tool. (Courtesy of Buick)

To install a new bearing, you should:

1. Lubricate the new bearing with gear lubricant.

2. Position the bearing in the housing bore, and using a driving tool that contacts the entire side of the bearing, drive the bearing inward until it contacts the shoulder in the housing. (Fig. 6-44)

3. Lubricate the lip of the seal with gear lubricant. (Fig. 6-45)

FIGURE 6-45.
Before a new seal is installed, it should be prelubed with the same type of grease that it is going to seal. (Courtesy of CR Industries)

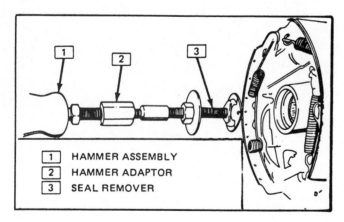

1	HAMMER ASSEMBLY
2	HAMMER ADAPTOR
3	SEAL REMOVER

FIGURE 6-43.
This adapter allows a slide hammer to be attached to the bearing so a faulty bearing can be pulled from the axle housing. (Courtesy of Buick)

4. If the outside edge of the seal case does not have a sealant coating, apply a thin film of **Room Temperature Vulcanizing (RTV) silicone rubber** or nonhardening gasket sealer around the seal case.

5. Position the seal in the axle housing with the seal lip facing inward toward the lubricant. Using a driving tool that contacts the entire side of the seal case, drive the seal inward until it is flush with the edge of the axle housing. (Fig. 6-46)

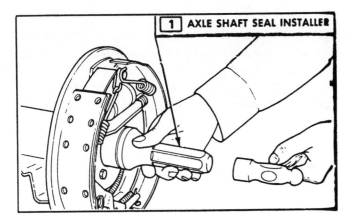

FIGURE 6-46.
A flat, correctly sized installer should be used to drive the new seal into the housing. (Courtesy of Buick)

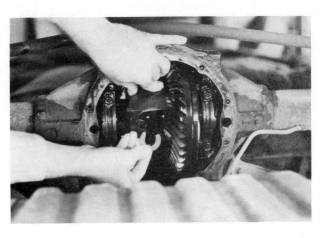

FIGURE 6-47.
After the axle has been carefully slid into the housing, the "C" locks are replaced, the differential pinion shaft is installed, and the shaft lock bolt is tightened to the correct torque. (Courtesy of CR Industries)

6.6.6   Installing a "C" Lock Axle

Installing a "C" lock axle follows a procedure that is essentially the reverse of the removal procedure.

To install a "C" lock axle, you should:

1. Lubricate the bearing and seal area of the axle.

2. Slide the axle into the axle housing and through the bearing and seal, being careful not to let the axle splines or rough axle surface drag across the seal.

3. When the axle stops a few inches from being fully installed, grip the axle flange so the inner end is lifted enough to enter the axle gear in the differential. It might be necessary to rotate the axle slightly to align the splines. Gently slide the axle inward as far as possible.

4. Slide the "C" lock into the groove at the inner end of the axle. (Fig. 6-47)

5. Slide the axle outward so that the "C" lock seats completely in the recess in the axle gear.

6. Slide the differential pinion shaft into the differential case with the shaft hole aligned with the hole for the locking bolt.

7. Install the locking bolt, and tighten it to the correct torque.

8. Install a new cover gasket and the rear axle cover, and tighten the retaining bolts to the correct torque. If RTV sealant is used in place of a solid, paper gasket, thoroughly clean the gasket surfaces on the cover and the back of the axle housing, and apply a bead of RTV sealant 1/16 to 1/8 inch (1.5 to 3 mm) wide around the cover or housing gasket surface, circling each of the bolt holes. (Fig. 6-48)

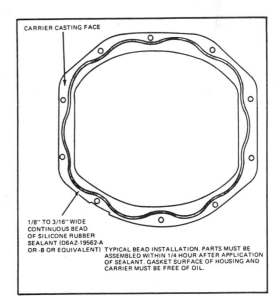

FIGURE 6-48.
Older axle covers were sealed with gaskets. Many newer covers use a formed-in-place gasket, usually of a silicone rubber material. (Courtesy of Ford Motor Company)

9. Add lubricant through the filler hole to bring the gear oil to the correct level in the axle housing.

10. Replace the brake drum, with the marks aligned, and replace the tire and wheel.

6.7 REPAIRING SERVICEABLE FWD FRONT WHEEL BEARINGS

There are several different styles of FWD front wheel bearings: ball bearings, tapered roller bearings, and sealed ball or roller bearings. Most of these do not require periodic maintenance. The major reason for servicing is to replace a damaged hub or to eliminate rough, loose, or noisy operation. (Fig. 6-49)

Servicing the front wheel bearings makes it necessary to remove the outer end of the axle or CV joint housing from the hub. In cases where a prevailing torque hub nut is used, a new nut should be installed during replacement.

This nut can usually be removed using an air-powered, impact type of wrench, but it should be installed by hand to prevent damage to the bearings. This allows a good feel of the nut's condition. Care should also be taken to prevent damage to the CV joint or the CV joint boot. Servicing procedures for the CV joint and boot will be described in Chapter 7.

The repair procedures for the various FWD wheel bearings vary among different makes of cars. It is highly recommended that the specific repair steps for a particular car be followed while servicing the wheel bearings.

In general, to remove the front wheel bearing from an FWD car, you should:

1. With the tires on the ground, the transmission in park or first gear, and the parking brake applied, remove any locking devices, and loosen the hub nut. (Fig. 6-50)

2. Raise and support the car on a hoist or jack stands.

3. If available, install a boot protector over the CV joint boots.

4. Remove the hub nut, the tire and wheel, and the brake caliper. Using wire, suspend the brake caliper from some point inside the fender.

5. Remove the hub from the steering knuckle. This operation often requires the use of a puller. The steering knuckle must be disconnected from either the lower control arm or the strut. In some cars, the tie rod should be

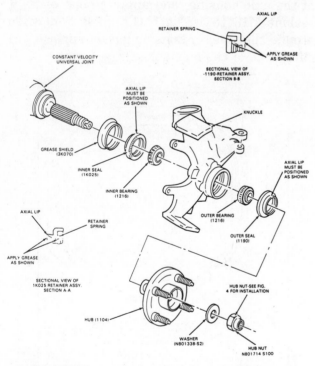

FIGURE 6-49.
A disassembled view of the front knuckle and hub of an FWD car. This particular car uses tapered roller bearings; some cars use ball bearings. (Courtesy of Ford Motor Company)

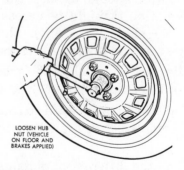

FIGURE 6-50.
The hub nut is usually very tight; it should be loosened with the tires on the ground. This nut must be removed for access to the front-wheel bearings on an FWD car. (Courtesy of Chrysler Corporation)

FIGURE 6-51.
This puller is being used to pull the knuckle from the front hub. (Courtesy of CR Industries)

FIGURE 6-53.
A socket, sized to fit against the outer bearing race, is being used to press the bearing out of the steering knuckle. (Courtesy of CR Industries)

disconnected from the steering arm. Disconnecting these parts allows the steering knuckle freedom to move outward so the hub can be separated from the CV joint housing. (Fig. 6-51)

6. Remove the bearing and seal from the steering knuckle. On some cars the bearing needs to be removed from the hub. This operation can also require the use of a special puller. (Figs. 6-52 & 6-53)

Like the disassembly procedure, bearing installation should follow the manufacturer's procedure.

In general, to install front wheel bearings on an FWD car, you should:

1. Install the new bearing and seal into the steering knuckle. Lubricate the seal lips and the bearing, if required, with the proper amount and type of lubricant.

2. Install the hub into the steering knuckle. (Fig. 6-54)

3. Install the hub and steering knuckle onto the axle end with the CV joint housing.

4. Replace the hub nut and brake caliper. Tighten the hub nut and the caliper mounting

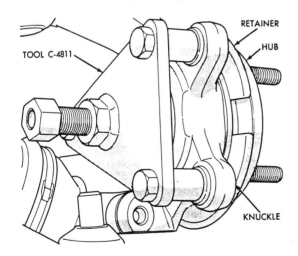

FIGURE 6-52.
Tool C-4811 is being used to remove the hub from the knuckle. (Courtesy of Chrysler Corporation)

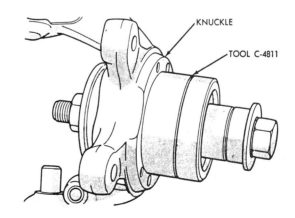

FIGURE 6-54.
Tool C-4811 is pushing against the new bearing to install it in the knuckle. (Courtesy of Chrysler Corporation)

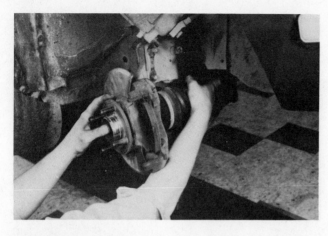

FIGURE 6–55.
The hub is being pushed back onto the drive shaft after the bearing and hub have been installed in the knuckle. (Courtesy of CR Industries)

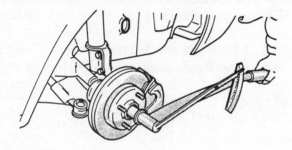

FIGURE 6–56.
The hub nut should be tightened to the torque recommended by the vehicle manufacturer; this will be somewhere between 100 and 250 foot pounds. (Courtesy of Chrysler Corporation)

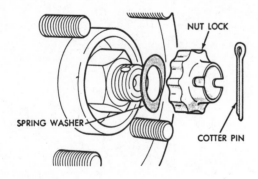

NUT LOCK

SPRING WASHER

COTTER PIN

FIGURE 6–57.
Various locking devices are used on hub nuts; if used, these should be replaced in the correct manner. (Courtesy of Chrysler Corporation)

bolts to the correct torque. Lock the hub nut in place as required. (Figs. 6–55 to 6–57)

5. Replace the tie rod, the lower end of the steering knuckle, and the steering knuckle-to-strut bolts if any of them were disconnected during disassembly. Tighten all of these nuts and bolts to the correct torque. Remove the CV joint boot protector, if used.

6. Replace the tire and wheel.

6.8 REPAIRING NONSERVICEABLE FWD FRONT WHEEL BEARINGS

Servicing of this style of bearing is limited to removing and replacing the hub and bearing assembly, much like the procedure used on nonserviceable wheel bearings. This operation makes it necessary to remove the axle end with the CV joint housing from the hub. A boot protector or cover should be used to prevent damage to the CV joint boot. A damaged, cracked, or torn CV joint boot should be replaced; this operation will be described in Chapter 7.

To remove a nonserviceable wheel bearing, you should:

1. With the tires on the ground, the transmission in park or first gear, and the parking brake applied, remove the locking device from the front hub nut, if used, and loosen the hub nut.

2. Raise and support the car on a hoist or jack stands.

3. If available, install a boot protector over the CV joint boot. (Fig. 6–58)

4. Remove the hub nut and the wheel.

5. Remove the brake caliper. Using wire, suspend it from some point inside of the fender, and remove the rotor.

6. Remove the hub and bearing mounting bolts and the rotor splash shield. (Fig. 6–59)

7. Install a hub puller onto the hub flange, and pull the hub and bearing assembly off of the end of the CV joint housing and out of the steering knuckle. (Fig. 6–60)

8. Some car models will also have a steer-

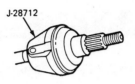

INSTALL DRIVE AXLE COVER

FIGURE 6–58.
It is recommended that a boot protector be installed around the CV joint boot to keep from damaging the boot while working on the front-wheel bearings or drive shaft. (Courtesy of Pontiac Division, GMC)

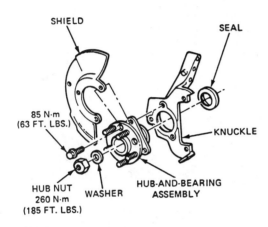

FIGURE 6–59.
The hub and bearing assembly on this FWD car is permanently sealed, lubricated, and adjusted. It is replaced as an assembly if there are problems. (Courtesy of Pontiac Division, GMC)

ing knuckle seal that should be removed at this time.

To replace a nonserviceable type of front wheel bearing, you should:

1. Clean the bore of the steering knuckle and the splined area of the CV joint housing.

2. Lubricate the lip of the new steering knuckle seal, if used, and install the seal in the steering knuckle.

3. Position the new hub and bearing assembly over the axle and into the steering knuckle. Install the hub nut and the hub and bearing assembly mounting bolts. Tighten them alternately to move the hub and bearing assembly into the correct location. Tighten the hub nut temporarily to about 70 foot pounds (100 N·M) of torque, and tighten the mounting bolts to the correct torque. A long bolt can be used in place of one of the mounting bolts to keep the hub from turning while tightening the hub nut. (Fig. 6–61)

4. Remove the boot protector.

5. Replace the rotor and caliper. Tighten the caliper mounting pins to the correct torque.

6. Install the wheel.

7. Lower the car to the ground, and tighten the hub nut to the correct torque. Lock the hub nut in place as required.

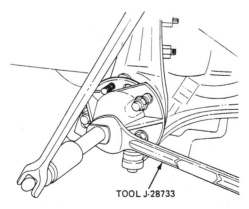

REMOVE HUB-AND-BEARING ASSEMBLY

FIGURE 6–60.
Tool J–28733 is being used to pull the hub and bearing assembly off of the drive shaft. (Courtesy of Pontiac Division, GMC)

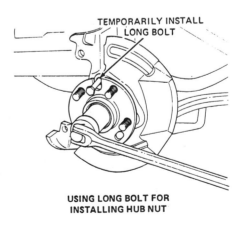

USING LONG BOLT FOR INSTALLING HUB NUT

FIGURE 6–61.
Some cars have a provision for installing a bolt through a notch in the hub to keep the hub from turning while the hub nut is tightened. (Courtesy of Pontiac Division, GMC)

REVIEW QUESTIONS

1. Technician A says that a faulty wheel bearing will usually change noise levels as the car is turned in different directions. Technician B says that faulty wheel bearings will usually change noise levels under different throttle and brake conditions. Who is right?

   a. A only   **c.** both A and B
   b. B only   **d.** neither A nor B

2. When checking the rear axle bearings of an RWD car, in all cases there should be
   A. no vertical play.
   B. no in and out play.

   a. A only   **c.** both A and B
   b. B only   **d.** neither A nor B

3. Technician A says that wheel bearing play is checked by gripping the tire at the top and bottom and trying to rock it. Technician B says the front wheel bearing of an RWD car should have a slight amount of play when rocked. Who is right?

   a. A only   **c.** both A and B
   b. B only   **d.** neither A nor B

4. A faulty wheel bearing will _____ when spun during the checking process.
   A. be noisy
   B. operate with a roughness

   a. A only   **c.** either A and B
   b. B only   **d.** neither A nor B

5. The strength grade of a bolt is indicated by
   A. the amount of embossed lines on the head.
   B. a number embossed on the head.

   a. A only   **c.** either A or B
   b. B only   **d.** neither A nor B

6. A nut can be secured onto a bolt by

   a. a lock washer.
   b. an anaerobic liquid.
   c. a distortion of the nut.
   d. any of these.

7. Technician A says that a prevailing torque nut is worn out if it turns too easily on the bolt. Technician B says that lock washers should always be used with prevailing torque nuts. Who is right?

   a. A only   **c.** both A and B
   b. B only   **d.** neither A nor B

8. Technician A says that the best way to remove the inner wheel bearing from a front hub is to drive the bearing and seal out with a wooden dowel. Technician B says that the best tool to remove a cotter pin from the spindle is either a cotter pin puller or diagonal cutting pliers. Who is right?

   a. A only   **c.** both A and B
   b. B only   **d.** neither A nor B

9. When cleaning wheel bearings, spinning the bearing with an air blast is
   A. an effective way of cleaning the bearing.
   B. not an unsafe shop practice.

   a. A only   **c.** both A and B
   b. B only   **d.** neither A nor B

10. Wheel bearings are repacked by forcing clean grease into the bearing using
    A. hand pressure.
    B. a bearing packer.

    a. A only   **c.** either A or B
    b. B only   **d.** neither A nor B

11. Technician A says that front wheel bearings of the tapered roller type should be

adjusted to a slight end play. Technician B says that after the bearings have been seated, the spindle nut should be tightened finger-tight. Who is right?

   a. A only   **c.** both A and B
   b. B only   **d.** neither A nor B

12. Technician A says that you must remove the axle to repair a leaky axle seal on the rear axle of a RWD car. Technician B says that the rear axle housing cover must always be removed in order to remove an axle. Who is right?

   a. A only   **c.** both A and B
   b. B only   **d.** neither A nor B

13. On "C" lock axles,
   A. a faulty axle bearing can easily ruin the axle.
   B. the axle bearing must be pressed off of the axle.

   a. A only   **c.** either A or B
   b. B only   **d.** neither A nor B

14. As an axle shaft is slid into the housing, the

   a. shaft should be kept from contacting the seal.
   b. shaft splines must be aligned with the gear in the differential.
   c. bearing should be lubricated with gear oil.
   d. all of these.

15. Lip seals should always be _____ before installing them.
   A. coated with sealant
   B. lubricated on the seal lips

   a. A only   **c.** both A and B
   b. B only   **d.** neither A nor B

16. Technician A says that it is necessary to remove the CV joint from the front hub when servicing the front wheel bearings of an FWD car. Technician B says these front wheel bearings should be serviced every 15,000 to 20,000 miles. Who is right?

   a. A only   **c.** both A and B
   b. B only   **d.** neither A nor B

17. Technician A says that it is a good idea to loosen the front hub nut of an FWD car with the tires on the ground and the parking brake applied. Technician B says that it is a good idea to place a boot protector over the CV joint boot when servicing the front wheel bearings of an FWD car. Who is right?

   a. A only   **c.** both A and B
   b. B only   **d.** neither A nor B

18. Nonserviceable wheel bearings are repaired by
   A. removing and replacing the entire hub and bearing assembly.
   B. removing, replacing, and adjusting the sealed bearing unit.

   a. A only   **c.** either A or B
   b. B only   **d.** neither A nor B

19. Technician A says that the front hub nut of some FWD cars is secured in place by bending the lip of the nut into a groove at the end of the threads. Technician B says that this nut is secured by a cotter pin. Who is right?

   a. A only   **c.** either A or B
   b. B only   **d.** neither A nor B

20. The front hub nut of an FWD car must be
   A. tightened using an air-impact wrench.
   B. tightened enough to obtain the correct bearing preload.

   a. A only   **c.** either A or B
   b. B only   **d.** neither A nor B

Chapter 7

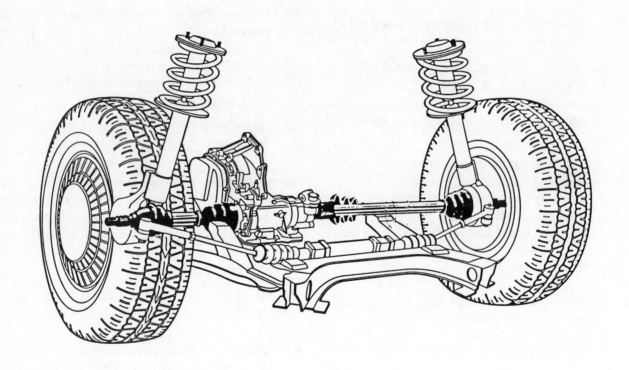

FRONT WHEEL DRIVE DRIVE SHAFTS, THEORY AND SERVICE

After completing this chapter, you should:

- Be familiar with the terms commonly used with the drive shafts on front wheel drive cars.

- Be able to determine the cause of a drive shaft or CV joint problem on these cars.

- Be able to remove and replace a drive shaft on a front wheel drive car.

- Be able to remove and replace a CV joint or boot on these drive shafts.

- Be able to rebuild a CV joint.

7.1   INTRODUCTION

In the late 1970s, the evolution of the automobile progressed to front wheel drive (FWD). This design is very efficient in terms of fuel mileage and passenger space per pound of vehicle. Compact, small-engined cars commonly use **transverse** (crosswise to the car) mounted engines with a **transaxle** (transmission and final drive with differential) at the end of the engine or right alongside of it. Two drive shafts extend from the differential in the transaxle to the front wheel hubs in the steering knuckles. These drive shafts are fitted with a **constant velocity (CV) universal joint** at each end. These drive shafts are also called halfshafts or axles. (Fig. 7–1)

Replacement or service of the drive shafts, CV joints, or boots requires some front suspension disassembly and occasionally the disassembly of the front hubs and wheel bearings. This is when the front-end technician becomes involved with the drive train. Traditionally, in our world of specialization, the front-end technician had very little to do with the car's drive train, and the power train technician had very little to do with front suspensions. As mentioned earlier, FWD has blurred or mixed up more than one aspect of car repair.

7.2   DRIVE SHAFT CONSTRUCTION

Most FWD drive shafts consist of two CV joints connected by a small diameter, solid steel shaft. CV joints have the ability to transfer power without a speed fluctuation that would cause a noticeable vibration or steering wheel shake on corners.

Normal **cross and yoke** or **cardan universal (U-) joints**, as used on RWD drive shafts, can transfer power while allowing a change of angle. But when they run at an angle, they will cause a speed fluctuation. The driven end will speed up and slow down twice in each revolution. Cross and yoke U-joints are usually used in pairs, and the two joints are run at equal angles and are timed so that the speed fluctuations generated by one joint are cancelled out by those generated by the other one. (Fig. 7–2)

CV joints are specially designed and constructed so they can operate at sharp angles and transfer power smoothly without fluctuations in speed. The output of these joints will turn at a constant velocity relative to the input. Also, a CV joint will allow about 40° of operating angle. This large angle is necessary to provide sufficient turning angles and suspension travel. Most current, outboard CV joints are of the **Rzeppa** design or variations of it. This

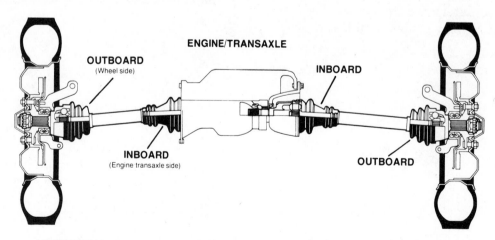

FIGURE 7–1.
A front wheel drive car uses two axles or drive shafts. Each drive shaft has an inboard and an outboard CV joint. (Courtesy of Neapco)

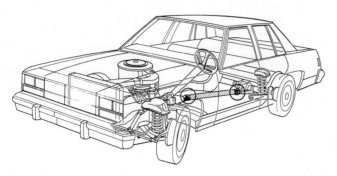

FIGURE 7-2.
A conventional RWD drive shaft using two universal joints (circled). These two joints are arranged so the vibrations developed by one joint are canceled out by the other. (Courtesy of Dana Corporation)

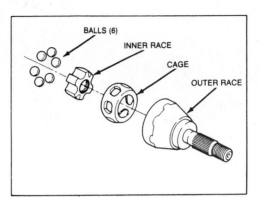

FIGURE 7-3.
A Rzeppa design CV joint. This is the most popular style of outboard, fixed joint. (Courtesy of AC-Delco, General Motors Corp.)

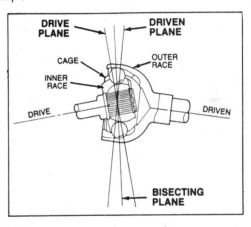

FIGURE 7-4.
In operation, the balls in a Rzeppa joint will bisect the operating planes of driving and driven shafts. (Courtesy of AC-Delco, General Motors Corp.)

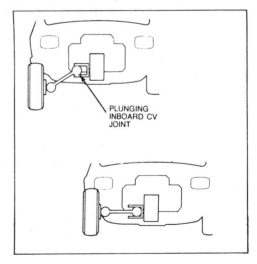

FIGURE 7-5.
The inboard, plunge joint allows the drive shaft to change length as the suspension travels up and down. (Courtesy of AC-Delco, General Motors Corp.)

joint usually uses six precision balls that operate in precisely sized and shaped grooves in the inner and outer races of the joint. All types of CV joints share a high degree of precision machining and all are relatively expensive. They must be kept clean and lubricated or they will soon fail. (Figs. 7-3 & 7-4)

The two joints on an FWD drive shaft are usually of two different types. Suspension movement causes the steering knuckle and hub to travel in an arc, relative to the car, as they move up and down. The path of this arc is determined by the geometry of the suspension. The outer end of the drive shaft also travels in an arc. The inner CV joint is the center point, and the length of the drive shaft is the radius of this arc. The arcs of the drive shaft and the hub are slightly different so the drive shaft must change length while the tire and wheel move up and down. If it does not, there will be a binding of the drive shaft or suspension or both. The inner CV joint is usually of a **plunging** or **plunge** design. The drive shaft can move in and out of this joint while it is operating. This joint is often referred to as a **tripot, tripod, tripode,** or **tulip joint**. This style of joint uses three precision grooves in which needle-bearing equipped balls or rollers mounted on a three-legged cross operate. (Figs. 7-5 & 7-6)

The outer joint is often called a **fixed joint** because it does not allow any side motion of the drive shaft. This keeps the hinge point of the joint directly in line with the steering axis.

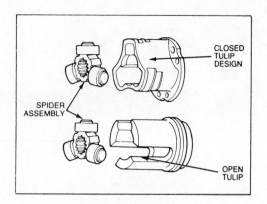
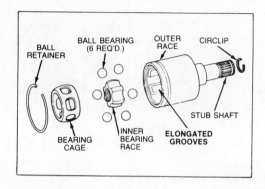

FIGURE 7-6.
The tripot (left) and the double offset (right) designs are the most popular styles of plunge joints. (Courtesy of AC-Delco, General Motors Corp.)

Steering or CV joint bind would occur on turns if the CV joint axis and the steering axis were not in exact alignment. (Fig. 7-7)

The drive shaft itself is usually a small diameter, solid steel shaft that is splined at each end to accept the CV joints. In most FWD designs, one drive shaft, usually the left, is shorter. This is because the differential is not in the center of the car. Having one drive shaft longer than the other can cause **torque steer**; during a hard acceleration, engine torque tries to twist the two drive shafts. The longer shaft allows more twist, so torque will reach the tire on the shorter shaft first. This will cause the car to steer away from that tire. The car will steer straight when there is a steady throttle but will pull to one side under heavy throttle. Some cars will use a larger diameter tubular shaft on the longer side to balance out the shaft twisting tendencies that cause torque steer. Another cure is to use equal length drive shafts of the same size. A short, intermediate shaft is added to the long side to make up for the differences in length. Equal length shaft installations require the use of a third U-joint, usually a simple cross and yoke type, and a support bearing. (Figs. 7-8 & 7-9)

CV joints are kept clean by a flexible, pleated, sealed **boot**. The boot also keeps the lubricant in the joint. The boot is made from a synthetic rubber and of an accordian or bellows type of construction for flexibility. It is usually retained in place on the CV joint housing and drive shaft by metal bands. The boot is usually

the first part of a joint to fail. A failing boot will often show up as a grease **spray**. Centrifugal force will throw the grease out of the joint and onto the area around it. It is said that the joint will last for about 10 to 20 hours of operation after it starts spraying grease. A bro-

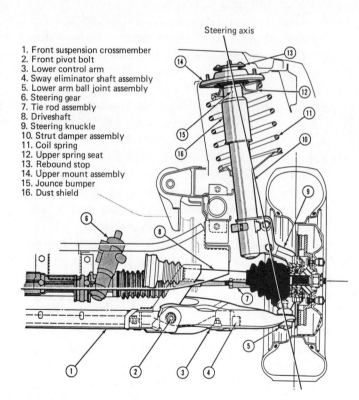

1. Front suspension crossmember
2. Front pivot bolt
3. Lower control arm
4. Sway eliminator shaft assembly
5. Lower arm ball joint assembly
6. Steering gear
7. Tie rod assembly
8. Driveshaft
9. Steering knuckle
10. Strut damper assembly
11. Coil spring
12. Upper spring seat
13. Rebound stop
14. Upper mount assembly
15. Jounce bumper
16. Dust shield

FIGURE 7-7.
The fixed, outboard joint must be centered with the steering axis to prevent binding during turns. (Courtesy of Chrysler Corporation)

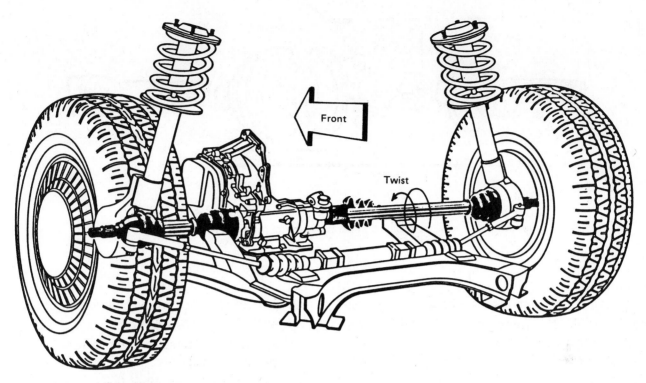

FIGURE 7–8.
Torque steer can occur on a hard acceleration because torque can twist the longer and weaker right-side drive shaft. Torque arrives at the left tire earlier than the right, and the car steers to the right. (Courtesy of AC-Delco, General Motors Corp.)

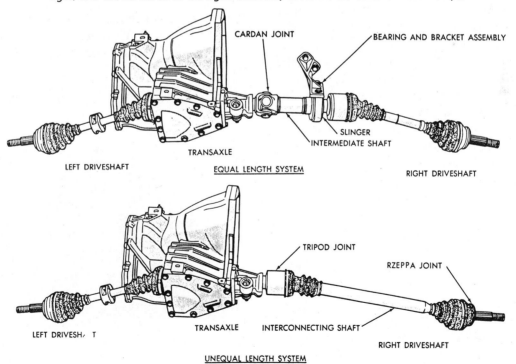

FIGURE 7–9.
To reduce torque steer, some designs use equal-length drive shafts that require an intermediate shaft and support bearing. Other designs will use an oversize shaft on the long side. (Courtesy of Chrysler Corporation)

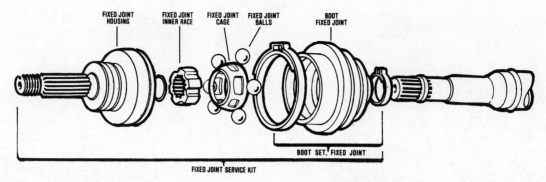

FIXED JOINT HOUSING · FIXED JOINT INNER RACE · FIXED JOINT CAGE · FIXED JOINT BALLS · BOOT FIXED JOINT · BOOT SET, FIXED JOINT · FIXED JOINT SERVICE KIT

FIGURE 7-10.
CV joints are enclosed in an accordion type bellows or boot. A replacement boot set includes a new boot plus clamps. (Courtesy of Moog Automotive)

ken or torn boot can also let water and dirt enter the joint, which will also cause early failure. (Fig. 7–10)

7.3 CHECKING CV JOINTS

CV joint boots are easily checked by raising the car on a hoist and visually inspecting them. First, check for grease spray in the area around the boot. Turn the front of the wheel outward, and rotate the tire and axle to get a good look at the condition of the outboard boot. Pull the accordian pleats apart while rotating the tire and axle to get a good look into the creases and folds of the inboard boot. A cut, torn, or cracked boot should be replaced. (Fig. 7–11)

A faulty CV joint can sometimes be felt as loose or rough operation. The first indication of possible failure is often noisy operation on turns. Most CV joint problems can be found on a road test. Listen to the operation of the joints as you drive the car slowly in right and left circles. Then drive the car in a straight line while alternately applying and releasing the throttle. Listen for clunking noises during the power change. Apply the throttle in such a way that the front end bounces up and down. A faulty inner plunge joint can often be made to react during a power change while the front end is bobbing up and down. One or more faulty joints is indicated if any of the following conditions show up:

- a clicking, clacking or chattering noise on turns (outboard joint);
- a clunking noise during power change (plunge joint);
- a vibration or shudder during acceleration (either); or
- a vibration at highway speeds (either).

If any of these faults show up and if it is believed that the CV joints are faulty, the drive shaft should be removed so you can physically feel the operation of the CV joint and so you can clean and inspect the joints. Cleaning the

FIGURE 7-11.
A torn or broken boot will allow the grease to spray out of the joint or dirt and water to enter the joint. It should be replaced immediately. (Courtesy of AC-Delco, General Motors Corp.)

FIGURE 7–12.
A split or quick CV joint boot. This style of joint is slid over the clean CV joint from the side, and the seam is sealed with a special adhesive. (Courtesy of Specialized Products)

joint requires repacking it with new grease and installing a new boot.

The service that can be done on a drive shaft will vary. Some drive shafts cannot be disassembled. In those cases, the only repair that can be made is to remove and replace the entire drive shaft. It is possible, though, to install a replacement boot of the split type. This procedure will be described in the next section. On drive shafts that can be disassembled, boot replacement with either an O.E.M. or split type, CV joint replacement, and CV joint rebuilding are the various service operations that can be done. (Fig. 7–12)

7.4   SPLIT CV JOINT BOOTS

Split type CV joint boots are available from aftermarket suppliers. They are made with interlocking, zip-lock type edges that are assembled with an adhesive. This boot is said to be grease- and water-tight. It can be installed on a drive shaft that is still mounted in the car. Removal of the drive shaft for boot installation might be a real advantage, but it is not necessary.

To install a boot of this type, cut off the old clamps and the old boot. Then slip the new boot over the shaft, apply the adhesive to the lock groove, lock the boot into itself, position the boot over the CV joint, and install the new clamps. It is absolutely necessary to keep the seam clean before and after applying the adhesive; a little grease or dirt can ruin the seal. The shaft and boot have to remain stationary until the adhesive cures, which takes about an hour. Cure time can be reduced by heating the boot with a heat lamp or hair dryer.

Some technicians recommend caution in the use of split boots because it is very difficult to clean out any water, dirt, or grit that might have entered through the old, torn boot. It is difficult to clean and inspect the internal portion of the CV joint for damage while the joint is assembled, especially when it is on the car. It is a waste of time and money to replace the boot on a dirty or worn CV joint.

7.5   DRIVE SHAFT REMOVAL

An FWD drive shaft is held in place by the outer splines passing through the hub that are secured by the hub nut and by a connection of the inner end to the transaxle. The drive shaft floats between these two points. On most domestic and some **imported cars (made outside of the United States)**, the inner end is splined into the differential side or axle gear and is held in place by a **spring clip**, which is a weak internal circlip or snap ring that will release under pressure. On some cars, the splined, inner end is retained in the differential gear by a conventional snap ring. This ring must be removed or installed from the inside of the differential using snap ring pliers. At least one manufacturer spring-loads the inner joint. The internal spring provides a constant pressure to keep the splined end of the drive shaft in the differential. Some imported cars use bolts or a spring clip to connect the inner CV joint to a flange extending from the differential. (Figs. 7–13 & 7–14)

Care should be taken during removal of the drive shaft so the boot does not get torn or damaged and the inner joint does not get overly compressed or extended. When the outer end of

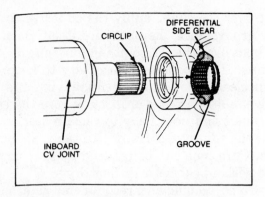

FIGURE 7-13.
A common method of securing the inboard joint into the transaxle is a circlip on the joint splines that snaps into a groove in the differential side gear. (Courtesy of AC-Delco, General Motors Corp.)

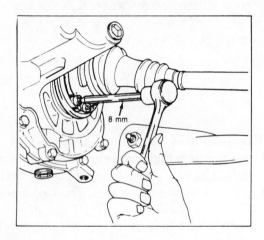

FIGURE 7-14.
Some inboard joints are bolted onto a flange, which is the part of the shaft that extends out of the transaxle differential. (Courtesy of AC-Delco, General Motors Corp.)

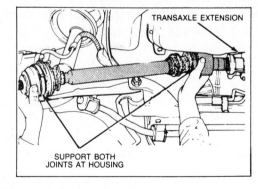

FIGURE 7-15.
When a drive shaft is removed from the car, it should be kept in a horizontal position as it is carried or handled. (Courtesy of AC-Delco, General Motors Corp.)

the shaft is removed, the shaft should be supported so it does not hang on the plunge joint. When the shaft is removed it should be carried, handled, and stored in a horizontal position. (Fig. 7-15) Drive shafts will vary. It is a wise practice to follow the procedure listed in the service manual when servicing them.

To remove an FWD drive shaft, you should:

1. Loosen the hub nut. This is usually done with the tires on the ground, the transmission in park or first gear, and the parking brake set. Staked hub nuts are usually removed by merely unthreading them; new nuts should be used for replacement.

2. Raise and support the vehicle on a hoist or jack stands.

3. Remove the hub nut and the wheel.

4. Remove the tie rod end from the steering arm. This operation usually requires the use of a special puller, which will be described in Chapter 18.

5. Disconnect the steering knuckle from the lower control arm. On most cars, this is simply a matter of removing the ball joint stud pinch bolt and prying the control arm downward until the ball joint stud is out of the steering knuckle boss. (Fig. 7-16) Cars using a tapered ball joint stud with a nut at the end of the stud require the use of a special puller (see Fig. 7-17). Usually, it is necessary to discon-

FIGURE 7-16.
Many cars use a pinch bolt to lock the ball joint stud into the steering knuckle. After the stud is removed, the control arm can be pried away from the steering knuckle. (Courtesy of Moog Automotive)

PLACE J 29330 INTO POSITION AS SHOWN. LOOSEN NUT AND BACK OFF UNTIL . . .

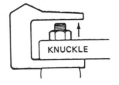

KNUCKLE

. . . THE NUT CONTACTS THE TOOL. CONTINUE BACKING OFF THE NUT UNTIL THE NUT FORCES THE BALL STUD OUT OF THE KNUCKLE.

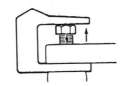

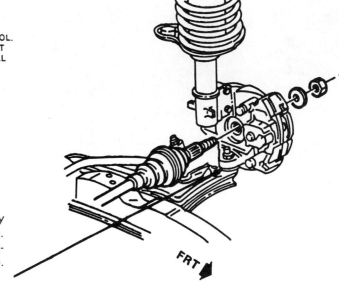

FRT

FIGURE 7–17.
Some cars use a tapered ball joint stud that is secured by a nut that locks the ball joint stud into the steering knuckle. A special tool is required to break the tapers loose and allow separation of the control arm from the steering knuckle. (Courtesy of Pontiac Division, GMC)

nect the stabilizer bar end links so the control arm can be moved downward far enough to move the ball joint stud out of the steering knuckle boss.

6. It is usually necessary to install a puller onto two of the wheel studs and tighten the puller screw to push the end of the CV joint housing into the hub. Occasionally, a tap on the end of the CV joint housing from a hammer and brass punch is sufficient. Pull outward on the steering knuckle to help remove it from the CV joint housing splines. (Fig. 7–18)

7. Pull the drive shaft out of the differential and out of the car. On some cars, an attachment can be installed onto a slide hammer to help pop the inner joint out of the differential. On other cars, a pry bar can be inserted between the inner joint housing and the differential case to pry the inner joint out of the differential. (Figs. 7–19 & 7–20)

CAUTION: Some drive shafts will pop out rather suddenly. Be prepared to catch it.

NOTE: Some manufacturers recommend reversing Steps 6 and 7. Also, those drive shafts that are attached to the differential by other methods will require slightly different disassembly procedures. Always check a service manual if you are not sure of the operation.

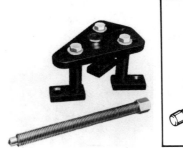
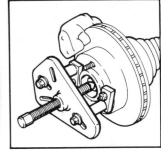

FIGURE 7–18.
A puller attached to the wheel studs is often required to push the CV joint inward through the hub. (Courtesy of Brannick Industries)

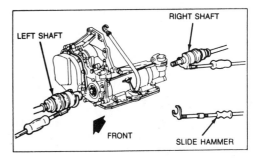

RIGHT SHAFT

LEFT SHAFT

FRONT

SLIDE HAMMER

FIGURE 7–19.
A special tool can be attached to a slide hammer to pull the inboard joint out of the differential. (Courtesy of AC-Delco, General Motors Corp.)

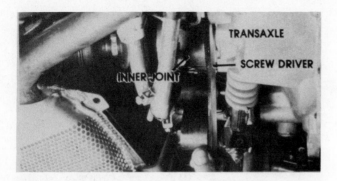

FIGURE 7-20.
On some cars, a pry bar or large screwdriver can be used to pry the inboard joint out of the differential. (Courtesy of Moog Automotive)

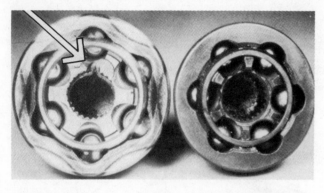

FIGURE 7-21.
The fixed joint is secured onto the drive shaft by either a visible snap ring (left) or circlip at the end of the drive shaft that cannot be seen. (Courtesy of Moog Automotive)

FIGURE 7-22.
Before removing a boot, a scratch mark should be made on the drive shaft so the new boot can be positioned correctly. (Courtesy of Moog Automotive)

NOTE: On some cars, if both drive shafts are removed at the same time, the differential gears can rotate inside of the differential in such a way as to make it impossible to reinstall the inner joint. It is highly recommended that these gears be held in alignment with a special tool.

7.6 DRIVE SHAFT DISASSEMBLY

The drive shaft must be partially disassembled in order to replace a boot using the standard O.E.M. type, to replace a CV joint, or to rebuild a CV joint. Two methods are commonly used to secure the fixed CV joint to the shaft. One method uses a visible snap ring; the other method uses a hidden spring clip or snap ring. Looking at the inner race of the CV joint will tell you which method is used. (Fig. 7-21)

To disassemble an FWD drive shaft, you should:

1. Mark the position of the boot on the drive shaft, and cut the boot clamps and the boot so they can be removed. Discard them. You should always use a new boot and clamps when assembling a drive shaft. (Figs. 7-22 & 7-23)

2. After removal of the boot, clean your fingers and rub a sample of the grease from inside of the joint between your fingers. Dirt or metal particles will cause a gritty feeling. This is a sign of a worn joint. A CV joint with metal or dirt particles in the grease will probably need to be replaced. (Fig. 7-24)

3. Clean the side of the inner race and look for the snap ring and recess to determine which style of retainer is used. If a snap ring is visible, expand the snap ring and slide the CV joint off the end of the shaft. If the hidden spring clip is used, clamp the shaft in a vise and tap the inner race of the CV joint sharply with a brass drift punch and hammer. The joint should pop loose and slide off the end of the shaft. Some manufacturers recommend replacement of the spring clip. If you replace this clip, do not expand it any farther than necessary to install it.

FIGURE 7-23.
A boot is removed by prying or cutting off the old clamps and then cutting off the old boot. (Courtesy of Moog Automotive)

A tool is available to separate the CV joint from the drive shaft on some car models. (Figs. 7-25 to 7-27)

4. Repeat Steps 1 and 2 on the remaining joint. On many inner, plunge joints the outer race or tulip will slide off the joint once the boot has been removed. The tripod and balls are held in place by a visible snap ring.

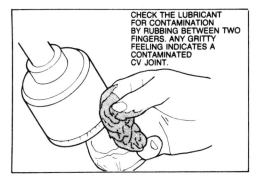

FIGURE 7-24.
When replacing a CV joint boot, the lubricant should be checked. If grit is felt, the joint should be disassembled, cleaned, and inspected or replaced. (Courtesy of AC-Delco, General Motors Corp.)

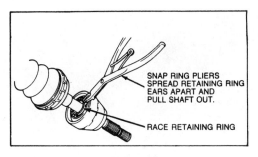

FIGURE 7-25.
If a visible snap ring is used to secure the joint to the drive shaft, the joint can be slid off the shaft after the snap ring has been spread. (Courtesy of AC-Delco, General Motors Corp.)

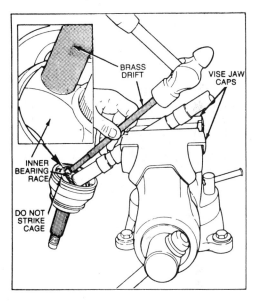

FIGURE 7-26.
If a snap ring is not visible, a brass drift is used to force the joint off the shaft. A sharp, hard blow is usually required to pop the circlip loose. (Courtesy of AC-Delco, General Motors Corp.)

7.7 SERVICING CV JOINTS

Service parts for CV joints are available as boot kits that include a boot, grease packet, and clamps. Any or all of the CV joint parts, including the outer race or housing, inner race, ball set, and cage, are available separately. Complete CV joint assemblies and complete drive shaft assemblies are also available. If the joint is noisy, worn, or rough operating, the old joint should be rebuilt or a new joint installed. (Fig. 7-28)

(A)

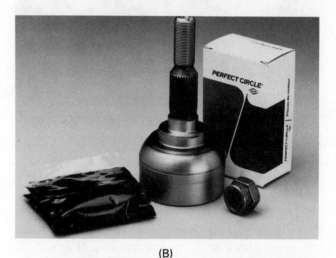

(B)

FIGURE 7–27.
This tool can be used to force the inboard and outboard CV joints off the drive shaft on some cars. (Courtesy of OTC)

Boot replacement is simply a matter of cutting off the old boot, removing the CV joint from the shaft, sliding the new boot onto the shaft, replacing the joint onto the shaft, and clamping the new boot in the proper location. This job becomes a little more difficult because the joint must have a supply of clean grease before the new boot is installed. After the joint has been removed from the shaft, it is a fairly easy job to thoroughly clean it up using solvent. After cleaning, it is important to thoroughly dry the joint and remove all traces of solvent from the inside. Compressed air can be used to blow out all of the solvent. Some shops use a hair dryer to heat the joint and help evaporate out all traces of solvent. After the joint is clean and dry, check all of the balls and race grooves for wear or damage.

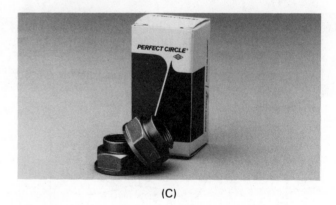

(C)

FIGURE 7–28.
A replacement CV joint boot kit (A), CV joint (B), and retaining nut (C). Some suppliers also market the inner parts for a CV joint. Note that this boot kit contains a grease packet and clamps. (Courtesy of Dana Corporation)

A packet of the correct amount of grease comes with most boot kits. This grease should be worked into the joint so it is thoroughly distributed. Normally, as much grease as possible is packed into the joint, and any remaining grease is spread around the inside of the boot. If the grease packet is not available, it is best to purchase CV joint grease from the vehicle manufacturer and use the exact type and amount of grease that is specified by that manufacturer.

7.7.1 Disassembling an Outer, Fixed-type CV Joint

If a joint has a worn or damaged cage, balls, inner race, or outer race, it can be disassembled for replacement of these parts. Often, it is less expensive to replace the entire joint assembly because a fractured or pitted ball will often be accompanied by fractured or pitted inner and outer races. If only one of the two parts is damaged, it will probably be less expensive to rebuild the joint.

Removing the inner race and cage from some CV joints is much like a puzzle. Parts have to be positioned a certain way to allow disassembly or reassembly. Look for a difference in the widths of the lands between the ball grooves of the outer race or a pair of longer windows in the cage as clues for the disassembly or reassembly procedure. The differences you find are there specifically for this purpose. (Fig. 7–29)

To disassemble a fixed CV joint of the Rzeppa type, you should:

1. Using a brass punch, push or tap inward on one side of the inner race causing the cage and inner race to tilt. A special tool is available to tilt the inner race. After they have tilted far enough, lift the exposed ball from the cage. (Fig. 7–30)

2. Repeat Step 1 on the remaining balls until all of them have been removed.

3. Pivot the cage and inner race 90° so the windows of the cage are aligned with the ball race lands of the outer race. Observe and mark the cage, if necessary, so you know which side goes inward. Withdraw or swing the cage and inner race up and out of the CV joint housing. (Fig. 7–31)

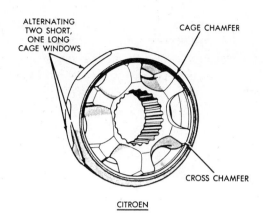

FIGURE 7–29.
Before disassembling a Rzeppa joint, check the parts for differences in the inner race, cage, and outer race. (Courtesy of Chrysler Corporation)

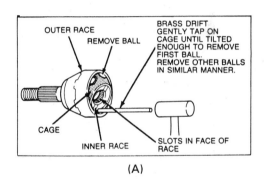

(A)

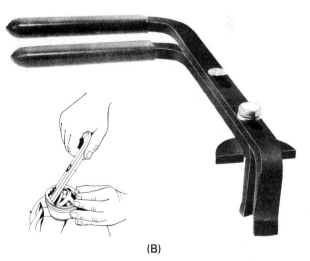

(B)

FIGURE 7–30.
Begin disassembly of a Rzeppa joint in the manner shown here. The cage and inner race are tilted enough to allow removal of each ball by driving inward on the inner race with a brass punch (A) or by gripping the inner race with a special tool (B). (A is courtesy of AC-Delco, General Motors Corp.; B is courtesy of Lisle)

4. Observe and mark the inner race, if necessary, so you know which side goes inward. Pivot the inner race 90° in the cage so one of the lands aligns with a window—the longer ones if there are any—and swing the race out of the cage. (Fig. 7-32)

5. Clean the parts in solvent, dry them with compressed air, and inspect them for damage. (Fig. 7-33)

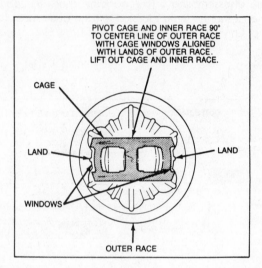

FIGURE 7-31.
After the balls are removed, the cage and inner race can be turned to this position and then pivoted up and out of the outer race. (Courtesy of AC-Delco, General Motors Corp.)

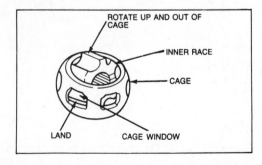

FIGURE 7-32.
The inner race can be removed from the cage by turning it to this position and then pivoting it up and out of the cage. (Courtesy of AC-Delco, General Motors Corp.)

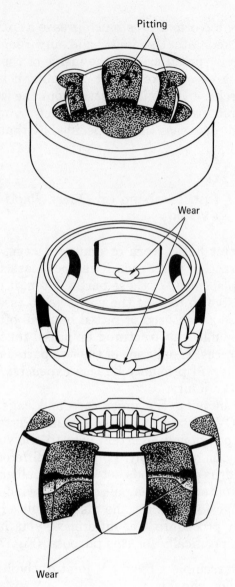

FIGURE 7-33.
After a CV joint has been disassembled and cleaned, the ball grooves and cage should be inspected for pits, wear, and cracks. If any of these faults are found, the joint or the individual parts should be replaced. (Courtesy of AC-Delco, General Motors Corp.)

7.7.2 Assembling an Outer, Fixed-type CV Joint

Reassembly of a CV joint is normally a reversal of the disassembly procedure. On most joints, it is important that the parts end up in the same relative position as they were before disassembly. The inside of all the parts should face toward the inside.

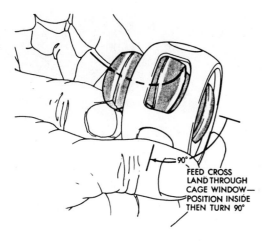

FIGURE 7-34.
To reassemble a Rzeppa joint, first replace the inner race into the cage. (Courtesy of Chrysler Corporation)

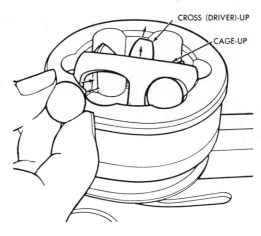

FIGURE 7-36.
Assembly of a Rzeppa joint is completed as the balls are replaced, one at a time. (Courtesy of Chrysler Corporation)

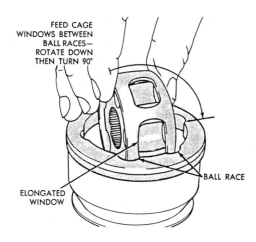

FIGURE 7-35.
The second assembly step is to replace the cage and inner race into the outer race. (Courtesy of Chrysler Corporation)

To reassemble a fixed CV joint of the Rzeppa type, you should:

1. Place the inner race into the cage, align one of the race lands with a cage window, and swing the race into the cage. Pivot the inner race 90° so the race and cage are parallel, and check the two parts for correct positioning. (Fig. 7-34)

2. Place the inner race and cage into the outer race so one of the cage windows is aligned with a race land. Swing the cage and inner race inward, and pivot them 90°. All three of the

parts should be parallel and in the correct position. (Fig. 7-35)

3. Tilt the inner race and cage so a ball can be inserted into one of the cage windows and into an inner and outer race groove. Using a brass punch and hammer, tap the inner race so the ball enters the outer race, or use the special tool to tilt the inner race so another cage window is exposed. (Fig. 7-36)

4. Repeat Step 3 until all of the balls have been inserted.

7.7.3 Disassembling an Inner CV Joint

Some inner joints are disassembled in a manner very similar to an Rzeppa-type joint. A tripod joint is much simpler. Most technicians consider a tripod type of joint much easier to work on.

To disassemble an inner joint of the tripod type, you should:

1. Mark the shaft at the edge of the boot, remove the boot clamps, and cut off the old boot. (Fig. 7-37)

2. Slide the outer race/tulip off the drive shaft. On some joints, it will be necessary to remove a retaining clip. (Fig. 7-38)

FIGURE 7-37.
Before removing the boot from a plunge joint, the shaft should be marked to ensure replacement in the correct position. (Courtesy of Moog Automotive)

FIGURE 7-38.
After the boot has been removed, the outer race or tulip can be pulled off the drive shaft in many cases. (Courtesy of Moog Automotive)

3. Remove the retaining clip from the end of the drive shaft. Note which way the tripod is positioned, and slide the tripod and rollers off the shaft. (Figs. 7-39 & 7-40)

4. Clean the parts in solvent, dry them with compressed air, and inspect them for damage.

7.7.4 Assembling an Inner CV Joint

Assembling a tripod joint follows a procedure that is essentially the reverse of the disassembly process.

To assemble an inner joint of the tripod type, you should:

1. Slide the new boot and clamps onto the drive shaft. (Fig. 7-41)

FIGURE 7-39.
Removal of this snap ring allows removal of the tripod or spider assembly from the shaft. (Courtesy of Moog Automotive)

2. Slide the tripod and rollers onto the drive shaft with the correct side inward. Secure it in position with the snap ring. (Figs. 7-42 & 7-43)

3. Distribute the proper amount of grease over the tripod and rollers and in the tulip.

4. Slide the outer race or tulip over the tripod and rollers, and install the retainer, if used. (Fig. 7-44)

5. Install the boot and clamps as described in the next section.

7.8 INSTALLING CV JOINT BOOTS

New boots are installed as the CV joints are replaced onto the drive shaft. Normally, the boot is slid onto the shaft, the joint is installed and packed with grease, and the boot then is clamped in place. It is important that the boot be installed without twists or unusual wrinkles and that the clamps make a tight seal between

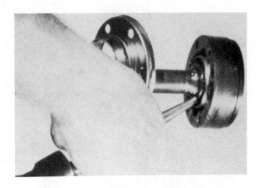

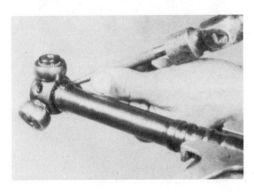

FIGURE 7-40.
It is often necessary to use a brass drift and a hammer to remove the tripod assembly from the shaft. (Courtesy of Moog Automotive)

FIGURE 7-42.
The tripod assembly can be tapped into position using a brass drift or a socket and hammer. (Courtesy of Moog Automotive)

FIGURE 7-43.
After the tripod is installed, it is secured with a snap ring. (Courtesy of Moog Automotive)

FIGURE 7-41.
The new boot and clamp must be slid onto the shaft before assembling a plunge joint. (Courtesy of Moog Automotive)

FIGURE 7-44.
Spread the correct amount of grease through the tulip and around the tripod assembly before replacing the tulip. (Courtesy of Moog Automotive)

FIGURE 7–45.
The new boot and clamp should be slid onto the shaft before replacing a fixed joint. (Courtesy of Moog Automotive)

FIGURE 7–46.
The fixed joint is tapped onto the shaft until you hear or feel the circlip slip into position. (Courtesy of Moog Automotive)

the boot and the drive shaft or CV joint housing. Most cars will use a different boot on the inner and outer joint.

To install a CV joint boot, you should:

1. Slide the small end of the boot with the clamp in place onto the shaft. (Fig. 7–45)

2. On fixed, outer joints, distribute the proper amount of grease through the joint, and position the joint onto the drive shaft splines. Tap the joint onto the shaft until you hear or feel the spring clip snap into place. Pull outward on the joint to ensure that it is locked in place. On inner joints, refer to Steps 2, 3, and 4 of Section 7.7.4. (Fig. 7–46)

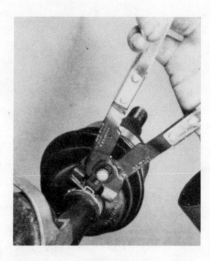

FIGURE 7–47.
With the joint in position, the new boot is aligned with the scratch mark and clamped in place using special pliers. (Courtesy of Moog Automotive)

3. Line up the small end of the boot with the marked location on the shaft, and tighten the small clamp. (Fig. 7–47)

4. Spread any remaining grease into the boot, if required, and slide the large end of the boot into position on the CV joint housing. Check to make sure that the boot is not twisted or collapsed. A collapsed boot can usually be cured by **burping the boot,** sliding a small screwdriver under the lip of the boot and letting trapped air pressure equalize between the inside and the outside of the boot. (Fig. 7–48)

5. Tighten the large clamp.

NOTE: *Some joints use a boot retainer ring in place of a clamp; this ring is pressed onto the CV joint housing along with the boot. (Figs. 7–49 to 7–51)*

7.9   INSTALLING AN FWD DRIVE SHAFT

Installing an FWD drive shaft is another operation that is the reverse of removal. One additional check to make is that many joints have a wear or deflector ring on the outer joint on which a wheel bearing seal fits. If the ring or seal is damaged, it should be replaced. It can usually be tapped off using a punch and ham-

FIGURE 7–48.
If the boot is twisted or collapsed, it should be straightened out and burped or vented (left). (Courtesy of Moog Automotive)

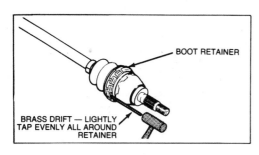

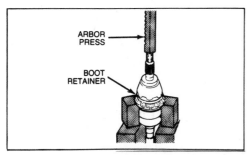

FIGURE 7–50.
A retainer is used to secure the large end of the boot on General Motors cars. This retainer can be removed by tapping it with a punch and hammer and installed by using a press and two blocks. It must be kept straight and even during installation. (Courtesy of AC-Delco, General Motors Corp.)

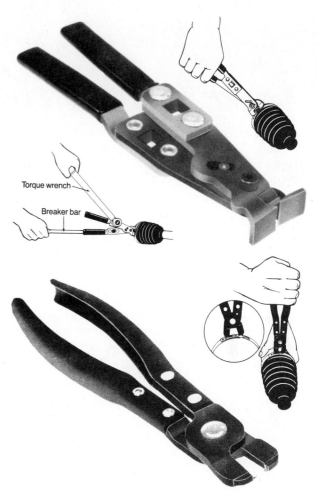

FIGURE 7–49.
The clamp on the large end of the boot is also secured using special pliers. One manufacturer requires that the clamp be tightened to a specific torque. (Courtesy of Lisle)

mer to allow replacements to be installed. Care should be taken during installation of these parts to ensure that they are not bent or damaged. Be sure to follow the procedure listed in the service manual. (Fig. 7–52)

To install an FWD drive shaft, you should:

1. Slide the inner end of the drive shaft into the differential until you feel the spring clip snap in place; test its connection by trying to pull the CV joint back out. It should be locked. (Fig. 7–53)

OR Connect the inner CV joint to the differential flange and install the locking ring or retaining bolts. If bolts are used, tighten them to the correct torque. (Fig. 7–54)

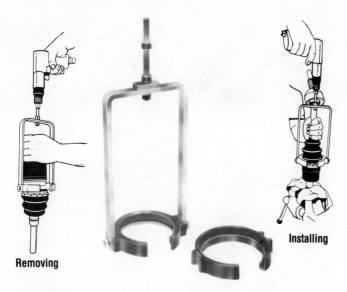

Removing

Installing

FIGURE 7-51.
This special tool is used with an air hammer or can be used with an ordinary hammer to remove or install boot retainer rings. (Courtesy of Lisle)

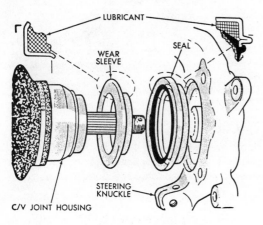

FIGURE 7-52.
A wear sleeve is installed on some outboard CV joints. This wear sleeve runs in a seal used to protect the front hub bearing. If either is damaged, it should be replaced. (Courtesy of Chrysler Corporation)

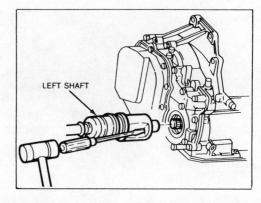

FIGURE 7-53.
A circlip-retained axle is slid into the differential. If necessary, a punch or screwdriver can be used to tap it inward. It should slide inward until the circlip is engaged. (Courtesy of AC-Delco, General Motors Corp.)

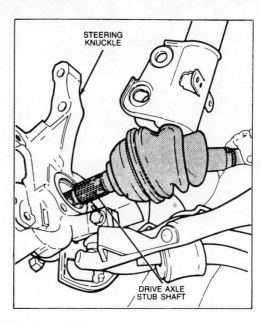

FIGURE 7-54.
The fixed joint is slid into the hub. Make sure the seal or wear ring, if used, is not damaged. (Courtesy of AC-Delco, General Motors Corp.)

2. Slide the hub and steering knuckle over the outer end of the drive shaft. (Fig. 7-55)

3. Install the washer and hub nut, using new ones if required, onto the CV joint housing. Tighten the nut to pull the hub into position on the CV joint housing. Pretighten the nut to about 50 to 75 foot pounds (70 to 100 N·M) of torque. If necessary, tap the hub onto the drive shaft splines so the nut can be started on the

threads using a soft hammer to strike the wheel mounting face on the hub. (Fig. 7-56)

4. Fit the ball joint stud into the steering knuckle, install the pinch bolt and nut (some manufacturers require new ones) or ball joint stud nut, and tighten the nut to the correct torque. If a prevailing torque nut is used, make sure it offers sufficient resistance. If a cotter pin is used, install and secure it. (Fig. 7-57)

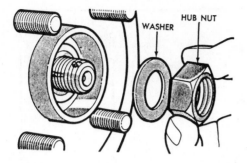

FIGURE 7-55.
The outer nut and washer can be installed and tightened to pull the drive shaft into the hub. (Courtesy of Chrysler Corporation)

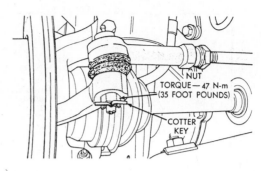

FIGURE 7-57.
If the tie rod end was disconnected, it should be replaced, the nut tightened to the correct torque, and secured with a cotter pin. (Courtesy of Chrysler Corporation)

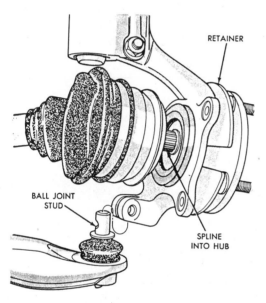

FIGURE 7-56.
The ball joint stud is slid into the steering knuckle boss and locked in place by tightening the bolt to the correct torque. If the ball joint stud has a single notch for the bolt, be sure to align the notch to the bolt hole. (Courtesy of Chrysler Corporation)

FIGURE 7-58.
The hub nut must be tightened to the correct torque and secured in place, usually with a cotter pin. This nut in particular should be tightened to somewhere between 100 and 250 foot pounds (136 and 340 N-M) of torque, as specified by the manufacturer. (Courtesy Chrysler Corporation)

5. Install the wheel, and tighten the lug nuts.

6. Lower the car to the ground, set the parking brake, and tighten the hub nut to the correct torque. (Fig. 7-58) Lock the hub nut as required. If staking is required, a staking tool can be made by regrinding the point of a chisel. Stake the nut as illustrated in Figure 7-59.

SAFETY NOTE
Make sure that all nuts and bolts are properly tightened and secured and that the brake pedal has a normal feel before moving the car.

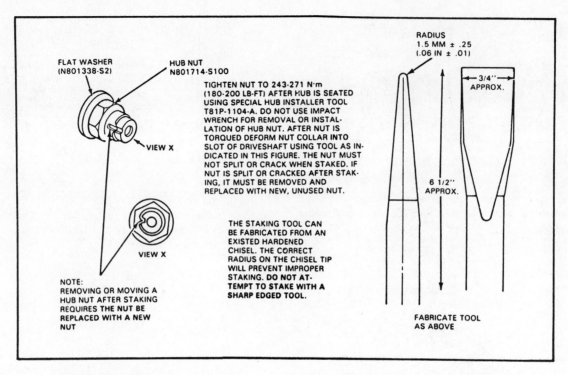

FIGURE 7-59.
After it has been tightened to the correct torque, the retaining nut is secured by staking on some cars. A staking tool can be made by regrinding a chisel. (Courtesy of Ford Motor Company)

REVIEW QUESTIONS

1. The term CV joint is short for

   a. continuous volume joint.
   b. constant volume joint.
   c. constant velocity universal joint.
   d. continuous velocity universal joint.

2. Most CV joints will
   A. allow a large amount of angle change between the two shafts.
   B. transmit power at an angle with no speed fluctuation.

   a. A only   c. both A and B
   b. B only   d. neither A nor B

3. Statement A: The outer CV joint is usually of the plunge type. Statement B: The inner CV joint is usually a fixed type. Which statement is correct?

   a. A only   c. both A and B
   b. B only   d. neither A nor B

4. Technician A says that most CV joints are provided with a special lube fitting so they can be lubricated. Technician B says that all FWD CV joints are protected by an accordian boot. Who is right?

   a. A only   c. both A and B
   b. B only   d. neither A nor B

5. A Rzeppa joint is
   A. a type of fixed joint.
   B. also used on RWD drive shafts.

   a. A only   c. both A and B
   b. B only   d. neither A nor B

6. Many plunge joints are of the _____ type.

   a. tripod joint   c. tripot joint
   b. tulip joint   d. any of these

7. FWD designs with one long drive shaft and one short drive shaft are prone to have
   A. torque steer.
   B. vibrations on sharp corners.

   a. A only   c. both A and B
   b. B only   d. neither A nor B

8. Technician A says that a fixed joint that is failing will often make a clunking noise during a load change. Technician B says that a faulty inner joint will often make a clicking noise during sharp turns. Who is right?

   a. A only   c. both A and B
   b. B only   d. neither A nor B

9. Technician A says that a faulty CV joint boot can be replaced with the drive shaft still in the car. Technician B says that the joint should be clean and repacked with wheel bearing grease before installing a new boot. Who is right?

   a. A only   c. both A and B
   b. B only   d. neither A nor B

10. Technician A says that the ball joint or strut must be disconnected from the steering knuckle in order to remove an FWD drive shaft. Technician B says that many FWD drive shafts are secured into the transaxle by a spring clip. Who is right?

    a. A only   c. both A and B
    b. B only   d. neither A nor B

11. Technician A says that many fixed CV joints can be removed from the drive shaft by driving outward on the inner race. Technician B says that some fixed CV joints are secured to the drive shaft by a set screw. Who is right?

    a. A only   c. both A and B
    b. B only   d. neither A nor B

12. When replacing a boot on a CV joint, it is a good practice to
    A. feel the old grease for signs of grit.
    B. clean and inspect any dirty joints.

    a. A only   c. both A and B
    b. B only   d. neither A nor B

13. Technician A says that it is necessary to disassemble the fixed CV joint in order to replace the boot. Technician B says that a plunge joint is easier to disassemble than a fixed joint. Who is right?

    a. A only   c. both A and B
    b. B only   d. neither A nor B

14. CV joint boots are secured in place using
    A. RTV and mechanics wire.
    B. a metal retainer ring.

    a. A only   c. either A or B
    b. B only   d. neither A nor B

15. When replacing an FWD drive shaft, it is usually necessary to
    A. drive the inner joint into the transaxle using a special driving tool.
    B. lower the tire onto the ground before final tightening of the hub nut.

    a. A only   c. both A and B
    b. B only   d. neither A nor B

Chapter 8

FRONT SUSPENSION TYPES

After completing this chapter, you should:

- Be familiar with the different styles of automotive front suspensions.

- Be familiar with the terms commonly used with front suspensions.

- Understand how a short-long arm suspension operates.

- Understand how a strut suspension operates.

- Understand the operating differences between rear wheel drive and front wheel drive front suspensions.

8.1 INTRODUCTION

The front suspension of a vehicle is designed so the steering knuckle and spindle can pivot on the **steering axis** to allow steering of the vehicle. The spindle must also rise and fall, relative to the body, to allow the springs and shock absorbers to reduce bump and road shock from the vehicle's ride. The suspension system allows the springs and shock absorbers to absorb the energy of the bump so we can have a smooth ride. While doing these two jobs, the suspension system must not allow loose, uncontrolled movement of the tire and wheel and must keep the alignment of the tire as correct as possible.

There are essentially four types of front suspensions used on cars, pickups, and trucks: **short-long arm (S-L A), MacPherson strut (strut), swing axle,** and **solid axle.** Of these four, only the first two are commonly used on passenger cars. All of these suspension types can be used with FWD, RWD, or 4WD vehicles.

These suspension types have several differing aspects: simplicity or complexity, size, cost, weight, alignment stability, ride and handling characteristics, and how well they adapt to automated, robot-aided construction. When a car design is developed, these aspects are carefully compared when choosing a particular design. Often, the type of suspension and drive system at one end of the car has to be balanced with the same handling characteristics at the other end. For example, a rear engine, RWD car tends to oversteer. If a front suspension with superior tire adhesion characteristics were put on this car, it would be dangerous to drive in some conditions. Handling characteristics such as oversteer and understeer will be discussed in more detail in Chapter 19.

As will be described in Chapter 17, wheel alignment is controlled by the tilt of the spindle, the angle of the steering axis, and the angle of the steering arms. Briefly, these alignment angles are

1. **camber, positive** or **negative**—the vertical angle of the tire and wheel (viewed from the front); caused by a horizontal—downward, or upward—tilt of the spindle.

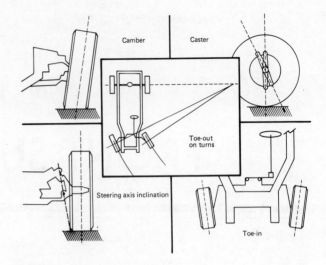

FIGURE 8-1.
The front tires and steering axis must be held in correct alignment as the car operates; these are the five alignment angles or factors. (Courtesy of Chrysler Corporation)

2. **toe-in** or **toe-out**—inward or outward angle of the tires and wheels (viewed from above); caused by a forward or rearward angle of the spindles and controlled by the length of the tie rods.

3. **caster, positive** or **negative**—an angle of the steering axis (viewed from the side); controlled by the position of the ball joints (S-L A), ball joint, and upper pivot damper unit (strut) or king pin. The steering axis is the imaginary line on which the steering knuckle pivots for turns.

4. **steering axis inclination**—an angle of the steering axis (viewed from the front); controlled by the same things as caster.

5. **toe-out on turns**—the relative position of the tires while turning; controlled by the angle of the steering arms. (Fig. 8-1)

The first three of these angles are adjustable in many cars. The last two are not.

8.2 SHORT-LONG ARM, S-L A SUSPENSIONS

Short-long arm, or S-L A, is the typical RWD car's front suspension. It is also called an **unequal arm suspension.** It consists of two con-

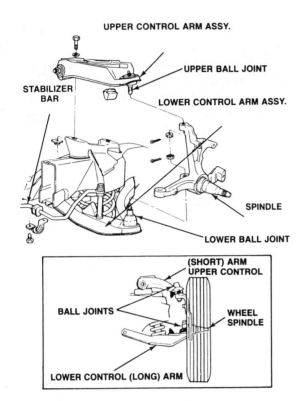

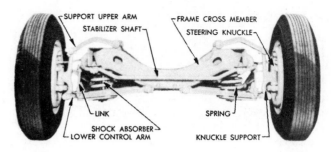

FIGURE 8–3.
Older cars with S-L A suspensions used king pins to connect the steering knuckle or spindle to the knuckle support; metal bushings connected the knuckle support to the control arms. (Courtesy of Ammco)

FIGURE 8–2.
Two views of a short-long arm, S-L A, suspension system. (Courtesy of Ford Motor Company)

trol arms (a short upper and a longer lower arm), a steering knuckle with spindle, and the necessary bushings and ball joints. (Fig. 8-2) The control arms are usually triangular, A-arm shapes with two inner pivot bushings mounted on the car's frame or reinforced body structure. These control arms are also called **A-arms** or **wishbones**. Another type of lower control arm uses a single inner pivot bushing along with a **trailing strut rod** or a **leading strut rod**. A trailing strut rod is also called a tension rod because it has the pivot bushing at the front and the control arm end at the rear. A leading strut rod is also called a compression rod; it has the pivot bushing at the rear. The strut rod prevents forward or rearward movement of the free (outer) end of the control arm. The outer end of both control arms connects to the steering knuckle, which includes the spindle, through a ball joint. At one time, before the development of ball joints, a king pin was used to connect the spindle to the spindle support. The ball joints or king pin form the steering axis. (Fig. 8-3)

The length of the control arms and the mounting location of the pivot points are selected by engineers with the following factors in mind:

1. Camber change during tire and wheel movement to reduce track change and tire scrub.

2. **Instant center and roll center** location for roll axis considerations and the camber change and resulting tire scrub during leaning motions of the car. Instant and roll centers will be described in detail in Chapter 19. (Fig. 8-4)

3. **Anti-dive**, a lowering of the front end during braking, is reduced by mounting the inner control arm pivots in a plane that creates a

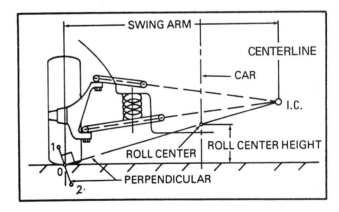

FIGURE 8–4.
If imaginary lines are drawn along the control arms pivots, they will meet at a point called the instant center, IC. If a line is drawn from the IC to the center of the tire, it will cross the center of the car at the roll center, RC. (Courtesy of Chevrolet)

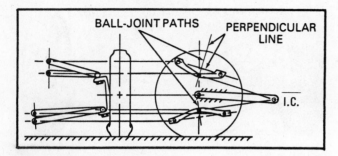

FIGURE 8-5.
Resistance to brake dive can be generated by the front suspension. The lengthwise angles of the control arms form an imaginary lever arm. Lines drawn along the planes of the control arms show us the length and location of the end of this lever arm. (Courtesy of Chevrolet)

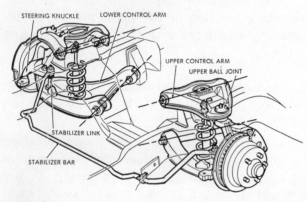

FIGURE 8-6.
The inner control arm bushings are located in what appears to be rather odd angles, but when we consider the roll center, antidive, and smoother operation over bumps, their relationship begins to make more sense. (Courtesy of Chevrolet)

lever arm converging near the car's center of gravity. This will be described in Chapter 19. (Fig. 8-5)

4. Bump severity is less pronounced if the path that the tire travels when it goes over a bump is toward the rear of the car; this path is affected by the mounting location of the inner control arm pivots. (Fig. 8-6)

8.2.1 Control Arm Geometry

Control arm design is matched with the size of the spring to provide optimum control arm position, tire travel over bumps, and comfortable car ride characteristics. In a static position (i.e., no bump condition), the lower control arm will

be horizontal or, in a new car, slightly lower at the outer end.

The ball joint at the end of this arm travels in an arc, pivoting at the inner bushing as the tire and wheel move up and down. The center point of this arc is the inner bushings; the radius of the arc is the length of the control arm. A longer control arm has a longer radius with a flatter arc. This arc is not only up and down but also slightly horizontal, depending on the midpoint and end locations of the arc (see Fig. 8-7). If the control arm is horizontal at the midpoint of the arc, there will be a minimum amount of side motion. The more the control arm is inclined, the more side motion that will occur. In a car, this side motion causes tire side scrub. If the springs sag, the lower control arms incline more and more upward at the outer end, resulting in more tire scrub and wear.

The amount of side scrub resulting from movement of the lower control arm can be reduced by the length and mounting angle of the upper control arm. The arc of travel for the upper ball joint is designed not to match that of the lower ball joint. Because of this difference,

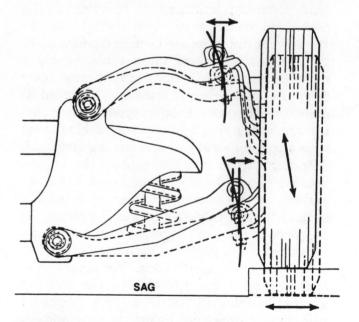

FIGURE 8-7.
As the front suspension moves up and down, the end of the lower control arm travels in an arc; side scrub of the tire can occur if the control arm gets too far above horizontal. (Courtesy of Dana Corporation)

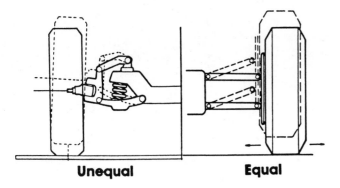

FIGURE 8–8.
If the control arms were parallel and of an equal length, a sideways scrub would occur at the tire tread during suspension movement; unequal length arms that are mounted at angles can move through bump travel without tire scrub. (Courtesy of Moog Automotive)

the steering knuckle changes vertical position during wheel travel, and, because of this, the camber of the tire will change.

When a car goes over a bump, the tire moves upward, compressing the spring. The upward motion at the end of the control arm will tend to move the tire inward, shortening the track and trying to scrub the tire. But, at the same time, the upper control arm is on a more severe arc that causes the tire to move to a negative camber position (inward at the top). As the camber changes, the bottom of the tire moves outward enough to compensate for the track change; the road contact point of the tire travels straight down the road with little or no side scrub. This is the major reason for S-L A suspensions. There are operational limits to any design. When you do a wheel alignment and lower a car's tire onto a turntable, a certain amount of side motion, which is evidence of tire scrub, can be noticed. (Fig. 8–8)

At one time, in the 1950s and 1960s, almost every passenger car used an S-L A front suspension. By the 1980s, most cars used strut front suspensions, mainly because of the lower cost and weight and the increased space provided. S-L A suspensions are still used on those cars in which superior handling and tire wear characteristics are desirable.

8.2.2 S-L A Springs and Ball Joints

The car's weight is transferred to the front wheels through a spring. On S-L A suspensions, this is usually a coil spring, but torsion bars, leaf springs, or air springs can be used. The spring is commonly mounted between the car's frame and the lower control arm, but it can be mounted between the upper control arm and the fender well, which is reinforced to accept this load. (Fig. 8–9)

The ball joint attached to the control arm with the spring is called the **load-carrying** ball joint because it must be constructed to carry a portion of the car's weight. Since a substantial amount of load is on this joint, the joint can be constructed with running clearance, and the vehicle load will hold the joint tightly together, eliminating any free play. (Fig. 8–10)

Some load-carrying ball joints are arranged with the control arm positioned above the boss or mount on the steering knuckle so the load squeezes the ball joint together. These are called **compression-loaded** ball joints. Most cars position the control arm under the steering knuckle mount so the load tries to pull the ball joint apart. These are called **tension-loaded** ball joints. (Figs. 8–11 & 8–12)

The other ball joint is called the **follower, friction-loaded, steering** or **dampening ball joint.** The follower joint carries no vertical load; its major job is to keep the tire and wheel in alignment. This joint is built with a slight internal preload to prevent any looseness or free

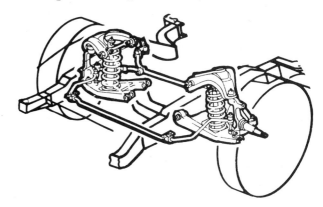

FIGURE 8–9.
Most cars with an S-L A suspension use a coil spring mounted between the frame and the lower control arm. (Courtesy of Moog Automotive)

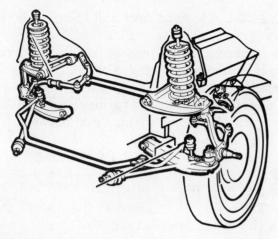

FIGURE 8–10.
Cars without frames, unibody cars, usually mount the spring between the upper control arm and the reinforced fender well. (Courtesy of Moog Automotive)

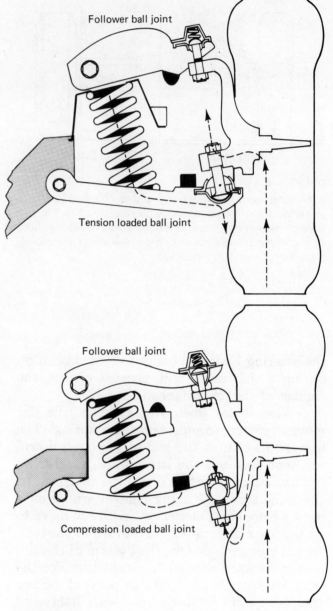

play; this preload is not great enough to cause steering drag. The follower and load-carrying ball joints form the steering axis, the pivot points for steering the front wheels. (Fig. 8–13)

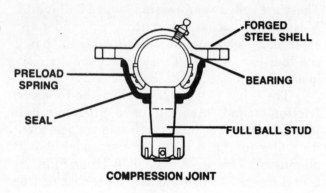

COMPRESSION JOINT

FORGED STEEL SHELL

PRELOAD SPRING

SEAL

BEARING

FULL BALL STUD

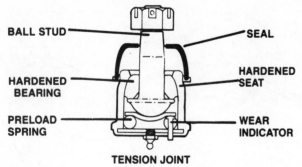

TENSION JOINT

BALL STUD

HARDENED BEARING

PRELOAD SPRING

SEAL

HARDENED SEAT

WEAR INDICATOR

FIGURE 8–11.
The vehicle's load travels from the frame, through the spring, along the control arm, through the ball joint, through the steering knuckle, and across the wheel bearings to the hub, wheel, and tire. (Courtesy of Ammco)

FIGURE 8–12.
Depending on whether the control arm is above or below the steering knuckle boss, the load-carrying ball joint will be of a compression or tension design. Note the different bearing locations inside the joint. (Courtesy of Dana Corporation)

8.2.3 S-L A Wear Factors

If the front springs sag too far from their original length, the geometry of the control arms will change. Static and dynamic camber will no longer be correct, and side scrub during bumps

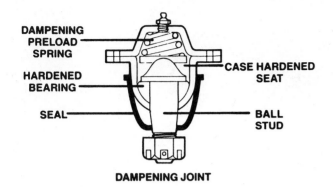

FIGURE 8-13.
The follower ball joint uses a preloaded spring to keep the joint tight so it will have no play or clearance. (Courtesy of Dana Corporation)

will increase. Vehicle driveability will deteriorate and the rate of tire wear will increase.

Time and wear will also tend to loosen the pivot points. The torsilastic, rubber bushings commonly used at the inner pivot points tend to harden, crack, and then break apart as they get older, harder, and less resilient. Rubber bushings will also tend to sag as they get old, letting the control arm change position. This will result in a change in alignment angles. Ball joints tend to wear and cause looseness at the outer pivot points. This can occur fairly fast if the boots tear or rupture. If these pivot points wear, the control arms and steering knuckle will change position causing a change in alignment,

increased tire wear, and reduced driveability. Inspection of these wear points will be covered in Chapter 14.

8.3 STRUT SUSPENSIONS

A **strut suspension**, often called a **MacPherson strut**, has no upper control arm or upper ball joint. The steering knuckle connects to a spring and shock absorber assembly, which is the strut. The upper end of this assembly connects to the car body through a pivot-damper unit. A lower control arm is used; it serves the same purpose as the lower arm of an S-L A suspension. Many strut systems use a control arm with a single inner pivot along with a strut rod. Some strut systems mount the stabilizer bar so the end of the stabilizer bar can serve as the strut rod. (Figs. 8-14 & 8-15)

The strut is basically a coil-over-shock, or a shock absorber with a coil spring mounted around it. The strut becomes shorter when the tire moves upward over a bump and longer if the tire drops into a hole. An oversize shock piston rod is required to withstand sideways bending forces; vertical loads on the tire and wheel result in sideways loads on the strut. This side load also tends to put a bind on the strut motion. Some manufacturers mount the spring

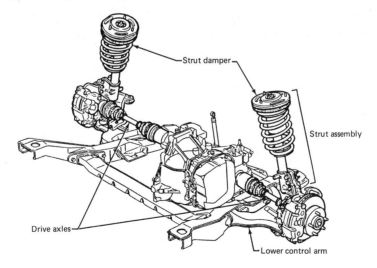

FIGURE 8-14.
A strut suspension system from an FWD car. Note that this lower control arm uses two bushings allowing the control arm to maintain the wheelbase as well as the track position of the front tire. (Courtesy of Chevrolet)

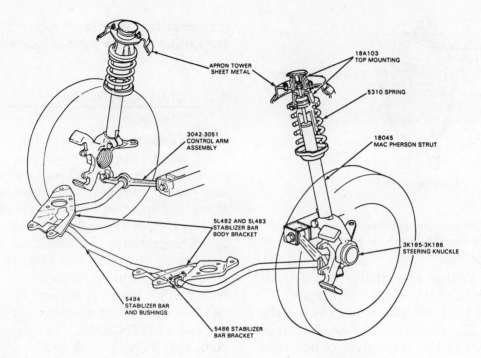

APRON TOWER
SHEET METAL

18A103
TOP MOUNTING

5310 SPRING

3042-3051
CONTROL ARM
ASSEMBLY

18045
MAC PHERSON STRUT

5L482 AND 5L483
STABILIZER BAR
BODY BRACKET

3K185-3K186
STEERING KNUCKLE

5494
STABILIZER BAR
AND BUSHINGS

5486 STABILIZER
BAR BRACKET

FIGURE 8–15.
This strut suspension uses a control arm with a single inner bushing to control the track position of the tire. The wheelbase position is controlled by the end of the stabilizer bar acting as a trailing strut rod. (Courtesy of Ford Motor Company)

seats at an angle or off center to try and reduce strut rod bind. (Fig. 8–16)

As the tire and wheel move up and down, the ball joint in the lower control arm will travel in an arc similar to that of an S-LA suspension. As this occurs, the lower end of the strut must move inward and outward relative to the car; this movement changes the vertical position of the strut and, therefore, the camber angle. The strut's angle change is allowed by the design flexibility of the upper strut mount or pivot-damper assembly. The damper is also called an insulator. (Fig. 8–17)

The **upper strut mount** or **damper** serves several purposes, including the following:

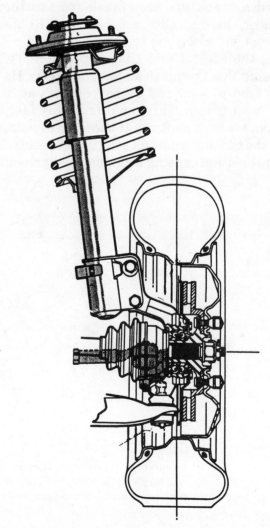

FIGURE 8–16.
Many struts have the spring seats located off center to reduce the side load on the strut piston rod bushing and piston rod. Vehicle load tends to bend the strut sideways; the spring load offsets that. (Courtesy of Chrysler Corporation)

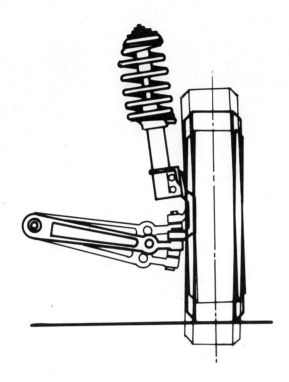

FIGURE 8–17.
During suspension travel, the strut changes length. The outer end of the lower control arm will travel in an arc, much like the lower control arm on an S-L A suspension.

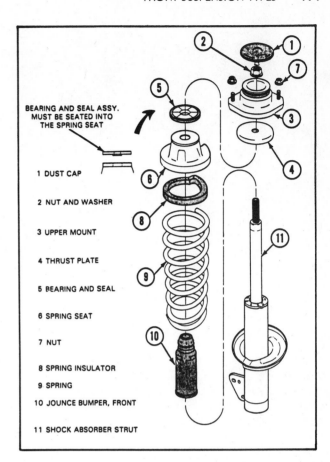

BEARING AND SEAL ASSY.
MUST BE SEATED INTO
THE SPRING SEAT

1 DUST CAP

2 NUT AND WASHER

3 UPPER MOUNT

4 THRUST PLATE

5 BEARING AND SEAL

6 SPRING SEAT

7 NUT

8 SPRING INSULATOR

9 SPRING

10 JOUNCE BUMPER, FRONT

11 SHOCK ABSORBER STRUT

FIGURE 8–18.
The upper mount parts of a strut assembly. (Courtesy of Ford Motor Company)

1. The flexibility of the mount allows the strut angle to change to follow the travel of the lower ball joint.

2. The rubber in the mount dampens or reduces road vibrations so they are not transmitted to the body of the car.

3. A bearing is built into the mount to serve as the upper end of the steering axis.

When the front wheels are turned, the entire strut assembly, from the pivot bearing to the ball joint, pivots or turns. The upper mount also transfers the load of that corner of the car into the strut and spring. In most cars, the lower control arm carries no load; a friction loaded ball joint is used. (Fig. 8–18)

Some struts are made so they bolt to the steering knuckle. On some of these cars, there is a provision for adjusting camber at this connection. Other struts are welded directly onto the steering knuckle. (Fig. 8–19)

8.3.1 Modified Struts

Some cars use a **modified strut**, in which the spring is mounted on the lower control arm instead of the strut. This is also called a **single control arm** suspension. In this suspension, the strut is essentially the same as a standard strut without the spring mounts; the upper strut mounts or dampers and pivots are the same. The lower control arm and ball joint are essentially the same as the lower control arm on an S-L A suspension, and the lower ball joint is a load-carrying ball joint. A torsion bar can also be used with a strut. On these cars, the lower control arm usually becomes the lever arm for the torsion bar. (Figs. 8–20 & 8–21)

One import manufacturer has modified a strut design still further by adding a control

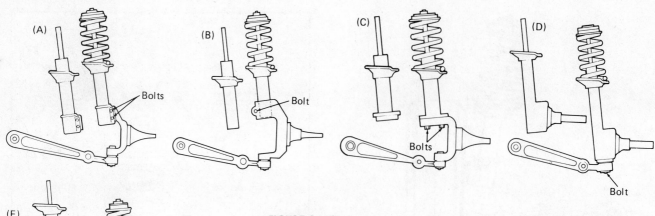

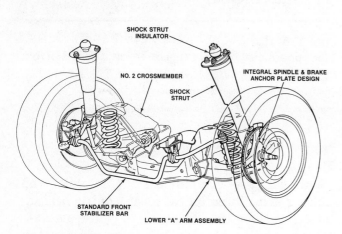

FIGURE 8-19.
There are five general methods of attaching the strut to the lower control arm. Type A, B, and C connect the strut to the steering knuckle; Type D connects the strut and steering knuckle to the steering arm, or ball joint boss; and Type E is a single assembly from the ball joint to the upper bushing. All of these styles can be found on both FWD and RWD front struts.

FIGURE 8-20.
A modified strut. Note the spring on the lower control arm much like an S-L A suspension. The strut serves as the upper end of the steering axis and the shock absorber. (Courtesy of Ford Motor Company)

FIGURE 8-21.
In a strut suspension, the load of the vehicle travels from the body to the upper strut mount through the spring to the lower strut mount to the steering knuckle and then through the wheel bearings to the hub, wheel, and tire. In a modified strut, the vehicle load travels through the spring to the lower control arm, much like an S-L A suspension.

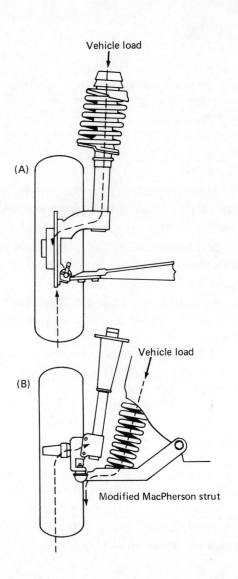

arm and upper ball joint. This design is really an S-L A, only it looks like a strut at first glance. It blends some features of a strut with improved tire and wheel position control of the S-L A suspension. (Fig. 8–22)

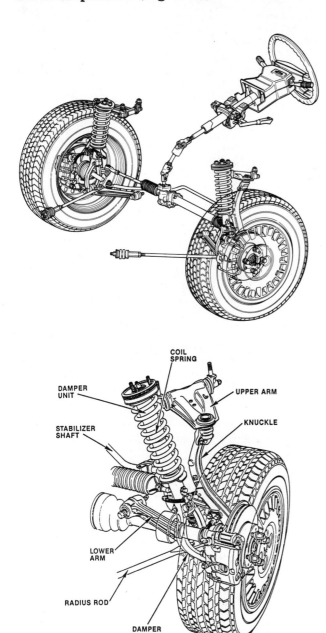

FIGURE 8–22.
This double wishbone suspension uses a spring and damper assembly much like a strut suspension. It uses a long, curved knuckle connected to a rearward angled upper control arm. This suspension provides excellent tire angle control during bump travel, an increased caster angle for stability during bump travel, and a minimum of intrusion into the engine compartment. (Courtesy of American Honda)

8.3.2 Advantages of Struts

In comparing a strut suspension to an S-L A suspension, the strut is simpler to build and is also lighter in weight and less expensive. Because there is no upper control arm, this suspension allows more room in the engine compartment for the transverse engine mounting plus transaxle and driveshafts of an FWD car. The under-hood area of an FWD, transverse engine car is very crowded. There is not enough room for the upper control arm of an S-L A suspension.

8.3.3 Strut Wear Factors

Strut suspensions are also affected by spring sag and bushing wear. In some cases, they are affected more than an S-L A type. Many strut systems do not have any or have a very limited provision for camber or caster adjustment to compensate for wear. Most S-L A systems are adjustable, so a readjustment can be made for misalignment caused by spring sag or bushing wear.

If the shock absorber wears out on a strut suspension, the strut needs to be removed for service. Depending on the strut or the availability of parts, the shock absorber is rebuilt, the shock absorber cartridge is replaced, or the entire strut is replaced. These operations will be described in Chapter 12. (Fig. 8–23)

8.4   SOLID AXLES

As mentioned earlier in this text, solid axles are not used on passenger cars because of their harsher ride and inferior handling characteristics on uneven roads. They are commonly used on trucks, 4WD, and some pickups because of their simpler, stronger, and less expensive construction. They generally require less maintenance because of their minimum number of wear points. A **solid axle** is simply a strong, solid beam of steel (usually I-shaped) with a king pin at each end to connect to the steering knuckle. This axle is called a **mono-beam** by one manufacturer. (Fig. 8–24)

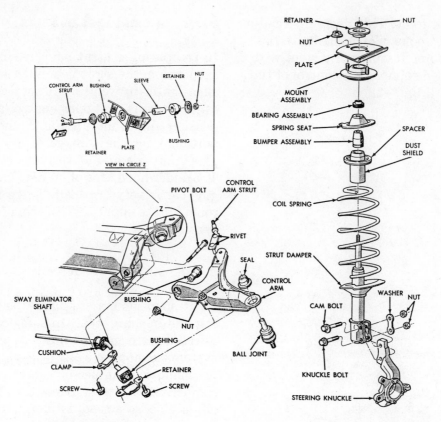

FIGURE 8-23.
A disassembled strut assembly from an FWD car showing the various parts. (Courtesy of Chrysler Corporation)

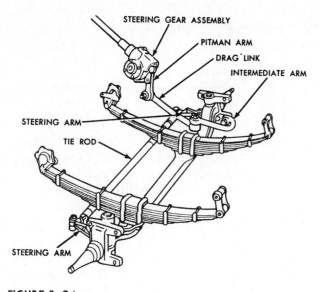

FIGURE 8-24.
A solid axle with springs and steering linkage. (Courtesy of Ford Motor Company)

A solid axle is usually connected to the vehicle's frame by a pair of leaf springs. The springs used are sturdy enough to position the axle and control wheelbase and caster. Camber, steering axis inclination, and track are controlled by the length and shape of the axle beam and the steering knuckles. These parts are usually durable enough to maintain proper wheel alignment for a relatively long period of time. (Fig. 8-25)

Solid axle suspensions have more potential design problems than independent suspensions. Because each style of independent suspension attaches each suspension unit separately to the frame and because each wheel moves separately from the other, the only wheel travel an engineer is concerned with is bounce travel. A solid axle is normally attached to the frame through a flexible coupling, that is, the springs. In addition to bounce travel, the axle can move in the following directions:

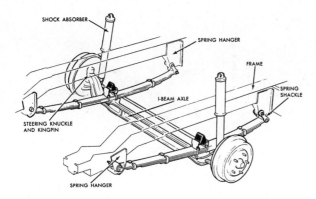

FIGURE 8–25.
The springs control the position of this solid axle and, therefore, control the wheelbase of the vehicle. The track is controlled by the length of the axle. (Courtesy of Chevrolet)

1. **windup**, a twisting of the axle when the brakes are applied; windup can also occur in a drive axle during acceleration;

2. **side shake**, a sideways motion of the axle relative to the frame;

3. **yaw**, a rotation of the axle—forward on one end and rearward on the other; and

4. **tramp**, a rotation of the axle—upward on one end and downward on the other; tramp is often called **shimmy**.

These potential problems can be cured with additional control arms, panhard rods, or radius arms, but these additional parts increase the complexity, cost, and number of wear points.

8.5 SWING AXLE, TWIN I-BEAM AXLES

Swing axles, also called twin I-beam axles, are used by one manufacturer on its pickups, 4WDs, and light trucks. Twin I-beam axles combine some of the sturdiness and simplicity of a solid axle with some of the improved ride and handling characteristics of an independent suspension. A twin I-beam axle is a compromise between these two suspension types. (Fig. 8–26)

A twin I-beam axle consists of two shortened I-beams that have a king pin boss at one end to connect to the steering knuckle just like one end of a solid axle. The inner end of these

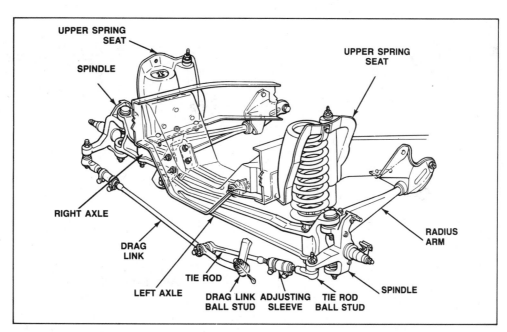

FIGURE 8–26.
A twin I-beam front suspension. This particular swing axle design uses two ball joints to connect the steering knuckle to the axle. (Courtesy of Ford Motor Company)

axles has a bushing boss to accept the bushing that connects the axle to the frame. This rubber bushing allows the wheel end of the axle to pivot or swing up and down for suspension movement. A radius arm is at the free or outer end of each axle beam to control the wheelbase. It also controls the vertical position of the king pin and therefore the caster. The radius arm is bolted securely to the axle and connected to the frame using a rubber bushing. (Fig. 8–27)

Camber and steering axis inclination, along with track, are controlled by the length and shape of the I-beams and the steering knuckle and by the height of the spring. As the spring changes length, the free end of the axle

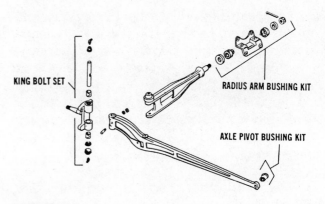

FIGURE 8–27.
An older twin I-beam axle that uses a king pin. Note the rubber bushing used at the axle pivot and the end of the radius arm. (Courtesy of Moog Automotive)

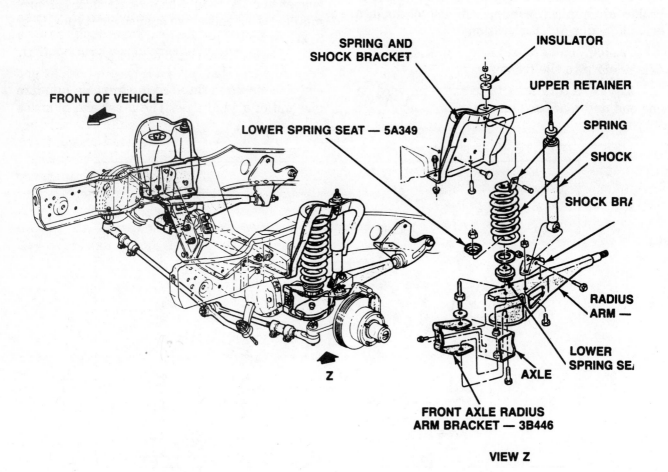

VIEW Z

FIGURE 8–28.
The radius arm controls the front end of the wheel base on this twin I-beam suspension. Track is controlled by the length and angles of the two axles. (Courtesy of Ford Motor Company)

moves up and down relative to the pivoting end. As this occurs, camber and track will change. In actual operation, the free wheel end stays about one-half the tire diameter off the ground while the pivoting end moves up and down along with the frame of the vehicle. Caster and wheelbase, controlled by the radius arm, will also change slightly during vertical vehicle movement. (Fig. 8–28)

Because of the four additional pivots, there are more wear points on a twin I-beam axle than on a solid axle. Wear at these points, plus camber and track change at different spring heights, can cause the tire and wheel alignment to change and cause tire wear or driveability changes. (Fig. 8–29)

8.6 MISCELLANEOUS SUSPENSION TYPES

At least two other front suspension types have been used on cars, but these are not in current use or commonly used by any major manufacturer. They are the following:

1. Trailing arm. The steering knuckle support is attached to one or a pair of trailing arms. During bump travel, the tire and wheel swing upward at the ends of the trailing arms. There

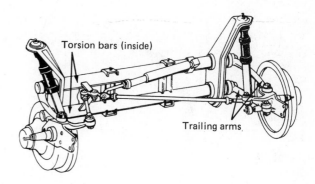

FIGURE 8–30.
A twin trailing arm front suspension. The steering knuckle is supported by the two trailing arms. Torsion bars are enclosed in the two transverse tubes to function as the front springs. (Courtesy of Moog Automotive)

is zero camber or track change. This suspension has a limitation in that vehicle lean will cause an equal change in camber. This usually results in poor front tire adhesion while cornering. (Fig. 8–30)

2. Sliding pillar. The steering knuckle slides vertically on an overly long king pin or pillar. The front springs are built into this pillar. During vertical tire and wheel travel, there is zero camber or track change. This suspension has limitations in that vehicle lean causes an adverse camber change, like the trailing arm suspension, and also the steering knuckle tends to bind on the pillar, which causes an interference with vertical travel. This design also tends to be fairly heavy. (Fig. 8–31)

8.7 FRONT-WHEEL DRIVE AXLES

All four major front-wheel suspension types described in this chapter have been adapted to front-wheel drive by various manufacturers. The biggest change required is using a steering knuckle with a hollow spindle for the driveshaft. In some vehicles the wheel bearings and the hub are mounted over the spindle while in others, wheel bearings and a spindle that will fit into the steering knuckle are used. Both of

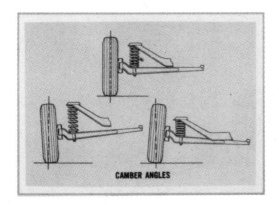

FIGURE 8–29.
Camber and track will change as the swing axle moves up and down. (Courtesy of Ford Motor Company)

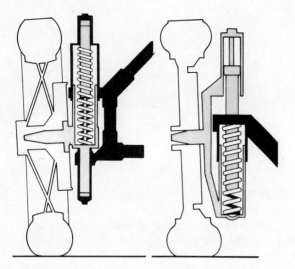

FIGURE 8–31.
A sliding-pillar suspension provides a means for the steering knuckle to move up and down on the king pin or steering pivot.

these methods have been described in Chapters 5, 6, and 7. Most FWD cars use a strut suspension.

On S-L A suspensions, the control arm length, angles, and mounting are essentially the same as on RWD cars. The major difference, other than the hollow spindle in the steering knuckle, is the common usage of torsion bars for springs. A coil spring would interfere with the driveshaft. (Fig. 8-32)

Front-wheel drive, solid axles, and twin I-beam axles cannot use a king pin; it would be in the way of the drive axle. A pair of ball joints are used above and below a hollow section of the axle where the CV joint or universal joint is positioned. These ball joints form the steering axis. Some 4WDs use a universal joint in place of a CV joint, which usually limits the use of 4WD to off-pavement use only. A universal joint is less expensive than a CV joint, but it can cause vibrations or pulsations during turning maneuvers. (Figs. 8–33 to 8–35)

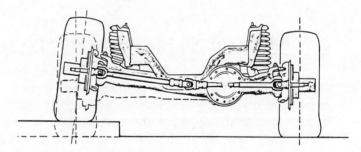

FIGURE 8–33.
A 4WD front suspension. This particular axle system is not used on two-wheel drive vehicles. (Courtesy of Dana Corporation)

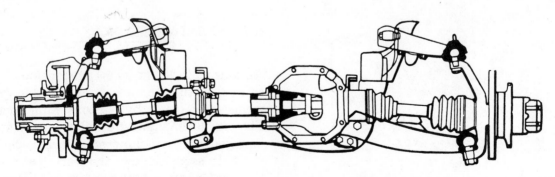

FIGURE 8–32.
An FWD, S-L A suspension system, this suspension can be used with two- or four-wheel drive. (Courtesy of Chrysler Corporation)

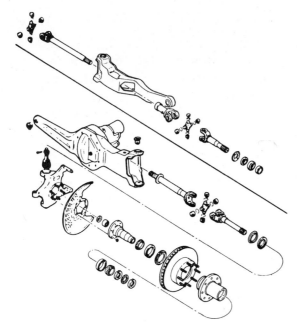

FIGURE 8–34.
An exploded view of a front wheel drive, twin I-beam front axle. The manufacturer calls this a Twin Traction Beam (TTB) axle. (Courtesy of Dana Corporation)

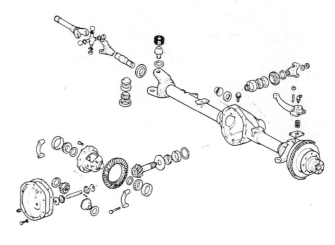

FIGURE 8–35.
An exploded view of an FWD, solid axle; this axle is normally used on 4WD vehicles. (Courtesy of Ford Motor Company)

REVIEW QUESTIONS

1. A car's front suspension system allows the front wheels to
   A. turn for steering maneuvers.
   B. rise and fall to provide for a smooth ride.

   a. A only   **c.** both A and B
   b. B only   **d.** neither A nor B

2. A control arm for a front suspension has a single pivot at the outer end and
   A. a single pivot at the inner end.
   B. two pivots at the inner end.

   a. A only   **c.** either A and B
   b. B only   **d.** neither A nor B

3. Statement A: An S-L A suspension system will cause the front end alignment to change as the wheel travels up and down. Statement B: These two arms are always parallel. Which statement is correct?

   a. A only   **c.** both A and B
   b. B only   **d.** neither A nor B

4. Statement A: An A-arm always has two inner pivot bushings. Statement B: A strut rod is never used with an A-arm. Which statement is correct?

   a. A only   **c.** both A and B
   b. B only   **d.** neither A nor B

5. A major advantage of the S-L A suspension design is that
   A. the camber angle remains constant during suspension travel.
   B. very little side scrub of the tire occurs during suspension travel.

   a. A only   c. both A and B
   b. B only   d. neither A nor B

6. Statement A: The load-carrying ball joint is on the lower control arm of most RWD cars with S-L A suspension. Statement B: The ball joint used in strut suspensions is also a load-carrying design. Which statement is correct?

   a. A only   c. both A and B
   b. B only   d. neither A nor B

7. The inner metal sleeve of a torsilastic bushing is

   a. designed to rotate on its mounting bolt.
   b. designed to allow the rubber insert to rotate on it.
   c. locked in place when the mounting bolt is tightened.
   d. none of these.

8. In a strut suspension, the shock absorber piston rod
   A. provides a turning pivot point for steering.
   B. holds the steering knuckle in alignment during suspension travel.

   a. A only   c. both A and B
   b. B only   d. neither A nor B

9. Statement A: A strut suspension will allow suspension travel with no change in camber angle. Statement B: This suspension will allow suspension travel with no change in track. Which statement is correct?

   a. A only   c. both A and B
   b. B only   d. neither A nor B

10. The upper strut-to-fender well mount provides

    a. a bearing for the steering axis.
    b. a damper to isolate suspension vibrations.
    c. flexibility so the strut can change angle.
    d. all of these.

11. Statement A: A modified strut uses a load-carrying ball joint. Statement B: There are no spring seats on a modified strut. Which statement is correct?

    a. A only   c. both A and B
    b. B only   d. neither A nor B

12. A strut suspension
    A. is slightly heavier but less complex than an S-L A suspension.
    B. allows more room in the engine compartment than an S-L A suspension.

    a. A only   c. both A and B
    b. B only   d. neither A nor B

13. Statement A: Spring sag in a strut suspension will cause the camber to change. Statement B: Shock absorber replacement is more difficult and expensive on a strut suspension than on an S-L A suspension. Which statement is correct?

    a. A only   c. both A and B
    b. B only   d. neither A nor B

14. Which of the following axle motions is not a potential problem on vehicles using a solid front axle?

    a. windup that can cause an alignment change
    b. camber change because of spring sag
    c. tramp that can result from an imbalance or worn parts
    d. yaw as a result of vehicle lean

15. A major advantage of a solid axle is that it

A. is sturdy with few wear points.
B. is lightweight and simple.

  a. A only   **c.** both A and B
  b. B only   **d.** neither A nor B

16. A solid axle can use _____ for the steering axis pivot.
  A. ball joints
  B. a king pin

  a. A only   **c.** either A or B
  b. B only   **d.** neither A nor B

17. Statement A: Suspension travel will cause a camber change on swing axle suspensions. Statement B: A track change will result from suspension travel on swing axles. Which statement is correct?

  a. A only   **c.** both A and B
  b. B only   **d.** neither A nor B

18. A swing axle suspension
  A. is classed as an independent suspension.

B. uses a pivot bushing connecting the inner ends of the two axles.

  a. A only   **c.** both A and B
  b. B only   **d.** neither A nor B

19. Statement A: The radius arm used with a swing axle helps control the track of the front suspension. Statement B: This radius arm prevents axle windup during braking. Which statement is correct?

  a. A only   **c.** both A and B
  b. B only   **d.** neither A nor B

20. Which of the following is not a front suspension type?

  a. unequal length control arm
  b. semi-trailing arm
  c. swing axle
  d. sliding pillar

Chapter 9

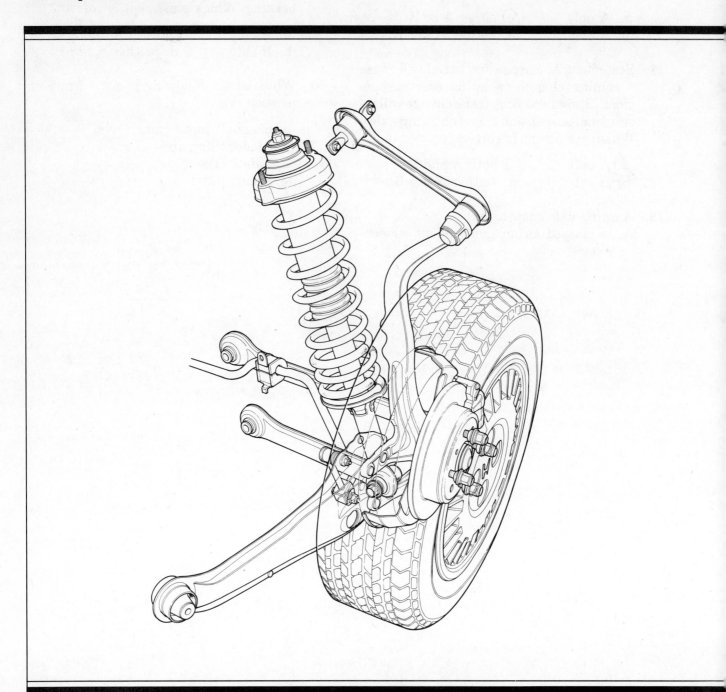

REAR SUSPENSION TYPES

After completing this chapter, you should:

- Be familiar with the different styles of automotive rear suspensions.

- Be familiar with the terms commonly used with rear suspensions.

- Understand the purpose for the various components of a rear suspension.

- Understand the operating differences between rear wheel drive and front wheel drive rear suspensions.

9.1 INTRODUCTION

Rear suspensions are very similar to front suspensions in that they allow vertical tire movement. However, they do not allow steering. Rear tires and wheels are normally set at or near zero camber (straight up and down) and zero toe (straight forward). Excessive camber or toe would cause tire wear and, possibly, a car that **dog tracks** and goes down the road slightly sideways.

An imaginary line drawn midway between the two rear wheels is called the **thrust line**. This thrust line should run down the middle of the car. The car's path during straight-ahead driving follows this thrust line. If the thrust line runs at an angle to the car's center line, the rear tires will not track or follow in direct alignment with the front tires. A rear tire and wheel thrust line that points to the right of the vehicle's center line (at the front) will cause the rear tires to steer to the right. The front tires will have to be steered to the right in order to go straight down the road, and the rear tires will leave tracks to the right side of the tracks left by the front tires. (Fig. 9-1)

Rear wheel alignment has traditionally been taken care of by the strength of the solid rear axle housing. More of today's cars have independent rear suspension or lighter weight rear axles with spindles that bolt on. A greater need has developed for rear wheel alignment to correct tire wear or thrust problems because of these changes. Rear wheel alignment will be discussed in Chapters 17 and 18.

9.2 SOLID AXLE, RWD SUSPENSION

Cars, pickups, and trucks have traditionally used driven, solid rear axles that are sometimes called live axles. This sturdy assembly holds the tires and wheels in alignment, transfers the vehicle load from the springs to the tires and wheels, and provides the gearing necessary to transfer the power from the drive shaft to the tires and wheels. The bearings used to transfer the vehicle weight were described in Chapters 5 and 6. (Fig. 9-2)

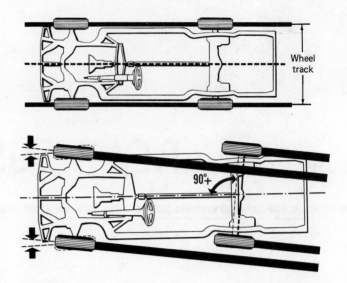

FIGURE 9-1.
A vehicle follows a path parallel to the rear tires. If the rear tires are parallel to the car's center line (top), the rear tires will follow the path of the front tires. If the rear tires are at an angle (bottom), the rear tire track will be to the side of the front tracks. (Courtesy of Hunter Engineering)

A solid rear axle is subject to the same movements possible from a front axle: windup, side shake, yaw, and tramp. Tramp is normally a much greater problem on steerable axles than nonsteerable axles. **Axle windup** is a greater problem on live axles than dead axles. Physics tells us that for every action there is an equal and opposite reaction. In a live axle, the action of sending power to the tires and wheels and causing them to turn in a forward direction produces a reaction of the axle housing trying to rotate it in the opposite direction. If we could see it, a hard acceleration of the car produces a turning force at the axle that tries to lift the

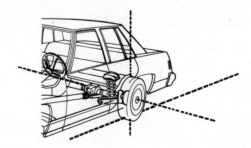

FIGURE 9-2.
On RWD cars, the camber and toe angles of the rear tires are controlled by the axle housing. (Courtesy of Dana Corporation)

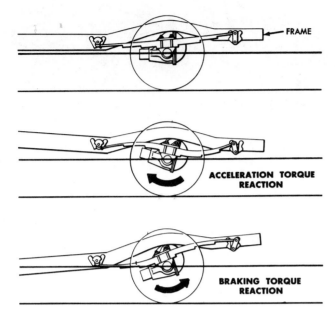

FIGURE 9-3.
The rear axle of an RWD car tends to windup or rotate in response to acceleration and braking torque. (Courtesy of Ford Motor Company)

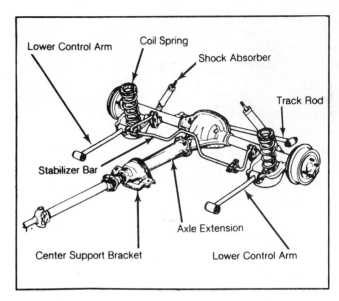

FIGURE 9-4.
The center support bracket and the axle extension prevent axle rotation from acceleration or brake torque. The drive shaft and axle extension are a type of torque-tube drive shaft. (Courtesy of Mitchell Information Services, Inc.)

front tires and wheels. A "wheelie" by a drag race car is a dramatic example of this. (Fig. 9-3)

Two styles of drive shaft design have been used with live rear axles: **torque tube** and **Hotchkiss.** Torque tube drive shafts enclose the drive shaft in a steel tube that is bolted solidly onto the axle housing and connected to the car's transmission or frame member through a flexible pivot. This tube acts as a long lever preventing axle windup. Normally, only a single universal joint is used at the forward end of the drive shaft. This style of drive shaft has seen limited use since the mid-1950s because it tends to be heavy and cumbersome. (Fig. 9-4)

Hotchkiss drive shafts are the most commonly used RWD drive shaft. This design usually has two universal joints, one at each end, and simply transfers power from the transmission to the rear axle. Axle windup must be controlled by other methods, usually the rear leaf springs or control arms. (Fig. 9-5)

One manufacturer uses a **torque arm** to control axle windup on some car models. This long, stamped steel beam bolts solidly to the axle housing and attaches to the transmission

using a rubber bushing. The torque arm functions much like the tube of a torque tube drive shaft.

Rear wheel alignment on solid axles is controlled as follows. Camber, toe, and track are controlled by the strength and length of the axle and axle housing. The rear wheel end of the wheelbase is controlled by the rear springs or control arms. (Fig. 9-6)

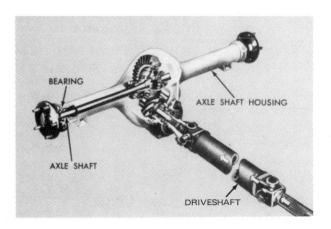

FIGURE 9-5.
A Hotchkiss drive shaft is the common drive shaft with a universal joint at each end; it transmits torque from the transmission to the rear axle. (Courtesy of Ford Motor Company)

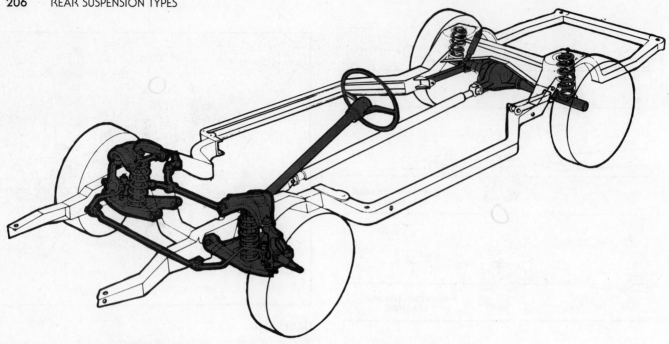

FIGURE 9-6.
On most RWD cars, the rear axle controls the camber and toe angle of the rear tires; the wheelbase and the crosswise position of the axle are controlled by the rear leaf springs or suspension members. (Courtesy of Moog Automotive)

9.2.1 Solid Axle, Leaf Spring Suspension

The solid axle, leaf spring suspension is the simplest form of rear suspension. At one time, it was also the most common. A pair of leaf springs attach to the frame through a rubber bushing at the front and through rubber bushings and a shackle at the rear. The rear axle housing bolts solidly to the center of the spring with a set of U-bolts. The front portion of the spring acts as a control arm, positioning the rear axle housing and establishing the wheelbase. (Fig. 9-7)

The front portion of the spring is often shorter than the rear to provide better axle position control. If this portion were made longer, it would be more flexible, and axle yaw and windup would increase. Severe windup can cause **wheel hop**, a rapid vertical bouncing of the tire and wheel. Yaw, usually a result of vehicle lean, will cause the rear of the car to steer right or left as the rear thrust line changes. Rear axle windup and yaw will be discussed more completely in Chapter 19.

9.2.2 Solid Axle, Coil Spring Suspension

Because a coil spring does not have the ability to locate an axle, coil sprung solid axles require more parts. A coil sprung rear end usually uses two lower suspension arms (sometimes called links) to control the rear axle end of the wheelbase and one or more upper suspension arms to control axle windup and side motion. When two upper arms are used, they are usually skewed diagonally (mounted at an angle) so as to be

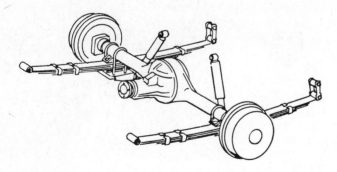

FIGURE 9-7.
A leaf spring rear suspension consists of the springs and the shock absorbers that are fitted with them. (Courtesy of Moog Automotive)

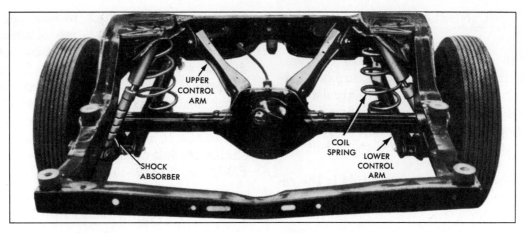

FIGURE 9–8.
The two lower control arms position the axle to control wheelbase. The two upper control arms prevent axle rotation, and, because they are mounted at an angle, they also prevent side movement. (Courtesy of Chevrolet)

able to control side motion as well as windup. The two lower arms are usually in a trailing and parallel position so as to allow bind-free travel. Some designs use only one upper arm to control axle windup plus a **Panhard rod**, also called a **track bar**, to control side motion. A Panhard rod connects to the axle, runs across the car, and connects to the frame of the car. A flexible rubber bushing is used at each end. Sometimes a

Watt's link is used in place of a Panhard rod. Panhard rods and Watt's links will be discussed more completely in Chapter 19. Some coil spring suspensions use two lower control arms and a Panhard rod with a torque arm. (Figs. 9–8 to 9–10)

The control arms are usually connected to the car's frame and the rear axle through rubber bushings. These bushings are very similar to those used on front ends. They offer maintenance-free service, a high degree of compliance, and the ability to dampen road vibrations.

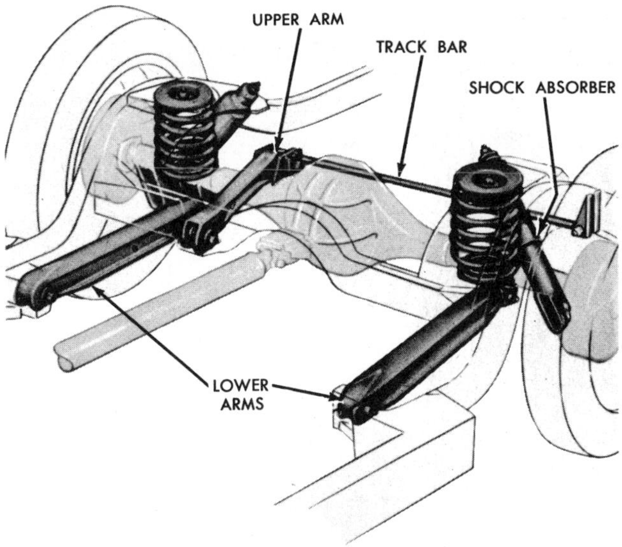

FIGURE 9–9.
This rear suspension has a single upper arm to prevent axle rotation. Side motion is controlled by the track bar, also called a Panhard rod; the track bar connects to the axle behind the upper arm. (Courtesy of Ford Motor Company)

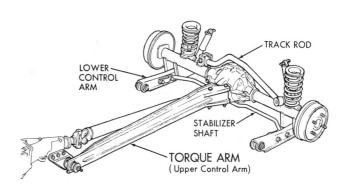

FIGURE 9–10.
This suspension uses a torque arm to prevent axle windup. The track rod prevents side motion, the stabilizer shaft reduces vehicle lean on corners, and the lower control arms control the wheelbase. (Courtesy of Chevrolet)

9.3   INDEPENDENT REAR SUSPENSION, IRS

Independent rear suspension (IRS) on rear-wheel drive has been used only on cars of a luxury or sporty nature. IRS is more expensive to build and has more wear points than a solid axle, but it provides better ride qualities and usually better camber and toe control of the rear tires.

One reason for an improved ride is the reduction of unsprung weight. The housing and final drive gears, which include the ring and pinion plus differential, are mounted on the frame. Their weight is carried by springs, not directly by the tires and wheels. Reducing the unsprung weight allows better spring and shock absorber control of tire and wheel movement. Power is transmitted from the differential to the tires and wheels by a pair of drive shafts or half shafts, much like those on an FWD car. (Fig. 9–11)

9.3.1   IRS, Semi-Trailing Arm Suspension

In this design, each rear tire and wheel is mounted on a single control arm, which is usually mounted to the frame through one or a pair of rubber bushings. A single pivot can be used if an additional control arm or strut rods are used. A pure trailing arm has the pivot mounted at a 90° angle to the center line of the car. The path that the tire follows as it pivots will be exactly parallel to the car's center line. If the car leans, there will be an equal change in camber angle. A 5° lean as the car corners will cause a 5° change (more positive on the outside tire) in camber. This would reduce the amount of traction and tire adhesion. (Figs. 9–12 & 9–13)

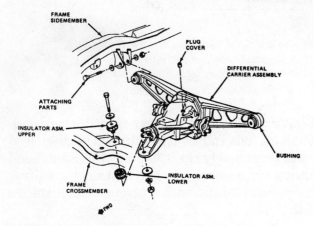

FIGURE 9–11.

The differential assembly portion of the rear axle of an independent rear suspension (IRS) car is mounted to the car's frame or body through rubber bushings. (Courtesy of Chevrolet)

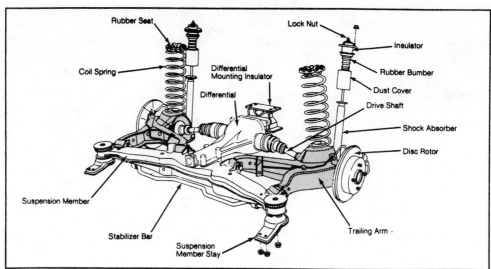

FIGURE 9–12.

Semi-trailing arms support the rear wheel bearings to control camber and toe angles as well as control the wheelbase. Note the angles of the front pivot bushings, which makes these semi-trailing arms. (Courtesy of Mitchell Information Services, Inc.)

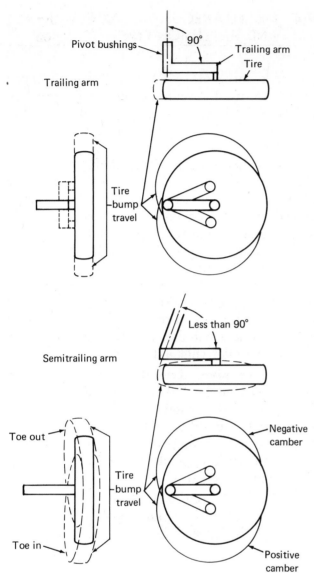

FIGURE 9-13.
The camber and toe angles of a tire will not change as a trailing arm suspension moves through its bump travel. The tire of a semi-trailing arm suspension will have more toe-out and a more negative camber during upward travel and more toe-in and a more positive camber during downward travel.

Semi-trailing arms have their pivots angled or skewed about 10 to 15° from those of a pure trailing arm. With this change, tire angle will change with suspension movement. Camber will remain closer to vertical as the vehicle leans on a corner and the control arm sweeps through this different arc of travel. This is an improvement, but, along with this camber change, there is also a toe change that can cause

a change in the rear tire and wheel thrust line. Rear wheel steering will occur when the thrust line changes. This suspension design is relatively simple, inexpensive, and lightweight.

9.3.2   IRS, Trailing Arm Suspension

Chevrolet Corvettes use a trailing arm IRS, but additional control links have been added to provide better tire angle control. Early IRS designs used a single trailing arm on each side. Newer models use a pair of trailing links. In the early design, the trailing arm controlled the toe angle and the rear end of the wheelbase. In the newer models, the trailing links control the wheelbase, and a separate link controls toe. In both styles, the half shaft and a lateral strut rod control camber. These three or five links produce very good tire and wheel control in the three-link suspension and excellent control in the five-link suspension. A few other cars use a similar type of rear suspension. (Fig. 9-14)

9.3.3   IRS, Strut Suspension

This suspension closely resembles the strut suspension used on front ends. RWD strut suspensions are sometimes called Chapman struts. Rear struts do not use steering knuckles, and the A-arms used often have a wide spread where they connect to the strut to provide good toe control. Acceleration or braking forces try to change rear wheel toe. Toe should be held very close to zero during acceleration and braking. One or two rubber bushings are used at this connection between the control arm and the strut. The upper mount or damper does not have a pivot bearing. It is mainly used to allow strut angle changes and to dampen road vibrations. (Fig. 9-15)

Strut suspensions are relatively inexpensive, fairly lightweight, and offer fairly good tire and wheel control. But, they tend to take up space in the car's storage area or in part of the space used by the rear seat.

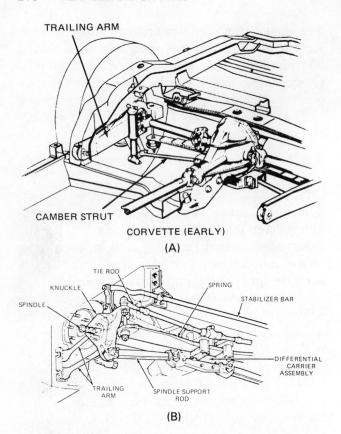

FIGURE 9-14.

In early Corvettes (A), a single trailing arm was used to control wheelbase, brake torque, and toe angle; the camber angle was controlled by the axle shaft and the lower strut. In newer Corvettes (B), a five-link suspension is used that consists of two trailing arms to control wheelbase and braking torque, a tie rod to control toe angle, and a spindle support rod and axle shaft to control camber angle. (Courtesy of Chevrolet)

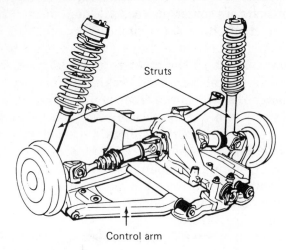

FIGURE 9-15.

A strut rear suspension, also called a MacPherson or Chapman strut. The lower arm and the strut support the axle bearings. (Courtesy of Moog Automotive)

9.4 MISCELLANEOUS RWD AXLE AND SUSPENSION TYPES

At least three other RWD suspension types have been used on cars. Their use is very limited at the present time. They include the following:

1. Swing axle. The axle shafts pivot from the universal joint at the inner end. A single or double trailing arm is usually used to control wheelbase and to absorb brake loads. This design has definite drawbacks in that camber and track change with vehicle height. The swing axle length is fairly short, so this camber and track change is fairly severe. Potentially dangerous suspension **jacking** can also occur. Hard cornering forces tend to tuck or fold the inside rear tire under the car, which tends to lift the car up and over that axle. (Fig. 9-16)

2. Low pivot axle. This is essentially a solid axle that bends in the middle. A single pivot is placed at the bottom of the axle housing near the middle, and a universal joint is built into the axle shaft at this pivot point. The pivot allows independent, vertical tire and wheel movement. (Fig. 9-17)

3. de Doin axle. This is a solid axle with a separate gear housing and axle shafts. This is not an IRS design; the tires and wheels are

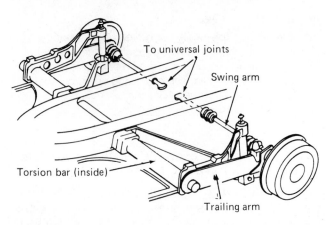

FIGURE 9-16.

A trailing arm with swing axle rear suspension. The axles that control camber connect to universal joints at the differential assembly. Torsion bars are used for the rear springs. (Courtesy of Moog Automotive)

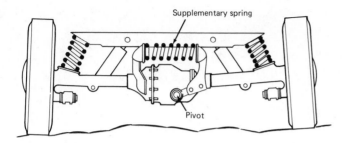

FIGURE 9-17.
A low pivot or split housing rear axle. A pivot and a universal joint allow this axle assembly to bend, giving IRS. (Courtesy of Mitchell Information Services, Inc.)

mounted on the ends of a solid axle. The rear axle gears are mounted to the frame in a manner similar to IRS cars. A de Doin axle provides the simplicity and tire angle control of a solid axle and reduces the unsprung weight for better tire and wheel control. (Fig. 9-18)

9.5 FWD REAR AXLES

FWD cars use rather simple rear axles. All that is necessary is to allow the tire and wheel to move up and down while staying in alignment. The suspension design can be a variation of any of the types previously discussed, but most

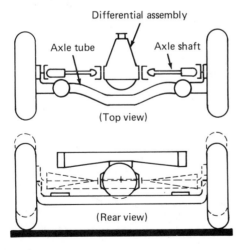

FIGURE 9-18.
A de Doin rear axle uses a solid axle to locate and support the wheel bearings while the differential assembly is mounted to the frame, similar to an IRS design. During suspension travel, the axle and tires will move up and down relative to the frame or body and the differential assembly.

manufacturers use variations of three styles: the solid axle, trailing arm, and strut.

With many FWD cars, the rear suspension carries a minor portion of the car's weight. The front tires carry most of the weight plus the driving and steering forces. The rear tires have a very limited traction because of the lower weight they carry. They also have a fairly low wear rate. They should stay in a nearly perfect alignment position like the front tires do.

Because of this, rear tires will not wear out at the same rate that front tires do. The rear tires on FWD cars often show an odd, scalloped wear pattern, probably because of this lighter loading. Regular tire rotation is usually necessary to obtain an even wear rate.

9.5.1 FWD, Rear Solid Axle Suspension

A solid axle beam, usually of stamped steel, connects the two tires and wheels. A pair of trailing arms are bolted rigidly or welded onto the axle beam and connect to the frame through rubber bushings. The axle and trailing arms are somewhat flexible to allow for slight twists when the car leans on corners. Side motion of the axle is controlled by a Panhard rod. Some designs use two lower links with rubber bushings at both ends and one or two upper links to control axle windup during braking. Both of these designs normally use coil springs. These axles can also use two leaf springs to locate the axle and provide the spring action. (Figs. 9-19 to 9-21)

Camber, toe, and track are controlled by the length and strength of the axle beam. Wheelbase is controlled by the trailing arms, links, or leaf springs. This is a simple, relatively lightweight, inexpensive design. Like other solid axles, it is not an independent suspension.

9.5.2 FWD, Rear Trailing Arm Suspension

Each rear tire and wheel is attached to a trailing arm. The trailing arm is attached to the frame through a pivot bushing at the front. Most FWD, rear trailing arm designs connect the two trailing arms to each other with a cross

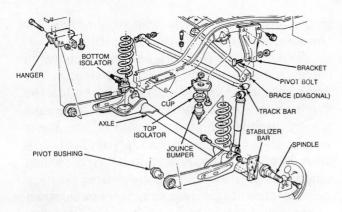

FIGURE 9-19.
A trailing arm, solid axle rear suspension from an FWD car. The axle controls camber and toe angles, the trailing arms control wheelbase, and the track bar controls side motion. (Courtesy of Chrysler Corporation)

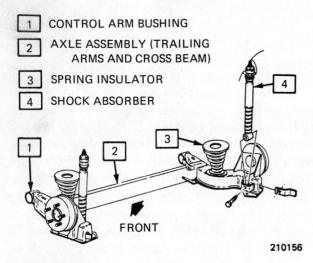

1	CONTROL ARM BUSHING
2	AXLE ASSEMBLY (TRAILING ARMS AND CROSS BEAM)
3	SPRING INSULATOR
4	SHOCK ABSORBER

FIGURE 9-20.
A trailing arm with twisting cross beam rear suspension. Twisting action of the cross beam allows semi-independent suspension action. (Courtesy of Pontiac Division, GMC)

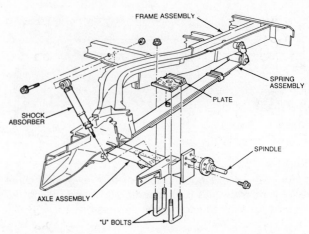

FIGURE 9-21.
A solid axle, leaf spring rear suspension. The leaf springs control wheelbase and side motion. (Courtesy of Chrysler Corporation)

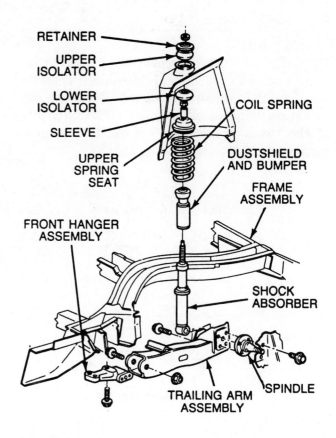

FIGURE 9-22.
A trailing arm rear suspension. A lateral beam (not shown) connects the two trailing arms. (Courtesy of Chrysler Corporation)

beam. This cross beam strengthens the trailing arms (in a crosswise direction) and tends to reduce body roll; it has to twist during body lean. Without the cross beam, the trailing arms would operate completely independently. (Fig. 9-22)

If we imagine a trailing arm suspension with the cross beam moved to the rear, in line with the wheel centers, we would have a solid axle with trailing arms as described in the previous section. Designers can place the cross beam at the front, rear, or anywhere in between, depending on the ride and handling characteristics they desire.

Another style of trailing arm rear suspension is used on one import car model. It is called a **double wishbone** suspension by the manufacturer. This trailing arm pivots on a rubber bushing at the front and is bolted solidly to the knuckle to which the hub and wheel are attached. This knuckle is located laterally by two lower arms and one upper arm. Vehicle loads are carried by the damper unit and spring, which is fitted between the knuckle and the body. The trailing arm establishes the wheelbase and absorbs braking torque. The upper and lower arms control camber, and the two lower arms control toe. The geometry of the knuckle and the control arms produce a controlled camber change and toe change during suspension travel. (Fig. 9–23)

9.5.3 FWD, Rear Strut Suspension

Nonpowered, rear struts normally use a strut, control arm, and strut rod to control rear tire

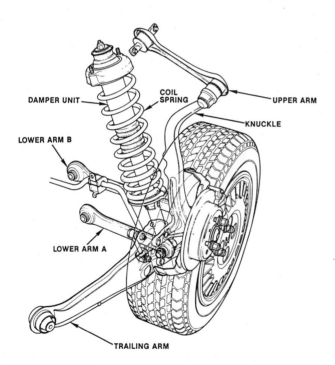

FIGURE 9–23.
The manufacturer calls this a double wishbone rear suspension. It combines a trailing arm with a long, curved knuckle. The top of the knuckle is located by an upper control arm. Two lower arms locate the rear hub laterally plus combine to provide almost zero toe change during bump travel. (Courtesy of American Honda)

and wheel movement. The spring can be mounted on the strut or between the car body and the lower control arm to save mounting room. In some designs, the lower control arm is replaced by a pair of lateral links. (Fig. 9–24)

With strut suspensions, camber is controlled by the strut, toe and track are controlled by the control arm, and wheelbase is controlled by the strut rod or lower control arm. Strut suspensions are completely independent, lightweight, fairly simple, and fairly inexpensive to manufacture.

9.5.4 FWD, Rear Short-long Arm Suspension

An S-L A rear suspension is used on some Ford Motor Company station wagons. This is a very compact system. Although it is more complex than the strut suspensions used on other car models, it provides a maximum of cargo space in the rear area. Variable rate coil springs are used to combine load-carrying capacity with a comfortable ride and small space. (Fig. 9–25)

9.6 REAR-WHEEL STEERING

Several developments are underway to improve a car's handling and maneuverability by steering the rear wheels. Two major approaches are being taken: (1) to actively steer the wheels by an input from the car's steering wheel and (2) to let road forces and suspension geometry realign the toe and camber angles. Both of these systems add more moving parts, complexity, weight, and cost to the rear suspension. Many people feel that these disadvantages outweigh any gain that rear-wheel steering offers the average passenger car.

The second style of rear-wheel steering, called **Dynamic Tracking Suspension System** by its manufacturer, is in full production. The dynamic tracking system has a toe control mechanism located in a unique two-part rear hub system and a camber control mechanism located with the semi-trailing suspension arm. In both of these two assemblies, a com-

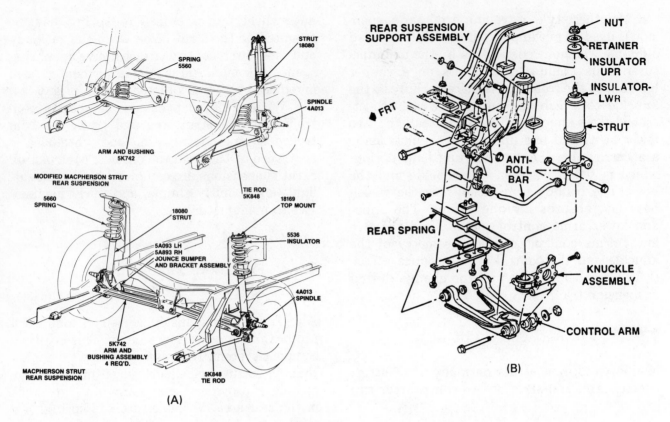

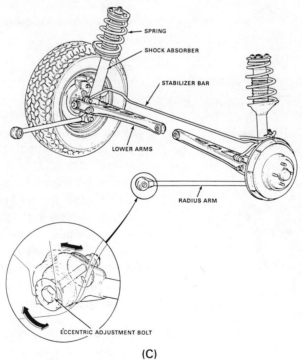

FIGURE 9-24.
Several styles of rear strut suspensions for FWD cars. Note the type and location of the spring and the different styles of lower control arms. (A is courtesy of Ford Motor Company; B is courtesy of General Motors Corporation; C is courtesy of American Honda)

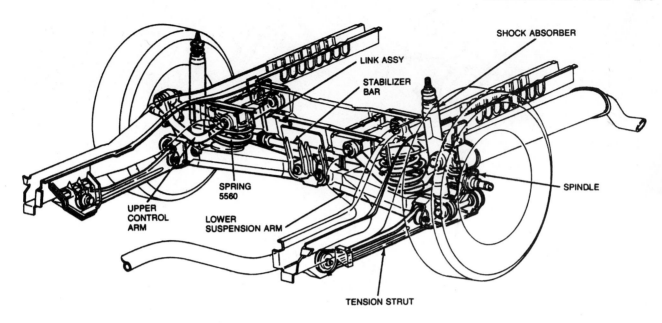

FIGURE 9-25.
An S-L A rear suspension for a FWD car. This particular suspension was designed for a station wagon and provides a flat cargo space. (Courtesy of Ford Motor Company)

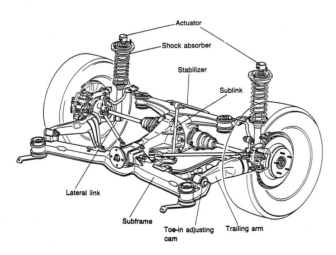

FIGURE 9-26.
A Dynamic Tracking Suspension System (DTSS). This rear suspension is designed to produce carefully planned toe and camber changes during braking, acceleration, and cornering maneuvers. Note the use of a double rear hub, the semi-trailing arm mounts, and the lateral links. (Courtesy of Mazda Motors of America)

bination of rubber and metal or ball and socket bushings is used. The rubber bushings are designed to allow a controlled amount of movement in specific directions. The metal bushings are used as pivot points. Each bushing is carefully positioned to take advantage of pressures generated by turning maneuvers, braking and acceleration forces, and vehicle lean to produce the following rear wheel reactions:

- braking or deceleration—toe-in;
- acceleration—toe-in;
- cornering—initial toe-out of the outer wheel that changes to toe-in at a lateral force of 0.4 to 0.5 g; and
- cornering or bounce—positive camber during jounce or compression travel and negative camber during rebound or extension travel. (Figs. 9-26 to 9-28)

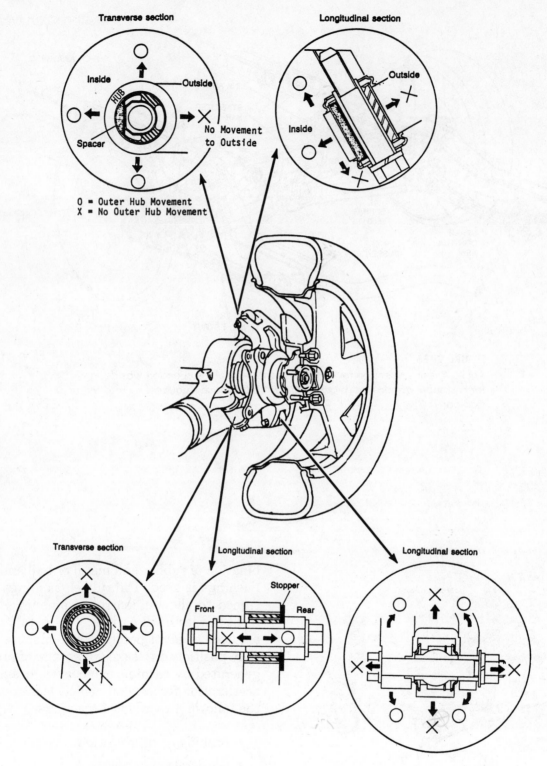

Transverse section

Inside · Outside

HUB

Spacer

No Movement to Outside

O = Outer Hub Movement
X = No Outer Hub Movement

Longitudinal section

Outside

Inside

Transverse section

Longitudinal section

Front · Stopper · Rear

Longitudinal section

FIGURE 9–27.
The three bushings between the inner and outer DTSS hubs allow specific motions so the inner hub can change position and allow a toe change during certain maneuvers. (Courtesy of Mazda Motors of America)

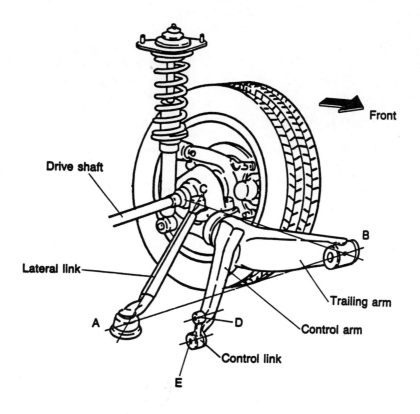

FIGURE 9–28.
The DTSS uses a rubber bushing at point B and ball joints at points A, C, D, and E along with the lateral link, or trailing arm, and control link to produce a specific camber change during vehicle lean. (Courtesy of Mazda Motors of America)

REVIEW QUESTIONS

1. Statement A: The alignment of the rear wheels is not important because they do not do any steering. Statement B: Straight-ahead in a car is a line that is parallel to the rear tires. Which statement is correct?

   a. A only   **c.** both A and B
   b. B only   **d.** neither A nor B

2. The rear axle of most RWD cars
   A. is a solid axle.
   B. provides driving action to the rear tires as well as support the car.

   a. A only   **c.** both A and B
   b. B only   **d.** neither A nor B

3. Torque tube drive shafts or torque arms are designed to prevent rear axle
   A. tramp.
   B. windup.

   a. A only   c. both A and B
   b. B only   d. neither A nor B

4. Statement A: A solid axle with leaf spring suspension uses the leaf springs to position the axle under the car. Statement B: Coil spring rear suspensions also use the spring to position the axle. Which statement is correct?

   a. A only   c. both A and B
   b. B only   d. neither A nor B

5. A Panhard rod
   A. controls the fore–aft position of the axle relative to the frame.
   B. uses rubber bushings at the mounting points.

   a. A only   c. both A and B
   b. B only   d. neither A nor B

6. Three or four control arms are used with coil sprung rear suspensions to
   A. position the axle lengthwise in the chassis. To ESTABLISH WHEEL BASE
   B. prevent rear axle yaw, windup, and tramp.

   a. A only   c. both A and B
   b. B only   d. neither A nor B

7. Rear axle windup in a coil sprung solid axle can be controlled by

   a. one or two upper links.
   b. a single lower link.
   c. a Panhard rod.
   d. any of these.

8. The major purpose of the two lower arms on a coil sprung solid rear axle is to

   a. prevent side sway.
   b. prevent axle windup.
   c. control the wheelbase.
   d. none of these.

9. Cars that use independent rear suspension generally offer
   A. better traction on rough roads.
   B. an improved ride quality.

   a. A only   c. both A and B
   b. B only   d. neither A nor B

10. Statement A: A semi-trailing arm rear suspension causes a slight camber change during suspension travel. Statement B: This suspension design usually uses a single inner pivot bushing on the lower control arm. Which statement is correct?

    a. A only   c. both A and B
    b. B only   d. neither A nor B

11. Which of the following is not a type of IRS?

    a. trailing arm   c. sliding pillar
    b. swing arm   d. strut

12. Statement A: A trailing arm rear suspension can use as many as five control links for each rear wheel. Statement B: The major purpose for the trailing arm is to absorb braking loads and wheel hop during acceleration. Which statement is correct?

    a. A only   c. both A and B
    b. B only   d. neither A nor B

13. Statement A: A de Dion rear axle can be used on FWD cars. Statement B: RWD swing axles can have problems with jacking of the rear suspension. Which statement is correct?

    a. A only   c. both A and B
    b. B only   d. neither A nor B

14. Statement A: FWD cars with a solid axle front suspension must use a solid axle rear suspension. Statement B: An independent front suspension with a solid axle rear suspension is never used on FWD cars. Which statement is correct?

    a. A only   **c.** both A and B
    b. B only   **d.** neither A nor B

15. The rear axle type used in many FWD cars is a

    a. solid axle with trailing arm.
    b. strut suspension.
    c. trailing arm.
    d. any of these.

Chapter 10

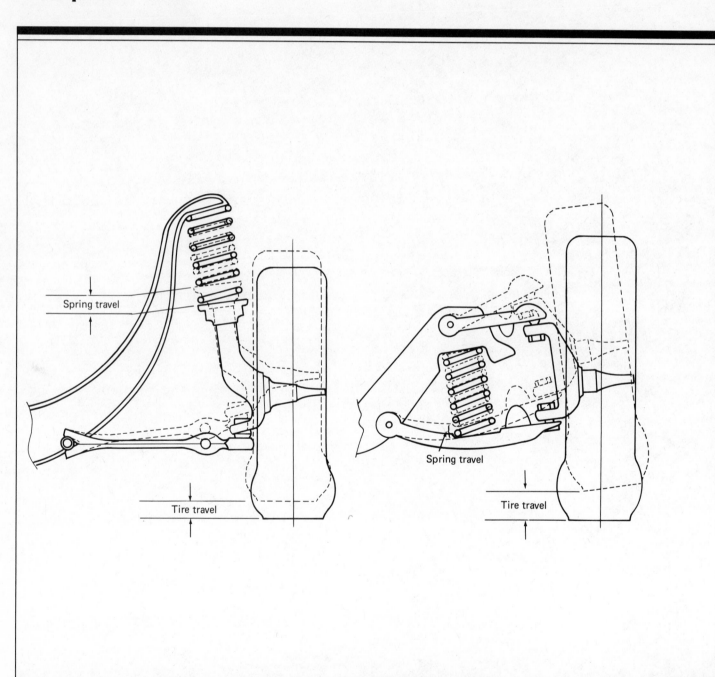

SPRINGS

After completing this chapter, you should:

- Be familiar with the terms commonly used with springs.

- Understand the purpose for springs as a suspension component.

- Understand the various types of spring systems and what suspension components are required with each spring type.

- Be familiar with electronically controlled air suspension systems, and be able to determine the correct diagnosis and repair procedure to correct faults with this system.

10.1 INTRODUCTION

Springs are the elastic portion of the suspension that allow vertical tire and wheel movement. They carry the weight of the car, but they also have the ability to absorb more weight. When additional load is placed on a spring because of more passengers or objects or because a tire meets a bump, the spring absorbs this load by compressing further. In reality, a spring stores energy by deflecting rather than absorbing weight. The springs are the most important suspension component that provides comfort in a car ride. Good spring action keeps the tires in contact with the road surface, adding greatly to the handling and stopping ability of the car. (Fig. 10–1)

When discussing springs, the term **bounce** refers to the vertical movement of the suspension, either up or down. Upward suspension travel that compresses the spring is called **jounce**. A lowering of the tire and wheel that extends the spring is called **rebound**.

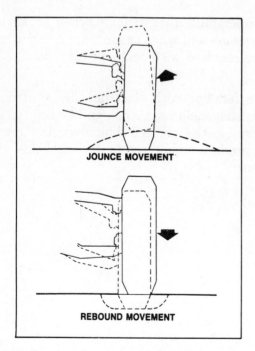

FIGURE 10–1.

A tire and suspension have two directions of bounce travel. The upward motion is called jounce, and the downward motion is called rebound. (Courtesy of Ford Motor Company)

To dramatize the need for springs, let's imagine a car traveling down the road at 55 MPH (88.5 KPH) or 4,480 feet per minute. This vehicle is traveling at 968 inches per second (24,587 cm per sec). If the road has a bump in it that is 1 inch (2.54 cm) high, the tire will only have a small part of a second to rise 1 inch and pass over this bump. Part of this rapid impact will be absorbed by the tire, deflecting the tread and part of the sidewall. Much of the bump will be absorbed by the suspension. That portion of the impact that is absorbed by the car body will lift the body slightly. Ideally, the bump is absorbed by compressing the springs. Now imagine the same bump, but on a vehicle without springs and, worse yet, with solid metal tires and wheels. When this vehicle meets the same bump at the same velocity, all of the bump will be absorbed by lifting the whole vehicle the height of the bump. The entire vehicle will have to rise 1 inch in a fraction of a second. The vertical motion will be very fast, and the inertia involved will place a large amount of pressure on the tire when it impacts the bump. Also, the inertia of the rapidly rising vehicle will tend to cause the vehicle to keep rising after it is past the bump. The tire will leave the road, and the vehicle will probably fly for a certain distance after leaving the bump.

10.2 SPRUNG AND UNSPRUNG WEIGHT

The **sprung weight** of a car is weight that is carried by the springs in the car's suspension. If the sprung weight is increased, the springs will compress further. **Unsprung weight** is not carried by the springs; it consists of those components that support or carry the springs. All the weight of the car is carried by the tires; the sprung components transfer their weight through the springs to the unsprung components and then through the tires to the ground. The tires, wheels, brake assemblies, and any other suspension part that travels directly with the tires and wheels are unsprung weight. The frame, body, engine, transmission, and any other parts that move directly with the frame and body are sprung weight. The parts that connect the sprung weight to the unsprung weight, the suspension components, are partially sprung and partially unsprung, depending on how much movement they have during suspension travel. (Fig. 10-2)

In reality, the tires are springs. Some off-road vehicles rely entirely on the elastic nature of their tires for suspension travel. The spring action of the tires normally softens most of the smaller bumps that the tires meet. Those of you who have had the opportunity to ride in a car before and after switching the tires to a different type have probably noticed this. A change in ride quality from bias tires to belted radials is usually quite noticeable. Most discussions of car springs and sprung and unsprung weight disregard the spring action of the tires. (Fig. 10-3)

The springs allow the frame and body to ride relatively undisturbed while the tires and suspension members follow the bumps and holes in the road surface. The lower weight of the unsprung parts allows them to be less affected by inertia. A high sprung weight to a low unsprung weight provides improved ride and

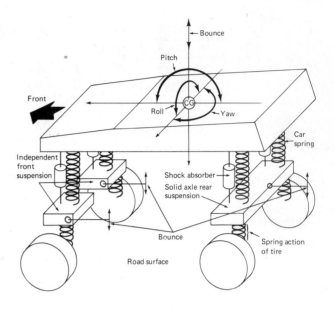

FIGURE 10-3.
This is an engineer's model of a car's suspension system showing the various types and directions of motions.

also improved tire traction. The springs have an easier job to do in absorbing bumps and keeping the tires in contact with the road.

10.3 SPRING RATE AND FREQUENCY

Rate is the system used to measure spring strength. It is the amount of weight that is required to compress the spring a certain distance, usually 1 inch. Traditionally, spring rates have been measured in pounds per inch or Newtons (N) per millimeter. A Newton is a force equal to 0.225 pounds. If it takes a force of 120 pounds (54 kg or 533.76 N) to compress a certain spring 1 inch (25.4 mm), that spring would have a rate of 120 pounds per inch. It would take another 120 pounds of force to compress the spring one more inch, and 120 pounds of force for each additional inch of compression until the spring bottomed out or used up all of the available travel. (Fig. 10-4)

Most springs have a constant rate of resistance from the point where compression begins—the free, uncompressed length—to the point where the spring bottoms out—the coils touch each other. Springs can be made with a

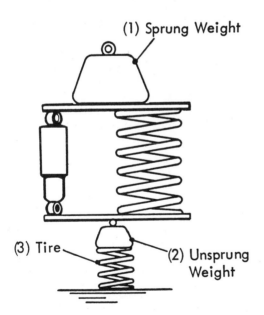

FIGURE 10-2.
The tires and axle support the springs. They are classified as unsprung weight even though the tires are really a type of pneumatic spring. The springs support the frame, body, and other car parts that are classified as sprung weight. (Courtesy of Monroe Auto Equipment)

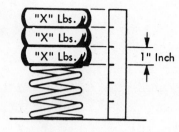

DEFLECTION RATE

FIGURE 10–4.

Spring rate, also called deflection rate, refers to the amount of weight that it takes to compress the spring one inch. In this example, weight "X" compresses the spring one inch; each additional amount of weight "X" will compress the spring an additional inch. (Courtesy of Monroe Auto Equipment)

variable rate; that is, the rate changes during part of the travel. Normally, variable rate springs will increase in rate as they are compressed. Variable rate is usually accomplished by constructing the spring from material having different widths or thicknesses or by winding the spring so the coils will progressively bottom out. A variable rate spring is more ideal in that it can be fairly soft at ride height and progressively stiffer as the suspension gets closer to the end of its travel. This would give us the softest ride possible on smooth roads yet not crash out the suspension stops on rough roads. (Fig. 10–5)

Spring **frequency** is closely related to spring rate. Frequency refers to the speed of the natural oscillations of a spring. If we hung an object from a spring and let it drop, stretching the spring, we would see the object bounce up and down. Each travel period from the uppermost point to the lowest point and back to the highest point is a called a cycle. The number of cycles completed in one second is called the **cycles per second** or **CPS**. This is the normal spring frequency measuring system. A soft spring (low spring rate) will have a fairly slow natural frequency; a strong spring will bounce at a much higher rate. (Figs. 10–6 & 10–7)

Springs of different rates can be used in a particular installation depending on the ride quality or load-carrying ability that is desired. Let's say a particular installation requires a spring that is 10 inches (25.4 cm) long when it

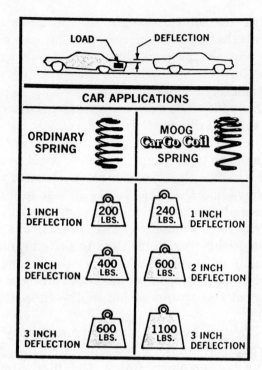

FIGURE 10–5.

A variable rate spring will accept larger and larger loads as it compresses. In the example shown, the rate of the variable spring is 240 pounds per inch for the first inch and 500 pounds per inch for the third inch of compression. (Courtesy of Moog Automotive)

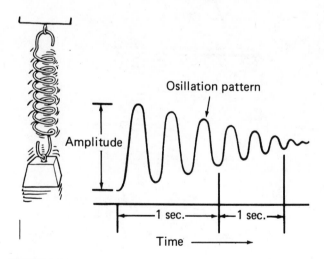

FIGURE 10–6.

A spring will try to bounce at its natural frequency when a load is released. This frequency is the number of bounce oscillations per second. As the energy in the spring is used up, the amplitude or size of the bounce will decrease; the frequency does not change.

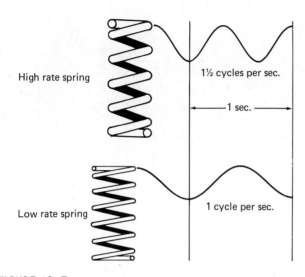

FIGURE 10-7.
A strong, high rate spring will have a higher natural frequency than a soft, low rate spring.

is at normal ride height and has a static, installed load on it of 500 pounds (227.3 kg). A spring with a free or unloaded length of 15 inches (38.1 cm) and a rate of 100 pounds per inch (57 N per mm) would compress to the right length when installed. An 11 inch (free length) spring with a rate of 500 pounds per inch (285 N per mm) would also compress to 10 inches long when it is installed. In comparing the two springs, the first one would give a softer ride at a slower frequency but would allow more leaning on turns and bottom out easier. The

second, stronger spring could carry more load and would give a firmer and harsher ride and would bounce at a higher frequency. (Fig. 10-8)

10.4 WHEEL RATE AND FREQUENCY

Wheel rate is a measurement of the spring's strength at the tire and wheel instead of directly at the spring. In many cases, the spring does not operate in direct relationship with the tire and wheel. The spring is compressed through the leverage of a control arm. An S-L A suspension might have the spring mounted halfway between the tire and the inner pivots on the control arm. This would produce about a 2:1 ratio between the tire and wheel motion and the spring motion. The tire would move vertically two inches while the spring moved one inch. This leverage would also require a spring with twice the strength of the load it carried. A 400 pound (182 kg) vehicle load would make the spring support 800 pounds (364 kg), but this spring needs to be able to travel only half as far as the tire and wheel. (Fig. 10-9)

The springs in most strut suspensions operate much closer to a direct relationship with the tire and wheel. In this case, a 400 pound spring would carry about 400 pounds of vehicle load, and one inch of tire and wheel movement

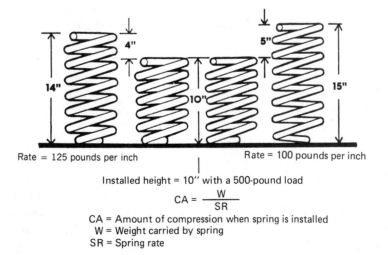

Rate = 125 pounds per inch   Rate = 100 pounds per inch

Installed height = 10″ with a 500-pound load

$$CA = \frac{W}{SR}$$

CA = Amount of compression when spring is installed
W = Weight carried by spring
SR = Spring rate

FIGURE 10-8.
At least three different springs can be fitted in a particular car; they would produce three different ride qualities and bounce frequencies.

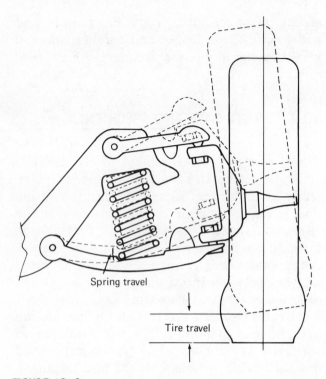

FIGURE 10-9.
Because of the leverage of the lower control, the wheel rate of the spring will be different from the spring rate. Note that the tire travel is about double that of the spring.

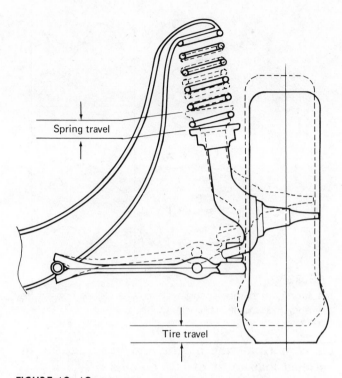

FIGURE 10-10.
The wheel rate on a strut suspension will be almost the same as the spring rate because the amount of travel of the spring is almost the same as the travel of the tire.

will cause almost one inch of spring travel. (Fig. 10-10)

It should be noted that the track width also has an effect on the tire-to-spring-lever ratio. Installing wheels with an increased negative offset and a reduced back spacing that moves the tire outboard and increases the track width will increase the tire travel relative to the spring. This change tends to reduce the wheel rate of the springs, and usually causes the car to sit lower.

Suspension or wheel frequency varies depending on the spring's frequency and the amount of sprung weight carried by that spring. Suspension frequency is the natural speed at which the body and frame would bounce if the shock absorbers were disconnected. It is also called the natural frequency. Increasing the sprung weight will lower the natural frequency of the suspension. A typical suspension is designed to have a natural frequency of about 1 CPS. The natural frequency of a racing car suspension is about 2 CPS. Most drivers would find that higher suspension fre-

quencies produce an annoying, disagreeable ride quality. A car with no suspension will bounce at about 10 CPS on the tire alone; 10 CPS is the natural frequency of most passenger car tires.

Many manufacturers design their cars so the front and rear suspensions have slightly different natural frequencies to reduce or eliminate **pitching** motions. This is called **flat ride tuning.** Pitch is an unpleasant oscillation of the car characterized by a lowering at the front while rising at the rear and then rising at the front while lowering at the rear. Pitch often results when a car goes over a bump. The front tires strike the bump first, and then, a short time later, depending on the car's speed and wheelbase, the rear tires strike the same bump. Pitch begins when the rear suspension is in the jounce portion of its travel while the front suspension is in the rebound portion. It can be very disturbing when regularly spaced bumps, such as highway expansion joints, keep this pattern going. If the vehicle has two different natural frequencies, the bouncing motions will get out

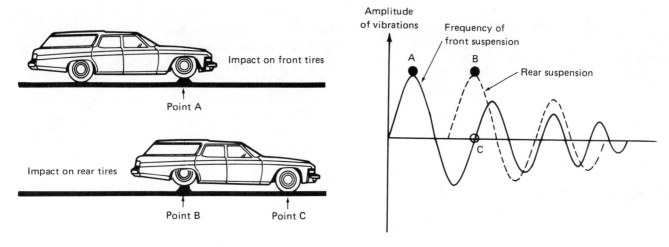

FIGURE 10-11.
Flat ride tuning reduces vehicle pitch by using a different suspension frequency at the front and rear. The bounces of the front and rear suspensions synchronize after a few oscillations.

of synchronization with each other (level out) after a few cycles, and pitch will soon become a simple bouncing up and down. Flat ride tuning adjusts the natural frequency of the front suspension to about 80 percent of that of the rear. (Fig. 10-11)

10.5 SPRING MATERIALS

Traditionally, springs have been made from steel alloy. This material, which is very strong to begin with, is heat treated to be more resilient. Springs flex through thousands of compression and expansion cycles without breaking and still retain their original shape. The most common spring failure is **sag**, a gradual reshaping of the spring that lowers the car.

Tempering a spring is a spring maker's art. It requires heating the metal to certain temperatures and then cooling the metal at a carefully controlled rate. Cooling the metal too slowly might cause annealing. An annealed spring is soft and will bend or sag very easily. Cooling the spring too quickly might cause brittleness or hardening. An overly hard spring will snap because it will be too hard to bend.

Composite leaf springs are being used by a few manufacturers. These springs are made from **fiberglass reinforced plastic (FRP)** or **graphite reinforced plastic (GRP)**. Advantages of composite springs are that they are lighter weight (about one-quarter to one-third the weight of steel, mostly unsprung), free from corrosion, and can be easily made with a variable spring rate that produces an improved ride. Composite springs are usually constructed to give a variable rate by varying the width or thickness of the spring's cross section. They are usually more expensive than steel springs. A composite leaf spring is often mounted in a transverse position with one end used for the right suspension and the other end for the left suspension. It is found on the rear suspension of some General Motors luxury cars and the front and rear of Corvettes. (Fig. 10-12)

Cars with electronically or computer-controlled suspensions use air trapped in a flexible chamber for a spring. The chamber is made partially from metal with a reinforced rubber diaphragm or reinforced rubber cylinder to form the flexible section. Air is a very good variable rate spring; suspension travel compresses the air which, in turn, increases the spring rate. The spring rate and length of air springs is easily adjustable. All that is needed is more air pressure. In most electronic systems, one or more height sensors note when the car is too low; this will cause the computer to either start up an on-

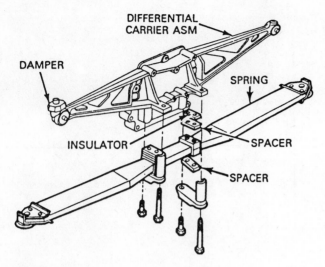

FIGURE 10-12.
This rear spring from a Corvette is made from a fiberglass reinforced plastic. Note the varied cross section that gives this spring a variable rate. (Courtesy of Chevrolet)

board air compresser or open an air valve to increase the air pressure in the spring chambers. The computer will close the control valves to

trap the correct amount of pressure in the chambers when the sensors note that the car is at the correct height. Air suspensions will be discussed more completely later in this chapter. (Figs. 10-13 & 10-14)

Rubber is one of the better elastic materials because of its ability to store a lot of energy relative to its weight. At this time, pure rubber springs are only used for suspension travel stops. As the suspension members approach the end of their upward and downward travel, they run into a **strike out bumper** or **bump stop** instead of crashing into a metal frame or body member. The bump stops are of a conical or partially hollow shape so as to cause a **rising rate.** The further the bump stop is compressed, the harder it becomes to compress. Occasionally, one or both of the bump stops are built into the shock absorber. In the future, rubber will possibly be used for the major suspension springs. Rubber is lightweight and has good variable rate characteristics. Drawbacks

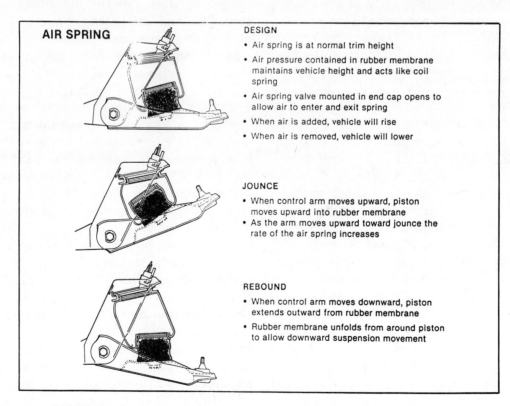

FIGURE 10-13.
An air spring is usually a flexible rubber container or membrane containing pressurized air; the trapped air compresses and expands during suspension movement. (Courtesy of Ford Motor Company)

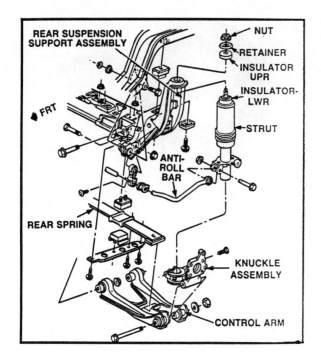

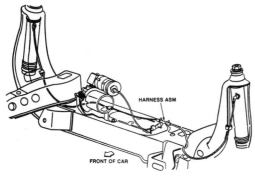

FIGURE 10–14.
This suspension combines a transverse, fiberglass leaf spring with an air spring in the strut; the air spring allows for automatic level control. (Courtesy of General Motors Corporation)

FIGURE 10–15.
The jounce travel bump stop (lower arrow) and the rebound travel bump stop (hidden by frame at upper arrow) are common examples of rubber springs; they serve to keep the control arms from crashing into the frame during extremes of suspension travel.

of rubber springs are that rubber is affected by various chemicals, and it will change spring rate with temperature changes. (Fig. 10–15)

10.6 LEAF SPRINGS

Most leaf springs serve a dual function as an axle-control member as well as a load-carrying member. They are normally of a **semi-elliptical** shape (i.e., shaped like a portion of an ellipse), have a main leaf with an eye at each end, and are assembled from a group of different length leaves. Some cars use a single spring for both ends of the axle mounted transverse or across the car. Springs mounted in this manner are sometimes called buggy springs. A few cars have used quarter-elliptic springs which look like a spring that has been cut in half. The thick end of the spring is bolted solidly to the car's frame, and the free end is attached to the axle through a bushing. (Figs. 10–16 to 10–18)

As the spring absorbs loads and deflects, the leaves change from a curved shape to a flat shape and then recurve in the other direction. The leaves also appear to change length as they flatten out and bend. A shackle is used with most leaf springs to allow for this change; there is also a certain amount of sliding of the ends of one leaf over the next leaf. This sliding, often called **interleaf friction**, could cause binding in the spring action. Pads of low friction material are usually put between the leaves to reduce this friction. Interleaf friction is difficult to control on older cars and often causes a harsh ride. Instead of using a shackle, some overload

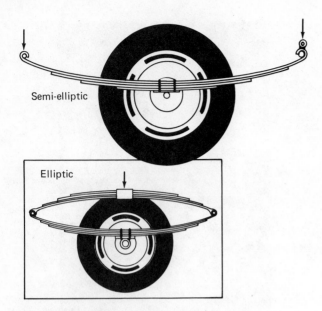

Semi-elliptic

Elliptic

FIGURE 10-16.
Most common, curved leaf springs are of a semi-elliptical shape; some very early cars used an elliptic spring. (It was usually a pair of semi-elliptical springs.) (inset)

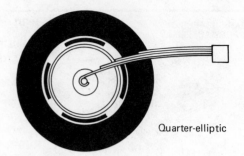

Quarter-elliptic

FIGURE 10-18.
A quarter-elliptic spring is half of a semi-elliptic spring; the thick section is secured solidly to the frame while the suspension attaches to the spring eye.

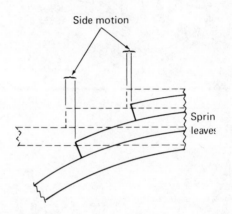

Side motion

Spring leaves

FIGURE 10-19.
As a leaf spring flattens, the end of one leaf must slide over the leaf next to it; this generates interleaf friction that adds to the spring rate.

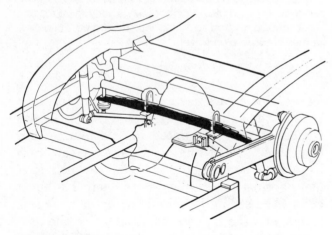

FIGURE 10-17.
A transverse leaf spring is positioned across the car. The center is securely attached to the frame or body, and each end provides spring travel for each wheel on the axle. (Courtesy of Dana Corporation)

spring installations on trucks use a mount that sits on top of the spring and allows the spring to slide. **Mono** or **single leaf springs** are used in some installations. When only one leaf is used, it is often tapered to produce a variable rate. A mono leaf spring has no interleaf friction. (Fig. 10-19)

Leaf springs are made stronger by making the leaves thicker or wider or both. They are made more flexible by increasing the length of the leaves. The formula for calculating the spring rate for a leaf spring is given in Figure 10-20. Adding leaves will increase the spring rate, and if the leaves are of different lengths, the rate will be variable. The ends of the longest leaf will bend first, then the ends of the next longest leaf will bend with the main leaf, and so on. These leaves are usually bolted together using a **center bolt**. The center bolt is also used to align the axle to the spring. Metal clips are normally bent around the spring to hold the leaves in alignment with each other. (Fig. 10-21)

Rubber bushings are normally used in the spring eyes to allow the necessary spring motion. They also provide the compliance needed for slight side and twisting motions that occur

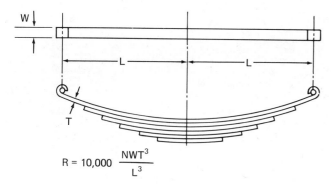

$$R = 10,000 \ \frac{NWT^3}{L^3}$$

R = Spring rate at point of force (pounds per inch)
N = Number of leaves
W = Width of leaf (inches)
T = Thickness of leaf (inches)
L = Length from eye to center (inches)

FIGURE 10-20.
This is the formula used to compute the spring rate of a leaf spring.

Suspension designers prefer to use a spring with the flattest curve possible and a shackle that is just long enough to allow the necessary spring travel. Excessive spring curve or

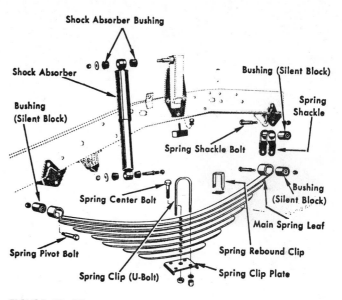

FIGURE 10-22.
A disassembled leaf spring. Note the center bolt and spring rebound clip that hold the leaves together and the spring eye bushings. (Courtesy of Mitchell Information Services Inc.)

during suspension travel and vehicle lean. The rubber bushings also dampen road vibrations from traveling into the frame and body. Some manufacturers place rubber pads between the spring and axle for this same purpose. Some vehicles use metal spring bushings, which require periodic lubrication. (Figs. 10-22 & 10-23)

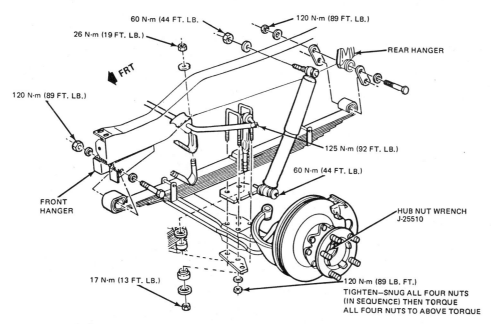

FIGURE 10-21.
A leaf spring suspension from a light truck that uses a solid front axle.
(Courtesy of Chevrolet)

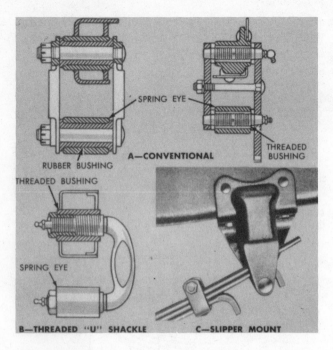

FIGURE 10-23.
Several types of bushings are used to connect the ends of a leaf spring to the frame bracket and shackle. Threaded bushings require periodic lubrication; the slipper mount is a sliding connection. (Courtesy of Ford Motor Company)

shackle length allow greater amounts of axle side and yaw motions, which produce roll oversteer and raise the roll center. These will be discussed in Chapter 19.

10.7 COIL SPRINGS

Coil springs are made from a round steel alloy wire that is wound into the common spiral or helical shape. Since there is no interleaf friction as in leaf springs, coil springs offer a good ride quality for an extended period of time. The strength of the spring is basically determined by the diameter and length of the wire. An increase in wire diameter will produce a stronger spring while an increase in length will make it more flexible. The formula for calculating the rate of a coil spring is given in Figure 10-24.

Coil springs differ in terms of wire diameter, coil diameter, coil free length, number of coils, coil winding direction, and the shape of the coil ends. The different coil end shapes are tapered, if the end of the wire is flattened; tan-

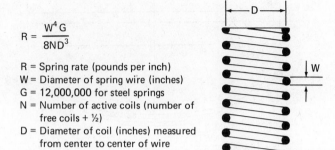

$$R = \frac{W^4 G}{8ND^3}$$

R = Spring rate (pounds per inch)
W = Diameter of spring wire (inches)
G = 12,000,000 for steel springs
N = Number of active coils (number of free coils + ½)
D = Diameter of coil (inches) measured from center to center of wire

FIGURE 10-24.
This is the formula used to compute the rate of a coil spring.

gential, if the coil is a continuous spiral; square, if the last coil is bent to be square with the coil; and pigtail, if the last coil is wound to a smaller diameter. These end treatments are shown in Figure 10-25. Coil springs are often wound with two different end shapes. When this occurs, the spring will have a top and a bottom, and each end will fit into a specific pocket. Coil springs can be wound from a tapered wire to get a variable rate, but this is more expensive. Sometimes coil springs are wound with a varying pitch between the coils. As the spring compresses some of the coils will bottom out and the spring rate will increase. (Figs. 10-26 & 10-27)

Coil springs are normally trouble-free; the common type of failure is sag. Sag can shorten the spring enough to allow the suspension to drop too far below design height for correct suspension geometry. When this occurs, tire

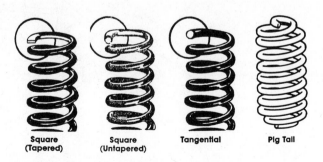

Square (Tapered)   Square (Untapered)   Tangential   Pig Tail

FIGURE 10-25.
Coil springs are made with different styles of ends depending on how the end of the wire is cut or bent. A square spring end is bent so the spring will stand up straight. A tangential end is simply a cut coil; it will stand off center. (Courtesy of Moog Automotive)

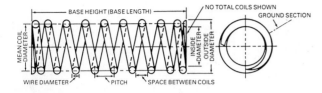

FIGURE 10–26.
These are the commonly used coil spring dimensions. (Courtesy of Chevrolet)

wear or handling difficulties will be the result. Sag will also lower the car to the point where the suspension members are constantly hitting the bump stops.

Coil spring breakage is fairly rare. If one breaks, it is usually the result of a **stress raiser**. A stress raiser is a flaw in the metal surface, sometimes just a scratch, nick, or pit caused by rust or corrosion. Many newer springs are painted or coated with epoxy to retard rust ac-

tion and reduce the possibility of a stress raiser occurring. The stress raiser weakens the metal so that future bending or twisting motions tend to concentrate and bend or twist this portion of the spring more than the other areas. If it is serious enough, the increased working will cause work hardening, also called metal fatigue, which hardens the metal to the point that it will no longer bend, and the spring will break. (Figs. 10–28 & 10–29)

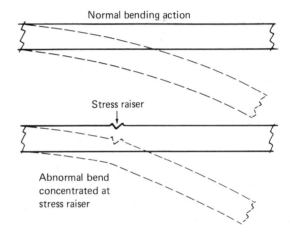

FIGURE 10–28.
A stress raiser begins when corrosion or a nick removes metal from one portion of a spring. This portion, which is now weaker, will bend easier than the rest of the spring, and the increased and concentrated bending action will cause the spring to fail.

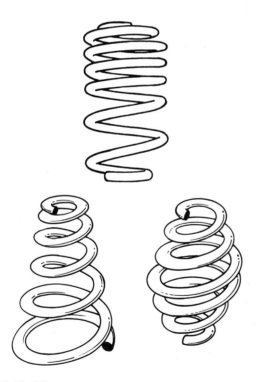

FIGURE 10–27.
Springs are normally wound with an even spacing between the coils; a variable rate can be produced by winding the coils tightly at one end and wider spaced at the other (A). Variable rate springs can also be produced by winding the spring from a tapered wire (not shown) or by winding the spring in a semi-conical shape (B). (A is courtesy of Moog Automotive)

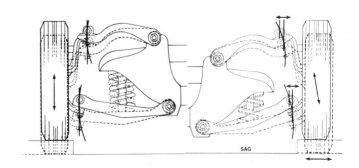

FIGURE 10–29.
If a coil spring sags an excessive amount, the front suspension geometry will also change. Wheel alignment, tire wear, and vehicle handling will be affected. (Courtesy of Dana Corporation)

The correct repair for a sagged spring is to replace it. This can be expensive, and in cases of older, high-mileage cars, which are usually the ones with sagged springs, the cost can be very expensive relative to the value of the car. One less expensive cure for sagged springs is to place a shim at the end of the spring. The shims are made from rubber or soft metal, usually aluminum. The drawbacks with using a shim are that the labor cost of installing a shim is the same as replacing the spring and that the shortening of the spring by the shim can cause coil clash. Coil clash occurs when the coils bottom and touch each other. When this occurs, the coil becomes solid, with a drastic increase in spring rate. Normally, coil clash does not occur because the suspension will bottom out on the stops before the coil bottoms. Another cure for sagged springs is to install spacers between the coils. Solid metal spacers are not recommended because they eliminate spring action in the coil where the spacer is. The spring rate will increase because of this, and a severe bending stress is placed on the active coils right next to the spacer. Rubber spacers are less severe and can be used without serious effects. Some styles of rubber spacers have a problem of staying in place and might fall out. Spacers can usually be installed in a few minutes so they offer cost savings in terms of labor as well as parts. (Figs. 10–30 to 10–32)

FIGURE 10–30.
A shim can be placed between the control arm and a coil spring to adjust the car to the correct trim height. (Courtesy of Arn-Wood)

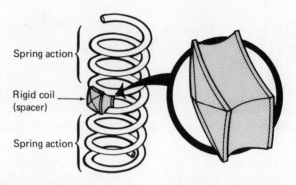

FIGURE 10–31.
Several styles of solid spring spacers and boosters, designed to be inserted between the coils, are available. These spacers are not recommended because they place a severe bending load on the spring next to the spacer.

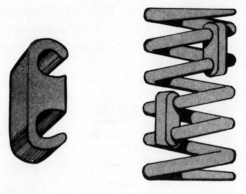

FIGURE 10–32.
Rubber spacers can be inserted in a coil spring to restore a car to the correct trim height.

10.8 TORSION BARS

A torsion bar is a bar of steel that is stationary at one end and forced to twist at the other. It can be made from a single piece of metal or a group of laminations. Some cars use an L-shaped torsion bar. The rate or strength of a torsion bar can be calculated using the formula given in Figure 10–33. If you have ever placed a large amount of torque on a long extension bar while using a socket and handle to either tighten or loosen a bolt, you possibly noticed a twisting of the extension bar. While you are doing this, the extension bar acts like a torsion bar. Most torsion bar suspensions anchor one end of the torsion bar solidly to the frame and

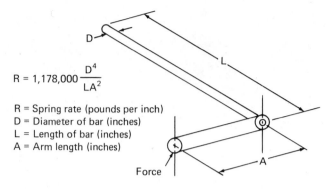

$$R = 1{,}178{,}000 \frac{D^4}{LA^2}$$

R = Spring rate (pounds per inch)
D = Diameter of bar (inches)
L = Length of bar (inches)
A = Arm length (inches)

Force

FIGURE 10–33.
This is the formula used to compute the rate of a torsion bar.

connect the other end to a suspension control arm or trailing arm. The control arm becomes the lever arm for the torsion bar. L-shaped torsion bars provide their own lever arm. As the suspension allows vertical tire and wheel motion, the torsion bar will twist tighter or looser. If you think about it, you will realize that the metal wire in a coil spring also twists, much like a torsion bar, as the spring compresses and extends. (Fig. 10–34)

An adjustable connection is usually placed at one of the ends of the torsion bar to provide an adjustment to compensate for bar sag. If the bar sags, the adjustment allows twisting the bar a little tighter, restoring the vehicle to correct ride height. (Fig. 10–35)

The long, skinny shape of a torsion bar gives the suspension engineer an alternative to the shorter, fatter coil spring in some tight mounting installations. A torsion bar can easily be used with S-L A, trailing arm, or strut-type FWD and RWD suspensions. Another possible advantage of a torsion bar is that a large portion of the spring load is transmitted to the fixed end of the bar. This, depending on how the torsion bar is mounted, can move some of front suspension load from the weaker suspension area to the stronger body cowl area. (Fig. 10–36)

Torsion bars are trouble-free. With an occasional adjustment, they will normally last the life of the car. They can break, but this is rare. Like coil springs, breakage is usually the result of a stress raiser. All metal springs should be handled somewhat gently to avoid scratches in the protective painted or plastic covering or in the metal that might become stress raisers.

If a torsion bar needs to be replaced, it should be noted that the bars on most cars are not interchangeable between the right and left sides. Many bars are prestressed during manufacture. The bars are usually marked or coded in some way to tell them apart, and a left-side bar should only be used on the left side of the car and vice versa.

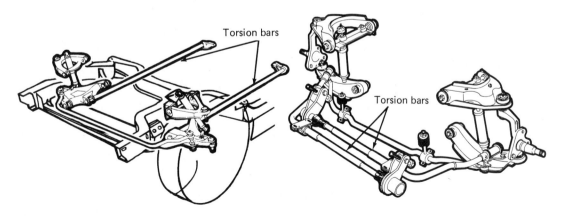

Torsion bars

Torsion bars

FIGURE 10–34.
Torsion bars are usually straight, round bars of spring steel, but they are also bent, L-shaped, so the end of the bar is also the lever. Torsion bars can also be flat or laminated or both. (Courtesy of Moog Automotive)

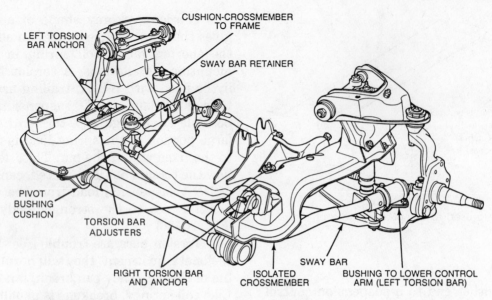

FIGURE 10–35.
Torsion bars are normally designed with an adjustment so that spring sag can
be compensated for. (Courtesy of Chrysler Corporation)

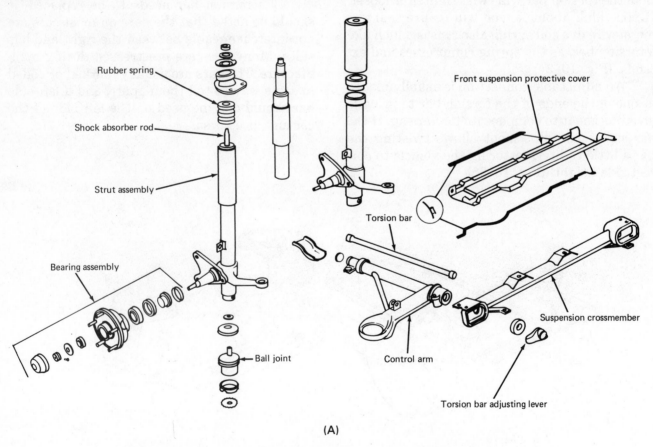

(A)

FIGURE 10–36. *(Continued on next page)*
Torsion bars are sometimes used with strut suspensions (A) and trailing arms suspensions (B). (Courtesy of Mitchell Information Services Inc.)

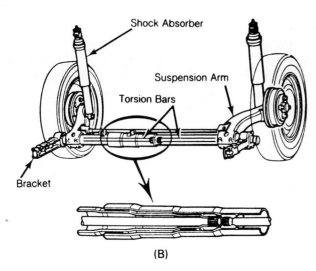

(B)

FIGURE 10-36 (Continued).

10.9 ELECTRONICALLY CONTROLLED AIR SUSPENSION

Air suspensions are complex and expensive, but they have the ability to stay at a constant height, regardless of load. In some cases, air suspensions provide a variable height that is controlled by the driver or a computer-controlled variable rate and height for improved handling or vehicle aerodynamics. Air suspensions use the compressibility of air for a spring. The spring rate is determined by the size of the air chamber and the amount of air pressure. If the car is too low, the air pressure in the spring chambers is increased; if the car is too high, some of the air pressure is released. The no-load spring rates are usually lower than if steel springs are used because air suspensions have the ability to compensate for heavier loads. Metal springs of the same lower rate would bottom out when more load is added. (Fig. 10-37)

Depending on the manufacturer, there are several versions of electronically controlled air suspensions. Nearly all of the systems at the present time are computer-controlled. The computer monitors the height of the car through three or four **height sensors**, one for each wheel or one for each front wheel and one for the rear axle. Some systems use brake or door sensors so automatic height adjustments are cancelled

during stopping and while passengers are getting in or out of the car. Some systems use speed sensors so that the car can be lowered at higher speeds for aerodymanic improvement. In some systems, the spring rate is increased at the same time. Some systems also use a "g" sensor and a steering shaft velocity sensor so the spring rate can be increased during fast cornering, braking, and acceleration maneuvers. Some systems use dual chamber air springs or coil or leaf springs plus air springs. Some systems use single air springs only. The flexible air chamber for the air spring can be built into a strut or mounted in place of the standard coil spring. A rigid air chamber can be connected into the flexible chamber to provide a lower spring rate. All systems use an on-board **air compressor** to maintain a supply of compressed air. Some systems will pressurize and bleed down the air chamber using a valve mounted

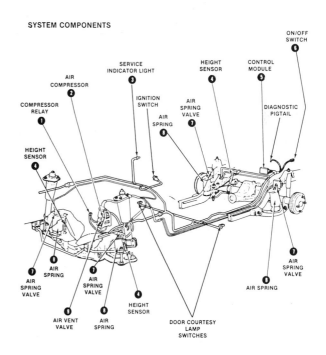

FIGURE 10-37.

A computer-controlled, air suspension system. Air springs (8) support the car's load and the air compressor (2) provides the air pressure. When the height sensors (4) sense a change in the car's height, the control module (5) will cause the air spring valves (7) to change the pressure in the springs. (Courtesy of Ford Motor Company)

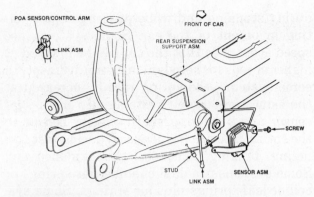

FIGURE 10-38.
Height sensors are connected between the body and a suspension member; they send a signal to the control module if a change in car height occurs. (Courtesy of General Motors Corporation)

between the compressor and the air chamber. Other systems will pressurize the air chamber using the compressor and bleed down the

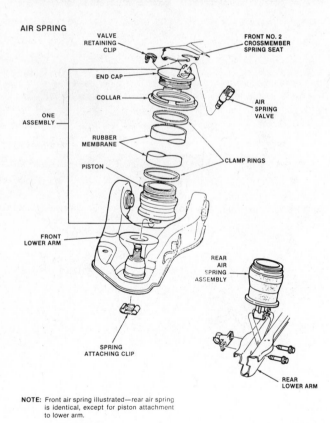

NOTE: Front air spring illustrated—rear air spring is identical, except for piston attachment to lower arm.

FIGURE 10-39.
This air spring consists of a reinforced, tubular rubber membrane that has an end cap clamped onto each end; the end caps fit into the body and the suspension member as well as provide a connection to the air line or valve. (Courtesy of Ford Motor Company)

chamber using a valve mounted on the air chamber. Some systems use a two-wheel-only system, mounted at the rear; the air chambers are built into the rear shock absorbers. Some two-wheel systems are a dealer-installed option. (Figs. 10-38 & 10-39)

The actual air chamber for the spring can take several shapes, the most common of which are somewhat tubular- or diaphragm-shaped. Reinforced rubber is used for the flexible membrane. Various ply materials, much like the sidewall of a tire, provide the flexible reinforcement. Each air chamber is connected to the air compressor or control valve through a plastic tube. (Fig. 10-40)

Most present-day systems use on-board air compressors that are driven by a 12-volt electric motor. A **desiccant**, to remove moisture, is usually placed in the compressed air flow. Water vapor can enter the system with the incoming air. If this water vapor condenses in the system, it can cause freeze-up in cold weather or corrosion of the metal parts or both.

The height sensors electronically measure the distance between the control arm or axle

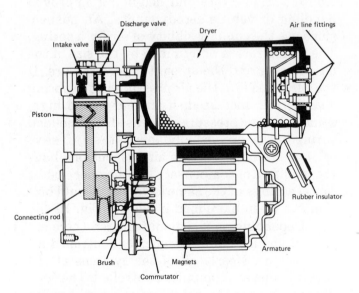

FIGURE 10-40.
A 12-volt motor drives this compressor to provide the air volume and pressure needed for an air suspension system. The dryer contains a desiccant to remove moisture from the compressed air. (Courtesy of Ford Motor Company)

and the frame. They signal the computer if this distance is longer or shorter than the design height. The computer normally ignores changes that occur over a short period of time; these are usually bumps. A height change that lasts longer than 5 or 6 seconds is responded to by changing the pressure in the air springs. Some systems will wait as long as 45 seconds during certain sequences of events before responding with a height adjustment. The computer uses the height sensor information to adjust the pressure in the air chambers to keep the car at the correct ride height.

Electronically controlled air suspension systems have many possibilities for failure. They are complex, and their operation varies from one manufacturer to another. Most systems have diagnosis features built into them so that by using the right equipment, usually very simple and inexpensive, a code, giving the nature of the problem, can be read out of the computer. This code gives the technician the cause of the problem. This text will not try to describe the process of diagnosing and curing electronic air suspension problems. This information is available from the vehicle manufacturer or the mechanic's service manuals. (Fig. 10–41)

Caution should be used when raising some cars with electronic air suspension. Some systems stay on all the time; they are not deactivated when the ignition is turned off. A separate switch is usually provided for that purpose. If the car is lifted on a frame contact hoist, which is recommended, the suspension will hang down, the height sensors will lengthen, and the computer will respond to a high-car condition and bleed the air pressure from the spring chambers. When the car is lowered, not only will the car be too low, but the air chambers might collapse and fold inward. This could damage the flexible rubber membranes and cause a need for replacement.

10.10 AFTERMARKET AIR SUSPENSIONS

Aftermarket air suspensions are add-on systems that are sold by various aftermarket suppliers. They are two-wheel systems designed for use on the rear of the car; the air chambers are built into the shock absorbers. They are commonly called **air shocks**. The air chambers are used to supplement the standard, existing springs on the car. The air chambers in the shock absorbers are connected to an inflater by small plastic tubing. When the added load-car-

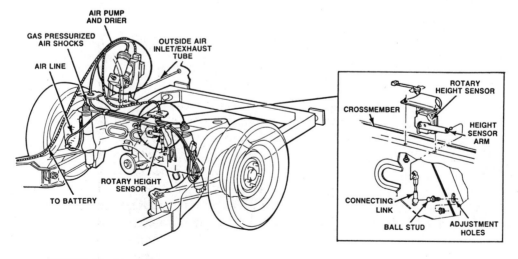

FIGURE 10–41.
A two-wheel, automatic leveling system. When weight is added to the back of this car, the air pressure in the shock absorbers will be increased to raise the car back to the normal height. (Courtesy of Ford Motor Company)

FIGURE 10–42.
An aftermarket, air adjustable shock absorber. Adding air pressure will allow this shock absorber to carry some of the weight of the car. (Courtesy of Monroe Auto Equipment)

rying ability is needed, air pressure is added through the inflater. Some systems use a manual, driver-controlled, 12-volt air compressor to provide the air pressure adjustment; others are inflated just like a tire. (Fig. 10–42)

When adding air shocks to a vehicle, it is important to ensure that the frame mounts are strong and secure. These mounting points will be carrying more of the vehicle's load.

10.11 OVERLOAD SPRINGS

Add-on springs are available for many cars, pickups, and trucks. These are commonly called **overload springs**. These springs allow the vehicle to carry additional load without bottoming out the suspension.

The easiest system to install is a shock absorber with a coil spring mounted around it.

This unit is commonly called a **coil-over-shock.** Like air shocks, a coil-over-shock places a greater load on the shock absorber mounts. Overload springs can also be in the form of a leaf spring that is attached to the leaf springs already on the car, a coil spring that is mounted between the rear axle and the frame, or a flexible air chamber that is placed inside of the coil springs already on the car or ones that can be added. (Fig. 10–43)

Overload springs and air shocks increase the spring rate, which will tend to raise the vehicle or make the ride harsher. Don't forget, if you are adding overload devices to a vehicle, the maximum load to be carried should not exceed the capacity of the tires.

10.12 STABILIZER, ANTIROLL BAR

The stabilizer or antiroll bar is used with many front and rear suspensions. A stabilizer bar is not really a spring like those we have already studied; it acts like a spring only when the car leans. The ends of the stabilizer bar usually connect to the control arm for each tire or onto the outer ends of the axle. The center section of the bar connects to the frame through a pair of rubber bushings. The bushings allow the bar to rotate and even deflect vertically a small

FIGURE 10–43.
Aftermarket overload or helper springs are available that can be added to passenger cars (left) or pickups and light trucks (right). (Courtesy of Moog Automotive)

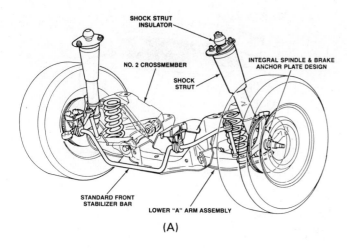

(A)

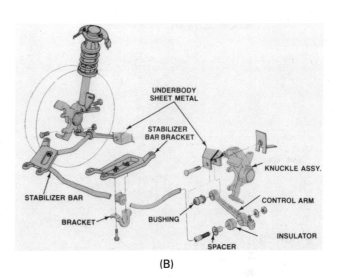

(B)

FIGURE 10–44.
A stabilizer bar, also called an antiroll bar, is connected to the suspension arm at each side of the car. It can be connected to the control arm through a link (A) or a rubber bushing or insulator (B). In (B), the end of the stabilizer bar also acts as a trailing strut to help locate the lower control arm. (Courtesy of Ford Motor Company)

amount. When both tires on an axle meet a dip or rise in the road, the bar merely rotates in the bushings. When only one tire meets a dip or rise or when the car leans on a corner, the bar is forced to twist, much like a torsion bar. (Figs. 10–44 to 10–46)

Stabilizer bars are used primarily to reduce body roll and lean on turns. The bar adds to the roll resistance of the springs and torsilastic control arm bushings. Stabilizer bars have a drawback in that they tend to increase the spring rate of one-tire bumps and, therefore, remove some of the independence of independent suspensions. The formula given in

Figure 10–47 can be used to calculate the strength of a stabilizer bar. The compliance of the rubber bushings in the bar end links and center bushings tends to soften and reduce the effects of one-tire bumps. (Fig. 10–48)

Stabilizer bars can also be built into the control arms and trailing arms on rear suspensions. The axle beam of an FWD, solid rear axle is often designed to function as a stabilizer bar. (Fig. 10–49) Stabilizer bars are trouble-free. The only problems that are normally encountered are the deterioration of the rubber pivot and end link bushings.

STABILIZER BAR

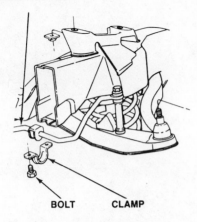

BOLT   CLAMP

STABILIZER BAR

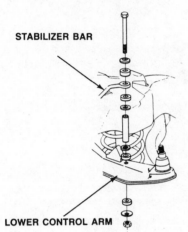

LOWER CONTROL ARM

FIGURE 10–45.
The stabilizer bar is often connected to the suspension member by an end link that consists of four rubber bushings, support washers, a spacer, and a bolt and nut. (Courtesy of Ford Motor Company)

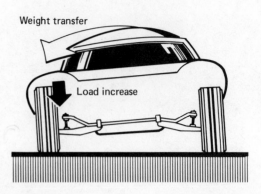

Weight transfer

Load increase

FIGURE 10–46.
When a car goes around a corner, body lean will raise the suspension on one side of the car and lower the suspension on the other side. This action will twist the stabilizer bar, and the strength of the bar will reduce the amount of lean.

Changing to stronger or weaker stabilizer bars should be done with caution. Many suspension designers use this bar to adjust the oversteer and understeer characteristics of the car. Changing the strength of the bar will definitely change these important handling characteristics. Stabilizer bar selection will be discussed more completely in Chapter 19.

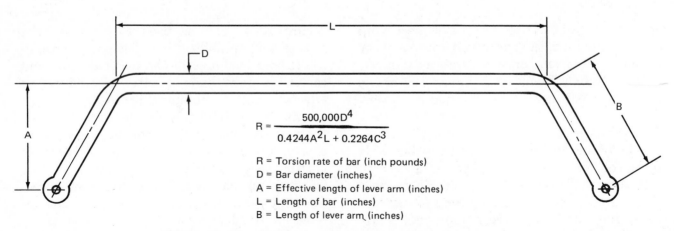

$$R = \frac{500{,}000D^4}{0.4244A^2L + 0.2264C^3}$$

R = Torsion rate of bar (inch pounds)
D = Bar diameter (inches)
A = Effective length of lever arm (inches)
L = Length of bar (inches)
B = Length of lever arm (inches)

FIGURE 10–47.
This is the formula used to compute the strength of a stabilizer bar.

FIGURE 10-48.
A rubber bushing is normally used as the pivot at each end of a stabilizer bar, and an end link set connects each end to the suspension. (Courtesy of Moog Automotive)

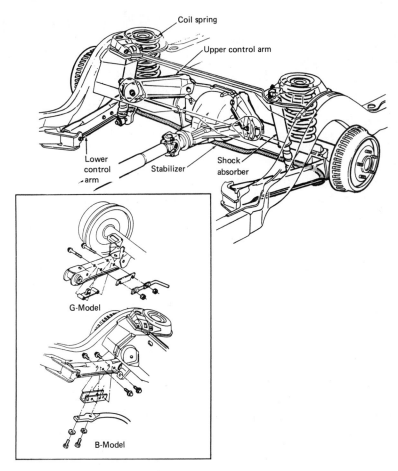

FIGURE 10-49.
This rear-mounted bar is bolted solidly to each control arm by one of two different methods (inset). (Courtesy of Chevrolet)

REVIEW QUESTIONS

1. Which of the following terms is correct?
   A. Vertical travel of the suspension is called jounce.
   B. Downward travel of the suspension is called rebound.

   a. A only
   c. both A and B
   b. B only
   d. neither A nor B

2. Statement A: The tires, wheels, and brake assemblies are all unsprung weight. Statement B: Increasing the unsprung weight of a car will place a greater load on the springs. Which statement is correct?

   a. A only
   c. both A and B
   b. B only
   d. neither A nor B

3. The car's springs
   A. are the flexible part of the suspension system that carries no load.
   B. provide a compressible link between the frame and the axle.

   a. A only
   c. both A and B.
   b. B only
   d. neither A nor B

4. The amount of force or weight that is required to compress a spring one inch is called

   a. spring rate.
   c. rebound weight.
   b. spring weight.
   d. any of these.

5. The speed at which a spring bounces
   A. is measured in cycles per second.
   B. increases in proportion to the strength of the spring.

   a. A only
   c. both A and B
   b. B only
   d. neither A nor B

6. Statement A: The spring on an S-L A suspension has a higher rate than the load at that corner of the car. Statement B: The spring on this suspension will compress about one inch for each two inches of vertical tire motion. Which statement is correct?

   a. A only
   c. both A and B
   b. B only
   d. neither A nor B

7. A variable rate spring
   A. has a rate that gets lower as the spring is compressed.
   B. provides a softer ride as more load is placed on it.

   a. A only
   c. both A and B
   b. B only
   d. neither A nor B

8. Which of the following is not a common spring type? ALL ARE

   a. coil
   c. composite
   b. multi-leaf
   d. torsion bar

9. Statement A: Air springs provide a softer ride because the system can compensate for high load conditions. Statement B: Air spring systems are not simple and inexpensive. Which statement is correct?

   a. A only
   c. both A and B
   b. B only
   d. neither A nor B

10. Which of the following spring types is the most difficult to make with a variable rate?

    a. coil
    c. leaf
    b. composite leaf
    d. torsion bar

11. The shape used for most leaf springs is the

    a. elliptical.
    c. quarter-elliptical.
    b. semi-elliptical.
    d. half-elliptical.

12. Multiple leaves are used in a leaf spring to provide
    A. interleaf friction.
    B. a variable spring rate.

    a. A only
    c. both A and B
    b. B only
    d. neither A nor B

13. Statement A: Rubber leaf spring bushings provide enough compliance for all of the various axle motions. Statement B: Rubber bushings keep road vibrations from traveling from the spring to the car body. Which statement is correct?

    a. A only
    b. B only
    c. both A and B
    d. neither A nor B

14. Statement A: When a coil spring has an end that is simply cut off with no reshaping, it is called a square end. Statement B: A spring with a pigtail end has the last coil wound tighter. Which statement is correct?

    a. A only
    b. B only
    c. both A and B
    d. neither A nor B

15. Variable rate coil springs are made by winding the spring
    A. from tapered wire.
    B. with tighter spaced coils at one end.

    a. A only
    b. B only
    c. both A and B
    d. neither A nor B

16. Coil spring rate is affected by the

    a. diameter of the wire.
    b. diameter of the coil.
    c. number of turns in the coil.
    d. all of these.

17. A torsion bar is always
    A. mounted so it is anchored at one end and free to turn at the other.
    B. a long, straight bar of steel.

    a. A only
    b. B only
    c. both A and B
    d. neither A nor B

18. A definite advantage of torsion bars is that they

    a. have a variable rate.
    b. can be adjusted to compensate for sag.
    c. offer a lower unsprung weight.
    d. all of these.

19. Electronically controlled suspensions are being discussed. Statement A: They use air springs. Statement B: Height sensors are used so the computer will know when the tire meets a bump. Which statement is correct?

    a. A only
    b. B only
    c. both A and B
    d. neither A nor B

20. Which of the following is not a part of an average air ride system?

    a. air compressor
    b. coil spring
    c. air bladder
    d. height sensor

Chapter 11

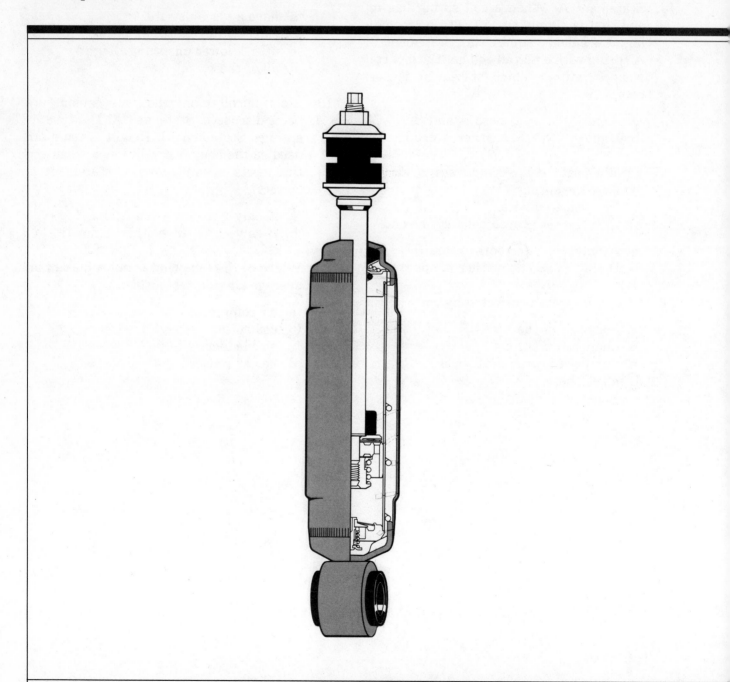

SHOCK ABSORBERS

After completing this chapter, you should:

- Be familiar with the terms commonly used with shock absorbers.

- Understand the purpose for shock absorbers as a suspension component.

- Be familiar with the internal operation of the two major shock absorber designs.

11.1   INTRODUCTION

Most technicians realize that shock absorbers do not absorb shock. The term **damper**, used in many countries outside of the United States, does a much better job in describing what they really do and that is to dampen the spring oscillations. When a spring is deflected, it absorbs energy; when it is allowed to, it will extend and release this energy. Spring inertia causes the spring to bounce too far and over-extend itself. It then recompresses, but it will travel too far again. The spring will continue to bounce back and forth and oscillate at its natural frequency until the energy that was originally put into the spring is used up by spring molecular friction. This usually takes quite a few oscillations. In a car, if shock absorbers are not fitted, a bump will cause the car to bounce up and down at the natural frequency of the suspension for an uncomfortably long period of time. With shock absorbers, the suspension is allowed to oscillate through one or two diminishing cycles. The shock absorber will absorb much of the energy from the spring. (Fig. 11-1)

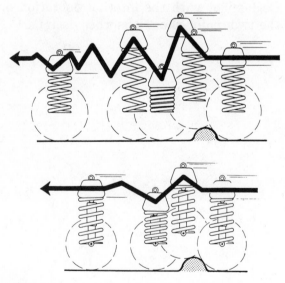

FIGURE 11-1.
If an undampened suspension meets a bump, the sprung weight will bounce at the frequency of the suspension until the energy of the bump is used up. The shock absorber will limit or dampen these vibrations to a few oscillations. (Courtesy of Monroe Auto Equipment)

Shock absorbers also give the suspension engineer another method of tailoring the suspension movement. Shock absorbers resist suspension movement, bounce, body roll, brake dive, or acceleration squat. The added resistance of the shock absorber during the compression or extension portions of the spring cycle can do much to change the action of the suspension travel. A firm shock absorber can add to the spring rate during the compression phase.

Some suspension designs use the shock absorber as the limiting member for suspension travel. **Bump stops** or **rubber cushions** can be installed on the piston rod just above (outside the body) or just below the upper main body bushing (inside the body). These bushings will stop the piston rod and suspension travel. One manufacturer incorporates a hydraulic extension or rebound travel stop in some of its models. Be aware that the removal of the shock absorber from some suspensions can cause unexpected surprises. It is possible for the car to be driven, but if the suspension drops from driving over a severe bump, the rear spring can fall out. (Fig. 11-2)

11.2   SHOCK ABSORBER OPERATING PRINCIPLES

The shock absorber must absorb energy in order to reduce the spring oscillations. Laws of physics tell us that energy cannot be created or destroyed. However, energy can be converted from one form to another. Another name for a shock absorber could be energy converter. The energy that is absorbed by the shock absorber is converted into heat and then dissipated into the surrounding air. As the heat leaves to the air, much of the energy absorbed by the shock absorber goes with it. A shock absorber is really a heat machine. A hard working shock absorber gets quite hot, sometimes reaching temperatures that cause changes in operation or even failure.

Some early styles of shock absorbers used friction to absorb the spring energy. They were constructed with one lever that connected to

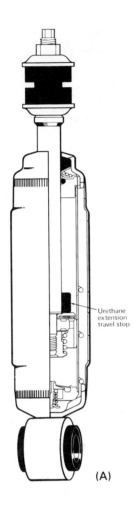

Urethane
extension
travel stop

(A)

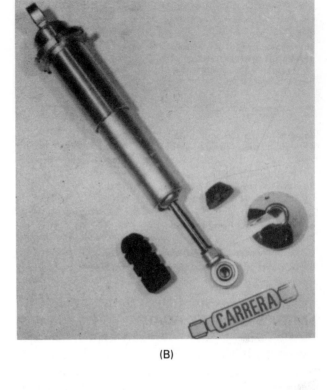

(B)

FIGURE 11-2.
Some shocks are designed with flexible travel stops to serve
as rebound (A) or jounce (B) travel stops. (A is courtesy of
Monroe Auto Equipment; B is courtesy of Carrera Shocks)

the axle and one lever that attached to the
frame. The levers were separated by a friction
material, similar to clutch or brake lining, and
tightened snugly against each other. Vertical
suspension movement caused a rotation and
rubbing of the levers; the friction produced by
the rubbing retarded the suspension and generated heat. The amount or rate of energy ab-
sorption was controlled by the tightness of the
levers against each other. If they were tightened too much, they would eliminate all motion
and lock up the suspension. Also, **friction shock
absorbers** were very prone to wear because of
the constant rubbing. (Fig. 11-3)

Modern shock absorbers operate **hydraul-
ically.** They are basically oil pumps that force
oil through small **openings** or **orifices.** Heat is
generated when a liquid is forced through a restriction. Because the internal friction is fluid
friction, hydraulic shock absorbers can operate
through many cycles without wearing. (Fig.
11-4)

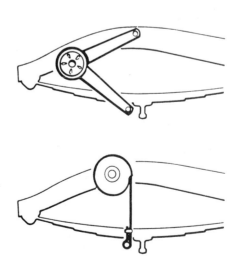

FIGURE 11-3.
A Hartford friction shock (top). This shock absorber could be
adjusted by tightening the friction disc to increase the resistance. A one-way, ribbon damper is shown at the bottom.
(Courtesy of Monroe Auto Equipment)

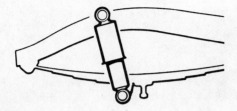

FIGURE 11-4.
All modern shocks are of the direct-acting, telescoping type; one end is connected to the frame and the other end is connected to the suspension. (Courtesy of Monroe Auto Equipment)

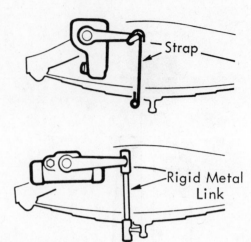

FIGURE 11-5.
Two older shock designs: a one-way (rebound direction control), lever shock (top) and a two-way, lever shock (bottom). (Courtesy of Monroe Auto Equipment)

All of the shock absorbers used today are **direct acting**. One end of the shock absorber is connected to the suspension and the other end is connected to the frame. Rubber bushings are used at these connections to offer compliance (i.e., allow for slight mounting angle changes) and to dampen vibrations. The shock absorber telescopes and changes length as the suspension moves up and down. At one time, **lever** shock absorbers were fairly common. The shock absorber body was bolted solidly to the frame with a lever extending from the body. The lever was connected to the suspension so that suspension movement operated the lever and worked the shock absorber. On some cars, the shock absorber lever was also the upper control arm for the front suspension. (Fig. 11-5)

11.3   SHOCK ABSORBER DAMPING RATIOS

Hydraulic shock absorbers are constructed so they offer different resistances during the two different operating directions of (1) compression or jounce and (2) extension or rebound. The differences in resistance are referred to as the **shock absorber ratio**. A friction shock absorber will have a ratio of fifty-fifty. It has the same resistance on compression as on extension. A hydraulic shock absorber can be constructed with whatever ratio the ride engineer desires. The ratios are normally given with the extension control resistance printed first. Confusion occurs because the ratio numbers are usually given in the reverse order by people involved with racing. This chapter will use ratios

as referred to by the engineer. A 90/10 shock absorber will have 90 percent of its control ability on extension and 10 percent on compression. This shock absorber will probably compress easily and be difficult to extend. A 10/90 would compress with difficulty (90 percent of control) and extend easily (10 percent). (Fig. 11-6)

The ratio is designed into a shock absorber by ride engineers with the requirement that the compression cycle should let the suspension system rise during jounce while the tire is going over a bump and fall back down after the tire passes the bump. If the shock offers too much resistance during the compression cycle, the bump will become more severe. If too much resistance occurs during extension, it might be possible to hold the tire off the road momentarily. The shock ratio and resistance must add enough resistance to dampen the spring oscillations yet allow free enough suspension travel to give a good ride. The actual ratio of most shock absorbers is known only by the ride engineers. The term has a very limited use in the repair field. Shock specifications are seldom printed in service manuals or shock absorber catalogs. It is very difficult and seldom necessary for a mechanic to determine the ratio of a particular shock absorber.

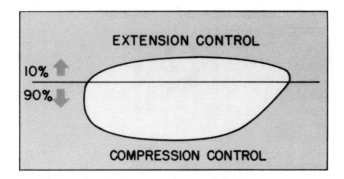

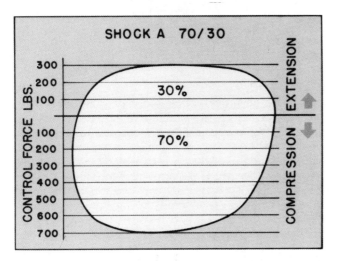

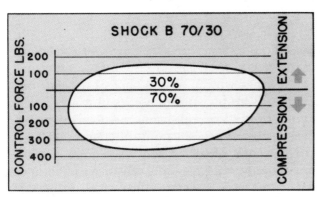

FIGURE 11–6.
A 90/10 shock ratio tells us that 90 percent of the shock absorber's control occurs during compression. The curves for two 70/30 shocks are also shown. Note that one of these has about twice the resistance of the other. (Courtesy of Monroe Auto Equipment)

11.4 SHOCK ABSORBER DAMPING FORCE

The amount of **damping** or **control force** that a shock absorber exerts is of more importance than the ratio. The ratio is just a comparison of the two control forces, compression and extension. A 50/50 shock absorber with a control force of 200 pounds (91 kg) would offer a softer but bouncier ride than a 50/50 shock absorber with a control force of 400 pounds (182 kg) on the same car. The amount of control force is designed into the shock absorber by ride engineers. Too much control force will give a harsh, jiggly ride with the possibility of tire hop on washboard roads. Too little control force can give a soft, bouncy ride with excessive roll on corners and excessive dive on braking. Some shock absorbers are constructed so that the control force can be adjusted for one or both directions of travel. This allows tailoring the shock absorber to a particular car and to the driver's preferences. (Figs. 11–7 & 11–8)

Shock absorber control is measured on a **shock absorber dynamometer**. This machine allows the resistance to be measured over various operating speeds. The resulting graph allows the ride engineer to see the various operating stages plus any weak spots or lags in operation. (Fig. 11–9)

Ride engineers design the rate of control force to change as the speed of suspension

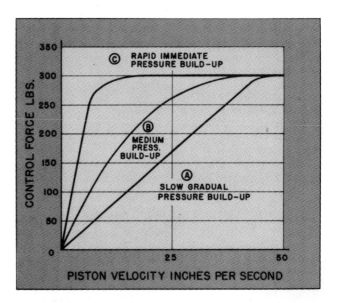

FIGURE 11–7.
The resistance of a shock increases with the speed of the piston movement. It is also controlled by the internal valving. These three curves show three different stage 1 valving. (Courtesy of Monroe Auto Equipment)

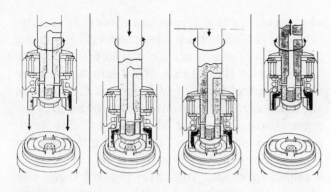

FIGURE 11–8.
This shock uses an adjustable piston orifice. Rotating the adjuster nut upward on the piston rod will increase both the compression and extension resistance. Compressing the shock completely engages the adjuster nut with the base of the shock and turning the piston rod will then turn the adjuster. (Courtesy of Koni)

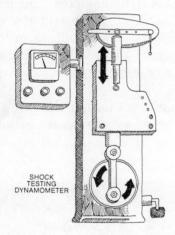

FIGURE 11–9.
A shock absorber dynamometer. This machine is used to measure shock resistance at various speeds. (Courtesy of Monroe Auto Equipment)

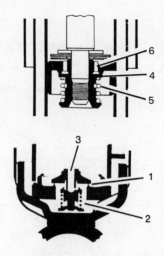

FIGURE 11–10.
Most shocks have three or more operating stages in each direction. During compression, stage 1 (slow speed), oil passes through a small orifice, # 1; stage 2 (faster speed), oil flow opens a spring controlled valve, # 2; and stage 3 (very high speed), oil flow is controlled by the size of the valve pin orifice, # 3. During extension, stage 1 (slow speed), oil flow is controlled by small orifices under the valve seat, in the piston, #4; stage 2 (faster speed), oil flow works against a spring, # 5; and stage 3 (very high speed), oil flow is controlled by the sizes of the oil passages, #6. (Courtesy of Monroe Auto Equipment)

travel changes. These changes are often referred to as **stages of operation**. A car operating at slow speeds over small bumps does not require or want much shock absorber control. The same car operating at high speeds over a rough road needs a lot more control; the shock absorber will need to absorb a lot more energy. Shock absorber valving is sensitive to the speed at which the suspension travels.

Hydraulic shock absorbers have a natural tendency to increase resistance with an increase in velocity. Laws of fluid dynamics state that a liquid's resistance to flow through an orifice increases with the square of the flow velocity. In other words, as the shock absorber piston speeds up because of a high bump rate, the higher fluid flow rate that results meets increased resistance at the fluid control orifices, so the shock absorber causes a higher resistance to motion. The shock absorber resistance will increase. The second and third stages of shock absorber operation open up additional orifices or flow paths to keep the resistance from becoming too great at high speeds. These stages are controlled by spring pressure and orifice size. (Fig. 11–10)

11.5 DOUBLE-TUBE SHOCK ABSORBERS

The traditional, direct-acting shock absorber is of a **double-tube** design. The shock absorber is made with two major steel tubes. The inner, pressure tube is the cylinder in which the pis-

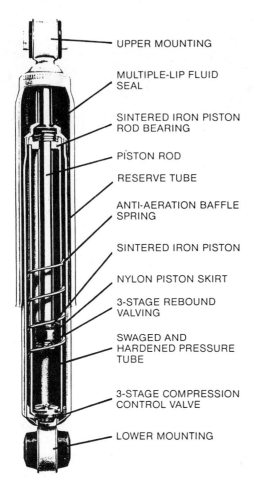

UPPER MOUNTING

MULTIPLE-LIP FLUID SEAL

SINTERED IRON PISTON ROD BEARING

PISTON ROD

RESERVE TUBE

ANTI-AERATION BAFFLE SPRING

SINTERED IRON PISTON

NYLON PISTON SKIRT

3-STAGE REBOUND VALVING

SWAGED AND HARDENED PRESSURE TUBE

3-STAGE COMPRESSION CONTROL VALVE

LOWER MOUNTING

FIGURE 11–11.
A double-tube shock. The piston travels in the inner, pressure tube while the reserve tube forms the outside of the reservoir. (Courtesy of Monroe Auto Equipment)

ton operates while the outer tube forms the outside of the reservoir. Many shock absorbers have an additional tube attached to the piston rod called a dust or stone shield. It protects the piston rod and upper seal from road damage. (Fig. 11–11)

11.5.1   Construction

The upper end of the piston rod connects to the car's frame. The piston, which has some of the control valves built into it, fits snugly into the pressure tube. Metal, teflon, or plastic sealing rings are often used to ensure a good piston-to-pressure-tube seal. The smoothly ground, chrome plated piston rod passes through a bushing and seal in the upper end of

the pressure tube. This seal is usually a neoprene or silicone rubber of a multilip design. The bushing and seal allow the piston rod to move freely in and out of the pressure tube while keeping all of the hydraulic oil inside. A drain hole, just below the seal, allows any oil that passes through the bushing to drain back into the reservoir. This hole also allows air to bleed out of the pressure tube during operation. Some shock absorbers place a valve in this opening to offer more control. (Figs. 11–12 & 11–13)

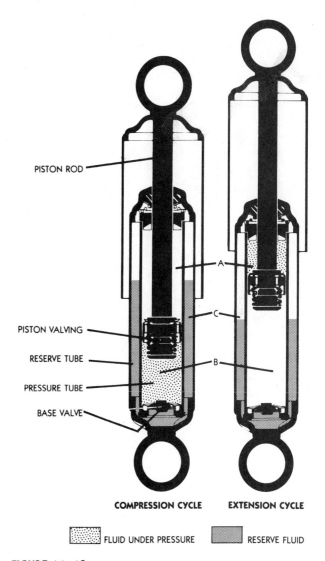

PISTON ROD

A

PISTON VALVING

C

RESERVE TUBE

B

PRESSURE TUBE

BASE VALVE

COMPRESSION CYCLE   **EXTENSION CYCLE**

FLUID UNDER PRESSURE   RESERVE FLUID

FIGURE 11–12.
During the compression cycle, the oil between the piston and the base valve is (b) under a high pressure, depending on the piston speed. During the extension cycle, the oil between the piston and upper bushing (a) is under pressure. (Courtesy of Monroe Auto Equipment)

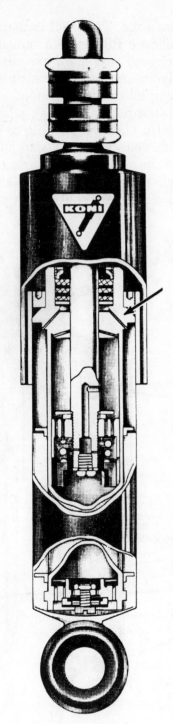

FIGURE 11-13.
As the piston rods strokes up and down, some oil flows through the upper bushing. This oil can run back into the reservoir through a passage (arrow). (Courtesy of Koni)

The base mount of the shock absorber serves several functions. Besides being the mount connection to the axle or suspension arm, it holds the **base** or **foot valve** assembly

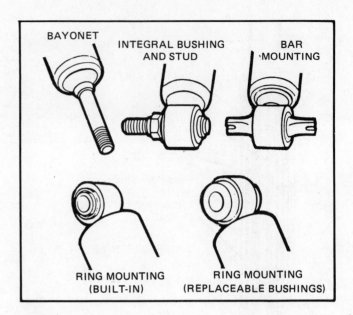

FIGURE 11-14.
Various types of shock mount types. All of these types use a rubber bushing between the shock absorber and the frame or body and suspension member. (Courtesy of Ford Motor Company)

that forms the bottom of the pressure cylinder and the reservoir chamber. Passenger car shock absorber mounts are usually of a **bayonet**, also called a **single-stud**, or **eye ring type**. Rubber bushings are used with each mount type to allow a somewhat flexible mounting so the shock absorber can align itself during suspension movement and also to reduce or dampen road vibrations. (Fig. 11-14)

11.5.2 Operation

As the suspension moves up and down, the shock absorber piston and rod strokes up and down in the pressure cylinder. There are essentially three different fluid chambers in a double-tube shock absorber: (1) a low pressure reservoir and the areas (2) above and (3) below the piston in the pressure cylinder. One of the cylinder areas will have a very high pressure and the other a very low pressure depending on the direction of piston travel. The pressures in the pressure cylinder are affected by the piston rod as well as the piston. The rod displaces a relatively large amount of fluid as it enters the shock absorber during compression travel, and

COMPRESSION CYCLE

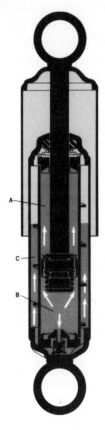

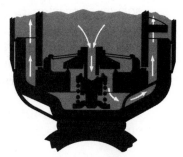

Compression Valve System

Piston Valve System

FIGURE 11–15.
During the compression cycle, oil must flow from chamber B into chambers A and C; shock control is provided by the orifices and spring-controlled valves in the piston and the base valve, mostly the base valve. (Courtesy of Monroe Auto Equipment)

at this time, an equal volume of fluid must leave the pressure cylinder and go into the reservoir.

During compression, the piston and rod are moving into the pressure cylinder. The piston and rod motion are generating a high pressure below the piston and a low or negative pressure above it. Oil flow between the chambers above and below the piston are controlled by valves and orifices in the piston. Oil flow from the lower pressure cylinder to the reservoir is controlled by valves and orifices in the base valve assembly. (Fig. 11–15)

These valves are usually staged. A small orifice provides the control for small bump operation. If the piston speeds are faster, the increased oil pressure that is generated will work against a coil or flat reed spring to open an ad-

ditional orifice. Higher pressures will open the valve farther and farther. Maximum flow is controlled by the size of this additional orifice passing through the valve.

During compression, most of the shock absorber control will be at the oil flow from the pressure tube to the reservoir. Excessive restriction at the piston valve might cause the upper cylinder pressure chamber to drop low enough to pull air in through the rod seal or the drain hole.

The amount of piston travel should be limited so it never is far enough for the piston to bump into the base valve. This can damage the base valve or piston or both, which would ruin the shock absorber. The compression travel is normally limited by external bump stops on the car's suspension or by rubber cushions between the upper shock absorber bushing and the upper shock absorber mount.

During extension, the piston and rod are moving upward in the pressure cylinder, and oil will flow back from the reservoir to the pressure cylinder. There will be a very high pressure in the chamber above the piston and a lower pressure below it. The flow into the lower chamber from the reservoir will be almost a free flow with little restriction. Nearly all of the extension control takes place in the piston valves and orifices as the oil moves downward through them. Because of the flow from the reservoir to the pressure cylinder, a double-tube shock absorber must be in a somewhat vertical position to operate. If it were mounted upside down, air, not oil, would flow into the pressure cylinder during the extension stroke. (Fig. 11–16)

Operation on rough roads will cause a certain amount of agitation in the reservoir oil as the body of the shock absorber is shaken up and down along with the suspension. This agitation tends to foam or aerate the oil. All double-tube shock absorbers leave a certain amount of expansion space in the reservoir to allow for heating of the oil; the most common expansion space is an air space above the oil.

Air bubbles that mix with the oil and enter the pressure cylinder interfere with shock absorber operation because air or any other gas can be compressed and, also, because a gas will flow through an orifice much faster than a liquid. Shock operation will be mushy. Normal shock absorber operation will eventually move this air into the upper chamber and then through the drain passage back to the reservoir, but while the air is in the pressure cylin-

EXTENSION CYCLE

Piston Valve System

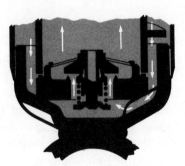

Compression Valve System

FIGURE 11–16.
During the extension cycle, oil must flow from chambers A and C into chamber B; shock control is provided by the valve stages in the piston. (Courtesy of Monroe Auto Equipment)

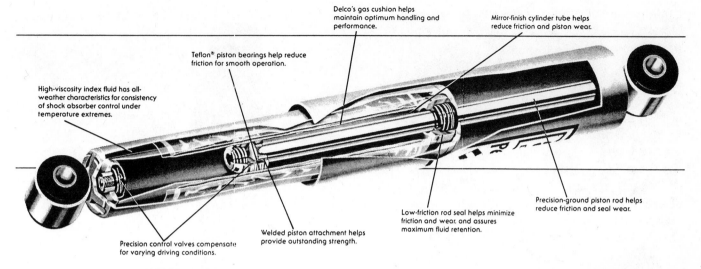

Delco's gas cushion helps maintain optimum handling and performance.

Mirror-finish cylinder tube helps reduce friction and piston wear.

Teflon® piston bearings help reduce friction for smooth operation.

High-viscosity index fluid has all-weather characteristics for consistency of shock absorber control under temperature extremes.

Precision-ground piston rod helps reduce friction and seal wear.

Low-friction rod seal helps minimize friction and wear, and assures maximum fluid retention.

Welded piston attachment helps provide outstanding strength.

Precision control valves compensate for varying driving conditions.

FIGURE 11–17.
The gas cushion in this shock is a gas-filled plastic bag that can compress or expand to allow for oil expansion and contraction. This allows complete filling of the reservoir with oil, which greatly reduces aeration of the oil. (Courtesy of AC-Delco, General Motors Corporation)

der, a loss of shock absorber control takes place. Many shock absorber designs use a spring-shaped baffle in the reservoir or a spirally grooved reservoir tube to reduce the air or gas that can work its way down to the base valve and into the pressure tube.

Some shock absorber designs place a gas-filled plastic bag in the reservoir and fill the reservoir completely with oil. The gas-filled bag compresses and expands during shock absorber operation or heat expansion of the oil. The plastic bag holds the gas, which is usually a freon gas under a pressure of about 150 psi (1,034 kPa), captive and prevents aeration. Another design pressurizes the reservoir with a nitrogen gas charge of about 100 to 150 psi (690 to 1,034 kPa). These are often called **gas-charged shocks**. This increased internal pressure increases the load placed on the piston rod seal. The seal has to keep in this added pressure. A shock absorber with internal pressure will tend to extend itself, much like a spring. In special installations, the reservoir can be mounted remotely and connected to the base valve and pressure tube with tubing. A remote reservoir has a much better chance of providing cool, air-free oil. (Figs. 11–17 to 11–19)

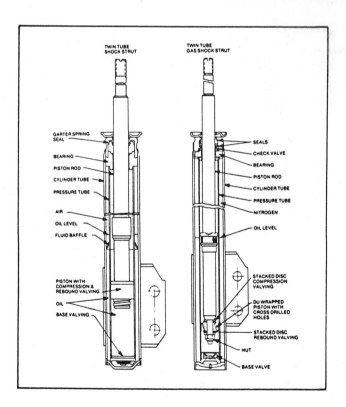

FIGURE 11–18.
These two shock struts are very similar. The one on the right has the reservoir gas pressurized, charged with nitrogen at a pressure of 100 psi (690 kPa). The gas pressure will prevent aeration of the oil. (Courtesy of Ford Motor Company)

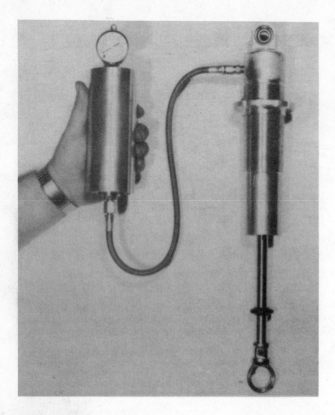

FIGURE 11–19.
This racing shock has a remote reservoir that can be run with various amounts of pressure. The remote mounting reduces vibration-caused aeration as well as produces improved cooling. Also note the adjustable spring seat. (Courtesy of Carrera Shocks)

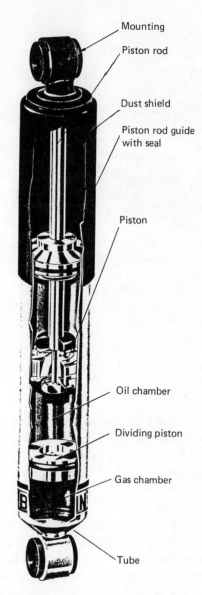

FIGURE 11–20.
A single-tube shock. Note that there is no reservoir and that there are two pistons in the pressure tube. (Courtesy of Bilstein)

11.6   SINGLE-TUBE SHOCK ABSORBER OPERATION

Single-tube shock absorbers are also called **de Carbon, monotube,** or **gas pressure** shock absorbers. There is only one shock absorber tube, the pressure tube, but there are two pistons, a **dividing piston** in addition to the **working piston.** This design offers an advantage in that the pressure tube is closer to the outside air so it will dissipate heat better and run cooler; plus, it can be mounted with the shock body to the car frame. As the oil in a shock absorber heats up, it will thin out. This can cause reduction in the control resistance, a lag or skip in the operation, or both. A potential problem with a single-tube shock absorber is that a bend or dent

in the tube can destroy the piston-to-cylinder seal and upset or destroy shock absorber operation. The reservoir of a double-tube shock absorber adds quite a bit of protection for the pressure tube. (Fig. 11–20)

11.6.1   Construction

The working piston and rod of a single-tube shock absorber are very similar to those of a double-tube shock absorber except that they are often upside down. A single-tube shock absorber will operate with either end up. Many designers prefer to mount the body of the shock absorber on the frame to help reduce unsprung weight and agitation of the major part of the shock absorber. The pressure tube is longer than needed for piston travel and is sealed at the top where it connects to the frame mount. A free-floating, dividing piston travels in the mount end of the pressure tube. The chamber between the dividing piston and the mount end is pressurized to about 75 to 350 psi (517 to 2,400 kPa) with nitrogen gas. This gas pressure will tend to move the dividing piston and the working piston toward the extended position.

11.6.2   Operation

Since this shock absorber does not use a base valve, both the extension and compression control valves must be built into the working piston. The dividing piston moves up or down in the pressure tube as the piston rod moves in or out of the shock absorber. Movement of the di-

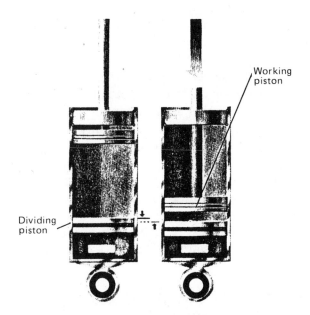

FIGURE 11-21.
The lower dividing piston separates a high pressure gas charge in the lower chamber from the oil in the rest of the shock. As the working piston and rod move in and out of the pressure tube or as the oil expands or contracts, the dividing piston will travel up and down to compensate. (Courtesy of Bilstein)

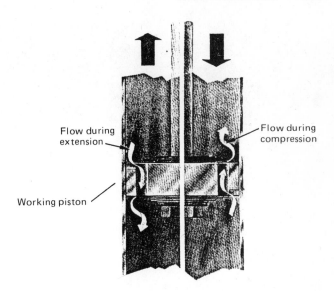

FIGURE 11-22.
All of the compression and extension control of a single-tube shock takes place in the piston valves. (Courtesy of Bilstein)

viding piston compensates for the oil displaced by the piston rod. (Fig. 11-21)

During compression, the piston rod and working piston move inward into the pressure tube and the dividing piston is forced toward the mount end of the pressure tube. This action tends to compress the gas charge. Shock absorber control takes place as the oil pressure increases in the chamber between the two pistons and forces oil through the orifices and valves in the working piston. Single-tube shock absorbers use staged valving much like that used in double-tube designs.

During extension, the piston rod and working piston move outward in the cylinder, and gas pressure moves the dividing piston toward the working piston. The oil pressure in the chamber on the rod side of the piston increases and forces oil to flow through the orifices and valves in the working piston. Shock absorber control occurs in the working piston while this oil flow takes place. (Fig. 11-22)

11.7   STRUT SHOCK ABSORBERS

The body of most struts is essentially a large shock absorber with an oversize piston rod and

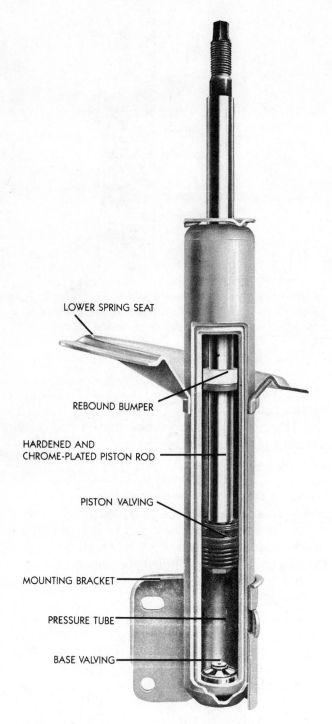

LOWER SPRING SEAT

REBOUND BUMPER

HARDENED AND
CHROME-PLATED PISTON ROD

PISTON VALVING

MOUNTING BRACKET

PRESSURE TUBE

BASE VALVING

FIGURE 11-23.
A strut. This unit is essentially an oversize shock with a spring seat and different mounts. Note that this particular strut cannot be disassembled. (Courtesy of Monroe Auto Equipment)

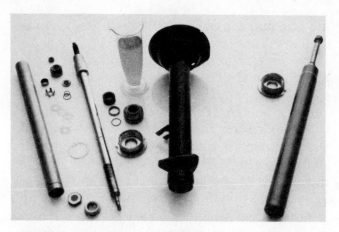

FIGURE 11-24.
Many struts can be disassembled (left) for service. It is common practice to replace the internal parts with a preassembled cartridge (right). (Courtesy of Monroe Auto Equipment)

is essentially the same as that of a shock absorber. (Fig. 11-23)

Some O.E.M. struts are constructed so they can be taken apart. The upper bushing and seal are retained by a **gland nut**. Some struts are welded together like most standard shock absorbers. Removal of the gland nut allows removal of the piston rod bushing and seal, the piston and rod, the pressure tube and base valve, and the oil. Any or all of these components can be replaced in order to rebuild this strut. Normal shop practice is to remove all of these parts and replace them with a new, sealed shock absorber cartridge. Some O.E.M. struts are welded together so they cannot be easily disassembled. When service is required, they are serviced by replacement of the strut assembly or, in some cases, by cutting the top off the strut, replacing the internal parts with a cartridge, and threading a gland nut into threads already in the outer tube. Replacement shock absorber cartridges can be either double-tube or single-tube designs. (Figs. 11-24 & 11-25)

11.8   COMPUTER-CONTROLLED SHOCK ABSORBERS

Several manufacturers have incorporated electronic control into the shock absorbers of strut suspensions. These systems use a computer module to electronically change the dampening

spring seats. Most O.E.M. struts are of a double-tube design, but some are single-tube. The operation and internal construction of a strut

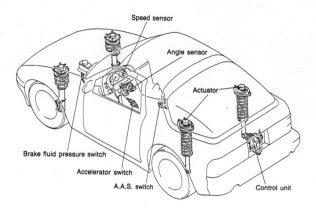

FIGURE 11–26.
A computer-controlled ride control system. An actuator at each of the four shock absorbers can adjust the shock valving to obtain different ride characteristics. The actuators are controlled by the control unit, which receives input from the speed sensor, angle sensor brake fluid pressure switch, accelerator switch, and the auto adjusting system (A.A.S.) switch. (Courtesy of Mazda Motors of America)

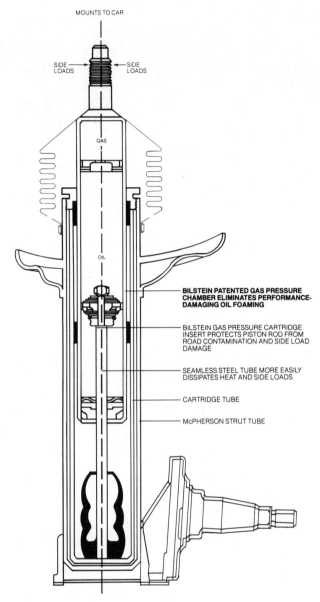

FIGURE 11–25.
A strut with a single-tube insert or cartridge. Strut cartridges are available in both single-tube and double-tube styles. (Courtesy of Bilstein)

rate of the shock absorber to suit various driving conditions.

The major parts of these systems are the four shock absorbers with their **variable valving**, an **actuator** for each shock absorber to change the valve setting, the **control module** that controls the actuators, the **sensors** that provide data needed for the control module, a switch that allows the driver to select the type of ride desired, and the wiring to connect these various parts. (Fig. 11–26)

The shock absorbers used with computer-controlled systems have a **variable sized orifice** that can allow fluid to bypass the usual extension and compression control valves. At the soft setting, the orifices are at the widest opening, which gives the least amount of dampening resistance. At the very firm setting, the bypass orifices are completely closed, which gives the greatest amount of dampening resistance. A third, intermediate setting is included in some systems, which allows some fluid to bypass and provides a dampening resistance between the other two. The valve settings are changed by either rotating or changing the height of a control rod that is located in the shock absorber's piston rod. (Figs. 11–27 & 11–28)

The actuators can be electronic, which receive their operating signal directly from the control module, or air-powered. Air-powered units use air pressure that is controlled by air valves switched on or off by the control module. The control module is programmed to operate the actuators under various driving conditions depending on the particular system and the position of the selector switch. Some of these conditions include the following:

1. Antidive. During braking, a brake fluid pressure switch will cause a firm to very firm

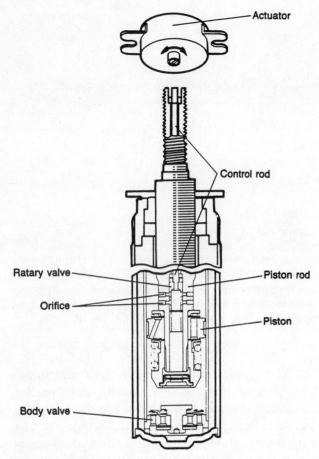

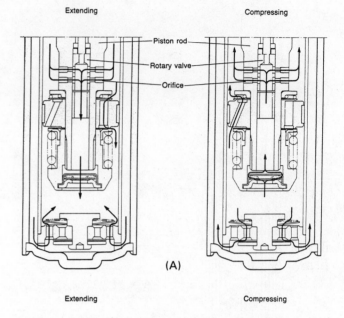

FIGURE 11-27.
An A.A.S. shock absorber. The actuator is a three-position, electric device that can rotate the control rod 120°. These three positions will change the amount of orifices that are open in the piston rod. (Courtesy of Mazda Motors of America)

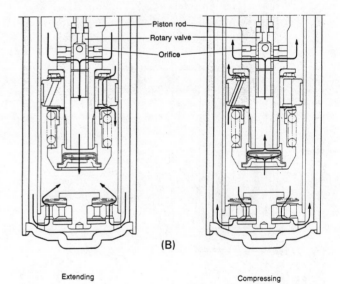

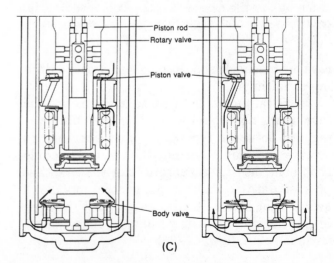

FIGURE 11-28.
With the actuator and control rod in the soft position, there are four orifices open to allow free flow of oil through the piston (A). With the actuator in the firm position (B), there are two orifices open. With the actuator in the very firm position (C), all of these orifices are blocked and the shock oil flows in the same manner as a standard shock. (Courtesy of Mazda of America)

setting depending on the speed of the car and the position of the selector switch.

2. Antiroll. During cornering, a steering wheel angle sensor, a steering shaft angular velocity sensor, or a "g" sensor can cause a firm or very firm setting depending on the car speed and the position of the selector switch.

3. Antisquat. During acceleration, the rate at which the throttle position sensor changes position can cause a firm or very firm setting depending on the car speed and selector switch setting.

4. Speed reset. Depending on the setting of the control switch, the setting can change to firmer settings above predetermined speeds. (Fig. 11-29)

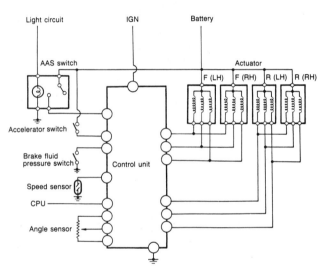

FIGURE 11-29.
The electrical circuit for an A.A.S. ride control system. Note that the control unit can adjust the front and rear shocks separately. (Courtesy of Mazda Motors of America)

11.9 LOAD-CARRYING SHOCK ABSORBERS

Shock absorbers are normally not designed to carry any vehicle load except for the slight spring rate caused by the internal gas pressure of a gas-filled or single-tube design. Removal or replacement of the shock absorbers does not affect the height of the vehicle. Aftermarket shock absorbers are available that can help the

FIGURE 11-30.
A load-lever or coil-over shock. A pair of these particular shocks at the rear will increase the load-carrying ability of the car by as much as 1,000 pounds; front units do not have the same load-carrying ability. (Courtesy of Monroe Auto Equipment)

springs carry additional load. These are commonly used on the rear ends of vehicles that carry heavier than normal loads. They are also available for the front end on some vehicles. It should be noted that the use of load-carrying shock absorbers places more load on the frame mounts for the shock absorbers. There have been cases of mount breakage and failure because of this added load. Also, some vehicles use a height-sensitive, brake-proportioning valve. Altering the height on these vehicles will upset the bias or balance of the braking action.

Coil-over shock absorbers mount a coil spring around the shock absorber body. The upper spring seat is built into the upper shock absorber mount, and the lower spring seat is built into the lower portion of the shock absorber body. Coil-over shock absorbers use a constant rate spring. One coil-over shock absorber design uses a spring that can act in tension as well as compression. In actual practice, this design performs much like the action of an antiroll or stabilizer bar. (Figs. 11-30 & 11-31)

Air shock absorbers connect a reinforced tubular shaped membrane to the piston rod dust shield and to the shock absorber body. The area inside of the dust shield becomes the air chamber for an air spring. This air chamber is connected to an air pressure source by a thin plastic tube. Air shock absorbers are variable rate air springs, depending on how much air pressure is used. A small amount of air pressure must be kept in the air chambers at all times, even when there is no load. Operation with no pressure might cause the rubber membrane to be folded and pulled inward if a rapid extension and contraction of the shock absorber occurs. (Figs. 11-32 & 11-33)

FIGURE 11–31.
The spring seats of this type of coil-over-shock are designed to allow the spring to act in tension as well as compression. Both the spring and shock absorber will offer resistance to extension travel as well as compression travel. (Courtesy of 4-Way Suspensions)

11.10 SHOCK ABSORBER QUALITY

Replacement shock absorbers vary substantially in price. The cost-conscious motorist is often concerned about paying more for one

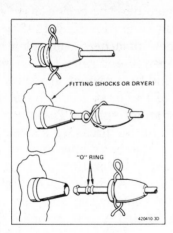

FIGURE 11–33.
The shock air lines on some O.E.M. units are sealed by "O" rings and held in place by a quick-disconnect system. (Courtesy of Chevrolet)

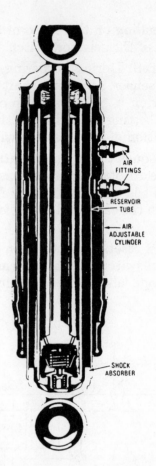

FIGURE 11–32.
A superlift or air shock. A flexible membrane connects the reservoir tube to the outer shield to form an air cylinder. This unit has one air fitting to connect to the other shock and one fitting to connect to the air supply. (Courtesy of Chevrolet)

product that looks just like another, much less expensive product. Some of the difficult-to-see features of a better quality shock absorber include the following:

1. **Piston rod.** A very smooth and hard finish is required to get maximum seal life and smooth operation. An oversize diameter is sometimes used to provide a stronger shock absorber plus to displace more oil during the compression stroke; the increased oil displacement provides better control capabilities.

2. **Upper bushing and seal.** A smooth precision bore is required for smooth operation, and a long-lasting seal is required to retain the oil for a longer operating life, especially with gas pressure designs. A shock absorber cannot function if it loses its oil.

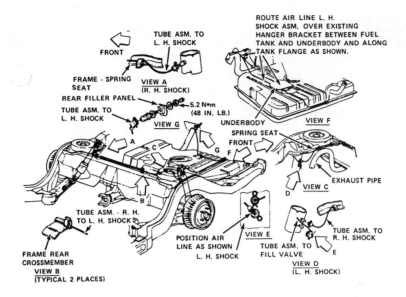

FIGURE 11-34.
The air line routing of a superlift or air shock system. Note that this system is filled from an outside air supply. (Courtesy of Chevrolet)

3. Piston and ring. The piston must be concentric with the piston rod and with the ring and must provide a good, fluid-tight, low-drag seal with the pressure tube.

4. Pressure tube. The pressure tube must be strong enough to withstand the fluid pressure without deforming and must be straight and smooth to provide a good seal with the piston and ring.

5. Valves. The valves must be able to operate through thousands of cycles.

6. Oil. The oil must not break down from age or high operating temperatures.

11.11   SHOCK ABSORBER FAILURES

Shock absorbers should be replaced when

1. a piston rod or mount has broken;
2. they are leaking oil;
3. they are noisy or give a rough operation; or
4. they do not provide the correct amount of resistance.

Most of these faults are easy to check. The first two can be done with a quick, visual inspection. The third requires disconnecting one end of the shock absorber and manually compressing and extending the shock absorber while feeling and listening to its operation.

Checking a shock absorber for correct resistance is a seat-of-the-pants, subjective operation. Some mechanics use a **bounce test**; that is, they bounce the car up and down and watch for an excessive number of oscillations, which indicates a weak shock absorber. Most mechanics feel the bounce test is almost useless because of the very slow bounce speed. All this test is really good for is to compare the first stages of two shock absorbers that are on the same axle. A road test over a rough road with a series of undulations is probably the best method of testing shock absorbers. Weak, worn shock absorbers will allow an excessive amount of bouncing (especially after the bump), sway, or front-end dive under hard braking. It is best to develop a standard test area and drive it with cars that have worn shock absorbers and then cars that have good shock absorbers so a comparison can be made. It also should be remembered that large, softly sprung cars have a much larger tendency to bounce and sway than smaller, lighter cars. On-car inspection of shock absorbers will be described in more detail in Chapter 12.

REVIEW QUESTIONS

1. Statement A: Shock absorbers are called dampers in many countries. Statement B: Shock absorbers help support the car's weight. Which statement is correct?

   a. A only   **c.** both A and B
   b. B only   **d.** neither A nor B

2. When driving on a rough road, a shock absorber will
   A. get hot.
   B. dampen excess spring oscillations.

   a. A only   **c.** both A and B
   b. B only   **d.** neither A nor B

3. A shock absorber with a 70/30 ratio will
   A. offer a different amount of resistance during compression and extension.
   B. allow the spring to travel easier in one direction than the other.

   a. A only   **c.** both A and B
   b. B only   **d.** neither A nor B

4. Statement A: Too much compression resistance in a shock can cause the tire to leave the ground as it passes over a sudden dip in the road at high speeds. Statement B: The amount of resistance is determined by the amount of oil in the shock absorber. Which statement is correct?

   a. A only   **c.** both A and B
   b. B only   **d.** neither A nor B

5. Which of the following terms does not describe a modern shock absorber?

   a. hydraulic
   b. direct acting
   c. multi-stage operation
   d. friction dampening

6. Internal resistance in a shock absorber changes as the _____ changes.
   A. size of the fluid orifice
   B. speed of the piston

   a. A only   **c.** both A and B
   b. B only   **d.** neither A nor B

7. Multi-stages are used in shock absorbers to provide
   A. an increased amount of resistance at ride height.
   B. a steadily reducing amount of resistance as the suspension moves into jounce or rebound travel.

   a. A only   **c.** both A and B
   b. B only   **d.** neither A nor B

8. Double-tube shock absorber operation is being discussed. Statement A: The amount of oil in the reservoir increases during the compression cycle. Statement B: There is a very high pressure between the piston and the base valve during the compression cycle. Which statement is correct?

   a. A only   **c.** both A and B
   b. B only   **d.** neither A nor B

9. Statement A: Oil flows from the reservoir into the pressure tube during the extension cycle of a double-tube shock absorber. Statement B: Aerated fluid in the reservoir will cause mushy shock operation. Which statement is correct?

   a. A only   **c.** both A and B
   b. B only   **d.** neither A nor B

10. A double-tube shock absorber
    A. cannot be operated upside down.
    B. will have the piston rod attached to the car body and the shock body attached to the suspension.

    a. A only   **c.** both A and B
    b. B only   **d.** neither A nor B

11. Fluid aeration in the pressure tube is reduced by

    a. placing a coil shaped wire in the reservoir.

b. pressurizing the reservoir with a nitro-gen gas charge.

c. filling the reservoir completely with a plastic envelope of freon gas and oil.

d. any of these.

12. Single-tube shock absorber operation is being discussed. Statement A: There are two pistons in the pressure tube. Statement B: This shock absorber will extend when disconnected. Which statement is correct?

a. A only   c. both A and B
b. B only   d. neither A nor B

13. Aeration is a lesser problem in single-tube shock absorbers designs because
A. there is no air in contact with the fluid in the pressure tube.
B. of the high pressure from the nitrogen charge.

a. A only   c. both A and B
b. B only   d. neither A nor B

14. Single-tube shock absorber operation is being discussed. Statement A: During the compression cycle, the dividing piston will move away from the working piston. Statement B: The pressure inside the pressure tube will increase during the compression cycle. Which statement is correct?

a. A only   c. both A and B
b. B only   d. neither A nor B

15. Shock absorber operation is being discussed. Statement A: Oil moves through the piston in a direction away from the dividing piston during the compression cycle in a single-tube shock absorber. Statement B: The oil flow in a double-tube shock absorber is downward through the piston during the compression cycle. Which statement is correct?

a. A only   c. both A and B
b. B only   d. neither A nor B

16. The shock absorber portion of a strut
A. is an oversize, standard shock absorber.
B. can be of a single-tube or a double-tube design.

a. A only   c. both A and B
b. B only   d. neither A nor B

17. Struts are being discussed. Statement A: The upper shock absorber piston rod bushing is retained by a gland nut in some shocks. Statement B: This bushing is welded in place in some struts. Which statement is correct?

a. A only   c. both A and B
b. B only   d. neither A nor B

18. Computer-controlled suspensions are being discussed. Statement A: Shock valving can be made softer or more firm by an electric actuator. Statement B: They will be made more firm during braking and hard cornering maneuvers. Which statement is correct?

a. A only   c. both A and B
b. B only   d. neither A nor B

19. Fluid loss in a shock absorber will

a. improve the shock and strut dampening characteristics.
b. make the shock cooler.
c. reduce the dampening capability.
d. none of these.

20. A _____ shock absorber can raise the height of the car.

a. coil-over   c. de Carbon
b. air   d. all of these

Chapter 12

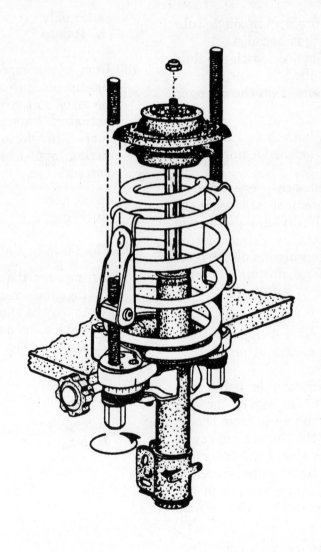

SPRING AND SHOCK ABSORBER SERVICE

After completing this chapter, you should:

- Be able to measure vehicle ride height and determine whether or not repairs are needed.

- Be able to determine the cause of noises, excessive body sway, or uneven ride heights and recommend the needed repairs.

- Be able to inspect springs, shock absorbers, and stabilizer bars for wear or damage.

- Be able to remove and replace shock absorbers.

- Be able to remove and replace a shock absorber cartridge in a front or rear strut.

- Be able to remove and replace front or rear springs and torsion bars.

- Be able to adjust a torsion bar suspension to the correct ride height.

12.1 INTRODUCTION

Shock absorbers are usually replaced on an as-needed basis by many technicians and do-it-yourselfers; springs, especially the front ones, are commonly serviced as part of a front-end rebuilding operation. Most shock absorbers can usually be easily removed and replaced using common hand tools and limited mechanical knowledge. Spring replacement often requires special tools and equipment, and it can be a very dangerous operation if the proper procedure is not followed or the normal cautions are not observed. Springs store energy, and they can release this energy very quickly.

Shock absorber replacement is much more complex on strut suspensions. It is no longer a driveway or shade tree operation. Special tools are required as well as a certain amount of knowledge.

Front spring or strut service is often done in connection with rebuilding a front end because the operation usually requires a partial disassembly of the front suspension. The ball joint stud is often disconnected when removing the spring from an S-L A suspension; as this is done, it is now a fairly easy operation to R&R the ball joint. The spring, as well as the other strut parts, are removed as part of the operation to service the shock absorber portion of a strut assembly; it is usually just as easy to install a new spring at this time as it would be to replace the old one.

Spring replacement is usually followed up with a wheel alignment. When the control arm or strut is removed to allow replacement of a spring on an S-L A or strut, the adjustable portions of the suspension are disconnected and reconnected; there is a good possibility that the alignment angles will be changed. Also, the change in height caused by the new springs changes the attitude of the vehicle and the relationship of the front suspension components, which changes some of the alignment angles. (Fig. 12–1)

Some of the steps needed to complete the operations given in this chapter, mainly the separation and connection of the ball joint stud

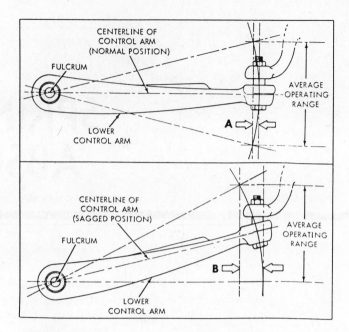

FIGURE 12–1.
Correct front suspension geometry is based on a horizontal lower control arm at ride height; a minimum amount of tire side scrub (A) will occur if this arm is horizontal. Sagging springs will cause a change in the operating range that will increase tire side scrub (B) and wear. (Courtesy of Moog Automotive)

on S-L A suspensions, are described in Chapter 15.

12.2 SPRING AND SHOCK ABSORBER INSPECTION

An inspection of the spring or the shock absorbers often begins with a customer complaint of noise, tire wear, low vehicle (one end, one side, or all over), excessive vehicle leaning on turns, or front end dive under braking. Any of these complaints might indicate weak or broken springs or shock absorbers. The best test for weak shock absorbers is a road test as described in Section 11.10. Begin your inspection by parking the car on a smooth, level surface so you can perform a bounce test and a height check.

12.2.1 Bounce Test

The bounce test is a simple and quick test that should give an indication of the condition of the suspension system.

To perform a bounce test, you should:

1. Grip one end of the bumper and alternately pull upward and push downward several times until you get that corner of the car bouncing up and down as far as you can. While the car is bouncing up and down, listen for any unusual noises that might indicate worn or broken parts.

2. With the car at the upper end of a bounce, release the bumper and watch the remaining oscillations until they stop. Two or more oscillations indicate the possibility of worn shock absorbers or, to a lesser degree, front end bushings. (Fig. 12-2)

3. Repeat Steps 1 and 2 at the other end of the bumper and compare the bouncing action of the two sides of the car. They should be the same; a difference would indicate a weak shock absorber or worn suspension bushings.

FIGURE 12-2.
A shock absorber bounce test is performed by pushing down and then pulling up on a corner of the car; after the car is bouncing as much as possible, the car is released and the bouncing action of the car is observed. If the car bounces more than 1 1/2 oscillations, the shock is probably weak. (Courtesy of Bear Automotive)

4. Repeat Steps 1, 2, and 3 at the other end of the car. Do not compare the number of bounces of the front with the rear; they are often different.

A car with no more than one or two bounce oscillations after releasing the bumper at each corner of the car and with smooth, quiet operation probably has good springs and shock absorbers, if the height and ride quality are good. Unusual or excessive noises, a differing number of bounce oscillations at each side of the car, or excessive bouncing indicate a need to follow up the bounce test with one of the remaining tests.

12.2.2 Suspension Ride Height Check

This is a simple, quick check to determine if the car is too low; weak, sagging springs let the suspension height drop. Excessive spring sag can cause excessive bottoming of the suspension (suspension members striking the bump stops severely) or tire wear and handling difficulties as the suspension system loses its proper geometry.

Ride height is a term meaning the standard no-load height of the vehicle; it is also called **trim height, curb height,** and **chassis height**. Ride height specifications are published by most vehicle manufacturers and are also available in technician's service manuals and specialized manuals available from some aftermarket spring manufacturers. (Fig. 12-3)

Ride height specifications are not published by all vehicle manufacturers; in these cases, it is possible to look at the position of the lower control arm and the condition of the bump stops to get an indication of whether the vehicle is too low. At ride height, the lower control arm of most independent suspensions should be level or slightly lower at the ball joint than at the inner pivot; sagging springs will cause the pivoting/frame end of the control arm to be lower. If the bump stops show a lot of activity or damage, the springs have probably sagged. (Fig. 12-4)

The measuring locations for ride height checks vary among manufacturers. Sometimes

MODEL	YEAR	BODY STYLE	MEASUREMENT POINTS & SPECIFICATIONS			
			FRONT		REAR #	
			Dim. Measurement Point	Specifications	Dim. Measurement Point	Specifications
CALAIS,	71-72	All (Exc. ALC)	"A"	$3\frac{7}{8} - 4\frac{5}{8}$	"B"	• $6\frac{1}{8} - 6\frac{7}{8}$
DeVILLE		ALC	"A"	$3\frac{7}{8} - 4\frac{5}{8}$	"B"	$4\frac{7}{8} - 5\frac{5}{8}$
	73-74	All (Exc. ALC)	"A"	$3\frac{7}{8} - 4\frac{5}{8}$	"B"	$5\frac{3}{8} - 6\frac{1}{8}$
		ALC	"A"	$3\frac{7}{8} - 4\frac{5}{8}$	"B"	$4\frac{7}{8} - 5\frac{5}{8}$
	1975	All (Exc. ALC)	"A"	$3\frac{1}{8} - 3\frac{7}{8}$	"B"	$5\frac{3}{4} - 6\frac{1}{2}$
		ALC	"A"	$3\frac{1}{8} - 3\frac{7}{8}$	"B"	$4\frac{7}{8} - 5\frac{5}{8}$
	1976	All (Exc. ALC)	"A"	$3\frac{7}{8} - 4\frac{5}{8}$	"B"	$5\frac{3}{8} - 6\frac{1}{8}$
		ALC	"A"	$3\frac{7}{8} - 4\frac{5}{8}$	"B"	$4\frac{7}{8} - 5\frac{5}{8}$
	77-79	Std.	"D" minus "E"	$2\frac{1}{8} - 2\frac{7}{8}$	"B"	$5\frac{3}{4} - 6\frac{1}{2}$
		ALC	"D" minus "E"	$2\frac{1}{8} - 2\frac{7}{8}$	"B"	$5\frac{3}{8} - 6\frac{1}{8}$

*With Radial Tires, Add ⅜" to Front and ½" to Rear.
\#With ALC and ELC depressurized.

FIGURE 12–3.
Suspension height specifications are available that can be used to determine if the springs have sagged and the car is too low. The commonly used measuring points are indicated in FIGURE 12–5. (Courtesy of Moog Automotive)

they will be easy-to-measure places like the bottom of the bumper to ground, the top of the fender opening to ground, or the bottom of the rocker panel to ground. If any of these measurements are used, remember that they are affected by the diameter and inflation pressure of

FIGURE 12–4.
This bump stop/strike out bumper is showing severe damage; the springs are probably sagged on this vehicle causing a ride height that is too low.

the tire and wheel; low tires will cause low height readings. Other measuring points can be the distance between the top of the axle and the bottom of the frame, between the control arm and the bump stop, or the difference in ground clearance between the ball joint and the inner control arm bushing. Making measurements of this second group of measuring points is more difficult, but they are not affected by tire diameter. Ride height measurements are usually minimum heights; a car with new springs will normally be somewhat higher than specifications. (Fig. 12–5)

To measure ride height, you should:

1. Park the car on a smooth, level surface; the ramps of a wheel alignment rack are ideal because they are level and allow easy access to the suspension members.

2. Check for unusual amounts of weight that might be in the trunk or back seat of the car. They should be removed or allowances made for any added weight; ride height specifications are given for unloaded cars.

3. Check the tire pressure and correct it, if necessary. Note whether the tires are of stock

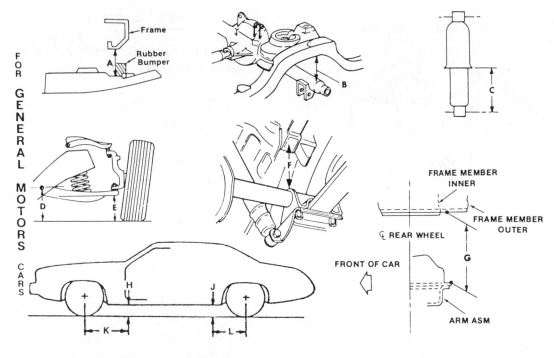

"A" Dim. — Vertical distance from top of front lower control arm in front of rubber strike-out bumper, to undersides of frame.
"B" Dim. — Vertical distance from top of rear axle housing to underside of frame side member.
"C" Dim. — Vertical distance from lower edge of front shock absorber dust shield to centerline of lower attachment stud.
"D" Dim. — Vertical distance from ground to centerline of front bushing bolt head.
"E" Dim. — Vertical distance from ground to underside of ball joint cover plate in board of and adjacent to lube fitting.

"F" Dim. — Vertical distance from top of rear axle housing to underside of bump stop bracket adjacent to rubber bumper.
"G" Dim. — Vertical distance from top of control arm flange, adjacent to shock absorber, to underside of frame outer side rail.
"H" Dim. — Ground to rocker panel at front.
"J" Dim. — Ground to rocker panel at rear.
"K" Dim. — Front wheel centerline to "H".
"L" Dim. — Rear wheel centerline to "J".

FIGURE 12-5.
The actual places where ride height is measured varies on different vehicles; the measuring location is normally indicated along with the specifications. (Courtesy of Moog Automotive)

size; if not, allowances must be made in the checking dimensions.

4. Obtain the ride height specifications and the locations of the measuring points.

5. Measure the distances at each measuring point and compare them to the specifications. Sagged springs are indicated if the measured distances are shorter or lower than specifications. (Fig. 12-6)

6. Compare the left and right measurements; they should be almost equal.

In cases where one side of the car sags more than the other, it is necessary to determine whether the lean is caused by a weak front spring, a weak rear spring, or both; either will cause this problem. The solution is simple;

FIGURE 12-6.
Ride height is usually measured with a tape measure and with the car on a flat, level surface. (Courtesy of Moog Automotive)

FIGURE 12-7.
This vehicle is probably leaning because of a weak front or rear spring; lifting the car in the exact center at front or rear will usually show if the front or rear spring is weak. (Courtesy of SKF Automotive Products)

merely lift one end of the car and see if the lean disappears. Carefully position a jack so it is exactly centered under a frame cross member at one end of the car and lift that end of the car until the tires are off the ground. Observe or remeasure the ride height; if the front end was lifted and a lean still exists, one of the rear springs is weak. If the lean goes away, a front spring is weak. (Figs. 12-7 & 12-8)

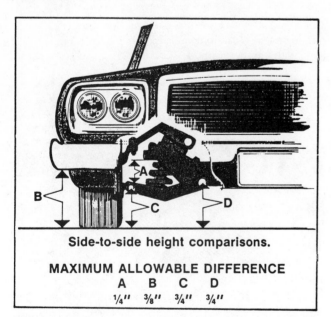

Side-to-side height comparisons.

MAXIMUM ALLOWABLE DIFFERENCE

A	B	C	D
¼"	⅜"	¾"	¾"

FIGURE 12-8.
Car lean is excessive if the difference between the right and left side is greater than the dimensions shown; note that all of these measuring points except A are affected by tire size. (Courtesy of SKF Automotive Products)

12.2.3 Visual Inspection

This is an important check in cases of noises; to make this check, you should:

1. Lift the vehicle on a frame contact hoist or a jack so the suspension will drop to allow a better view of the components.

2. Carefully inspect the springs and shock absorbers for these conditions:

    a. Excessive contact with the bump stops.

    b. Loose, worn, or broken shock absorber mounts.

    c. Worn, rough, or broken shock absorber piston rod.

    d. Badly dented shock absorber body.

    e. Leaky shock absorbers; disregard moist spots but condemn the shock if oil is dripping or running down the body of the shock absorber.

    f. Shiny, worn metal spots on the springs, shock absorber, or portions of the frame/body nearby.

    g. Improper positioning of the ends of coil springs.

    h. Loose spring or shackle mounting bolts.

    i. Worn spring or shackle bushings.

3. If any corner of the car was noisy during the bounce test, extra care should be spent while inspecting that suspension. If the cause of the noise cannot be seen, disconnect one end of the shock absorber and stroke it through its compression and extension travel to feel its internal condition.

For this check, the bottom end of the shock absorber is usually disconnected because it is easier to get to. After it is disconnected, pull down and push up on the shock absorber body as far and as fast as you can. You should only hear a faint swish of the oil through the valves, and you should feel a steady and even amount of resistance on each stroke. Hops or skips in the resistance or excessive noise indicate a faulty shock absorber. If the resistance feels high or low, compare it with the other shock absorber on the same axle.

12.3 REMOVING AND REPLACING SHOCK ABSORBERS

This is usually a rather simple job requiring the removal of only two or three nuts or bolts. Two types of problems sometimes occur to make shock absorber replacement difficult or possibly hazardous: removal of the nut on single-stud mounts and the possibility of a sudden dropping of the suspension as the shock absorber is removed.

FIGURE 12–9.
The rear axle on this car will probably drop several inches if the shock absorber (note the coil-over shock) was disconnected; note the support (arrow) positioned under the axle to prevent this from happening.

> **SAFETY NOTE**
>
> Before any job is begun, it is wise to look over the various parts to see how they are arranged and to develop a plan for the disassembly and reassembly procedures. It is a good idea to do this after the vehicle has been raised and before removing the shock absorbers. Be careful to note what devices are used for the extension or down-travel suspension stops. Many coil sprung rear axles use the shock absorber as the limiter for axle down-travel; when the shock absorber is removed the axle can drop several inches until it is stopped by the brake hose or other things under the axle. This will probably damage the brake hose and other vehicle parts as well as possibly injuring you. The axle should be supported by stands or some other lifting device before removing the shock absorber. (Fig. 12–9)

The nut on single-stud shock absorber mounts tends to rust and seize in place. Its removal is made more difficult because the stud and piston rod will often rotate during removal attempts. The polished piston rod should never be gripped with plain or locking pliers. The smooth surface will be damaged, which in turn will ruin the upper shock absorber bushing and seal, and the polished surface is too smooth to grip tightly. There are several approved ways of removing this nut; begin by applying penetrating oil to the threads. Next try one of these methods:

1. Often a fast operating air wrench will spin the nut off.

2. The shock absorber stud has two flats on the end, which can be held using a special wrench. (Fig. 12–10)

3. If the nut is rusted in place, the shock stud can be broken off by placing a deep-well socket with extension bar over the nut or thread, or a special tool made just for this purpose, onto the exposed threads of the stud. Bend the shock stud back and forth until it breaks off; this, of course, ruins the shock absorber. (Fig. 12–11)

4. A **nut breaker** can be placed over the nut and tightened to break or snap the nut open for easy removal. (Fig. 12–12)

5. An air impact hammer and chisel can be used to cut down each side of the nut until the nut snaps or is cut into two parts. (Fig. 12–13)

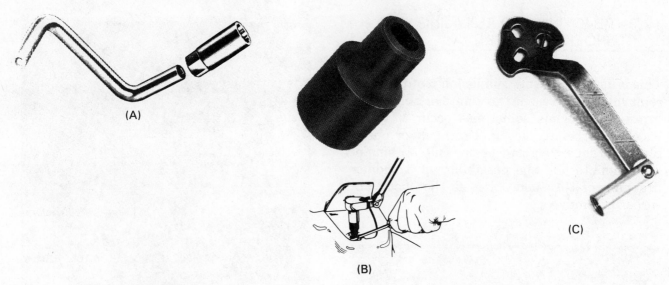

FIGURE 12–10.
This odd-shaped tool (A) has sockets in each end to fit the flats of a shock absorber stud; it will hold the stud from turning while the nut is being unscrewed using the deep-well socket. The socket (B) is shaped to fit the shock stud so the stud can be rotated while the nut is held with a wrench. Shock wrench C has three different shaped openings to fit different shock studs; the tubular shaped end can be threaded onto a shock stud to pull it into position. (A is courtesy of Specialty Products; B is courtesy of Lisle)

To remove shock absorbers, you should:

1. Raise and support the car securely on a hoist or jack stands.

NOTE: On those cars where the upper-front shock absorber mounts can be reached from under the hood, it is often faster to disconnect the upper shock absorber mount before raising the car.

2. Disconnect the upper shock absorber mount, if it is not already done.

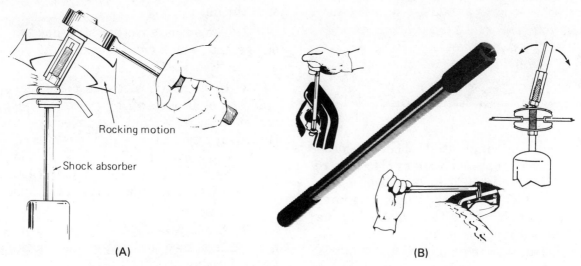

FIGURE 12–11.
Usually the quickest way to remove a stud-mount shock absorber is to break off the stud by bending it back and forth several times. This can be done using a deep-well socket (A) or a special tool (B). (B is courtesy of Lisle)

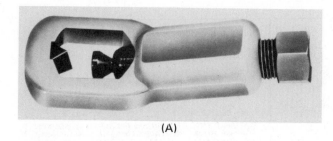

(A)

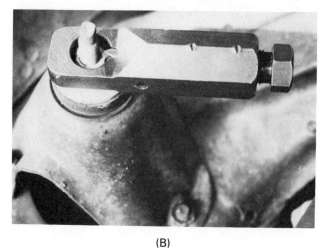

(B)

FIGURE 12-12.
A nut-splitting tool (A) can be placed over the stud nut (A) and tightened to split the nut open for easy removal. (A is courtesy of Specialty Products)

FIGURE 12-13.
An air hammer and chisel can be used to split the nut from the shock absorber stud. (Courtesy of Superior Pneumatic)

SAFETY NOTE
While removing the nut on stud-type mounts, observe the rubber bushing; if it stays under a compression load while the nut is being removed, the suspension will probably drop when the nut leaves the threads. In this case it is best to lift the suspension slightly or lower the body to remove this load.

3. Disconnect the lower mount and remove the shock absorber.

At times it is necessary to purge the air out of the pressure tube of a double-tube shock absorber; this is also called bleeding a shock absorber. If the shock absorber was stored on its side, air will enter the pressure tube. Air is compressible and also will pass through the shock valving much faster and easier than oil. A shock with air in the pressure tube will offer very little resistance for part or all of a stroke. When mounted on a car, a shock will purge itself after enough compression and extension cycles.

To purge the air from the pressure tube of a double-tube shock absorber, you should:

1. Compress the shock while holding it in an upside-down position.
2. Place the shock in an upright position and extend the piston rod.
3. Repeat Steps 1 and 2 several more times or until an even resistance is felt for the entire stroke. A shock absorber that does not develop a full stroke resistance is faulty.

To replace a shock absorber, you should:

1. Check the bushings and other pieces of shock absorber mounting hardware to make sure they are in good condition and assemble these parts in the proper relationship. (Fig. 12-14)
2. Place the shock absorber with bushings in position and connect the upper mount. On

SHOULDERED STEM

TIGHTEN NUT UNTIL RETAINER BOTTOMS ON STEM SHOULDER.

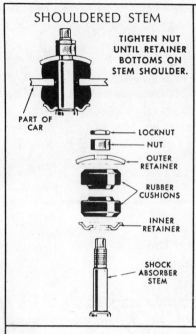

PART OF CAR

LOCKNUT
NUT
OUTER RETAINER
RUBBER CUSHIONS
INNER RETAINER
SHOCK ABSORBER STEM

NON-SHOULDERED STEM

TIGHTEN NUT UNTIL CUSHIONS BULGE ALMOST TO OUTER EDGE OF RETAINERS. DO NOT TIGHTEN EXCESSIVELY.

PART OF CAR

NUT
OUTER RETAINER

CUSHIONS HAVE A LARGE AND SMALL PILOT. USE THE PILOT THAT FITS THE HOLE IN THE CAR.

INNER RETAINER
SHOCK ABSORBER STEM

CROSS PIN MOUNTING

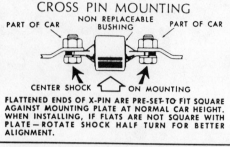

PART OF CAR   NON REPLACEABLE BUSHING   PART OF CAR

CENTER SHOCK   ON MOUNTING

FLATTENED ENDS OF X-PIN ARE PRE-SET-TO FIT SQUARE AGAINST MOUNTING PLATE AT NORMAL CAR HEIGHT. WHEN INSTALLING, IF FLATS ARE NOT SQUARE WITH PLATE — ROTATE SHOCK HALF TURN FOR BETTER ALIGNMENT.

USE NEW MOUNTING PIN PARTS TO REPLACE THE ORIGINAL PIN

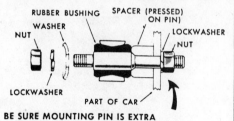

RUBBER BUSHING   SPACER (PRESSED) ON PIN)
WASHER   LOCKWASHER
NUT   NUT
LOCKWASHER   PART OF CAR

BE SURE MOUNTING PIN IS EXTRA TIGHT TO PREVENT IT FROM LOOSENING

MOUNTING WITHOUT INNER SLEEVE

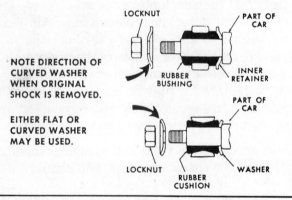

LOCKNUT   PART OF CAR
RUBBER BUSHING   INNER RETAINER
PART OF CAR

NOTE DIRECTION OF CURVED WASHER WHEN ORIGINAL SHOCK IS REMOVED.

EITHER FLAT OR CURVED WASHER MAY BE USED.

LOCKNUT   WASHER
RUBBER CUSHION

MOUNTING WITH INNER SLEEVE

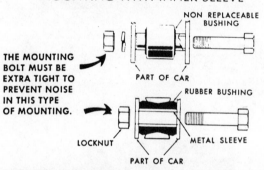

NON REPLACEABLE BUSHING
PART OF CAR

THE MOUNTING BOLT MUST BE EXTRA TIGHT TO PREVENT NOISE IN THIS TYPE OF MOUNTING.

RUBBER BUSHING
LOCKNUT   METAL SLEEVE
PART OF CAR

MAKE SURE METAL SLEEVE IS INSULATED IN RUBBER BUSHING. RE-USE MOUNTING BOLT, NUT AND LOCKWASHER.

DO

1. OBSERVE THE LOCATION of all mounting parts, washers, retainers, etc. as they are removed, and be sure to reassemble these parts in the same positions.
2. USE CARE when installing the new units. Be sure all nuts and bolts are securely tightened. Check clearances.
3. FOLLOW SPECIAL INSTRUCTIONS WHEN SUPPLIED WITH UNITS.

DO NOT

1. DO NOT TWIST OFF STUDS. If nuts are rusted, use penetrating oil on threads, allowing a few minutes for the oil to penetrate before unscrewing. If nut has rusted to a point where penetrating oil does not permit removal, it is recommended that a nut-splitter be used, taking care not to damage the threads on the mounting, if it is part of the car.
2. DO NOT GRIP POLISHED PISTON ROD with tools during installation. This will damage rod and cause unit failure.
3. DO NOT HANG WHEELS on cars having rear coil springs or air springs when changing shock absorbers.

FIGURE 12–14.

Different installation procedures are required for the different types of shock mounts. (Courtesy of Monroe Auto Equipment)

single-stud mounts, tighten the nut to compress the rubber bushing, but not too much. The correct amount of tightness will squeeze the rubber bushing to about the same diameter as the metal washer. If the nut has not tightened securely against a metal shoulder or sleeve at this point, or if a self-locking, prevailing torque nut is not used, a jam or pal nut should be used to lock the retaining nut in position.

3. Place the lower bushings in position and connect the lower shock absorber mount. Tighten the retaining nut(s) or bolt(s) to the correct torque.

Shock absorber service on strut suspensions is described in Section 12.7.

12.4   REMOVING AND REPLACING COIL SPRINGS

In some styles of rear suspensions using coil springs, the spring can easily be removed once the shock absorber is disconnected. In fact, on some cars, the spring will almost fall out; it is not recommended to operate a car that has rear coil springs with a shock absorber disconnected for this reason. On these cars, spring replacement is a matter of disconnecting the shock absorber, lowering the rear axle, or raising the car enough to lift out the old spring; checking to make sure the spring pads, if used, are in good condition; setting the new spring in place in the correct position; and replacing the shock absorbers. When the axle is raised to reinstall the shock absorber, the spring will be locked in place.

When installing new springs, be sure to get the correct replacement. Many cars use one of several different spring rates depending on the engine and the accessories that are on it. One manufacturer even uses a different spring on the right and left side on one model of car. The best method of getting the exact replacement is to use the part number, which is often a paper or plastic tag fastened to the old spring. Also note that some springs have a top and a bottom; they should be installed right side up.

If the car has a slight lean, a shim can be added to the spring on the low side so the car will sit level. (Fig. 12–15)

Springs are normally replaced in pairs; it is impossible to buy a new spring that will exactly match the rate and length of a used spring. Replacement of only one spring will usually cause a leaning of the vehicle or an odd ride quality.

Coil spring replacement on strut suspensions is described in Section 12.5

12.4.1   Removing and Replacing a Coil Spring, S-L A Suspension

To remove a coil spring from an S-L A suspension, you should:

1. Raise and support the car securely on a hoist or jack stands.

2. Remove the wheel, the stabilizer bar end-links, and the shock absorber.

3. Break the taper of the lower ball joint; this operation is described in Section 15.2. This step will not be necessary if you plan on disconnecting the lower control arm at the inner pivots.

FIGURE 12–15.
The original springs often have a tag on them that indicates the part number; this tab is very helpful in getting the exact replacement. This replacement spring set has a tag on one spring that specificies the spring for the driver's side. (Courtesy of Moog Automotive)

SAFETY NOTE

Removing a spring from this suspension requires a careful procedure and the use of special tools; this spring is very powerful and can cause injury. Depending on the equipment available, there are several ways of correctly performing this job; they begin in a similar fashion. It is wise follow the procedure recommended by the vehicle manufacturer and use the special spring tools in the manner recommended by their manufacturer.

If a spring compressor is used:

4. Install the spring compressor and tighten it until there is clearance at one end of the spring.

CAUTION: Observe the rebound bump stop; it will have very little load on it when the spring has been compressed sufficiently. (Fig. 12–16)

5. Disconnect the lower ball joint stud or remove the inner control arm pivot bolts and swing the lower control arm downward, releasing the compressed spring. Some sources recommend disconnecting the control arm pivot bolts at the inner bushings and pivoting the control arm downward at the ball joint; if you use this method, use care because the control arm can twist, slip, and allow the spring to escape. A jack adapter is available to make this job more secure. (Figs. 12–17 & 12–18)

6. Remove the compressed spring. If force is necessary, use a pry bar; be careful not to let your hand get trapped around a coil spring.

If spring clips are used:

4. Select the shortest clips that can be used that will capture as many coils as possible, and position the clips so they are toward the center of the car. It is possible to use a single clip, but most technicians recommend using two clips at a time for safety reasons. (Fig. 12–19)

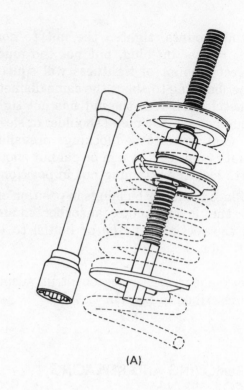

(A)

(B)

FIGURE 12–16.
Two of the commonly used types of spring compressors, used to take the pressure off the control arm for safe and easy spring removal. (A is courtesy of Moog Automotive; B is courtesy of SKF Automotive Products)

FIGURE 12-17.
A spring can be removed by disconnecting the ball joint stud and lowering the control arm downward with the aid of a jack. Note the use of a spring clip (arrow) to keep the spring partially compressed. (Courtesy of Moog Automotive)

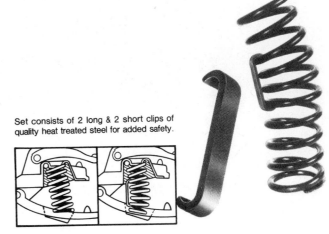

Set consists of 2 long & 2 short clips of quality heat treated steel for added safety.

FIGURE 12-19.
Spring clips (normally used in pairs) can be placed onto the spring while the control arm is still in the normal position; during removal, they will keep the spring partially compressed. Also, note the curve the spring takes as it tries to extend. (Courtesy of Moog Automotive)

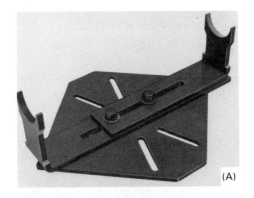

(A)

(B)

FIGURE 12-18.
A spring can be removed by disconnecting the inner pivot bolts and lowering the control arm down with the aid of a jack (A); the jack attachment (B) supports the control arm to keep it from slipping. (Courtesy of Kent-Moore)

HELPFUL HINT: Lifting the opposite rear corner of the car will place more load on the front corner; this will compress the spring farther, usually allowing more spring coils to be captured by the spring clip.

5. Position a jack under the control arm at the outermost position possible.

CAUTION: The jack must be able to travel inward a few inches as the control arm swings downward; the control arm might slip off the jack or the jack might tip if the jack cannot move inward. (Fig. 12-20)

6. Disconnect the lower ball joint stud, and carefully lower the jack to allow the control arm to swing downward.

7. After the control has swung downward, remove the spring; a pry bar is often needed to work the spring out, and it is also safer than placing your fingers between the spring and control arm.

While the spring is out, check the upper and lower spring mounts to determine how the spring is positioned. Note that many spring mounts have positioning locations for the ends of the coil as well as the sideways and lengthwise locations. (Fig. 12-21)

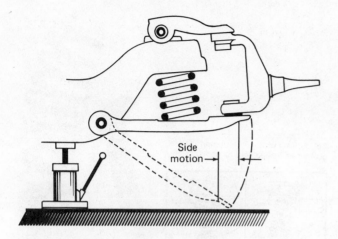

FIGURE 12-20.
As a control arm is lowered, the free end travels in an arc; note that the free end will travel sideways as well as downward. If a jack is placed under the control arm to lower it, the jack must be able to follow the movement of the control arm.

To prepare a spring for installation, you should:

If a spring compressor was used:

1. Measure the length of the old compressed spring, note the location of the spring compressor and compressor plates, and remove the spring compressor. (Fig. 12-22)

2. Position the spring compressor on the new spring in the same manner and position that it was on the old spring, and compress the new spring to the same length as the old one during removal. Failure to position the compressor and plates in the correct position might make them difficult or impossible to remove after the spring has been installed.

If spring clips were used:

1. Note the position of the clips on the old spring.

2. Install a spring compressor and compress the spring enough to allow removal of the clips. (Fig. 12-23)

3. Install the spring compressor on the new spring and compress the spring enough to

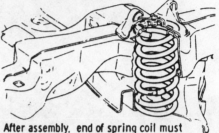

Spring to be installed with tape at lowest position. Bottom of spring is coiled helical, and the top is coiled flat with a gripper notch near end of wire.

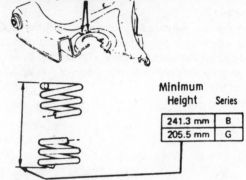

After assembly, end of spring coil must cover all or part of one inspection drain hole. The other hole must be partly exposed or completely uncovered.

Minimum Height	Series
241.3 mm	B
205.5 mm	G

FIGURE 12-21.
There is usually a pocket in which each end of the spring must fit during installation, and one end of the spring has a definite location where it must be placed. (Courtesy of Buick)

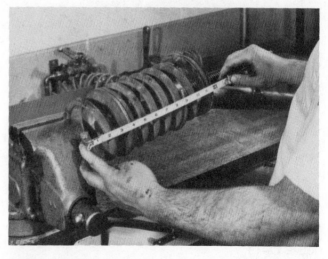

FIGURE 12-22.
A new spring is usually compressed to make it easier to install; it should be the same length as the old one was as it was removed. (Courtesy of Moog Automotive)

(A)

FIGURE 12–23.
When using spring clips to aid coil spring removal and installation, it is important that the clips be transferred to the same position on the new spring that they were on the old spring. Mark the clip location on the old spring and compress the spring to remove the clips (A), transfer the marks to the new spring (B), and then compress the new spring so the clips can be installed at the marked location. (C). (Courtesy of Moog Automotive)

(B)

(C)

allow installation of the clips in the same position as they were on the old spring during removal.

4. Remove the spring compressor.

To install a coil spring, you should:

If a spring compressor is used:

1. Place the compressed spring in the pocket and position it in the correct relationship with seats.

2. Swing the control arm upward, position the ball joint stud into the steering knuckle, and thread the retaining nut onto the ball joint stud. Tighten the ball joint stud retaining nut; the correct procedure for doing this is described in Section 15.7. (Fig. 12–24)

If the control arm was disconnected at the inner pivots, swing the control arm upward, align the holes, and install the bolts and nuts; a punch can be used to help align the holes in the control arm bushing with those in the frame/body. Final tightening of these bolts should be made with the vehicle weight on the tires. (Fig. 12–25)

3. Remove the spring compressor and reinstall the shock absorber, stabilizer bar endlinks, and wheel.

If spring clips are used:

1. Place the spring, with clips in place, in the spring pocket; the curve of the spring should be such that the ends of the spring approximately match the angles of the spring seats.

2. Using a jack, swing the control arm upward until the ball joint stud enters the steering knuckle boss, and the retaining nut can be placed onto the ball joint stud. As the control arm swings upward, check to make sure that

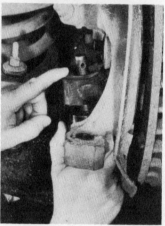

the ends of the spring are in the correct position in the spring seats.

3. Tighten the ball joint stud retaining nut to the correct tightness, as described in Section 15.7.

4. Remove the spring clips and reinstall the shock absorber, stabilizer bar end-links, and wheel.

In those cases where the coil spring is mounted above the upper control arm, a spring compressor is used to compress the spring enough to allow the spring to be lifted out of the spring pocket. If you are unsure of how to do this, consult a manufacturer's or technician's service manual. (Fig. 12–26)

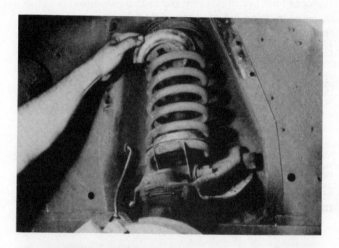

FIGURE 12–24.
If the ball joint stud was disconnected for spring removal, the spindle hole should be cleaned and inspected before reassembly; also, position the ball joint stud to allow access to the cotter pin hole. (Courtesy of Moog Automotive)

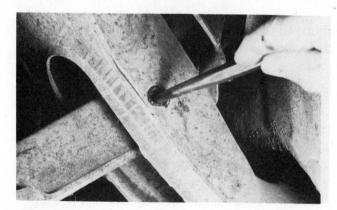

FIGURE 12–25.
If the inner pivot bolts were disconnected for spring removal, a punch can be used to pry the control arm (arrow) for easy bolt installation.

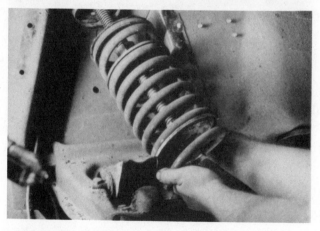

FIGURE 12–26.
When removing a coil spring that is mounted on the upper control arm, use of a spring compressor often allows spring removal with the control arm still in place. (Courtesy of Moog Automotive)

12.5 LEAF SPRING SERVICE

At this time, leaf springs are in limited use in passenger cars; only a small amount of service work is commonly done on them. The service operations that are occasionally done are: R&R of a spring, R&R of the spring and/or shackle bushings, and R&R of the spring center bolt.

To remove a leaf spring, you should:

1. Raise and support the car securely on a hoist or jack stands.

2. Remove the wheel.

3. Place a support under the axle and remove the axle to spring mounting "U" bolts.

SAFETY NOTE

Be sure that both the car and the axle are safely supported.

---

4. Depending on its position to the spring, raise or lower the axle so there is clearance between the axle and the spring.

5. Remove the front pivot bolt and allow the spring to hang down.

6. Remove the shackle bolts, the shackle, and the spring.

To replace a leaf spring, you should:

1. Assemble the shackle to connect the spring to the frame/body mounts; start the nuts onto the shackle bolts enough to hold the parts together.

2. Swing the spring up into position, place the pivot bolt in position, and thread the nut onto the bolt a few turns.

3. Bring the axle next to the spring, making sure that the axle is centered with the spring center bolt. Install and tighten the axle to spring mounting "U" bolts, and tighten them to the correct torque.

4. Tighten the pivot and shackle bolts to the correct torque.

5. Install the wheel.

Leaf spring bushings are of two basic types: a pair of straight or shouldered rubber bushings that are slid into the spring eye from each side or a one-piece metal-backed rubber bushing. The metal-backed bushings are pressed into the spring eye and can be replaced in a manner very similar to a control arm bushing; control arm bushing replacement is described in Chapter 15. The two-piece rubber bushings are simply slid into the spring eye; normally no special tools are required. The operation of installing rubber bushings can be made easier by using tire mounting, rubber lube. (Fig. 12-27)

Removing and replacing a spring center-bolt is merely a matter of taking the old bolt out and putting in the new one. If possible, a "C" clamp can be placed over the spring before removing the centerbolt to hold the leaves together while the centerbolt is changed. Normally, a centerbolt that is too long is installed to make it easier to pull the various leaves into

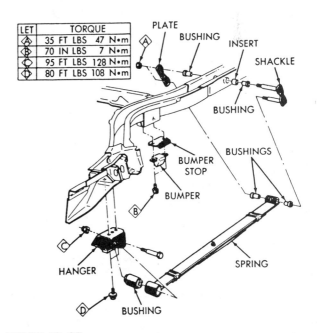

LET	TORQUE	
Ⓐ	35 FT LBS	47 N•m
Ⓑ	70 IN LBS	7 N•m
Ⓒ	95 FT LBS	128 N•m
Ⓓ	80 FT LBS	108 N•m

FIGURE 12–27.
Faulty leaf spring bushings can be removed and replaced with new ones. (Courtesy of Chrysler Corporation)

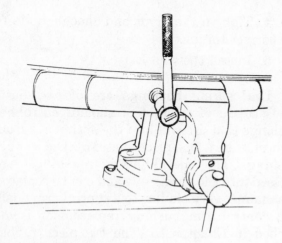

FIGURE 12-28.
When removing or replacing a spring center bolt, it is a good practice to use a vise or clamp to hold the spring leaves together. (Courtesy of Chrysler Corporation)

position; after the bolt is tightened, the excess length is cut off. (Fig. 12-28)

12.6   TORSION BAR SERVICE

Normal torsion bar service consists of adjusting the ride height and removing and replacing the torsion bar. The exact procedure to adjust or remove and replace a torsion bar varies with the manufacturer of the car. A manufacturer's or technician's service manual should be followed when performing these operations on a specific car model; the procedure described here is very basic and general.

To adjust a set of torsion bars, you should:

1. Park the car on a smooth level surface; an alignment rack is ideal because it allows easy access to the measuring and adjusting points and the side movement from the turntables will remove any binding which might result from tire scrub while the height adjustment is being made or when the car is lowered.

2. Locate the adjustment at the end of each torsion bar and lubricate the bolt threads. (Fig. 12-29)

3. Obtain the height specifications and measure the front ride height. If the measure-

FIGURE 12-29.
Turning the torsion bar adjuster screw changes the vehicle's ride height. (Courtesy of SKF Automotive Products)

ments are not within limits to each other or with the specifications, an adjustment is necessary. Gauges are available for some car models to make the height checks easier and more accurate. (Fig. 12-30)

4. Turn the adjusting screws as required to bring the heights into specification limits; it is a good practice to bounce the suspension to normalize the height after making an adjust-

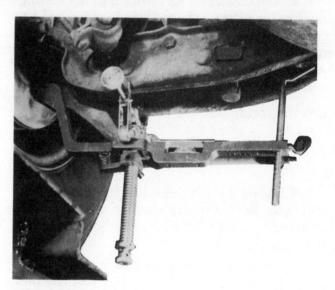

FIGURE 12-30.
This torsion bar height gauge (arrow) is clamped onto the control arm so one end is in contact with the ball joint. Then the gauge unit is adjusted to level and the measuring blade (right side) is moved to contact the inner pivot bushing so torsion bar height can be read from the gauge. (Courtesy of SKF Automotive Products)

ment and before taking a new measurement. The adjustment is complete when the heights are correct.

To remove a torsion bar, you should:

1. Raise and support the car securely on a hoist or jack stands.

2. Remove the wheel, the shock absorber, and the stabilizer bar end-links.

3. Turn the torsion bar adjuster screw all the way out to release as much of the torsion bar pressure as possible. It is a good practice to count and record the number of turns released so that during the replacement a quick, rough adjustment can be made by retightening the adjuster screw the same number of turns.

4. On some cars, the bar can be removed by removing the mounting bolts and then the bar. On many cars, the torsion bar is slid forward or backward, after removing a locating clip, so it leaves the hex-shaped opening or splines at one end; then the bar is slid the other way and out of the second opening. (Figs. 12-31 & 12-32)

To replace a torsion bar, you should:

1. Slide the torsion bar into position and replace the retaining clip. Or, place the bar into position and replace the mounting bolts. Tighten the bolts to the correct torque.

2. Retighten the adjusting bolt the same number of turns that it was loosened.

3. Replace the stabilizer bar end-links, the shock absorber, and the wheel.

4. Adjust the torsion bars to bring the car to the correct height.

12.7   STRUT SERVICE

Normal strut service includes rebuilding or replacing the shock absorber portion, removing and replacing the spring, or removing and replacing the upper mount/damper assembly. With experience and the proper equipment,

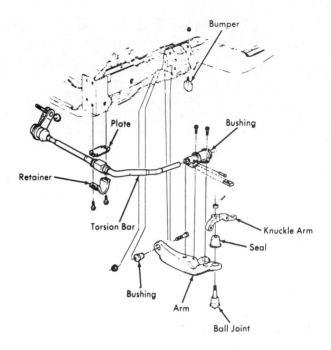

FIGURE 12-31.
A disassembled view of a later model Chrysler Corporation torsion bar. (Courtesy of Mitchell Information Services, Inc.)

some of these operations can be performed in the fender opening with the lower end of the strut still connected. This saves time and difficulty in disconnecting the lower end of the strut. In the average metropolitan garage or repair shop with labor rates of $25 to $50 per hour, time-saving steps become very important. Many technicians prefer to remove the strut assembly from the car for better access to the parts, and this is highly recommended. Many struts do not require much effort to remove them entirely, and the access and ease of working on the strut off the car helps ensure a better job. (Figs. 12-33 & 12-34)

The procedure to remove a strut varies depending on how the strut is attached at the bottom to the steering knuckle or control arm. Some manufacturers connect the strut to the steering knuckle using a pair of bolts or a telescoping connection and a pinch bolt. Some manufacturers weld the steering knuckle solidly to the bottom of the strut. On some of these, the steering arm is bolted to the bottom of the strut along with the ball joint boss; on others, the steering arm is forged as part of the

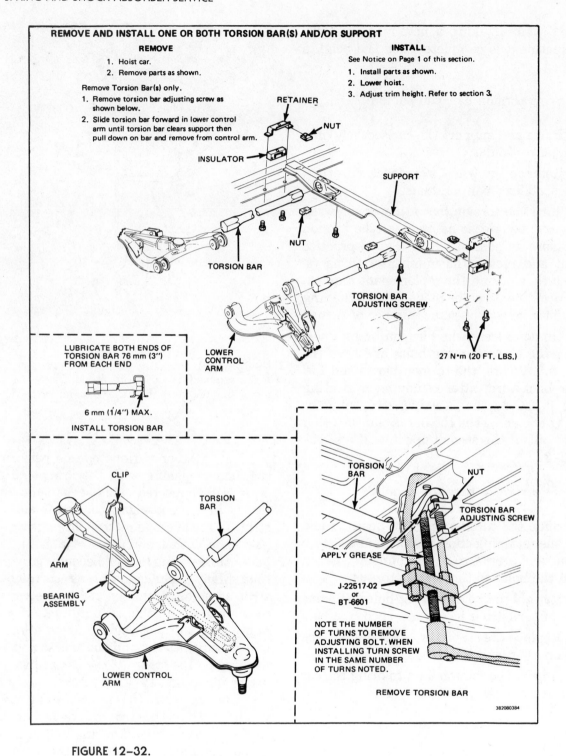

FIGURE 12-32.

The procedure to use when removing and replacing torsion bars from some General Motors cars. (Courtesy of Oldsmobile)

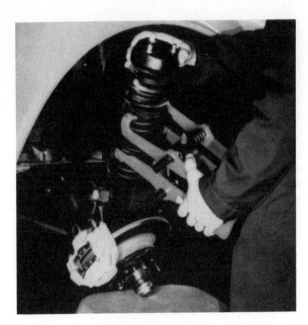
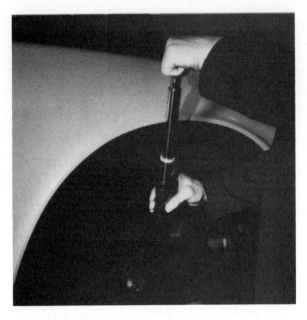

FIGURE 12–33.
On some cars, the top of the strut can be disconnected and pivoted outward enough to allow disassembly and replacement of the shock cartridge while the lower end is still connected. (Courtesy of Monroe Auto Equipment)

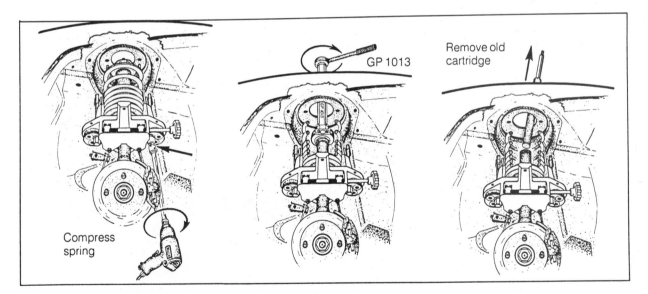

FIGURE 12–34.
A strut's upper spring mount and shock body nut can be removed while the bottom of the strut is still connected if you have the correct special tools. This sequence shows compressing the spring (left), removal of the strut body gland nut (center), and the removal of the old shock absorber cartridge (right). (Courtesy of GP Tools)

steering knuckle, and the ball joint boss is bolted to the bottom of the strut. Some newer cars locate the steering arm on the upper part of the strut, just under the spring seat. Some cars require disconnecting of the lower ball joint, the tie rod end, and the brake caliper. At this time, most FWD cars allow the separation of the strut from the steering knuckle, saving the time needed to disconnect the front hub and CV joint. More parts that are made into a strut assembly mean more parts that need to be disconnected for strut removal. (Fig. 12–35)

After the spring has been removed, many struts can be easily dismantled for internal service. Depending on the availability of parts and the condition of the old parts, the shock absorber portion of these struts can be rebuilt with new parts and new oil or have a cartridge installed. It is a common practice to remove all of the old parts and oil and install a replacement shock absorber cartridge. Some struts are welded together so a replacement strut assembly is required. This style usually is one that bolts to the steering knuckle, so it really is not too much more expensive than just the cartridge. Some of the welded struts can be cut open using a tubing or pipe cutter to allow removal of the old shock absorber parts and the installation of a new cartridge. Internal threads are already in the upper portion of the outer tube for installation of a gland nut. Be careful, if cutting a strut open, as some struts are charged with nitrogen gas under a high pressure; disassembly of these struts can cause injury. (Fig. 12–36)

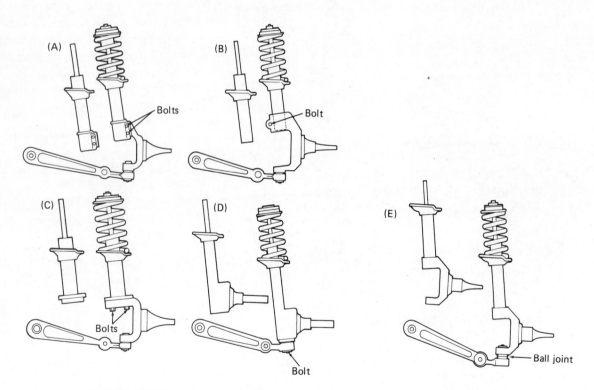

FIGURE 12-35.
There are at least five general ways to connect the strut to the steering knuckle or lower ball joint. Style A uses a pair of bolts to fasten the strut to the steering knuckle; style B slides the strut into the steering knuckle and locks it in place with a pinch bolt; style C uses three bolts to connect the strut to the steering knuckle; in style D, the strut is part of the steering knuckle and they are bolted to the steering arm or ball joint connector; and style E is a single, one-piece unit including the strut, steering knuckle, and steering arm.

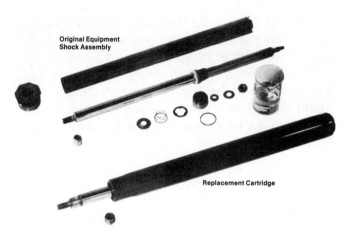

FIGURE 12-36.
The parts of the original shock absorber assembly from an imported car strut are shown at the top; these are normally replaced with a cartridge like the one at the bottom. (Courtesy of SKF Automotive Products)

SAFETY NOTE

Care should be exercised when compressing a strut spring to protect the spring as well as yourself. Many spring compressors have padded jaws to protect the painted or epoxy coating on the springs. This coating is to retard rust action, which can etch or pit the spring. Rust pitting can cause a stress raiser, which can lead to spring breakage. Spring breakage not only causes a sudden lowering of that corner of the car, but there have been cases where the broken end of the spring spiraled off the strut and into the tire. This can cut the tire, which could cause a loss of control of the car. Another potential hazard is caused by the normal tendency of a coil spring to twist as it is compressed or extended, and this tendency can cause it to slip loose. Many compressors have extra hooks or safety straps to contain the spring in case of slippage; they should be used if available. Also, many newer springs are fairly long and require a compressor with sufficient travel to allow the spring to completely extend without the compressor coming apart.

12.7.1   Special Strut Service Points

Some struts require special service procedures or precautions. Some cars use spring seats, which are positioned off-center with the ends of the spring at an angle to the strut. The lower spring seat is welded to the strut body, but the upper seat is held in place by the upper mount and can be turned. If this seat is not positioned correctly, the spring will be positioned improperly. It is a good practice to mark the upper mount and spring seat before disassembly to ensure the proper location during reassembly. (Fig. 12-37)

General Motors cars use springs that are coated with epoxy to retard rust. It is recommended that spring compressors used on these springs have jaws that are padded and free of any sharp edges that might cut the spring coating.

Most FWD Ford Motor Company cars use a strut-to-steering-knuckle connection similar to B in Figure 12-35; the bottom of the strut fits into the steering knuckle and is locked in place by a pinch bolt. It is recommended that the spring be partially compressed so as to re-

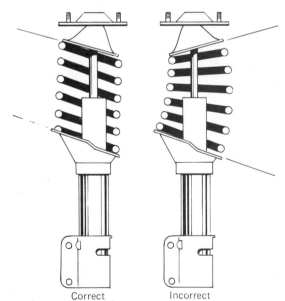

Correct   Incorrect
Note: Mark on the upper spring seat must line up with mark on top spring coil. Otherwise, the spring will bow.

FIGURE 12-37.
The upper spring seat on some struts is mounted off center or at an angle; it is a good practice to mark the upper spring seat to ensure its replacement in the correct position.

duce the length of the strut before removing the strut from the steering knuckle. Failure to compress the spring will require that the steering knuckle be pried downward an excessive amount to allow separation. This excessive travel can apply an excessive amount of force on the CV joints, possibly damaging them. An indication that this damage has occurred is a steering wheel shake, side-to-side rotation, under acceleration.

Some Chrysler Corporation "K" cars use a sintered metal bushing for the upper strut mount steering pivot. This bushing requires periodic lubrication to prevent the car from developing poor returnability, or **memory steer.** Memory steer causes a pull in the last direction that the car was turned. To lubricate this bushing you need to compress the spring, remove the top mount, and remove the bushing. Then a coating of grease can be put on the bushing and the bushing replaced. This style of bushing needs lubing every 15,000 to 20,000 miles (24,000 to 32,000 km); it can be replaced with a mount assembly that uses a ball bearing, as used on many "K" cars. (Fig. 12–38)

Some import cars use an anaerobic locking compound on the strut body gland nut to help ensure the retention of the nut. This compound makes the removal of the nut extremely difficult. It softens at temperatures of 200 to 275° F (93 to 135° C). Nut removal can be made easier by heating the top of the strut body and gland nut to temperatures slightly above the boiling point of water, using a propane or oxacetylene torch.

12.7.2 Strut Removal and Replacement

Strut removal and replacement procedures vary among different cars. It is wise to follow the manufacturer's recommended procedure.

To remove a strut assembly, you should:

1. Mark one of the strut upper mounting bolts so the upper mount can be replaced in the same position, loosen the nut on one of the mounting bolts so it is several turns from coming off, and remove the nuts from the remaining mounting bolts. (Figs. 12–39 & 12–40)

2. If you are planning on disassembling the strut, loosen the center retaining (piston rod) nut a couple of turns.

CAUTION: **DO NOT** *remove the nut at this time. Many technicians skip this step until the*

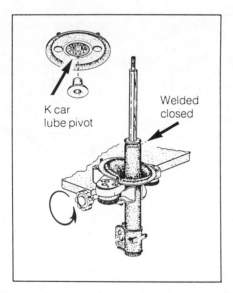

FIGURE 12–38.
Some Chrysler Product "K" cars use a sintered steering pivot bushing; this bushing requires periodic lubrication. Note that this strut is welded together at the upper bushing. (Courtesy of GP Tools)

FIGURE 12–39.
Before removing the upper strut mounting bolts, one of them (usually either the inner most or most forward bolt) is marked so the strut mount can be replaced in the same position. (Courtesy of SKF Automotive Products)

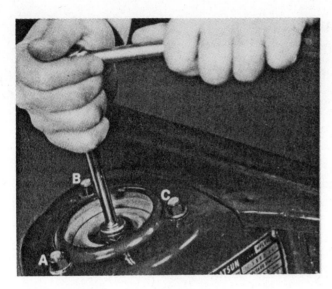

FIGURE 12–40.
Some mechanics prefer loosening the piston rod nut before removing the strut; DO NOT remove this nut at this time. The strut mounting nuts (A, B, & C) should be removed. (Courtesy of SKF Automotive Products)

strut is off the car and a spring compressor has been installed.

3. Raise and support the car securely on a hoist or jack stands.

4. Remove the wheel, stabilizer bar end-links, and strut rod if necessary.

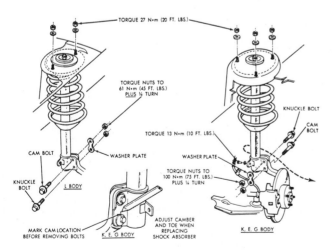

FIGURE 12–41.
Some struts can be removed from the steering knuckle by removing two or three bolts. (Courtesy of Chrysler Corporation)

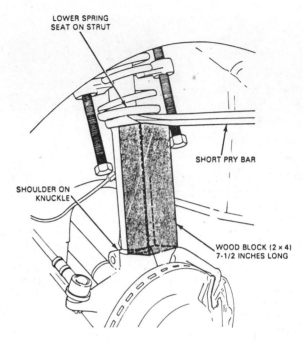

FIGURE 12–42.
Some struts are removed by removing a pinch bolt and sliding the end of the strut out of the steering knuckle. Note the spring compressor on the strut to shorten the spring for easy removal of the strut. (Courtesy of Ford Motor Company)

5. Disconnect any brake hose supports that attach to the strut. On cars where the steering arm is part of the strut, disconnect the tie rod from the steering arm.

6. As required by the strut design:
   a. Remove the strut-to-steering-knuckle nuts and bolts. (Fig. 12–41)
   b. Remove the strut-to-steering-knuckle pinch bolt. (Fig. 12–42)
   c. Remove the brake caliper from the steering knuckle and suspend it from some point inside of the fender and remove the steering-arm/ball-joint-boss-to-strut-assembly bolts. (Fig. 12–43)

   OR

   d. Remove the tie rod end from the steering arm, remove the brake caliper from the steering knuckle and support it from some point inside of the fender, and remove the ball-joint-boss-to-strut-assembly bolts. (Fig. 12–44)

FIGURE 12–43.
On some struts, the bottom of the strut bolts to the steering arm; the brake caliper needs to be disconnected and removed with the strut or removed from the strut. (Courtesy of SKF Automotive Products)

NOTE: Strut removal is usually easier if a spring compressor can be installed to compress and shorten the spring. This requires a small, compact compressor.

7. If necessary, pry downward on the lower control arm to separate it from the strut.

8. Remove the remaining nut from the up-

FIGURE 12–44.
On some struts, the ball joint stud (A), the steering arm (B), and the brakes need to be disconnected for strut removal. (Courtesy of Moog Automotive)

per mount, and remove the strut assembly from the car.

To install a strut assembly, you should:

1. Slide the strut into position with the previously marked mounting bolt in the correct hole and finger-tighten the nut.

2. If necessary, pry the control arm downward to allow the lower end of the strut to be placed in the correct position.

3. Reattach the lower end of the strut at those locations that were disconnected during removal; tighten all of these nuts and bolts to the correct torque.

4. If necessary, replace the brake caliper and brake hose clamps that were removed; tighten all of these nuts and bolts to the correct torque.

5. Replace the stabilizer bar end-links and/or strut rod, and tighten the nuts and bolts to the correct torque.

6. Replace the wheel, correctly tightening the lug nuts, and lower the car onto the ground.

7. Replace the remaining nuts on the upper mounting bolts and tighten them to the correct torque.

12.7.3 Strut Spring Removal and Replacement

To remove the spring from a strut assembly, you should:

1. Secure the strut assembly in a bench holding fixture or strut vise. (Fig. 12–45)

2. Following the directions of the tool manufacturer, install a spring compressor and compress the spring until there is clearance between the spring and one of the spring mounts.

CAUTION: **DO NOT** *overcompress the spring; there is no need to compress it to the point where the coils touch each other.*

3. Remove the center retaining (piston rod) nut; an air wrench can spin the nut off to make this a quick and easy operation. Many strut piston rods have wrench flats or an inter-

FIGURE 12–45.
In order to work on it, the strut should be secured in a strut vise to hold it securely without damage. (Courtesy of GP Tools)

nal hex opening so the piston rod can be held from turning if necessary. (Fig. 12–46)

4. Remove the upper strut mount. The steering pivot bearing should be checked at this

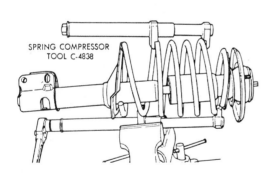

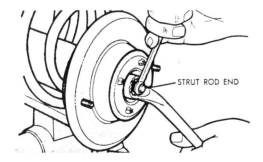

FIGURE 12–46.
There are several different styles of spring compressors that can be used to compress the spring on a strut. After the spring is compressed, the piston rod nut can be safely removed; if necessary, a wrench can be used to to hold the strut rod from turning. (Courtesy of Chrysler Corporation)

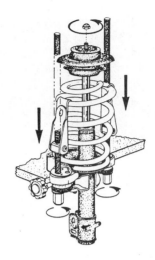

FIGURE 12–47.
After the piston rod nut has been removed, the upper mount assembly can be lifted off the strut. (Courtesy of GP Tools)

time to ensure its easy, smooth, and free but not sloppy rotation. Also, check the rubber damper portion to ensure that there is no excessive cracking or breakage. (Fig. 12–47)

5. Remove the spring and compressor. (Fig. 12–48)

To replace a spring onto a strut assembly, you should:

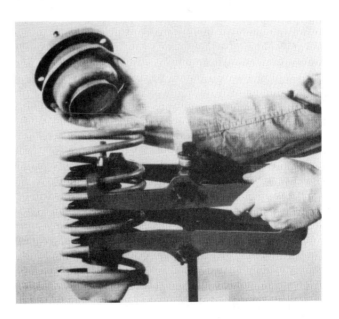

FIGURE 12–48.
With the piston rod nut removed, the upper mount and spring can be lifted off the strut body. (Courtesy of SKF Automotive Products)

FIGURE 12–49.
After the service work on the strut is completed, the compressed spring is replaced and aligned to the spring seat. (Courtesy of Moog Automotive)

1. Secure the strut in a strut vise.

2. Place the compressed spring and spring compressor onto the strut and turn the spring until it is correctly positioned on the lower spring seat. (Fig. 12–49)

FIGURE 12–50.
After the compressed spring has been installed onto the strut, the dust shield is installed (if used) and the upper mount assembly is replaced over the strut rod. The upper spring seat should be aligned with the spring during this operation. (Courtesy of Moog Automotive)

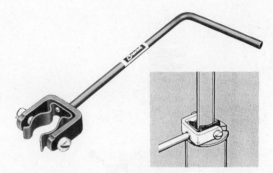

FIGURE 12–51.
This tool can be used to hold the strut rod in an extended position during replacement of the upper mount and rod nut. (Courtesy of Brannick Industries)

3. Pull the piston rod upward to the completely extended position.

4. Install the piston rod dust shield, if used, and the upper mount onto the piston rod. Align the upper spring mount with the top end of the spring, and in some cases, align the flat on the piston rod with the flat in the upper mount. A tool is available to hold the piston rod in the extended position, if necessary. If the tool is not available, a large rubber band can be placed over the piston rod to help hold it in an extended position. (Figs. 12–50 & 12–51)

5. Install the center retaining nut and tighten it to the correct torque. Some manufacturers recommend that the strut be installed in the car and turned to a straightahead position

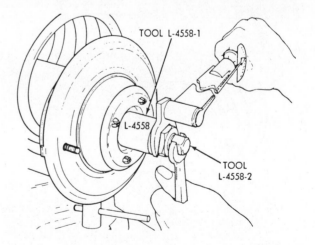

FIGURE 12–52.
During replacement, the strut rod nut must be tightened to the correct torque; some manufacturers prefer that this step be done with the strut mounted in the car. (Courtesy of Chrysler Corporation)

before final tightening of this nut. Many sources recommend that a new nut should be used. (Fig. 12–52)

6. Remove the spring compressor, making sure that the spring contacts the seats correctly.

12.7.4 Strut Cartridge Installation

To remove the shock absorber portion of a strut, you should:

1. Secure the strut in a strut vise and remove the spring as described in Section 12.7.3.

2. Remove the large gland nut from the top of the strut tube; special wrenches are available to fit the various styles of nut shapes. On all-welded struts, if a replacement cartridge and gland nut is available, use a pipe cutter to

FIGURE 12–54.
After the gland nut has been unscrewed, it can be lifted off the piston rod; many struts have an "O" ring seal (arrow), which should be removed also.

cut completely through the reservoir tube at the marked location. (Fig. 12–53)

3. Remove the "O" ring that was under the gland nut and pull upward on the piston rod to dislodge the upper shock absorber bushing. If necessary, use the piston rod and piston like a slide hammer to remove this bushing. Be ready to catch some shock absorber oil as the piston comes out. (Fig. 12–54)

4. Remove the piston rod and piston, pressure tube, and other internal shock absorber along with all of the old oil from the strut. (Fig. 12–55)

This text will not describe the procedure used to rebuild the shock absorber; if desired, consult the vehicle manufacturer's service manual for that information. (Fig. 12–56)

To install a shock absorber cartridge in a strut, you should:

1. Slide the shock absorber cartridge into the strut housing, adding any internal spacers required by the cartridge manufacturer. Many sources recommend pouring a couple of spoon-

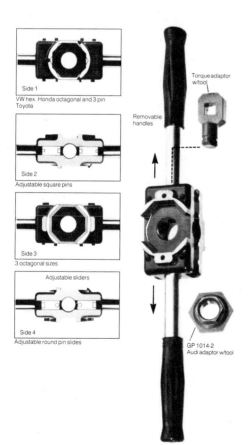

FIGURE 12–53.
There are many different sizes and shapes of gland nuts used on the strut body; a special wrench is recommended for their removal and installation. (Courtesy of GP Tools)

FIGURE 12-55.
With the gland nut and "O" ring removed, the piston rod with the pressure tube, rod guide/upper bushing, and lower valve can be lifted out of the strut. Be ready for an oil spill.

FIGURE 12-56.
After the gland nut and top bushing has been removed, the piston and rod, pressure tube with bottom valve, oil, and other internal parts can be removed from the strut body. (Courtesy of Moog Automotive)

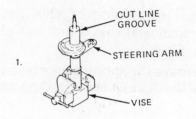

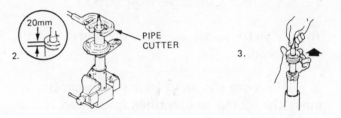

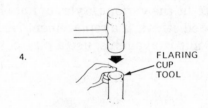

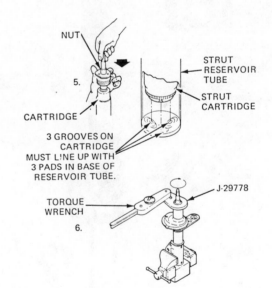

FIGURE 12-57.
On some General Motors struts, the top of the strut body can be cut off using a pipe cutter (2), the internal parts can be removed (3), the top of the strut body tube should be deburred (4) to allow installation of a shock absorber cartridge (5 and 6). (Courtesy of Pontiac Division, GMC)

fuls of lightweight oil into the housing before installing the cartridge; the oil helps transfer heat from the cartridge to the outer housing. If the strut was one that had to be cut open, smooth the cut portion of the reservoir tube before installing the cartridge to ensure that the gland nut will thread into it. (Figs. 12–57 to 12–59)

   2. Install a new gland nut, make sure that it has at least three threads of engagement, and tighten it to the correct torque. The cartridge should be locked tightly in the housing; there will often be a slight gap between the shoulder of the gland nut and the top of the strut housing. (Fig. 12–60)

12.8 STABILIZER BAR SERVICE

The common stabilizer bar service operations are the removal and replacement of the end-links and bushings and the center bushings. (Fig. 12–61)

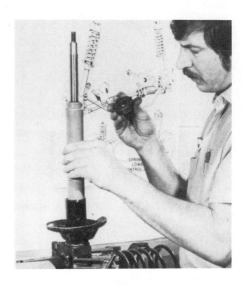

FIGURE 12–59.
Following the instructions that come with the new parts, slide the shock absorber cartridge into the strut housing. (Courtesy of Moog Automotive)

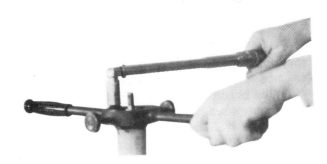

FIGURE 12–60.
With the new cartridge in place and using the correct spanner wrench, the gland nut should be tightened to the correct torque. (Courtesy of SKF Automotive Products)

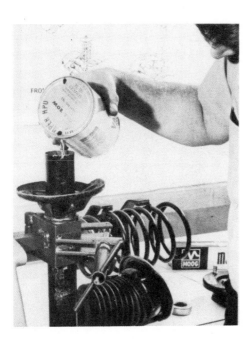

FIGURE 12–58.
It is recommended that a small amount of motor oil be poured into the strut housing to help keep the new cartridge cool; the oil should fill the reservoir to a few inches from the top when the cartridge is installed. (Courtesy of Moog Automotive)

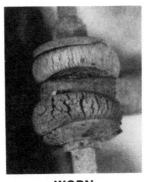 

WORN   **NEW**

FIGURE 12–61.
Worn and new stabilizer bar end-link bushings. (Courtesy of SKF Automotive Products)

To remove and replace the end-links and bushings, you should:

1. Raise and support the car on a hoist or jack stands.

2. Note the direction in which the bolt is positioned and remove the nut, bolt, and bushings from the end-link. (Fig. 12–62)

3. Install the new link bolt, bushings, washers, spacer (if used), and nut; the bolt should be positioned in the same direction as the old one.

4. Tighten the nut until it seats at the end of the bolt threads; if included with the new parts, install the jam/pal nut. (Fig. 12–63)

To remove and replace the center bushings, you should:

1. Raise and support the car on a hoist or jack stands.

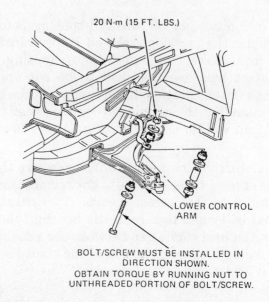

FIGURE 12–62.
On some cars, the bolt to retain the end-link bushings must be installed in a specific direction. (Courtesy of Oldsmobile)

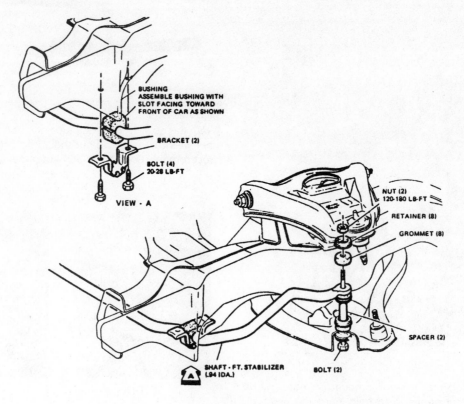

FIGURE 12–63.
The various stabilizer bushings and grommets. (Courtesy of Buick)

2. Remove the bolts securing the bushing brackets to the frame/body and remove the brackets and bushings.

3. Install the new bushings and brackets and install the retaining bolts finger tight.

4. Lower the car to the ground and tighten the retaining bolts to the correct torque.

12.9 COMPLETION

Several checks should always be made after completing any spring or shock absorber service operation. These are:

- Ensure proper tightening of all nuts and bolts, with all locking devices properly installed.
- Ensure that there is no incorrect rubbing or contact between parts or brake hoses.
- Ensure proper feel of the brake pedal.
- Check and readjust the wheel alignment if there is a possibility that the replacement parts have changed the alignment.
- Perform a careful road test to ensure correct and safe vehicle operation.

REVIEW QUESTIONS

1. Shock absorber tests are being discussed. Technician A says that a bounce test is a simple, quick, and very accurate test of the shock absorber's overall condition. Technician B says that a road test is the best way to test the second and third stages of shock absorber operation. Who is right?

   a. A only   **c.** both A and B
   b. B only   **d.** neither A nor B

2. Shock absorbers should be visually inspected for

   a. leakage.
   b. broken piston rods.
   c. broken or damaged mounts.
   d. all of these.

3. Which of the following does not indicate worn shocks?

   a. excessive dive during braking

   b. more than two cycles after being bounced
   c. cuppy tire wear
   d. tire wear at the center of the tread

4. Suspension ride height checks are important to see if the car has

   a. sagged springs.
   b. broken springs.
   c. weak shock absorbers.
   d. worn bump stops.

5. Suspension ride height checks are important because if the ride height is too low, the
   A. front end geometry will be wrong.
   B. car will bottom out too easily.

   a. A only   **c.** both A and B
   b. B only   **d.** neither A nor B

6. Ride height checks are being discussed. Technician A says that the tire size and inflation pressure can change the height and should be considered when comparing the measurements to the specifications. Technician B says these checks should be made on a flat, level surface. Who is right?

   a. A only   **c.** both A and B
   b. B only   **d.** neither A nor B

7. A car is leaning noticeably to the left. This is probably caused by a
   A. weak left front spring.
   B. weak left rear spring.

   a. A only   **c.** either A and B
   b. B only   **d.** neither A nor B

8. A car is a little too low at the front and rear. This is probably caused by
   A. weak shock absorbers.
   B. weak springs.

   a. A only   **c.** both A and B
   b. B only   **d.** neither A nor B

9. If the shock absorber nut is rusted in place on a bayonet-type shock mount, the shock can be removed by

   a. breaking the shock mounting stud.
   b. cutting the nut off with a nut breaker.
   c. breaking the nut in half with an air hammer and chisel.
   d. any of these.

10. Shock absorber replacement is being discussed. Technician A says that the rear axle can drop when the shock absorbers are disconnected. Technician B says that the shock absorber mounting bushings should be lubricated with motor oil during installation. Who is right?

    a. A only   **c.** both A and B
    b. B only   **d.** neither A nor B

11. The car has 25,000 miles on it, and the right front shock absorber is leaking. The

A. right front shock absorber should be replaced.
B. left front shock absorber should be replaced.

   a. A only   **c.** both A and B
   b. B only   **d.** neither A nor B

12. The car has 50,000 miles on it, and the right front spring has sagged out of specifications. The
    A. right front spring should be replaced.
    B. left front spring should be replaced.

    a. A only   **c.** both A and B
    b. B only   **d.** neither A nor B

13. In order to replace rear coil springs, it is necessary to remove the
    A. rear shock absorbers.
    B. rear control arms.

    a. A only   **c.** both A and B
    b. B only   **d.** neither A nor B

14. Two technicians are discussing the replacement of a front spring that is mounted on the lower control arm of an S-L A. Technician A says it is a good practice to install a spring compressor before disconnecting the ball joint or inner pivot bolts. Technician B says that the control can swing downward after the ball joint has been disconnected. Who is right?

    a. A only   **c.** both A and B
    b. B only   **d.** neither A nor B

15. A car equipped with torsion bars sits low at the front end. The torsion bars should be
    A. adjusted.
    B. replaced.

    a. A only   **c.** either A and B
    b. B only   **d.** neither A nor B

16. New front springs are being installed in a car with S-L A suspension. The
    A. ball-joint stud, inner control arm pivot bolts, stabilizer bar end-links, and the

shock absorber bolts should be tightened to the correct torque before lowering the car to the ground.

B. front end alignment should be checked and adjusted.

a. A only
b. B only
c. both A and B
d. neither A nor B

17. Rear-mounted leaf springs are being discussed. Technician A says that a broken center bolt is probably caused by loose axle-to-spring "U" bolts. Technician B says that the head of the center bolt is used to align the spring to the axle. Who is right?

a. A only
b. B only
c. both A and B
d. neither A nor B

18. Which of the following is important when compressing the spring for removal from a strut?

a. The compressor should have padded jaws.
b. The compressor must allow enough travel to let the spring extend fully.

c. The compresser must be able to contain the spring as it twists.
d. All of these.

19. A strut with a worn-out shock absorber is being discussed. Technician A says that the one repair method is to disassemble the strut and rebuild it with new components. Technician B says that you can install a new shock absorber cartridge in most struts. Who is right? *MISS USE OF ENGLISH LANGUAGE*

a. A only
b. B only
c. both A and B
d. neither A nor B

20. It is recommended to pour a small amount of oil into the strut housing as the new cartridge is being installed. The purpose of the oil is to

a. lubricate the cartridge.
b. keep the cartridge from rusting.
c. help keep the cartridge cooler.
d. all of these.

Chapter 13

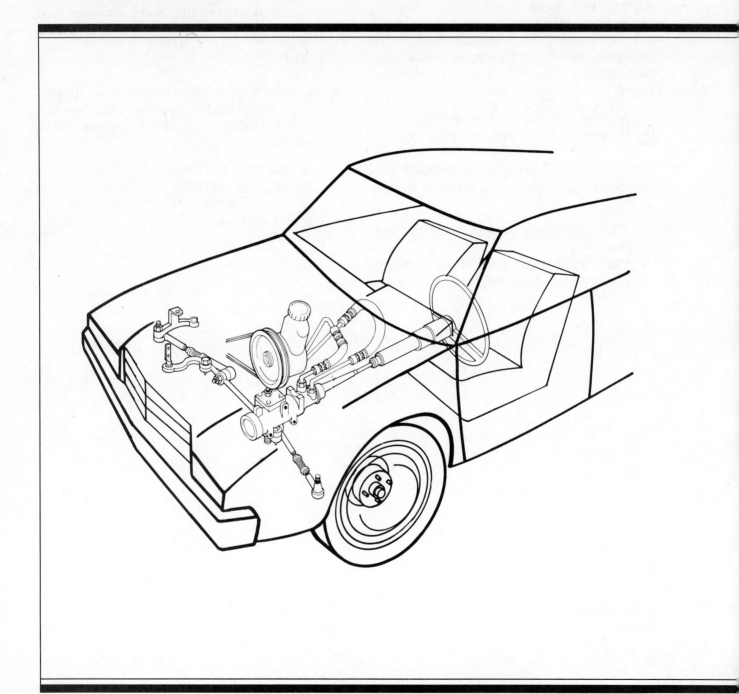

STEERING SYSTEMS

After completing this chapter, you should:

- Be familiar with the terms commonly used with automotive steering systems.

- Understand the operation of both major styles of steering gears.

- Understand the operation of a power steering system.

- Understand the operation of the different types of steering linkage.

13.1 INTRODUCTION

The steering system begins at the steering wheel, where the driver decides which direction the car should go. The steering motions are transmitted down the steering shaft and column, through one or more flexible couplings, to the steering gear. Two types of steering gears are used: **conventional/standard** and **rack and pinion**. The steering gear in turn connects to the tie rods. The steering system transfers the steering motions from the steering wheel to the steering arms. (Fig. 13-1)

Either type of steering gear has the function of changing the rotating motion of the steering wheel into a reciprocating, back and forth motion at the tie rods. They also provide a gear reduction to slow down the steering speed, reduce the amount of steering effort required, and make steering more precise. Automotive steering systems provide a **steering ratio** somewhere between 10:1 and 25:1. A steering ratio of 20:1 means that 20° of steering wheel motion will turn the front tires 1°. Steering ratios usually vary depending on the weight of the car and whether manual or power steering is used. Lighter cars or ones with

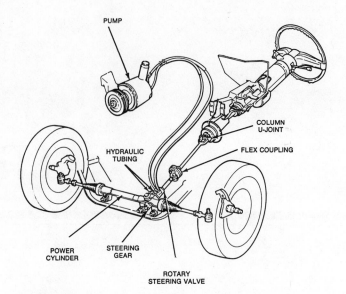

FIGURE 13-2.
A rack and pinion steering system, this particular one has a power steering rack. (Courtesy of Ford Motor Company)

power steering generally use faster ratios (toward 10:1) to allow quicker steering; heavier, manual steering cars use slower steering ratios (toward 25:1) to allow easier steering. Some power steering gears provide a variable ratio. Variable ratios are usually slower in the center position to provide precision steering while going in a straight line, and a faster ratio toward the ends, to provide quicker steering during cornering and parking maneuvers. (Figs. 13-2 & 13-3)

Steering ratios appear to the driver as the number of turns it takes to move the wheel from **lock to lock**, a term referring to the number of turns the steering wheel must move from one steering stop to the other. Faster ratios are about three turns lock to lock; slower ratios are four or five turns. (Figs. 13-4 & 13-5)

Overall steering ratios are controlled by the ratio of the steering gear plus the length of the steering arms and Pitman arm. Longer steering arms or a shorter Pitman and idler arm will produce slower steering ratios. A few cars have steering arms with two sets of holes for the tie rod ends. The holes closest to the steering axis provide the fastest steering and are normally used with lighter weight engines or power steering. The holes closest to the ends of

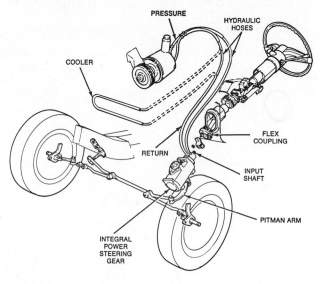

FIGURE 13-1.
A standard steering system, this particular one uses an integral power steering gear. (Courtesy of Ford Motor Company)

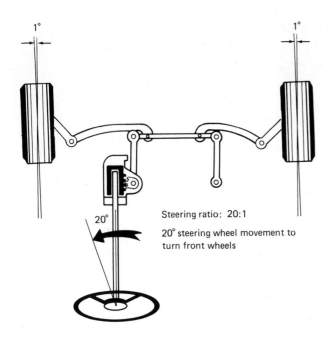

FIGURE 13-3.
A steering ratio of 20:1 tells us that it will require 20° of steering wheel rotation to produce a turn of 1°; a 30° turn will require 20 times 30 or 600° of steering wheel rotation.

the steering arms provide easier but slower steering. (Fig. 13-6)

Steering gears are connected to the steering arms by the steering linkage. On cars with rack and pinion gears, the steering linkage is the two tie rods. Several different styles of linkage are used with standard steering gears; the parallelogram style is the most popular. Sev-

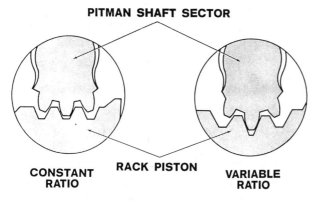

FIGURE 13-4.
The different profile of the steering gear teeth on the right provides a variable ratio. This design provides a slower ratio for the center one-half turn for precise steering and quicker ratio as the steering wheel is turned beyond that. (Courtesy of Saginaw Division, General Motors Corporation)

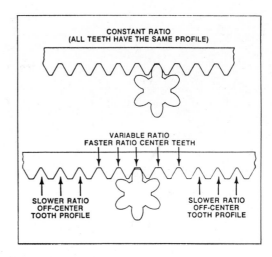

FIGURE 13-5.
The different profile of the rack and pinion teeth at the bottom provides a variable ratio. This design provides a faster ratio at the center for fast and responsive highway steering, and slower ratios near the ends of the steering rack travel for easy cornering and parking. (Courtesy of Ford Motor Company)

eral styles of linkage are used with trucks and pickups also; the exact style depends on the type of axle/suspension that is used. (Figs. 13-7 & 13-8)

13.2   STEERING COLUMNS

Traditionally, steering columns have been used to support the steering shaft. Bearings are pro-

FIGURE 13-6.
This Corvette steering arm has two tie rod mounting holes; the tie rod on this car is placed in the inner, faster position.

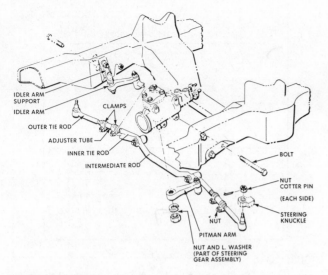

FIGURE 13-7.
This is a parallelogram style of steering linkage; it is commonly used with standard steering gears and S-L A suspension. (Courtesy of Saginaw Division, General Motors Corporation)

vided to allow free rotation of the shaft and wheel without sloppy side or end motions. As the years pass, the steering column is evolving into one of the more complex parts of the car.

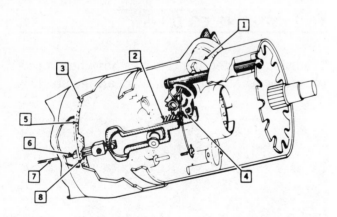

1. LOCK CYLINDER
2. RACK
3. BOWL PLATE
4. SECTOR
5. PARK POSITION
6. WEDGE SHAPE FINGER
7. ACTUATOR ROD ASSY.
8. NEUTRAL POSITION

FIGURE 13-9.
The upper end of a steering column showing the mechanism used to lock the steering wheel and part of the neutral start mechanism. (Courtesy of Oldsmobile)

1. CROSS STEER

2. HALTENBERGER LINKAGE

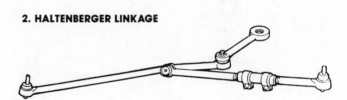

3. CENTER ARM STEER

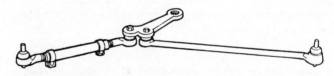

FIGURE 13-8.
Three different styles of steering linkage, the cross-steer linkage is used with solid axles, the Haltenberger linkage is used with swing axles, and the center arm steer has been used with several different independent suspensions. (Courtesy of Moog Automotive)

We are all familiar with the horn in the steering wheel and the turn indicator switch just in front of the steering wheel. Most of us are also familiar with the ignition switch, which not only controls the starting and running of the engine but also locks the steering shaft from turning and the transmission gear selector from moving when the ignition is locked and the key is removed. Some steering column ignition switches are also interconnected to the transmission linkage through an inhibitor system that requires a certain transmission gear before the key can be removed. Many newer cars have replaced this inhibitor linkage with a simple lever or button in the column, which must be moved before removing the key. (Fig. 13-9)

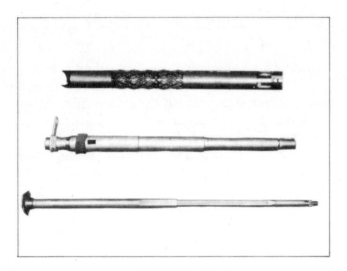
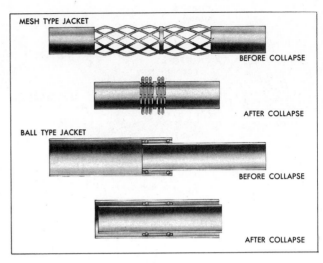

FIGURE 13-10.
The collapsible components of a steering column are the steering shaft, the gear selector tube, and the steering column jacket (left). At the right are two different styles of column jackets before and after collapsing. (Courtesy of Chevrolet)

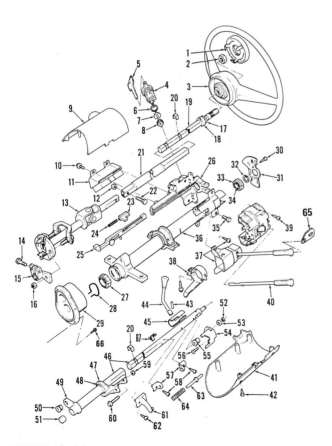

FIGURE 13-11.
A disassembled steering column showing the internal components. (Courtesy of Ford Motor Company)

Collapsible steering columns have become standard, and they are required by federal law. These columns and steering shafts are designed to shorten under impact, a feature that has probably saved many drivers from chest and upper body injury during front-end accidents. Care should be taken when working on steering columns or steering wheels to ensure that the collapsible feature remains in proper operation; excessive force by a mechanic can sometimes cause damage. Steering columns or shafts that have collapsed should be replaced. (Fig. 13-10)

Several more automobile controls have moved onto the steering column, including cruise control switches, windshield wiper and washer switches, headlight dimmer switches, and hazard switches. The control levers for these switches often combine several functions; movement of the lever one way controls the operation of one system, while movement in a different direction controls a different function. (Figs. 13-11 & 13-12)

Some columns are adjustable for steering wheel position, steering wheel angle, or forward-backward position. These adjustments are used to place the steering wheel in a more comfortable position, a feature benefiting the driver of a nonaverage size the most. (Fig. 13-13)

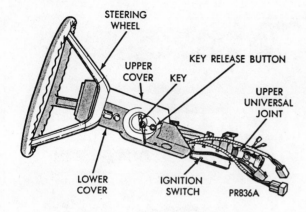

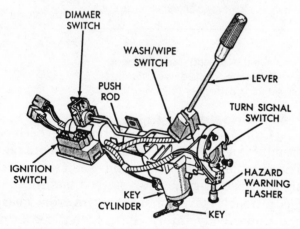

FIGURE 13-12.
A steering column showing how some of the switches are mounted. (Courtesy of Chrysler Corporation)

Flexible couplers are used to connect the steering shaft to the steering gear. These couplers allow for a slight angle difference between the two shafts and dampen road vibrations and noises from being transmitted up the shaft to the steering wheel. Most of the couplers are made from rubber, often reinforced with fabric. They can and do fail. A safety interconnection is made between the two portions of the coupler; in case of failure, the driver will still be able to steer. Coupler failure will introduce a large amount of looseness and sloppiness in the steering. (Fig. 13-14)

Steering column service is too complex to describe in a text of this type. An adequate manufacturer's or technician's service manual should be consulted before attempting to disassemble or reassemble a modern steering column.

13.3 STANDARD STEERING GEARS

All steering gears must provide a change in motion, a gear ratio, and operate as smoothly and easily as possible without excess free play. Most standard steering gears used in today's cars are of the **recirculating ball nut** type; two other design types are **worm and sector/worm and roller (Gemmer)** and **cam and lever (Ross)**. (Fig. 13-15)

Let's begin with the worm and sector steering gear because it is the easiest design to describe and understand. All steering gears use two shafts, an input connected to the steering wheel and an output connected to the linkage. The input shaft is commonly called the **steering shaft** or the **worm shaft** because it has a worm gear made into it; a worm gear resembles a large screw thread. The output shaft is commonly called the **Pitman shaft** because the Pitman arm connects to it, or the **sector shaft** because the gear that is made on it is a section of a gear. In some steering gears, the gear teeth on the sector have been replaced with a hardened steel roller that has the same side shape as gear teeth; this gear set is called a **worm and roller.** The roller is attached to the Pitman shaft through a set of ball bearings to help reduce gear friction. The sector gear meshes with the worm gear; when the worm gear turns, the sector gear and shaft are forced to swing one way or the other. This motion will cause the end of the Pitman arm to move one way or the other, giving the steering action. The steering shaft and the Pitman shaft are each supported by a set of bearings or bushings. (Figs. 13-16 & 13-17)

A recirculating ball nut steering gear also has a sector gear on the Pitman shaft; this gear meshes with the ball nut. The ball nut has an internal thread that resembles a spiral groove. It fits over the steering shaft, which also has a spiral thread groove. One or two sets of steel balls are fitted into these two grooves to form the interconnecting threads. As the shaft rotates, these balls force the ball nut to move up or down on the steering shaft, much like a nut on a bolt. The balls rotate and roll during this motion and circulate on a path provided by a

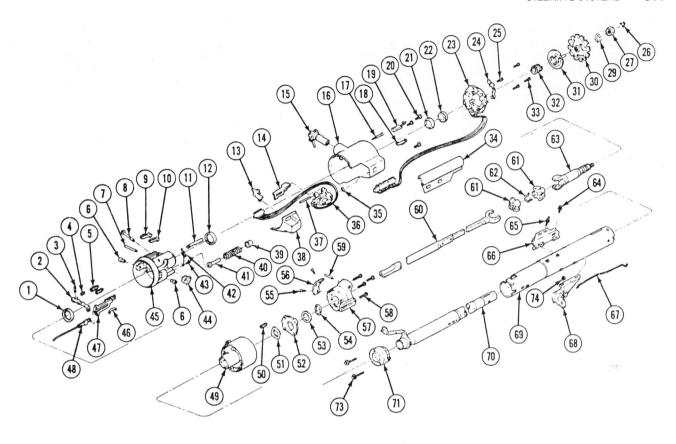

1. Bearing Assy.
2. Lever, Shoe Release
3. Pin, Release Lever
4. Spring, Release Lever
5. Spring, Shoe
6. Pin, Pivot
7. Pin, Dowel
8. Shaft, Drive
9. Shoe, Steering Wheel Lock
10. Shoe, Steering Wheel Lock
11. Bolt, Lock
12. Bearing Assy.
13. Shield, Tilt Lever Opening
14. Actuator, Dimmer Switch Rod
15. Lock Cylinder Set, Strg Column
16. Cover, Lock Housing
17. Screw, Lock Retaining
18. Clip, Buzzer Switch Retaining
19. Switch, Assy. Buzzer
20. Screw, Pan Head Cross Recess
21. Race, Inner
22. Seat, Upper Bearing Inner Race
23. Switch Assy. Turn Signal
24. Arm Assy. Signal Switch
25. Screw, Round Washer Head

26. Retainer
27. Nut, Hex Jam
29. Ring, Retaining
30. Lock Plate
31. Cam Assy. Turn Sig. Cancelling
32. Spring, Upper Bearing
33. Screw, Binding Hd. Cross Recess
34. Protector, Wiring
35. Spring, Pin Preload
36. Switch Assy. Pivot &
37. Pin, Switch Actuator Pivot
38. Cap, Column Housing Cover End
39. Retainer, Spring
40. Spring, Wheel Tilt
41. Guide, Spring
42. Spring, Lock Bolt
43. Screw, Hex Washer Head
44. Sector, Switch Actuator
45. Housing, Steering Column
46. Spring, Rack Preload
47. Rack, Switch Actuator
48. Actuator Assy. Ignition Switch
49. Bowl, Gearshift Lever
50. Spring, Shift Lever

51. Washer, Wave
52. Plate, Jacket Mounting
53. Washer, Thrust
54. Ring, Shift Tube Retaining
55. Screw, Oval Head Cross Recess
56. Gate, Shift Lever
57. Support, Strg. Column Housing
58. Screw, Support
59. Pin, Dowel
60. Shaft Assy. Lower Steering
61. Sphere, Centering
62. Spring, Joint Preload
63. Shaft Assy. Race & Upper
64. Screw, Wash. Hd.
65. Stud, Dimmer & Ignition Switch
    Mounting
66. Switch Assy. Ignition
67. Rod, Dimmer Switch
68. Switch, Assy. Dimmer
69. Jacket Assy. Steering Column
70. Tube Assy. Shift
71. Bearing Assy. Adapter &
73. Screw, Hex. Washer Head Tapping
74. Nut, Hex.

FIGURE 13–13.

A disassembled tilt-wheel steering column. (Courtesy of Chrysler Corporation)

INTERMEDIATE SHAFT INSTALLATION

1. COUPLING MUST BE FULLY ENGAGED WITH SPLINES OF STEERING GEAR BEFORE INSTALLING PINCH BOLT.

2. COUPLING SHIELD LATCH MUST BE SEATED AROUND THE RETURN PIPE NUT.

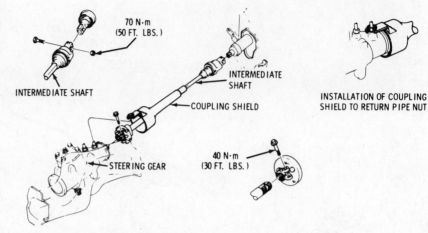

PIN & MARK MUST BE IN LINE

FIGURE 13–14.
An intermediate shaft is used on this car to connect the steering shaft to the steering gear; note the two different styles of couplers used at each end of the intermediate shaft. (Courtesy of Oldsmobile)

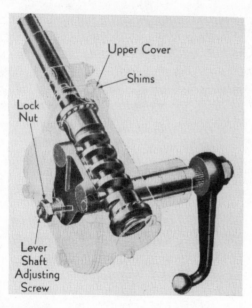

FIGURE 13–15.
A Ross, also called a cam and lever steering gear; note the two studs on the lever (sector gear) that engage the cam (worm shaft). (Courtesy of Bear Automotive)

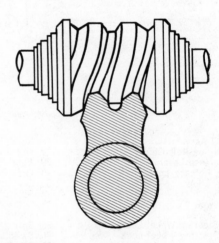

FIGURE 13–16.
Many early standard steering gears were of the worm and sector type; when the worm gear was turned by the steering wheel, the sector gear turned the Pitman shaft. (Courtesy of Saginaw Division, General Motors Corporation)

FIGURE 13–17.
The worm and roller steering gear is the evolution of the worm and sector; the three-tooth roller on the sector shaft of this gear reduces internal gear friction. (Courtesy of Ford Motor Company)

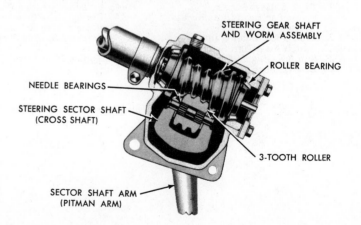

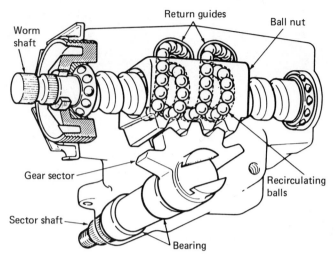

FIGURE 13-18.
The recirculating balls run in the grooves inside the ball nut and the worm shaft; as the worm shaft turns, the ball nut will be threaded up or down the worm shaft causing the sector gear to turn. (Courtesy of Ford Motor Company)

set of ball guides. The balls provide an almost frictionless connection between the steering shaft and the ball nut. When the steering shaft is rotated and the ball nut moves up or down, the sector gear, which is meshed with the ball nut, is forced to rotate or swing one way or the other producing the steering motion. (Fig. 13-18)

Both styles of standard steering gears allow freer motion to be transmitted through them from the steering shaft to the Pitman shaft than from the opposite direction. Steering motion is freely transmitted from the steering wheel to the Pitman arm, but road shock trying to move from the Pitman arm to the steering shaft is resisted. This action is the natural result of the type of gears used and tends to dampen road shock and provide steering that is free from kickback on rough roads.

All standard steering gears provide two adjustments to help ensure free turning without free play. These are **worm shaft bearing preload** and **gear lash**, sometimes called the **over center** adjustment. These adjustments are described in Chapter 16. (Fig. 13-19)

Standard manual (nonpower) steering gears contain their own supply of lubricant; the gear housing is normally filled through a filler plug or one of the sector gear cover retaining bolts. The lubricant type varies among manufacturers; it is usually gear oil or a thin, semi-fluid type of grease. Seals are used at the Pitman shaft and the steering shaft to keep the lubricant in the gear housing and dirt and water out.

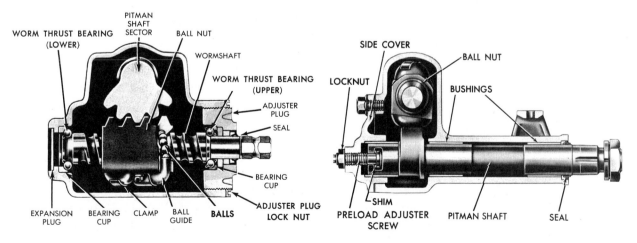

FIGURE 13-19.
Two views of the internal parts of a recirculating ball nut steering gear; note the adjuster plug used to adjust worm shaft preload (left) and the preload adjuster screw used to adjust the gear lash on the right hand gear. (Courtesy of Saginaw Division, General Motors Corporation)

13.4 RACK AND PINION STEERING GEARS

Rack and pinion steering systems are simpler and lighter in weight than standard systems. Many people credit rack and pinion systems with being more responsive and giving the driver a better feel of the road; this is probably largely due to the fact that they are often of a faster ratio then standard gears. Rack and pinion steering gears were traditionally fitted to smaller, lighter cars whereas standard steering gears were fitted to larger, heavier cars. (Fig. 13-20)

A disadvantage with rack and pinion steering gears is that they transmit more road shock to the steering wheel. Most rack and pinion gears are mounted through rubber bushings to help reduce this kickback; some systems use a steering damper to also reduce kickback. A **steering damper** is a unit built much like a shock absorber. The body of the damper is connected to the frame/body, and the piston rod is connected to the steering linkage. A damper adds turning resistance to the steering. The power steering piston in a power rack and pinion steering gear is very effective in dampening these motions. (Fig. 13-21)

The steering shaft in a rack and pinion gear set becomes the pinion gear; this pinion gear meshes with the teeth on the rack. When the pinion rotates, the rack is forced to move to the right or the left. The rack is supported by bush-

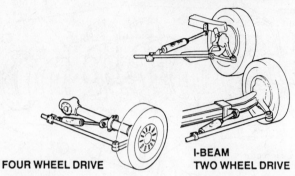

INDEPENDENT SUSPENSION

FOUR WHEEL DRIVE

I-BEAM
TWO WHEEL DRIVE

FIGURE 13-21.
When a steering damper is used, one end is attached to the frame/body and the other end is connected to the tie rod. (Courtesy of Dana Corporation)

ings in the gear housing, which allow it to slide sideways during turning maneuvers. Flexible rubber bellows are used to seal the ends of the gear housing to keep water or dirt from entering or gear lube from escaping. (Fig. 13-22)

Some rack and pinion gear sets provide a method of adjusting pinion gear bearings or the mesh clearance or lash between the rack and pinion gears. A rack and pinion gear is normally lubricated during assembly with gear oil or semifluid grease. Rack and pinion service is described in Chapter 16.

13.5 POWER STEERING

Automotive power steering units use conventional steering gears with a hydraulic assist. In addition to the steering gear, which is similar to those just described, there is a hydraulic pump, a control valve that is sensitive to the driver's turning effort, a pair of hoses, and a hydraulic actuator—piston and cylinder. The actuator can be mounted separately on the steering linkage, or it can be integral—built internally into either style of steering gear. (Fig. 13-23)

Most systems are of the integral type, with the piston being a modified recirculating ball nut or a modified rack and the cylinder being a modified gear housing. The control valve is also

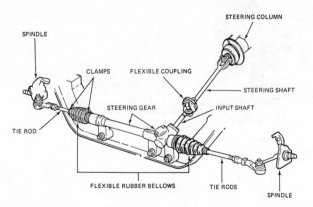

STEERING COLUMN

SPINDLE

CLAMPS   FLEXIBLE COUPLING

STEERING SHAFT

STEERING GEAR   INPUT SHAFT

TIE ROD

FLEXIBLE RUBBER BELLOWS   TIE RODS

SPINDLE

FIGURE 13-20.
A rack and pinion steering gear with tie rods and steering shaft. (Courtesy of Ford Motor Company)

Key No.	Part Name	Key No.	Part Name	Key No.	Part Name
1	FLANGE ASSY, COUPLING & STRG.	9	CLAMP, BOOT	17	SPRING, ADJUSTER
2	BOLT, PINCH	10	BOOT	18	PLUG, ADJUSTER
3	HOUSING ASSY, RACK & PINION	11	CLAMP, BOOT	19	NUT, ADJUSTER PLUG LOCK
4	BEARING ASSY, ROLLER	12	ROD ASSY, INNER TIE	20	GROMMET, GEAR MOUNTING (LH)
5	PINION ASSY, BEARING (ROTOR BRG)	13	NUT, JAM	21	GROMMET, GEAR MOUNTING (RH)
6	RING, RETAINING	14	ROD ASSY, OUTER TIE	22	BUSHING, RACK
7	SEAL, STEERING PINION	15	SEAL, TIE ROD	23	RING, RETAINING
8	RACK, STEERING	16	BEARING, RACK		

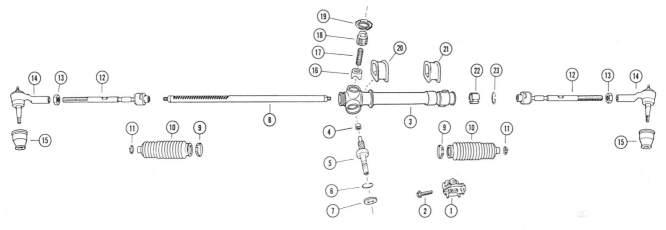

FIGURE 13-22.

An exploded view of a rack and pinion steering gear with tie rods, note that the rack is turned over so the gear teeth are visible. (Courtesy of Saginaw Division, General Motors Corporation)

built into the gear housing and is constructed into the steering shaft. When the driver turns the steering wheel, the torque/turning effort on the steering shaft moves the valve to redi-

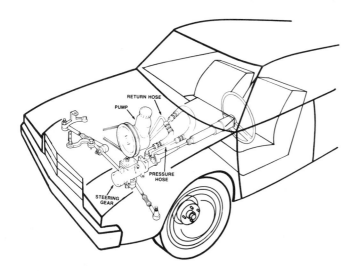

FIGURE 13-23.

A standard power steering system, the major parts are the pump, steering gear, and hoses. (Courtesy of Everco)

rect the pressurized hydraulic fluid to the proper side of the actuator to give a steering assist. The power steering fluid also serves to lubricate the gears and bearings in the steering gear. (Fig. 13-24)

The power for power steering comes from the engine; it is the turning effort exerted by the drive belt to the power steering pump. This pump delivers fluid under pressures up to about 600 to 1,300 psi (4,100 to 8,950 kPa) to the control valve in the steering gear or linkage. Two reinforced rubber hoses connect the pump to the control valve. A high pressure hose carries the fluid to the valve, and a low pressure hose returns the fluid to the pump reservoir. (Fig. 13-25)

To understand power steering, we need to have an understanding of some of the principles of hydraulics. **Hydraulics, fluid power,** is based on the fact that liquids are fluid and can flow easily, but cannot be compressed (squeezed into a smaller volume). Also, fluids can transmit pressure, and fluids under pressure will ex-

HOW IT WORKS

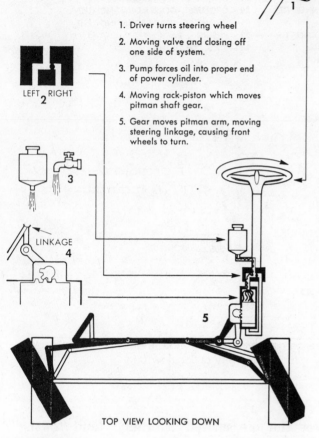

1. Driver turns steering wheel

2. Moving valve and closing off one side of system.

3. Pump forces oil into proper end of power cylinder.

4. Moving rack-piston which moves pitman shaft gear.

5. Gear moves pitman arm, moving steering linkage, causing front wheels to turn.

LEFT 2 RIGHT

3

LINKAGE 4

5

TOP VIEW LOOKING DOWN

FIGURE 13–24.
All power steering systems contain a valve (2), which is sensitive to steering wheel pressure (1), the pump (3), which provides the pressure, and the actuator/piston (4), where the pressure delivers the assist. (Courtesy of Saginaw Division, General Motors Corporation)

ert force equally in all directions. Fluid pressures do not come from just the pump. In fact, the pump does not pump pressure; it pumps volume or flow. Pressure is created when this flow is restricted, and then the pump's effort generates pressure. If we connected the output hose of a pump directly into the reservoir, with no restriction, there would be very little pressure in the hose. If we took this same pump and closed the end of the hose, there would be no place for the fluid to go and a lot of pressure would be generated. The amount of pressure would be controlled by the strength of the hose and pump, the amount of power driving the

1 Typical Integral Power Steering System

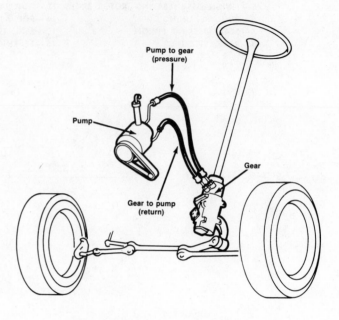

Pump to gear (pressure)

Pump

Gear

Gear to pump (return)

2 Typical Booster Type Power Steering System

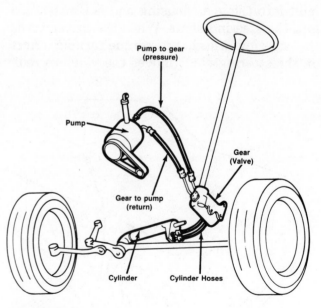

Pump to gear (pressure)

Pump

Gear (Valve)

Gear to pump (return)

Cylinder

Cylinder Hoses

FIGURE 13–25.
Two hoses connect the pump to the gear. One carries the high pressure fluid to the gear, and one returns the fluid back to the pump. Linkage booster type systems require two more hoses; a hose connects each end of the cylinder to the valve. (Courtesy of Everco)

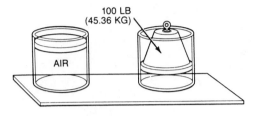

GAS CAN BE COMPRESSED

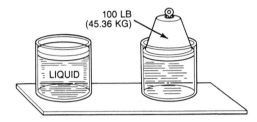

LIQUID CANNOT BE COMPRESSED

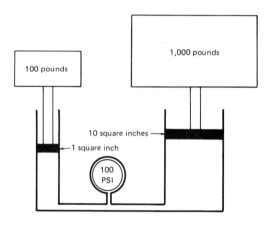

FIGURE 13–26.
Two important hydraulics principles are that liquids cannot be compressed and that liquids can transmit pressure equally in all directions. The amount of pressure is determined by the size of the piston and the amount of force on it.

pump, the ability of the drive belt to transmit this power, and possibly, by a relief valve. Hydraulics is a means to transfer energy from the engine to the actuator; energy is put into the fluid at the pump and removed at the actuator/piston and cylinder. The amount of power transmitted or consumed by a hydraulic system is determined by how much fluid flows and how much pressure is used. Hydraulic horsepower is equal to the flow times the pressure; this number is then adjusted by a small factor

to obtain the actual horsepower amount. (Fig. 13–26)

Small-engine, power steering equipped cars are often constructed with a pressure switch and computer controls that will increase the engine idle speed while the power steering is used. It is possible that the load of the power steering pump will cause engine stalling. Other new cars use computer controls on the power steering system to change the amount of assist relative to vehicle speed. The most assist is needed when the car is stopped or moving very slowly, and these driving conditions require the greatest turning effort from the driver. A reduced amount of assist is desired by many drivers so they can get a better feel of the road at higher speeds. Electronic controls activate a solenoid, which opens a fluid bypass above certain speeds. This bypass of fluid drops the pressure available for assist and, in turn, reduces the amount of assist.

13.5.1 Power Steering Pumps

Power steering pumps are usually **balanced vane** hydraulic pumps. They are built into an assembly that includes the pump, a **pressure and flow control valve**, and a **reservoir**; some pumps mount the reservoir remotely to provide more room on the engine. The valve limits the amount of pressure that the pump can generate. A vane pump has a series of slots in the center rotor in which the vanes fit; these vanes can slide in or out of these slots in order to follow the contour of the pump cam/chamber ring. Some pumps use slippers or rollers instead of vanes to serve the same function. The pump cam is somewhat elliptically shaped to create two pumping chambers, one on each side of the rotor, which balance the pump. Fluid pressure in the pumping chamber tends to force the rotor off center; a balanced pump is balanced by the same pressure on each side of the center rotor. (Figs. 13–27 & 13–28)

When the pump pulley is turned, the pump shaft turns the rotor and vanes. The vanes move into the rotor on the sections of the cam that are close to the rotor and outward on the

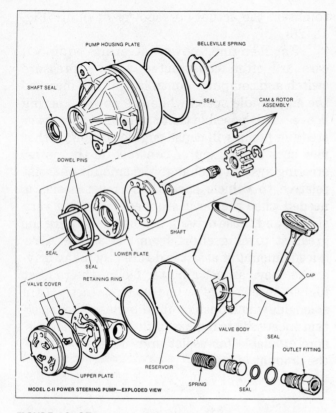

FIGURE 13-27.
An exploded view of a power steering pump showing the internal parts. (Courtesy of Ford Motor Company)

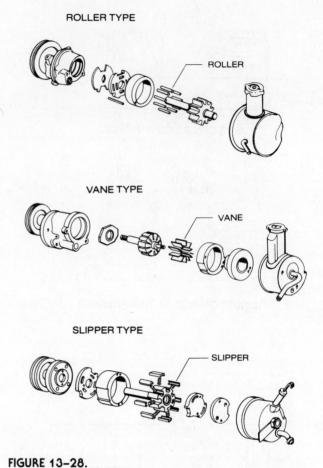

FIGURE 13-28.
The pumping member of a power steering pump can be a set of rollers, vanes, or slipper; the actual operation of each of these pumps is very similar. (Courtesy of Moog Automotive)

cam sections that are farther from the rotor. Outward movement of the vanes is caused by hydraulic pressure under the vanes and centrifugal force; some pumps use spring-loaded vanes or slippers. Pumping chambers are formed between two vanes and the rotor and outer cam surfaces. These chambers get bigger in the areas where the cam-to-rotor space is increasing and smaller where the space is decreasing. As the rotor chamber passes an area where a chamber is increasing in size, fluid flows into the pumping chamber through ports at the ends of the rotor; these ports are connected to the reservoir. After the rotation of about one-fourth turn (90°), the rotor and vanes enter a cam section that reduces the chamber size, and the fluid is forced out of the chamber through a second set of ports. These ports are connected to the control valve and pump outlet. (Fig. 13-29)

This style of pump is called a **positive displacement pump**; it pumps a specific volume of fluid on each revolution of the pump rotor. The displacement of a particular pump is determined by the length of the rotor, vanes, and cam ring, and the shape of the cam ring. Most automotive power steering pumps will pump about 1 to 1.5 gallons (3.8 to 5.7 l) per minute at idle speed at low pressure. The pump is normally designed to pump a large enough volume to supply the needs of the power steering gear with the engine idling. If the pump is too small, there is loss of hydraulic assist during faster steering manuevers; this is called pump **catch up**.

The reservoir of the pump is the plastic or stamped metal portion that encloses the back and most of the body of the pump. The filler neck and cap with dipstick are usually built into the reservoir. Passages in the pump body open

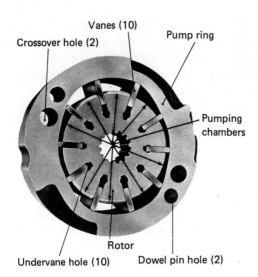

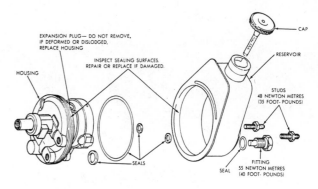

FIGURE 13-29.
The pumping elements from a vane pump, note the pumping chamber on each side of the rotor (top and bottom); their placement gives one complete pumping cycle each one-half turn of the rotor and balances the pumping pressures on each side of the rotor. (Courtesy of Saginaw Division, General Motors Corporation)

FIGURE 13-30.
In many power steering pumps, the reservoir is held onto the pump body by the high pressure fitting and one or two bolts and is sealed by rubber "O" rings. (Courtesy of Chrysler Corporation)

into the reservoir for the pump inlet. The pump outlet passage and pressure control valve pass through the reservoir surface and into the pump body. Remotely mounted reservoirs are usually mounted at the fender well and are connected to the pump inlet by a flexible hose and to the

steering gear by the return line. (Figs. 13-30 & 13-31)

The control valve on most power steering pumps serves two purposes: flow control and maximum pressure control. Since pumps must be sized to provide an adequate flow at engine idle speeds and because they are positive displacement pumps, they deliver more fluid than necessary at speeds above idle. As the engine speeds up, the increased flow will cause the flow control valve to open and bypass part of the flow back to the inlet port. This helps reduce the power needed to drive the pump, reduces

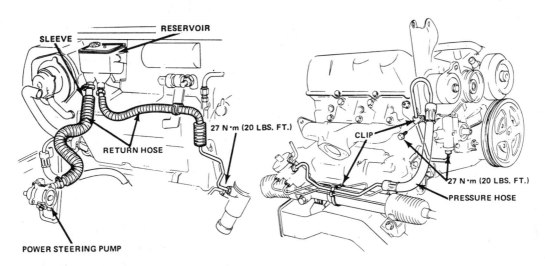

FIGURE 13-31.
The reservoirs on some of our modern cars with cramped engine compartments are remotely mounted, usually on the fender well or rear bulkhead. (Courtesy of Chevrolet)

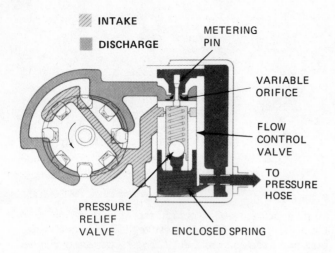

INTAKE

DISCHARGE

METERING PIN

VARIABLE ORIFICE

FLOW CONTROL VALVE

TO PRESSURE HOSE

PRESSURE RELIEF VALVE

ENCLOSED SPRING

PASSAGE IS OPENED

ORIFICE

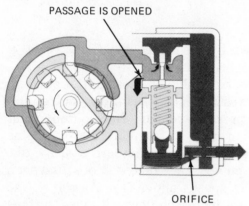

FIGURE 13–32.
When pump output becomes sufficient to create a pressure at the variable orifice, the flow control valve will move downward and allow some of the pump's output return to the intake cavities of the pump. (Courtesy of Ford Motor Company)

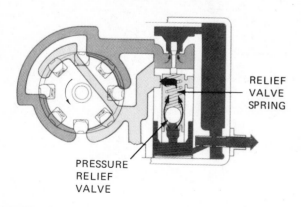

RELIEF VALVE SPRING

PRESSURE RELIEF VALVE

FIGURE 13–33.
If the pump's pressure becomes too high, the pressure relief valve ball will be pushed off its seat, and fluid will flow back to the intake cavity of the pump. (Courtesy of Ford Motor Company)

fluid temperatures, and increases the life of the fluid. The flow control valve function is in constant use while the car is driven at road speeds; the pressure control valve is seldom used. (Fig. 13–32)

As we will see later, the steering gear control valve allows a free flow of fluid to return to the pump reservoir when a steering maneuver is not being performed, but when there is a turning pressure on the steering wheel, the pressurized fluid is directed to one end of the actuator. Trapping the fluid coming from the pump generates the pressure needed to give us the steering assist, and the volume of the pump is used to move the actuator through its travel to produce a turn. When the actuator reaches the end of its travel, it will accept no more fluid, and additional pump output will instantly try to create excessive fluid pressures. At this time, the pressure release function of the control valve should occur. This valve will open and let part of the pump's output flow back to the pump inlet. The valve is noisy; we can usually hear it squeal or hiss when it is relieving pump pressure. Squeal along with a screeching noise is often caused by a slipping drive belt. (Fig. 13–33)

A turning pressure should not be kept on the steering wheel after the pressure control valve has started working; the high pressure can cause the fluid to heat and possibly damage the pump, high pressure hose, or other components of the system. Releasing the turning pressure on the steering wheel will let the control valve in the steering gear recenter itself and allow fluid to flow back to the pump. The car will turn just as sharply.

Power steering pump problem diagnoses and service operations are described in Chapter 16.

13.5.2 Steering Gear Control Valve

Most modern power steering gears use a rotary type of valve built into the steering gear to control the fluid flows. This valve is normally positioned where the steering shaft enters the steering gear. This style of valve uses a small torsion bar to control the amount of assist that is delivered by the power steering unit.

The valve assembly consists of an inner valve spool, an outer valve sleeve, a valve housing, and the torsion bar. The input shaft of the steering gear is built into or connected to the inner valve spool and is also connected to one end of the torsion bar. The other end of the torsion bar is connected to the pinion gear of a rack and pinion set or the worm shaft of a standard steering gear; it is also connected to the outer valve spool. The valve spool fits into the valve housing and has passages that connect to four separate fluid ports. Sealing rings are used so it can rotate without fluid leakage. The four fluid ports are pump pressure, pump return, and each end of the actuator. The sleeve and spool are machined to give very precise fit so as to control fluid leakage. Steering torque, being transmitted through the torsion bar, can cause the torsion bar to twist. This action will change the position of the valve spool relative to the sleeve and, in turn, change the fluid flows, producing or stopping hydraulic assist. The strength of the torsion bar controls the amount of assist and **steering feel**. A very weak torsion bar would twist easily; a lot of power assist would occur quickly and the steering would be very soft. An extremely strong torsion bar might not twist at all; we would get little or no assist and heavy steering. (Fig. 13–34)

In most valve assemblies, a projection of the valve spool fits into a slot in the worm shaft. The length of the slot allows enough spool-to-sleeve travel for valve operation along with a means to turn the worm shaft manually. Hard and quick steering maneuvers are done by manual force with power assist.

All power steering valves must have four ports: an inlet (pump pressure), an outlet (pump return), and a port to each end of the actuator. The actuator ports and passages are internal on standard steering gears and external on rack and pinion gears and linkage booster systems. When there is no turning effort and the torsion bar is relaxed, fluid enters the valve through the inlet port and passes through the valve to the outlet port; both actuator ports are open to inlet and outlet pressure, allowing low pressure fluid to fill both ends of the actuator. The fluid pressure, at this time, is controlled by any restriction in the steering valve or hose or anywhere between the pump outlet port and the pump reservoir. When a turning effort is exerted that is strong enough to twist the torsion bar, the valve spool and sleeve become realigned. Depending on the direction of twist/

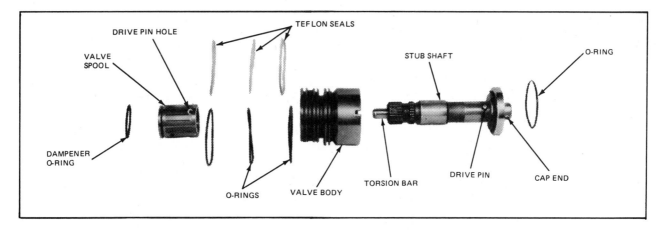

FIGURE 13–34.

A disassembled view of a power steering rotary valve, the torsion bar connects to the stub shaft at the left end and the cap end at the right. The valve spool will turn with the stub shaft, and the valve body will move with the cap end. (Courtesy of Ford Motor Company)

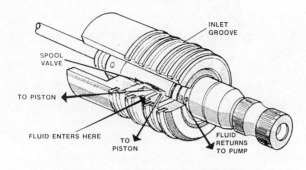

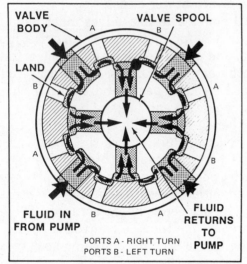

FIGURE 13-35.
While there are no turning pressures, the torsion bar is relaxed, and the passages in the valve spool and body are
aligned so the fluid flows in from the pump to both ports
for the steering gear and to the return port back to the
pump. (Courtesy of Ford Motor Company)

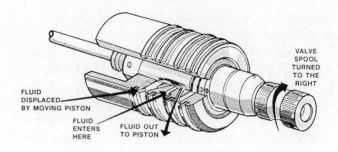

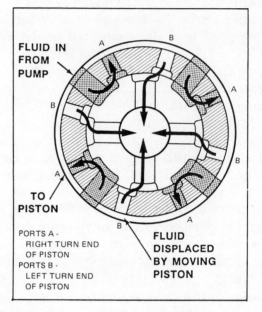

FIGURE 13-36.
Turning pressure from a steering motion can twist the torsion
bar; this motion will reposition the passages in the valve
spool and body. Pump pressure will now be sent only to
one end of the power steering piston; the other end of the
piston will be connected to pump return. This action will
reverse if the turning motions reverse. (Courtesy of Ford Motor Company)

turn, inlet fluid will be directed to one of the
actuator ports while the other actuator port is
connected to the return port. Reversing the direction of turn would reverse these two connections. As power assist occurs, the twisting
pressure on the torsion bar relaxes; movement
of the actuator tends to take the pressure off
the torsion bar, recenter the valve, and stop the
assist. This normally occurs as the front tires
reach the turning angle desired by the driver.
(Figs. 13-35 & 13-36)

The valves used with linkage-type power
steering and standard power steering gears
built by Chrysler Corporation are a sliding
valve type. The operation is essentially the
same as far as the fluid flows are concerned, but
the mechanical action is a spool valve sliding in

a bore instead of a spool rotating in a sleeve.
The valve spool is centered by a pair of reaction
springs when there is no steering action desired. These springs provide the feel of the
amount of steering assist. An important thing
to remember is that most sliding valves are adjustable for centering. If they are adjusted incorrectly, they can cause a power assist and
produce turning in one direction with no steering input. Check this by raising the front tires
off the ground and then starting the engine. If
the steering wheel turns one direction or the
other by itself, the sliding valve needs to be ad-

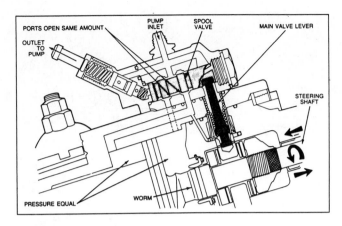

FIGURE 13-37.
A sliding power steering valve, turning motions of the steering shaft can cause the steering shaft to move slightly to one side or the other in the housing; the main valve lever transfers this motion to the valve spool. Movement of the valve spool redirects the fluid flows and produces the power assist. (Courtesy of Chrysler Corporation)

justed. The adjustment procedure is described in the manufacturer's and technician's service manuals. (Figs. 13-37 & 13-38)

13.5.3 Rack and Pinion Power Steering Gears

Power rack and pinion steering gears have become very common on most modern cars. The simple design matches the simplicity of the rack and pinion gear set. The rack gear is built

with a piston on the end opposite to the gear teeth; this piston runs in a bore in the rack housing with a sealed, sliding fit. Steel tubing is used to make the fluid connections between each end of the housing bore and the control valve. (Fig. 13-39)

During straightahead operation, the control valve allows a pressure equal to return line pressure to be exerted at each end of the piston bore. There will be no hydraulic assist because there will be an equal pressure at each side of the piston. A turning force on the steering wheel will move the control valve off center so that pump pressure is sent to one end of the piston bore while the other end of the bore is connected to pump return. This pressure difference on the rack piston will force the piston toward the side of lower pressure and produce the steering assist. It should be noted that once the rack reaches the end of its travel, there is no more room for additional fluid. Fluid flow will stop at this point, and the pressure will increase. (Fig. 13-40)

A power rack and pinion gear will also have a small vent tube interconnecting the two end bellows. During a turn, one of the accordion bellows extends while the other contracts; air must transfer from one of the bellows to the other to allow this to happen. In a manual rack, air can move through the center of the rack, but a power rack has three separate seals that pre-

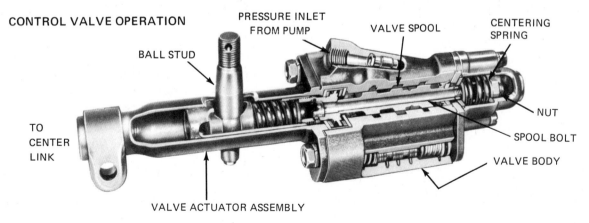

FIGURE 13-38.
A valve assembly from a linkage booster power steering system, sideways movement of the ball stud (connected to the Pitman arm of the steering gear) moves the valve spool in the body and produces the fluid flows necessary to produce a power assist. (Courtesy of Ford Motor Company)

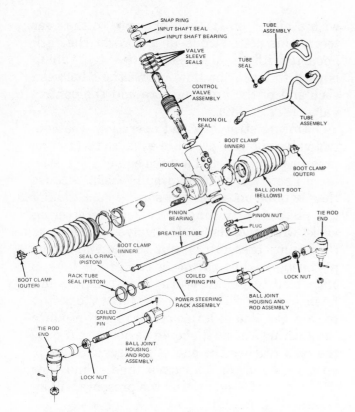

FIGURE 13-39.
An exploded view of power rack and pinion steering gear, note the raised flange near the center of the rack; this is the power assist piston. (Courtesy of Ford Motor Company)

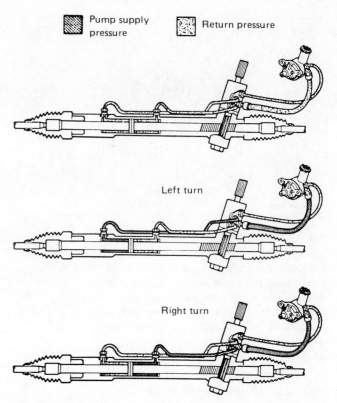

FIGURE 13-40.
The pressure exerted on each side of the rack piston will change from a straightahead position (top), a left-turn position (center), or a right-turn position (bottom). (Courtesy Ford Motor Company)

vent air movement, making the exterior vent necessary.

13.5.4 Standard Power Steering Gears (Integral)

Integral standard power steering gears use an actuator piston that is built into the ball nut, and the ball nut has a sliding, sealed fit in the gear housing. Two internal passages connect the actuator chambers to the ports of the control valve. The flows in this steering gear are similar to those in a rack and pinion gear. (Fig. 13-41)

When there is steering input and the control valve has all ports open to each other, both chambers will have the same (return) pressure, and there will be no assist. Steering input that is strong enough to twist the torsion bar and realign the valve spool and sleeve will direct pump pressure to one side of the ball nut piston and connect the chamber on the other side to the pump return. This difference in pressure across the piston will force the ball nut and sector gear in the correct direction to produce a power assist. Reversing the steering direction reverses the fluid flows and the assist direction. It should be noted that when the ball nut reaches the end of the chamber, it can move no farther, and there will be no more room for additional fluid. The fluid flow will stop at this point, and the pressure will increase. (Fig. 13-42)

13.5.5 Linkage Booster Power Steering

This style of power steering uses a normal standard steering gear; the control valve and actuator are mounted in the steering linkage. At one time this type of power steering was of-

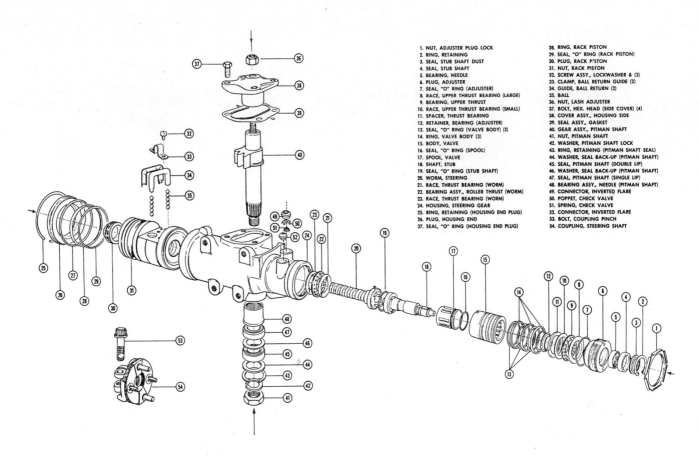

1. NUT, ADJUSTER PLUG LOCK
2. RING, RETAINING
3. SEAL, STUB SHAFT DUST
4. SEAL, STUB SHAFT
5. BEARING, NEEDLE
6. PLUG, ADJUSTER
7. SEAL, "O" RING (ADJUSTER)
8. RACE, UPPER THRUST BEARING (LARGE)
9. BEARING, UPPER THRUST
10. RACE, UPPER THRUST BEARING (SMALL)
11. SPACER, THRUST BEARING
12. RETAINER, BEARING (ADJUSTER)
13. SEAL, "O" RING (VALVE BODY) (3)
14. RING, VALVE BODY (3)
15. BODY, VALVE
16. SEAL, "O" RING (SPOOL)
17. SPOOL, VALVE
18. SHAFT, STUB
19. SEAL, "O" RING (STUB SHAFT)
20. WORM, STEERING
21. RACE, THRUST BEARING (WORM)
22. BEARING ASSY., ROLLER THRUST (WORM)
23. RACE, THRUST BEARING (WORM)
24. HOUSING, STEERING GEAR
25. RING, RETAINING (HOUSING END PLUG)
26. PLUG, HOUSING END
27. SEAL, "O" RING (HOUSING END PLUG)

28. RING, RACK PISTON
29. SEAL, "O" RING (RACK PISTON)
30. PLUG, RACK PISTON
31. NUT, RACK PISTON
32. SCREW ASSY., LOCKWASHER & (2)
33. CLAMP, BALL RETURN GUIDE (2)
34. GUIDE, BALL RETURN (2)
35. BALL
36. NUT, LASH ADJUSTER
37. BOLT, HEX. HEAD (SIDE COVER) (4)
38. COVER ASSY., HOUSING SIDE
39. SEAL ASSY., GASKET
40. GEAR ASSY., PITMAN SHAFT
41. NUT, PITMAN SHAFT
42. WASHER, PITMAN SHAFT LOCK
43. RING, RETAINING (PITMAN SHAFT SEAL)
44. WASHER, SEAL BACK-UP (PITMAN SHAFT)
45. SEAL, PITMAN SHAFT (DOUBLE LIP)
46. WASHER, SEAL BACK-UP (PITMAN SHAFT)
47. SEAL, PITMAN SHAFT (SINGLE LIP)
48. BEARING ASSY., NEEDLE (PITMAN SHAFT)
49. CONNECTOR, INVERTED FLARE
50. POPPET, CHECK VALVE
51. SPRING, CHECK VALVE
52. CONNECTOR, INVERTED FLARE
53. BOLT, COUPLING PINCH
54. COUPLING, STEERING SHAFT

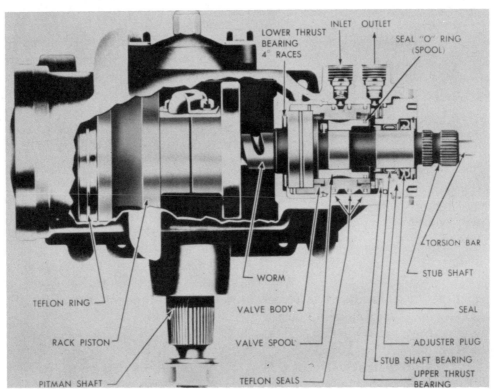

FIGURE 13-41.
A disassembled (top) and a cutaway (bottom) view of an integral, standard power steering gear, this gear is commonly found on older RWD cars. (Courtesy of Saginaw Division, General Motors Corporation)

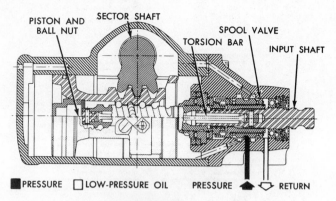

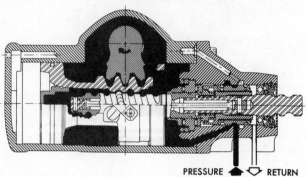

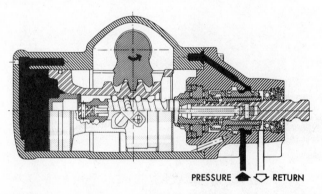

FIGURE 13-42.
The pressure exerted on each side of the piston and ball nut will change from a straightahead position (top), a left-turn position (center), or a right-turn position (bottom). (Courtesy of Ford Motor Company)

fered as a dealer-installed option. It is sometimes called a **hang-on** style. **Linkage booster power steering** was factory installed on several models of cars. (Fig. 13-43)

The sliding type of control valve is normally mounted at the Pitman arm end of the center steering link. The actuator is a separate hydraulic cylinder that has one end attached to the car's frame/body and the other end to the steering center link. Two high-pressure steel

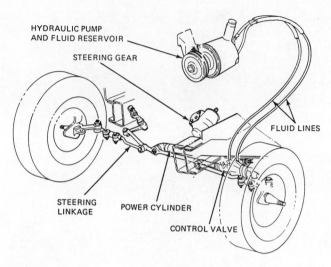

FIGURE 13-43.
A linkage booster or non-integral power steering system, the power cylinder is connected to the frame/body at the right end. (Courtesy of Ford Motor Company)

tubes or reinforced rubber hoses connect the ends of the actuator to the control valve.

When there is no steering input, the two reaction springs hold the control valve in the center position. Fluid pressure from the pump is directed to the pump return port and both actuator ports. When the driver turns the steering wheel, pressure between the Pitman arm and center link will move the control valve spool to an off-center position. Pump pressure will now be directed to one end of the actuator, and the other end will be connected to pump return. Power assist will occur because the pressures on both sides of the actuator piston are not equal. Like the other types of power steering, reversing the steering direction will reverse the fluid flows, power assisted movement will remove the valve spool pressure and recenter the valve, and fluid flow will stop and pressure will increase when the actuator reaches the end of its travel. (Figs. 13-44 & 13-45)

13.5.6   Power Steering Hoses and Fluid

Two reinforced rubber hoses connect the power steering pump to the control valve. They are the pressure hose and the return hose. The return hose connections are usually made by slip-

POWER CYLINDER OPERATION

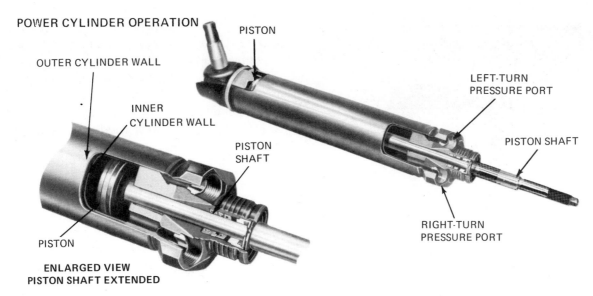

FIGURE 13-44.
The power cylinder used with linkage booster power steering, note that one fluid port enters directly into the cylinder; the other port connects to the far end of the cylinder using the area between the two cylinder walls as a passage. (Courtesy of Ford Motor Company)

ping the hose onto a connector and securing it in place with a band or screw-type clamps. The pressure hose is stronger and must use pressure-tight connections. High-pressure, swaged metal bands connect the hose to the steel end tubes, which connect to the pump and gear using flare or "O" ring seals. These hoses come in various lengths with various shaped end tubes to fit different installations. Some pressure hoses have built-in restrictions that resemble a reduction in hose diameter. (Figs. 13-46 & 13-47)

Some late model cars use **hydro-boost brakes**; the power assist for brake application comes from the power steering pump. These systems require at least two additional hoses. (Fig. 13-48)

Some systems use a cooler for the power steering fluid; this cooler can resemble a miniature radiator or just an extra length of metal tubing with or without fins. Hydraulic fluid tends to break down and oxidize at higher temperatures. Oxidized fluid tends to turn dark brown, smell like varnish, thicken, and leave gummy, varnish-like deposits on metal surfaces. Fluid life can be lengthened if it is kept

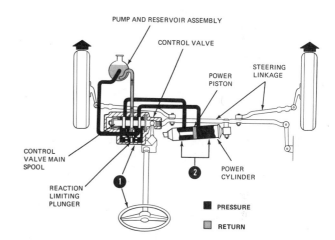

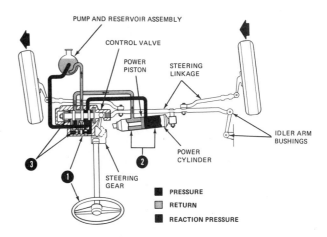

FIGURE 13-45.
The fluid flows for a right and left turn will be reversed from each other; during straightahead driving, pump return pressure will be at each end of the piston. (Courtesy of Ford Motor Company)

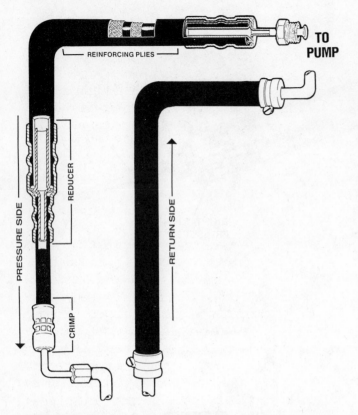

FIGURE 13-46.
The pressure hose for power steering is reinforced with several layers of fabric and uses crimped connectors; some pressure hoses will include a flow reducer. The return hose connections are usually secured with screw clamps. (Courtesy of Moog Automotive)

relatively cool. The cooler, if used, is mounted in the return line where it will be subject to low pressures. (Fig. 13–49)

Traditionally, power steering systems have used automatic transmission fluid/ATF for hydraulic fluid; the reddish color makes this easy to identify. Many later model systems are currently using power steering fluid; this fluid is dyed yellow or yellow-green. Power steering fluid has a lower viscosity (is thinner), which reduces some of the heat build-up and power loss in the steering system. The manufacturer's recommendations for fluid type should be followed when adding or changing fluid.

Power steering hose replacement and fluid changes are described in Chapter 16.

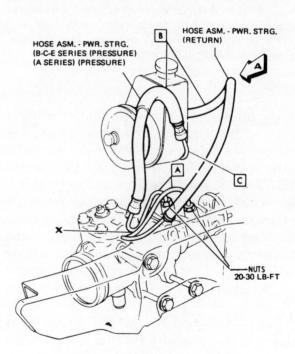

A — DO NOT BEND OR DISTORT PIPES TO FACILITATE INSTALLATION. AFTER TIGHTENING NUT, PIPE TO BE AGAINST GEAR AS SHOWN. RETURN HOSE PIPE TO BE AGAINST PRESSURE HOSE PIPE AS SHOWN AT "X"

B — HOSES MUST BE INSTALLED TO CLEAR EACH OTHER AND SURROUNDING PARTS. (RETURN HOSE MUST NOT BE TWISTED DURING INSTALLATION) (HOSE TO HOSE MINIMUM CLEARANCE 1.00" SURROUNDING PARTS MINIMUM CLEARANCE .50") HOSE ENDS MUST BE KEPT FREE FROM DIRT & OTHER CONTAMINATES.

C — AFTER TIGHTENING NUT, PIPE TO LIE AGAINST PUMP ASM AS SHOWN.

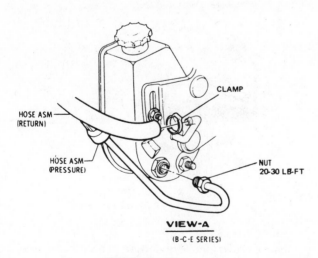

VIEW-A
(B-C-E SERIES)

FIGURE 13-47.
The power steering hose connections at the pump should be connected in the recommended manner. (Courtesy of Buick)

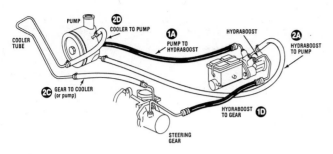

FIGURE 13–48.
Cars equipped with hydraboost braking systems require additional hoses and connections. Hoses 1A and 1D are pressure hoses; hoses 2A, 2C, and 2D are return hoses. (Courtesy of Moog Automotive)

13.6 STEERING LINKAGE

As mentioned earlier, two different styles of steering linkage are used on modern cars: a fairly simple system with rack and pinion gears and a more complex system with standard steering gears. The linkage transmits the side-to-side steering motions from the steering gear to the front wheels.

As the front tires raise and lower during normal suspension travel, the steering knuckle and arm will travel vertically in a slight arc relative to the frame/body. The outer tie rod end must travel in a similar arc or **bump steer** will result. Bump steer causes the car to turn right or left as the tires move up or down; this can cause potentially dangerous steering or tire wear. (Fig. 13–50)

Suspension designers try to match the arcs of travel of both the tie rod end and the steering arm so they will coincide. The arc of travel of the steering arm is controlled by the length or angles of the control arms. This is discussed in a little more detail in Chapter 19 when we describe instant centers and roll centers. The arc of travel of the tie rod end is controlled by the length of the tie rod and height of the inner tie rod end. This resulting arc and bump steer are discussed in Chapter 19. Some manufacturers recommend a steering linkage parallelism check to help ensure the correct inner tie rod end

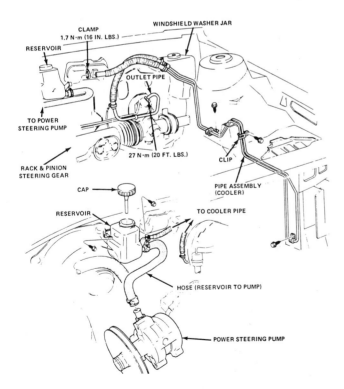

FIGURE 13–49.
Some power steering systems include a "cooler" loop in the return line to help reduce power steering fluid temperatures. (Courtesy of Chevrolet)

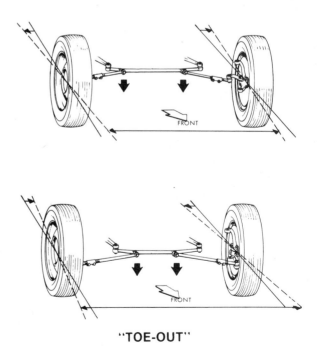

"TOE-OUT"

FIGURE 13–50.
Bump steer, also called toe change, can occur as the suspension moves up and down if the steering geometry does not match the suspension geometry. (Courtesy of SKF Automotive Products)

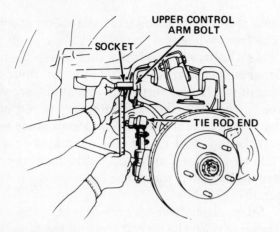

FIGURE 13-51.
One manufacturer recommends a steering linkage parallelism check and adjustment, if necessary, to ensure that
excessive toe change does not occur. (Courtesy of Oldsmobile)

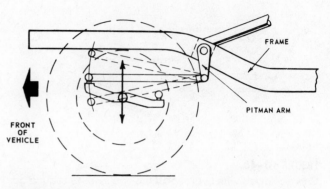

FIGURE 13-52.
The steering system used with some solid axles, turning on
bumps can occur if the bump path of the third arm does not
match the bump path of the drag link. (Courtesy of Ammco)

height and a minimum of bump steer effects;
this check is recommended when replacing idler
arms. (Fig. 13-51)

A type of bump steer can also occur on
solid axles if the vertical arc of travel of the axle
and the third arm does not match the arc of
travel for the forward end of the drag link.
Raising a vehicle very far above design height
specifications usually aggravates this situation; a drag link should be horizontal at ride
height. (Fig. 13-52)

13.6.1 Outer Tie Rod Ends

Most outer tie rod ends are a ball and socket
joint very similar to a ball joint. These tie rod
ends use a preload spring, either metal or rubber, to maintain a slight friction load between
the ball and socket to eliminate sideplay. The
stud of the ball joint is a locking taper so there
will be no sideplay when it is installed in the
steering arm. Most tie rod ends use a flexible
rubber boot to keep the lubricant in the joint
and dirt and water out. A tie rod will fail if the
boot is torn or broken. At one time, most tie
rod ends had some provision for lubrication;
this was either a grease/zerk fitting or a plug
for installing a fitting. Many modern tie rod
ends are lubricated and sealed during manufacture; there is no provision for additional lubrication. (Fig. 13-53)

One manufacturer has begun using a tie
rod end that uses a rubber bushing. The rubber
stretches or twists during steering manuevers
much like the rubber in a torsilastic control arm
bushing. Like control arm bushings, this tie rod
end requires no lubrication. (Fig. 13-54)

13.6.2 Tie Rods, Rack and Pinion Steering Gear

The inner end of the tie rods used with rack and
pinion steering becomes the ball for the ball and
socket joint; the socket encloses the ball and is
secured onto the end of the rack. In newer designs, a sealed, permanently adjusted socket is
built onto the tie rod. Some older designs used
an outer socket, which could be replaced separately from the tie rod. The older design required an adjustment for tightness during replacement. These joints are protected by the
flexible bellows/boot at the ends of the gear
housing. (Fig. 13-55)

The threads of the inner tie rod end socket
are locked onto the end of the rack by one of
five methods: a jam nut, a jam nut plus solid
pin, a roll or spiral lock pin, a set screw, and a
staked end. These locking methods help ensure
that the tie rod socket cannot come loose from
the rack. Each of these locking methods requires a different disassembly and reassembly
procedure, which is described in Chapter 16.
(Fig. 13-56)

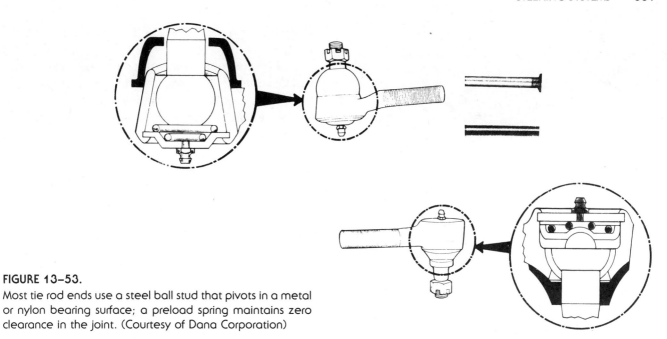

FIGURE 13–53.
Most tie rod ends use a steel ball stud that pivots in a metal or nylon bearing surface; a preload spring maintains zero clearance in the joint. (Courtesy of Dana Corporation)

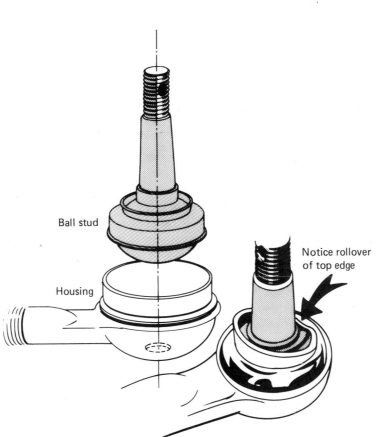

Ball stud

Housing

Notice rollover of top edge

FIGURE 13–54.
Rubber bonded tie rods use a steel ball stud encapsulated with rubber; the rubber twists and stretches during steering and turning manuevers. Note the rollover of the end socket which secures the ball stud in place. Since this joint is not greased, no protective boot is needed. (Courtesy of Replacement Parts Division, TRW, Inc.)

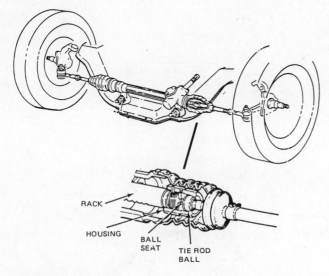

FIGURE 13–55.
The inner tie rod end used with rack and pinion steering is a ball-shaped inner end of the tie rod enclosed in a socket at the end of the rack. A flexible rubber bellows protects the tie rod end and the end of the steering gear. (Courtesy of Ford Motor Company)

Some General Motors FWD cars use a rack and pinion design in which the tie rod ends connect to the center portion of the rack. Replaceable rubber bushings are used between the ends of the tie rods and the tie rod mounting bolts. Center-mounted tie rods use a single, large bel-

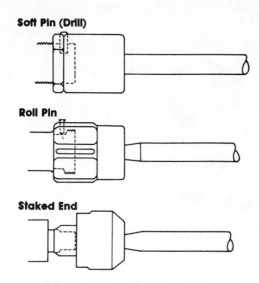

FIGURE 13–56.
Three of the most commonly used methods to secure the tie rod end socket to the end of the rack. The staked style squeezes a portion of the socket onto a pair of flats on the rack. (Courtesy of Moog Automotive)

lows/boot to protect the internal gear parts. (Fig. 13–57)

The tie rods themselves can be extensions of the inner tie rod end, which threads directly into the outer tie rod end (two-piece assembly) or use an adjuster sleeve, which threads into both the inner and outer ends (three-piece assembly). Mechanically, there is little difference between these two styles unless you are adjusting toe. The inner end of the tie rod is turned to make the adjustment on two-piece units while only the adjuster sleeve is turned on three-piece units. The toe adjustment is locked by a jam nut at the outer tie rod end or a clamp at the threads of the adjuster. Toe/tie rod adjustments are described more completely in Chapter 18.

13.6.3 Tie Rods, Standard Steering Gears

The inner tie rod ends used with standard steering gears are almost the same as the outer end. The major difference is that one has right-hand threads and the other left-hand threads, and the length of the threaded portion that connects to the threaded adjuster sleeve is usually much longer in the tie rod end. (Fig. 13–58)

All these tie rods are three-piece units, an outer end, an inner end, and the adjuster sleeve, which connects the two ends. The sleeve is locked to the tie rod ends by clamps that squeeze the slotted sleeve tightly onto the tie rod ends.

13.6.4 Idler Arm and Center Link

Parallelogram steering linkage is used with most standard steering gears. This style uses a center steering link with the inner tie rod ends connected to it in the correct location as far as bump steer is concerned. The center link is supported at one end by the Pitman arm on the steering gear, which also puts the steering motions into the steering linkage. The other end of the center link is supported by the idler arm. The idler arm is the same length and parallel to the Pitman arm and placed in exactly the same relative position on the opposite side of the car.

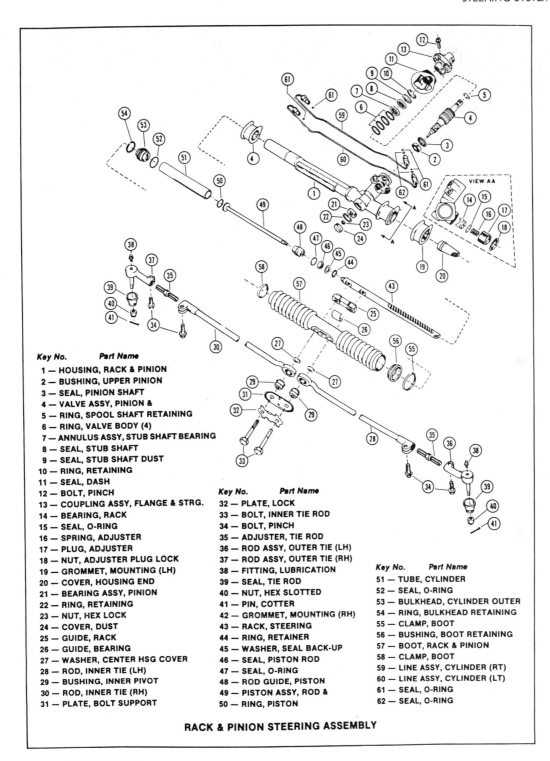

Key No.   Part Name

1 — HOUSING, RACK & PINION
2 — BUSHING, UPPER PINION
3 — SEAL, PINION SHAFT
4 — VALVE ASSY, PINION &
5 — RING, SPOOL SHAFT RETAINING
6 — RING, VALVE BODY (4)
7 — ANNULUS ASSY, STUB SHAFT BEARING
8 — SEAL, STUB SHAFT
9 — SEAL, STUB SHAFT DUST
10 — RING, RETAINING
11 — SEAL, DASH
12 — BOLT, PINCH
13 — COUPLING ASSY, FLANGE & STRG.
14 — BEARING, RACK
15 — SEAL, O-RING
16 — SPRING, ADJUSTER
17 — PLUG, ADJUSTER
18 — NUT, ADJUSTER PLUG LOCK
19 — GROMMET, MOUNTING (LH)
20 — COVER, HOUSING END
21 — BEARING ASSY, PINION
22 — RING, RETAINING
23 — NUT, HEX LOCK
24 — COVER, DUST
25 — GUIDE, RACK
26 — GUIDE, BEARING
27 — WASHER, CENTER HSG COVER
28 — ROD, INNER TIE (LH)
29 — BUSHING, INNER PIVOT
30 — ROD, INNER TIE (RH)
31 — PLATE, BOLT SUPPORT

Key No.   Part Name

32 — PLATE, LOCK
33 — BOLT, INNER TIE ROD
34 — BOLT, PINCH
35 — ADJUSTER, TIE ROD
36 — ROD ASSY, OUTER TIE (LH)
37 — ROD ASSY, OUTER TIE (RH)
38 — FITTING, LUBRICATION
39 — SEAL, TIE ROD
40 — NUT, HEX SLOTTED
41 — PIN, COTTER
42 — GROMMET, MOUNTING (RH)
43 — RACK, STEERING
44 — RING, RETAINER
45 — WASHER, SEAL BACK-UP
46 — SEAL, PISTON ROD
47 — SEAL, O-RING
48 — ROD GUIDE, PISTON
49 — PISTON ASSY, ROD &
50 — RING, PISTON

Key No.   Part Name

51 — TUBE, CYLINDER
52 — SEAL, O-RING
53 — BULKHEAD, CYLINDER OUTER
54 — RING, BULKHEAD RETAINING
55 — CLAMP, BOOT
56 — BUSHING, BOOT RETAINING
57 — BOOT, RACK & PINION
58 — CLAMP, BOOT
59 — LINE ASSY, CYLINDER (RT)
60 — LINE ASSY, CYLINDER (LT)
61 — SEAL, O-RING
62 — SEAL, O-RING

RACK & PINION STEERING ASSEMBLY

FIGURE 13–57.
Some General Motors cars attach the inner ends of the tie rods to the center of the rack; note the guide (25), which the mounting bolts pass through and the differently shaped boot (57), which seals the opening in the rack housing. (Courtesy of Pontiac Division, GMC)

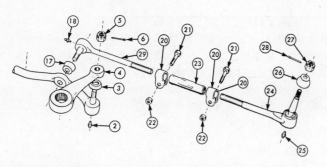

FIGURE 13–58.
Most tie rods used on RWD cars consist of two tie rod ends connected by an adjusting tube or sleeve; note the two clamps (20) that lock the three major parts together. (Courtesy of Saginaw Division, General Motors Corporation)

This ensures that the center link will move straight across the car. The center link is connected to the Pitman arm and idler arm by ball and socket joints, which are quite similar to tie rod end joints. These two pivots plus the center axis of the Pitman shaft and the idler arm pivot from the corner points of the parallelogram, which gives this linkage type its name. In a straightahead position, the idler arm, Pitman arm, and center link form three sides of a rectangle. (Fig. 13–59)

The idler arm pivots from a bracket that is bolted to the frame/body. The pivot bushing can be a rubber bushing, a threaded metal bushing, or a bearing type depending on the manufacturer. Each bushing style has different service requirements, which are described in Chapter 16. (Fig. 13–60)

13.7   DRAG LINKS

Many trucks and 4WD pickups use a solid axle and a single tie rod to connect the two steering arms. Use of two tie rods and parallelogram steering linkage would cause bump steer. A **drag link** connects the Pitman arm on the steering gear to the tie rod or an extension of the steering arm at the right steering knuckle, across the car.

Some trucks and pickups use a drag link to connect the Pitman arm to a **third arm** at the left steering knuckle. This design also uses a single tie rod to connect the two steering arms. A drag link uses a ball and socket joint at each end connection; this unit can be similar to a

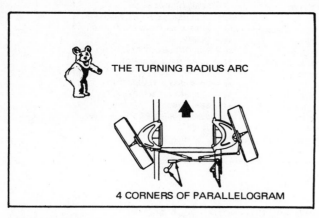

FIGURE 13–59.
The Pitman arm, center link, and idler arm form three sides of a parallelogram; this ensures a true side-to-side movement of inner ends of the tie rods. (Courtesy of Bear Automotive)

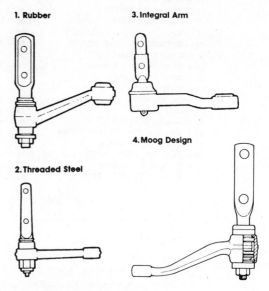

FIGURE 13–60.
Different styles of bearings are used between the idler arm and mounting bracket depending on the manufacturer. (Courtesy of Moog Automotive)

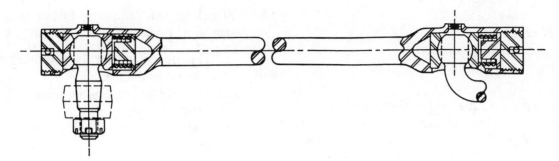

FIGURE 13-61.
A cutaway view of a drag link showing the internal drag link socket parts; note that a drag link can be removed from the ball stud by loosening the end screws. (© 1965, courtesy of Replacement Parts Div., TRW, Inc.)

normal tie rod end or a drag link socket. The ball stud of a drag link socket is solidly connected, sometimes riveted or welded, to the Pitman arm or third arm. Drag link sockets are disassembled or loosened to allow the socket to lift off the ball stud for removal; turning the adjuster screw enlarges the socket enough so that it can be slipped off the ball. Adjustment of a drag link socket is described in Chapter 16. (Fig. 13-61)

REVIEW QUESTIONS

1. Steering ratios are being discussed. Statement A: A 15:1 ratio will give one complete revolution of the tie rod for each 15 turns of the steering wheel. Statement B: This ratio would require 150° of turning the steering wheel to produce 10° of turning at the front wheels. Which statement is correct?

    a. A only   **c.** both A and B
    b. B only   **d.** neither A nor B

2. Which of these describe the effects of steering ratios?
    A. Fast ratios produce quick, maneuverable steering.
    B. Slow ratios produce stable, easy steering.

    a. A only   **c.** both A and B
    b. B only   **d.** neither A nor B

3. Besides containing the steering shaft, the steering column contains or holds

    a. the collapsible column and shaft features.
    b. several electrical switches for different circuits.
    c. a locking mechanism for the steering wheel.
    d. all of these.

4. The steering shaft is normally connected to the steering gear through

    a. a rigid, splined connector.
    b. a universal joint.
    c. a flexible coupler.
    d. all of these.

5. A worm and roller style of steering gear is also called a

a. Gemmer gear.
b. Ross gear.
c. recirculating ball-nut.
d. rack and pinion.

6. Steering gears are being discussed. Statement A: A steering shaft is also called a worm shaft. Statement B: The sector gear is part of the Pitman shaft. Which statement is correct?

a. A only   c. both A and B
b. B only   d. neither A nor B

7. The most commonly used style of standard, manual steering gear is the

a. cam and level.
b. worm and sector.
c. recirculating ball-nut.
d. worm and roller.

8. Statement A: The ball-nut moves up or down the worm shaft when the worm shaft is rotated. Statement B: The balls roll along grooves in the worm shaft and ball-nut when the worm shaft is rotated. Which statement is correct?

a. A only   c. both A and B
b. B only   d. neither A nor B

9. Which of the following is true about rack and pinion steering gears?
A. The rack is a long, straight gear.
B. The pinion is a small round, normally shaped gear.

a. A only   c. both A and B
b. B only   d. neither A nor B

10. Which of the following is true when comparing standard steering gears with rack and pinion steering gears?
A. Rack and pinion gears are heavier and more complex.
B. Rack and pinion gears do a better job of dampening road shock.

a. A only   c. both A and B
b. B only   d. neither A nor B

11. Which of the following is not true about standard power steering gears?

a. The power assist is provided by hydraulic pressure.
b. The control valve is mounted in the Pitman shaft.
c. The power piston is built into the ball-nut.
d. The steering gear is lubricated by power steering fluid.

12. Fluid flows through a power steering system are being discussed. Statement A: The control valve in the steering gear can cause a pressure change in the high-pressure hose. Statement B: The maximum pressure in the high-pressure hose is controlled by the relief valve in the pump. Which statement is correct?

a. A only   c. both A and B
b. B only   d. neither A nor B

13. Which of the following is true about power steering hoses?
A. The return hose has very little pressure in it.
B. The pressure hose contains a few hundred psi with no effort on the steering wheel and a pressure as high as the relief valve setting during turns.

a. A only   c. both A and B
b. B only   d. neither A nor B

14. Most power steering pumps are of the
A. sliding or roller vane type.
B. positive displacement type.

a. A only   c. both A and B
b. B only   d. neither A nor B

15. The control valve in the pump is being discussed. Statement A: This valve allows greater and greater flows as the engine speeds up. Statement B: This valve usually hisses or squeals when it relieves pressure back to the pump inlet. Which statement is correct?

a. A only   **c.** both A and B
b. B only   **d.** neither A nor B

16. The power piston section of the power steering can be mounted

    a. outside of the steering gear.
    b. on the ball-nut.
    c. on the rack gear.
    d. any of these.

17. The outer tie rod ends are
    A. usually ball and socket type joints.
    B. preloaded with an internal device to eliminate rotating motions.

    a. A only   **c.** both A and B
    b. B only   **d.** neither A nor B

18. An idler arm is

    a. used to support one end of the center link.
    b. the same length as the Pitman arm.

c. attached to the frame through a pivot bushing.
d. all of these.

19. Inner tie rod ends of a rack and pinion steering gear are being discussed. Statement A: They are always locked onto the end of the rack by a steel pin, jam nuts, or set screw. Statement B: They are enclosed by a flexible, accordion-like boot. Which statement is correct?

    a. A only   **c.** both A and B
    b. B only   **d.** neither A nor B

20. The tie rods
    A. connect the steering arms to the Pitman arm.
    B. are arranged so that they can move up and down at the outer end as well as laterally.

    a. A only   **c.** both A and B
    b. B only   **d.** neither A nor B

Chapter 14

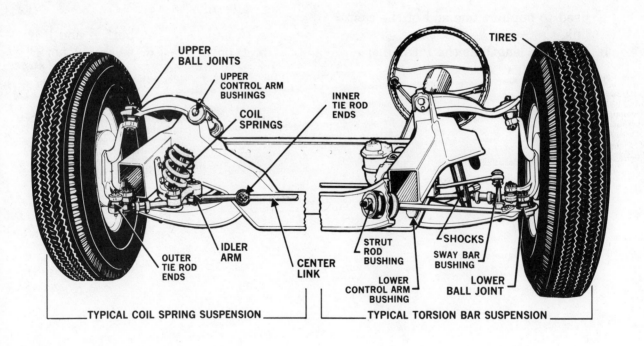

UPPER
BALL JOINTS

UPPER
CONTROL ARM
BUSHINGS

COIL
SPRINGS

INNER
TIE ROD
ENDS

TIRES

OUTER
TIE ROD
ENDS

IDLER
ARM

CENTER
LINK

STRUT
ROD
BUSHING

LOWER
CONTROL ARM
BUSHING

SHOCKS

SWAY BAR
BUSHING

LOWER
BALL JOINT

TYPICAL COIL SPRING SUSPENSION

TYPICAL TORSION BAR SUSPENSION

SUSPENSION AND STEERING SYSTEMS INSPECTION

After completing this chapter, you should:

- Be able to inspect and measure ball joints and king pins for excessive clearance, faulty seals, or binding.

- Be able to check control arm and strut rod bushings for wear or excessive clearance.

- Be able to check struts and strut mounts for wear or damage.

- Be able to check conventional steering gears and steering linkage systems for wear, excessive clearance, or binding.

- Be able to check rack and pinion steering gears and steering linkage for wear, excessive clearance, or binding.

- Be able to check power steering systems for abnormal noises, leakage, or improper operation.

- Be able to determine what repairs are needed to correct any suspension and steering faults located during this inspection procedure.

14.1 INTRODUCTION

Complaints that deal with tires or handling characteristics of a car indicate a need for a thorough inspection of the suspension and steering systems. Typical complaints include tire wear, vibrations, pull to one side, side pull during braking, wander, loose or sloppy steering, or noises. You should recognize some of these complaints from the earlier portions of this text in the sections dealing with tires, wheel bearings, springs, and shock absorbers. If any of these systems have a malfunction or failure, they affect the suspension system in their own particular way.

Remember that vehicle vibrations result from something that rotates, and that something, usually a tire, is either out-of-round or unbalanced. Also remember that weak springs and shock absorbers can cause poor ride quality and can cause the suspension system to sag out of the correct geometry. Since we have already dealt with these matters, this chapter is primarily concerned with the control arms, ball joints, and control arm bushing portions of the suspension system as well as the steering system components.

The purpose of an inspection is to determine the cause for the vehicle owner's complaint and to determine what steps will be needed to cure that complaint. It is a good practice to note any other parts that show signs of failing in the near future so the customer can be aware of them. It should be remembered that the suspension should operate for many miles and a year or so until the next time it is inspected; the average motorist does not check suspension components very frequently. (Fig. 14-1)

Sometimes an inspection will determine that a simple adjustment or realignment is all that is needed to correct the situation; often a worn bushing or ball joint will show up. Worn parts should be replaced before an alignment can be done. It does no good to do a wheel alignment if the suspension parts are sloppy. In most cases on an older car, when a realignment is necessary, that need is probably caused by worn parts or sagged springs. We should also re-

SAFETY NOTE
A suspension, steering, or brake failure can place the car and its passengers in a highly dangerous situation. While making an inspection, this fact is in the forefront of the front-end technician's mind. Any item that might fail in the near future and cause an accident is noted and brought to the car owner's attention.

member that the rear wheels also have a suspension system and that their parts also wear out. Rear suspension bushings and pivots are checked in the same manner as those at the front. (Fig. 14-2)

14.2 INSPECTION POINTS

As an inspection is being performed, it is a good practice to follow a set procedure to ensure that portions of the suspension and steering systems are not skipped or forgotten. A suspen-

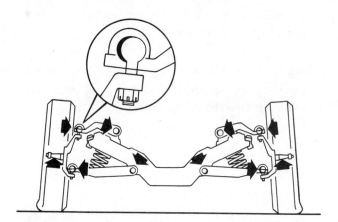

FIGURE 14-1.
As the suspension parts wear, they allow the front tires to have improper alignment angles as well as allow uncontrolled, sloppy tire movement. (Courtesy of SKF Automotive Products)

FIGURE 14–2.
Every moving part in the steering and suspension systems is a potential wear point. (© 1976, courtesy of Replacement Parts Div., TRW, Inc.)

sion and steering system inspection should include:

1. Steering wheel for excessive steering looseness or binding.

2. Tires for correct inflation (Chapter 4).

3. Tires for wear pattern to give an indication of incorrect alignment, balance, or worn parts, and also for physical defects that might cause failure (Chapter 4).

4. Vehicle for correct height and attitude (Chapter 12).

5. Vehicle for optional springs, shock absorbers, or overload devices that might change the ride quality or alignment (Chapter 10).

6. Tire spinning (by hand) for tire runout and wheel bearing condition (Chapters 4 and 6).

7. Tire and wheel shake (top and bottom) for wheel bearing looseness (Chapter 6).

8. Tire and wheel shake (side to side) for steering component looseness.

9. Ball joints for excessive looseness, boot condition, and binding.

10. Control arm bushings for wear or deterioration.

11. Strut rod bushings for wear or deterioration.

12. Stabilizer bar bushings and end-links for wear or deterioration (Chapter 12).

13. Springs for loose or broken parts (Chapter 12).

14. Shock absorbers or strut for leakage, loose or broken mounts, or broken parts (Chapter 12).

15. Tie rod ends for looseness or torn boots.

16. Steering gear, center link, and idler arm for loose pivot points or loose mounting bolts.

OR Steering rack and tie rods for loose mounting bushings, loose inner tie rod ends, or torn boots/bellows.

17. Power steering pump, hoses, and gear for leaks.

18. Power steering pump drive belt for wear and adjustment.

19. Power steering pump for correct fluid level.

OR Manual steering gear for correct lubricant level.

If the owner included a vibration complaint, the tires should be checked for excessive runout and spun up to speed to check for an unbalanced condition. Loose, worn, or misaligned suspension or steering parts will not normally cause vibrations; vibrations originate from a part that spins or rotates.

Some of the suspension checks are made with the tires on the ground; most of them are made with the vehicle raised on a frame contact hoist to provide good access to the tires and wheels, suspension components, and steering components. As indicated in some of the steps, some of the inspection procedures have already been covered in earlier chapters. This chapter describes the procedure for checking the remaining components.

Many technicians follow an inspection checklist such as the one illustrated in Figure 14-3. The checklist helps ensure that none of the checks is missed or forgotten, and it allows a more professional discussion with the car owner.

14.3 BALL JOINT CHECKS

As mentioned earlier, ball joints form the steering axis for S-L A suspensions. The steering axis for a strut suspension is the lower ball joint and the upper strut mount. These ball joints also form the pivot between the control arm and the steering knuckle to allow suspension travel. To perform these tasks, the ball joint must allow free movement without free clearance, which would allow sloppy tire and wheel or steering motions. Ball joint checks are to ensure that this free motion is within the wear tolerances set by the manufacturer. They also ensure that the ball joint is lubricated and that

the boot needed to retain this lubricant is in good condition.

Ball joint boots are checked visually; the area behind the boot and ball joint where you can't see can be checked by running your finger around the boot and feeling for problems. Look or feel for grease outside of the boot, which indicates breaks or tears. If the boot is torn, the ball joint will probably fail, if it has not already, and should be replaced. While checking the boot, squeeze it to ensure that there is grease inside of the boot. An empty boot indicates a need for lubrication.

Lubrication requirements for a ball joint vary among manufacturers. The lubrication intervals for modern joints are rather long; one manufacturer, for example, requires lubrication every three years or 30,000 miles (48,000 km). Long intervals such as this make it easy for the average motorist to forget about lubricating ball joints completely. Some ball joints are permanently sealed and require no further lubrication.

To lubricate a ball joint, you should:

1. Clean off all dirt or grease that might have accumulated around the grease/zerk fitting and boot. If a small plug is threaded into the ball joint instead of a grease fitting, remove the plug and install a grease fitting or use a grease gun attachment that threads directly into the joint. After lubricating the ball joint, some technicians leave the grease fitting in place; other technicians prefer to replace the plug for a ore positive seal. (Figs. 14-4 & 14-5)

2. Using a grease gun, pump enough of the correct type of grease into the joint until the boot begins to balloon or grease begins to flow from the bleed area of the boot.

It is a good practice while checking the boot to also check the visual condition of the ball joint and the control arm for cracks or breaks in the metal, which indicate a probable failure. These cracks will often show up as reddish-colored, loose rust streaks or dirt-free areas in otherwise dirty parts. A sudden separation

Moog problem solving parts help hold alignment, increase tire life, and improve steering and driving control . . .

UPPER BALL JOINTS
UPPER CONTROL ARM BUSHINGS
COIL SPRINGS
INNER TIE ROD ENDS
TIRES

OUTER TIE ROD ENDS
IDLER ARM
CENTER LINK
STRUT ROD BUSHING
LOWER CONTROL ARM BUSHING
SHOCKS
SWAY BAR BUSHING
LOWER BALL JOINT

TYPICAL COIL SPRING SUSPENSION

TYPICAL TORSION BAR SUSPENSION

YOUR CAR DESERVES A
WHEEL TO WHEEL SECURITY CHECK

Owner_____ Date_____ Phone_____

Make_____ Model_____ Year_____

Mileage_____ License Number_____ Engine Size_____

PARTS DESCRIPTION	OK	COMMENTS	PARTS	LABOR
SPRINGS				
CONTROL ARM BUSHINGS				
POWER STEERING				
LOWER BALL JOINT				
UPPER BALL JOINT				
WHEEL BRGS.				
BALANCE				
TIRES				
TIE ROD ENDS				
IDLER ARM				
PITMAN ARM				
CENTER LINK				
SWAY BAR BUSHINGS				
STRUT ROD BUSHINGS				
SHOCK ABSORBERS				
ALIGNMENT				

| REMOVED PARTS REQUESTED | YES | | | SUB TOTAL | |
| | NO | | | | |

TOTAL

CAR HEIGHT			
	Left	Right	Specs.
FRONT			
REAR			

BALL JOINT READINGS			
Load Carrier	Left	Right	Specs.
AXIAL			
RADIAL			

ALIGNMENT			
	Left	Right	Specs.
CAMBER			
CASTER			
TOE			

REMARKS:

Inspector
FORM 2000C • Litho in U S A

FIGURE 14-3.
Many mechanics follow a checklist like this to ensure that they don't skip any checks and to give the car owner a record of what was found during the inspection. (Courtesy of Moog Automotive)

FIGURE 14-4.
A cut or torn ball joint such as this one will let the grease escape and allow water and dirt to enter the joint; if the joint is not worn out already, it will soon fail. (Courtesy of SKF Automotive Products)

of a ball joint from a control arm can have catastrophic results. (Fig. 14-6)

A complaint of **"hard steering"** can be caused by a tight ball joint. This is not a common complaint, but it does occur. Place your fingers on the ball joint as you turn the steering knuckle; noisy operation or a rough feeling indicates a **dry/tight joint**. It should be checked further. Disconnect the tie rod end from the steering arm and rotate the steering knuckle by itself. The steering knuckle should rotate freely and smoothly, requiring only a slight pressure at the steering arm.

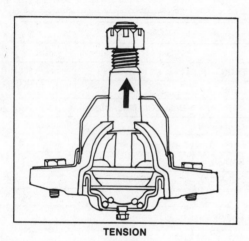

TENSION

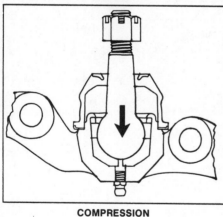

COMPRESSION

FIGURE 14-5.
Ball joints will have either a plug or zerk fitting in the housing so that the joint may be greased. (Courtesy of SKF Automotive Products)

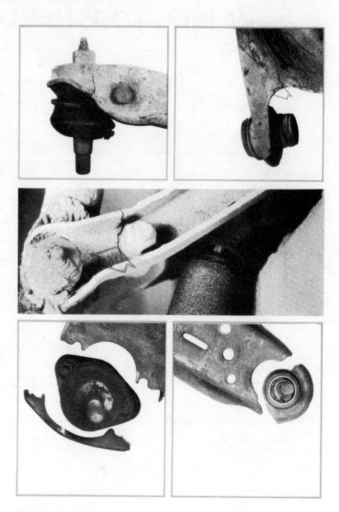

FIGURE 14-6.
When lubing ball joints or inspecting suspension systems, keep an eye out for cracks, which indicate a probable failure of the components; the cracks will often show up as clean or rusty brown streaks in otherwise dirty areas. (© 1978, courtesy of Replacement Parts Div., TRW, Inc.)

The most commonly encountered problem with ball joints is excessive clearance. The amount of clearance allowed in a ball joint varies depending on where the ball joint is used and the manufacturer. Generally speaking, a friction-loaded/follower style of ball joint (used on the control arm without the spring on an S-L A or strut suspension) should have no clearance/free play. A load-carrying ball joint is often manufactured with an operating clearance; the load on the joint removes this clearance and keeps the joint tight. When the load is taken off this style of joint, it often feels loose because of the clearance, which is now perceptible. If the clearance is less than the manufacturer's limits, the joint is still good. Some states require that the car owner be informed of the allowable clearance for that joint and how much clearance was measured in the worn ball joint during the inspection. Ball joint replacement cannot be sold without this.

Four different checking methods are used for ball joints depending on their style and location: wear-indicating ball joints, load-carrying ball joints on lower control arms, load-carrying ball joints on upper control arms, and friction-loaded ball joints. Specifications for the allowable ball joint clearance and the exact checking method for a particular ball joint can be found in the manufacturer's service manual, a technician's service manual, and specialized service manuals published by aftermarket suspension component manufacturers. A rule of thumb followed by many technicians when specifications are not available is that a load-carrying ball joint with 0.060 inch (15 mm) of vertical clearance or less with no visible damage should still be usable. (Figs. 14–7 & 14–8)

14.3.1 Checking a Wear Indicator Ball Joint for Excess Clearance

Wear indicator ball joints are designed to easily show if the ball joint has excessive clearance; they are used as load-carrying ball joints on S-L A and modified strut suspensions. They can often be identified by inspecting the lower face of the ball joint. Wear indicator joints have an

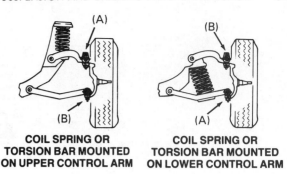

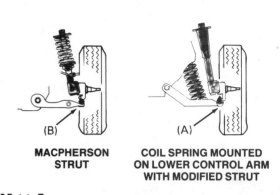

FIGURE 14–7
Depending on the suspension type and where the ball joint is located, a ball joint is either (A) a load-carrying or (B) a friction-loaded type. (Courtesy of Dana Corporation)

opening in the lower cover through which the grease fitting or a metal boss projects. The exact checking procedure varies slightly among manufacturers; the method described here is a general approach. (Fig. 14–9)

To check a wear indicator ball joint for excessive clearance, you should:

1. Park the car on a level surface, which allows access to the lower control arms and ball joints. The ramps of an alignment rack are ideal. The weight of the car should remain on the tires.

2. Wipe off any grease or dirt on the checking surface or the lower face of the ball joint.

3. On some styles, slide a plain, flat screwdriver or other flat, metal object about 1/4 to 1/2 inch (6 to 12 mm) wide across the bottom surface of the ball joint; it should bump into the

Model	Year	Vertical Movement	Model	Year	Vertical Movement
CHEVROLET (Cont'd)			**OLDSMOBILE (Cont'd)**		
Camaro	67-69*	.060″	Toronado	66-76	.125″
	70-73	.020″**	Omega	73-74	.0625″
	74-76	Wear Indicator♦		75-76	Wear Indicator♦
Nova (Chevy II)	62-67	See Table II	Starfire	75-76	Wear Indicator♦
	68-70*	.060″	**PLYMOUTH**	57-67	.050″
	71-74	.0625″		68-72	.070″
	75-76	Wear Indicator♦	Valiant, Barracuda	60-67	.050″
Vega,	71-74	.0625″		68-76	.070″
Vega, Monza	75-76	Wear Indicator♦	Volare	1976	.020″**
Chevette	1976	Wear Indicator♦	Fury	1973	.070″
CHRYSLER	57-64	.050″		74-76	.020″**
	65-73	.070″	Satellite	73-74	.020″**
	74-76	.020″**	**PONTIAC**		
Cordoba	75-76	.020″**	Catalina, Bonneville, Grandville, etc.	58-64	.060″
COLT/ARROW	71-76	.020″**		65-70	.050″
CRICKET	71-72	.020″**		71-72	.020″**
DODGE	57-67	.050″		73-76	Wear Indicator♦
	68-72	.070″	Tempest	61-63*	.093″
Dart	60-67	.050″		1964*	.060″
	68-76	.070″	LeMans (Tempest)	65-69	.050″
Challenger	70-73	.070″	Grand Prix, LeMans	70-72	.0625″
Aspen	1976	.020″**			

FIGURE 14–8.

Clearance specifications are printed for nonwear-indicating, load-carrying ball joints; the ball joint is considered worn out if the internal clearance exceeds this specification. (© 1976, courtesy of Replacement Parts Div., TRW, Inc.)

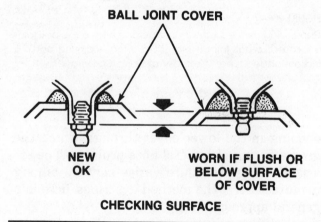

BALL JOINT COVER

NEW OK — WORN IF FLUSH OR BELOW SURFACE OF COVER

CHECKING SURFACE

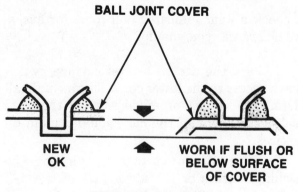

BALL JOINT COVER

NEW OK — WORN IF FLUSH OR BELOW SURFACE OF COVER

CHECKING SURFACE

FIGURE 14–9.

As a wear-indicator ball joint wears, the checking surface moves upward and into the joint. (Courtesy of Ford Motor Company)

checking surface. If the checking surface has moved up into the ball joint, the ball joint is excessively loose and should be replaced. (Fig. 14-10)

OR On some styles, grip the grease fitting with your fingers and try to rotate it; if the grease fitting can be rotated, the ball joint is excessively loose and should be replaced. (Fig. 14-11)

FIGURE 14-10.

When checking a wear-indicator ball joint, if a screwdriver blade catches on the checking surface, the joint is good. (Courtesy of SKF Automotive Products)

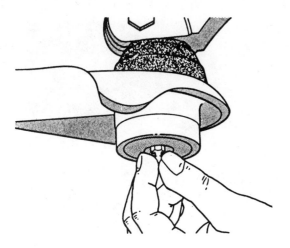

FIGURE 14-11.
If the zerk fitting can be rotated easily using finger pressure on some ball joints, the joint is worn and should be replaced. (Courtesy of Chrysler Corporation)

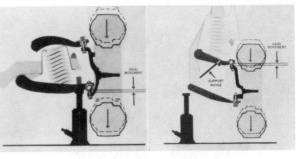

Lower   Upper

FIGURE 14-12.
When checking the clearance of a load-carrying ball joint, the load of the vehicle must be removed from the ball joint using a jack placed in the correct location depending on whether the load-carrying joint is the lower or upper one. Note the support wedge that is used when the load-carrying joint is on the upper arm (right). (Courtesy of Ammco)

14.3.2   Checking a Load-Carrying Ball Joint on a Lower Control Arm for Excessive Clearance

When the vehicle load passes from the spring and through the lower control arm to the steering knuckle, the lower ball joint is the load-carrying ball joint. This is true in cases where either a torsion bar, coil spring, or air spring is attached to the lower control arm. This load squeezes a compression-loaded ball joint tightly between the control arm and the steering knuckle or tries to pull a tension-loaded joint apart. The ball joint must be unloaded (vehicle and spring load removed) in order to measure the amount of clearance that is in the ball joint. This is usually accomplished by lifting the car by the lower control arm so the spring is compressed. If the car was lifted by the frame, the spring would push the lower control arm downward until the rebound/extension bump stop contacts the upper control arm. In this position the spring pressure will hold both ball joints tightly. The ball joints will appear to have zero clearance. (Fig. 14-12)

Wheel bearing clearance can be confused with ball joint clearance while shaking a tire. This problem can be eliminated by tightening the wheel bearing to eliminate its clearance or

applying the brakes to lock the rotor to the steering knuckle. Ball joint clearance is usually measured using a dial indicator; it can also be measured using vernier or dial calipers. Some states require that the amount of clearance must be measured by devices capable of measuring to the thousandths of an inch so the customer can be accurately informed as to the actual amount of clearance in the worn joints.

To check a load-carrying ball joint on a lower control arm, you should:

1. Place a jack or the lifting pads of a suspension contact hoist under the lower control arm. The jack or lifting pad should be positioned as close to the ball joint as possible to ensure that the spring is compressed and the ball joint is unloaded. (Fig. 14-13)

2. Raise the vehicle so the tire is off the ground and check the rebound/extension bump stop to ensure that it is not under a load. This ensures that the ball joints are unloaded. (Fig. 14-14)

3. Position a dial indicator so that it will read the **axial/vertical** motion of the ball joint. (Fig. 14-15)

OR Measure the distance from the lower face of the ball joint to the end of the ball joint stud using a vernier or dial caliper. (Fig. 14-16)

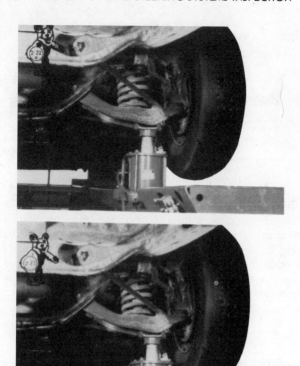

FIGURE 14-13.
The car should be lifted by a jack placed close to the ball joint to unload the car's weight from a lower load-carrying ball joint; without the car's weight, the clearance in the ball joint can now be measured. (Courtesy of Bear Automotive)

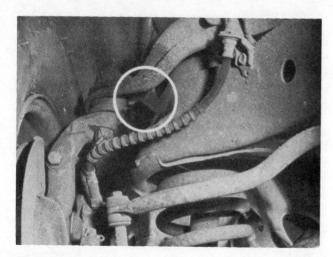

FIGURE 14-14.
As the load is removed from the lower ball joint, a gap will be evident between the extension bump stop and the frame. (Courtesy of Moog Automotive)

(A)

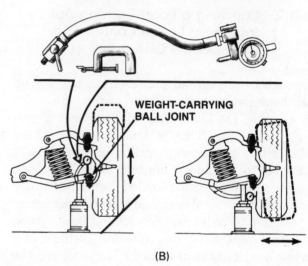

(B)

FIGURE 14-15.
Ball joint vertical (axial) clearance can be checked by mounting a dial indicator in the position shown. (A is courtesy of Moog Automotive; B is courtesy of Dana Corporation)

4. Raise the tire, wheel, and steering knuckle as far as you can without lifting the car; many technicians place a lever under the tire or steering arm to make this easier. Note the amount of travel on the dial indicator; this is the amount of clearance in the ball joint. (Fig. 14-17)

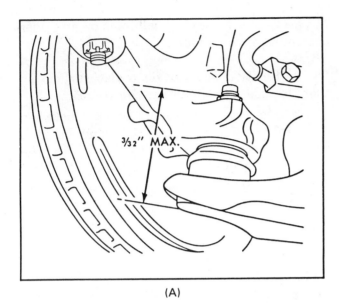

(A)

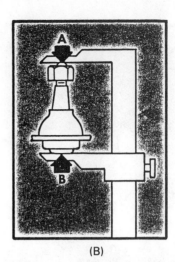

(B)

(C)

FIGURE 14–16.

Ball joint vertical (axial) clearance can be checked by measuring this distance (A) using calipers. Measure this distance with the vehicle's weight on the tires and then again with the vehicle raised and no weight on the tires; subtract one distance from the other to obtain the amount of clearance. C shows a measurement being made. (A is courtesy of Chevrolet; B is courtesy of Dana Corporation; C is courtesy of Volkswagen of America)

FIGURE 14–17.

When using a dial indicator to measure vertical clearance, use a bar to raise the tire and wheel while watching the amount of dial indicator movement. (Courtesy of Moog Automotive)

OR Remeasure the length of the ball joint with the calipers, and subtract the first measurement from the second. The result is also the amount of clearance.

5. Compare the amount of clearance measured to the specifications. An excessively worn ball joint is indicated if the measured clearance exceeds the specifications. It should be replaced.

6. Some manufacturers prefer to check ball joint clearance in a **horizontal/radial** direction. This is done by mounting the dial indicator in a horizontal rather than a vertical position and moving the tire, wheel, and steering knuckle in and out rather than up and down while making the check. If the radial clearance exceeds the specifications, the ball joint should be replaced. (Fig. 14–18)

FIGURE 14–18.
When checking ball joint horizontal (radial) clearance, the dial indicator is mounted in the position shown. (Courtesy of Moog Automotive)

14.3.3 Checking a Load-Carrying Ball Joint on an Upper Control Arm

This is essentially the same operation as described in Section 14.3.2 except for the procedure used to unload the ball joint. If the car is lifted by the lower control arm, vehicle load transmitted from the spring to the upper control arm will keep the ball joint tight between the steering knuckle and the upper control arm; both ball joints will have zero clearance. A prop or brace should be placed between the upper control arm and the frame to hold the control arm in near-to-normal position, and the vehicle should be lifted by the frame. The prop will transfer the spring load to the frame, unloading the ball joint while the control arm is held in a near-normal position. Both sides of the car should be braced and checked at the same time; if you try to check one side at a time, the stabilizer bar will twist and load the ball joints, making them appear tight.

To check a load-carrying ball joint on an upper control arm, you should:

1. Position the car on a flat surface so there is access to the upper control arm.

2. Place a prop/support bracket between the upper control arm and the frame on each side of the car so that when the car is lifted the control arm will be held upward. (Fig. 14–19)

3. Lift the car so the tire is off the ground.

4. Follow Steps 3 through 6 of Section 14.3.2. (Fig. 14–20)

FIGURE 14–19.
When checking a load-carrying ball joint on the upper control arm, a prop should be placed between the control arm and the frame to hold the control arm in its normal position as the car is lifted by the frame. (Courtesy of Moog Automotive)

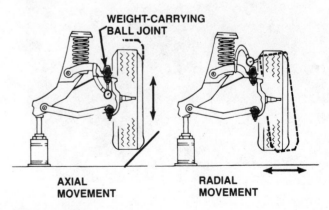

FIGURE 14–20.
Load-carrying ball joints are checked for axial or radial clearance with the dial indicator mounted in the positions shown here. (Courtesy of Dana Corporation)

14.3.4 Checking a Friction-Loaded/ Follower Ball Joint

These joints are manufactured with no clearance and in most cases have a slight preload. This joint is mounted on the control arm without a direct connection to a spring or torsion bar. There are two commonly used methods used to check a follower ball joint: free play and preload.

To check a follower ball joint for **free play**, you should:

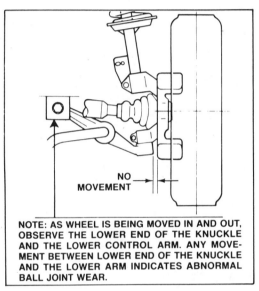

NOTE: AS WHEEL IS BEING MOVED IN AND OUT, OBSERVE THE LOWER END OF THE KNUCKLE AND THE LOWER CONTROL ARM. ANY MOVEMENT BETWEEN LOWER END OF THE KNUCKLE AND THE LOWER ARM INDICATES ABNORMAL BALL JOINT WEAR.

(A)

(B)

FIGURE 14–21.
When checking a nonload carrying/friction loaded ball joint, shake the tire in and out while watching for movement at the ball joint between the control arm and the steering knuckle. (A is courtesy of Ford Motor Company; B is courtesy of Bear Automotive)

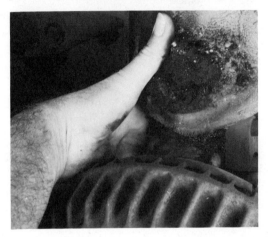

FIGURE 14–22.
As the tire is shaken in and out, there should be no apparent movement in an unloaded ball joint (circle). (Courtesy of Bear Automotive)

1. Raise the vehicle in such a way as to unload the load-carrying ball joint (described in Sections 14.3.2 and 14.3.3); on strut suspensions, lift the car by the frame.

2. Grasp the tire at the top or bottom, next to the ball joint being checked, and push in and out on the tire. Watch for any motion between the control arm and the steering knuckle that would indicate clearance. If you have trouble pushing on the tire while watching the ball joint, place your hand so your fingers span the gap between the steering knuckle and the control arm while the tire is being shaken. If you can feel or see any perceptible clearance in the ball joint, it has excessive clearance, and the ball joint should be replaced. (Figs. 14–21 & 14–22)

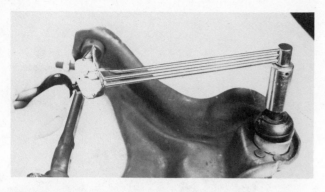

FIGURE 14-23.
Measuring the preload of a friction-loaded ball joint; it should take about 2 to 5 foot pounds of torque to rotate the ball joint stud. The torque wrench is attached to a nut on the ball joint stud.

To measure the **preload** of a follower ball joint, you should:

1. Raise and support the car on a hoist or jack stands.

2. Disconnect the ball joint stud from the steering knuckle; this operation is described in Chapter 15.

3. Replace the nut onto the ball joint stud and run it to the end of the threads or use a jam nut so the stud is forced to turn with the nut.

4. Use a low-reading torque wrench to measure the amount of torque that is required to rotate the stud under the internal ball joint resistance. Many manufacturers publish a torque specification for this check; a rule of thumb, if specifications are not available, is 2 to 8 foot pounds or 24 to 96 inch pounds (3 to 11 N·M). If the stud rotates too hard or too easily, the ball joint should be replaced. (Fig. 14-23)

14.3.5 Checking Ball Joint Clearance on an I-Beam or 4WD Solid Axle

These axles are used on some pickups and 4WD vehicles. The ball joints are used in pairs, with both of them carrying some of the vehicle load. There are two methods of checking these ball joints depending on the specifications given by the vehicle manufacturer, **turning torque** and **side clearance**.

To measure the turning torque, you should:

1. Raise and support the vehicle on a hoist or jack stands.

2. Disconnect the tie rod end from the steering arm; this operation is described in Chapter 16. (Fig. 14-24)

3. Attach a spring scale to the tie rod boss in the steering arm or a torque wrench onto a ball joint stud nut and measure the force needed to turn the steering knuckle. Excessive ball joint clearance is indicated if the steering knuckle turns too easily. (Fig. 14-25)

4. Some manufacturers use an adjustment sleeve at the ball joint to provide a means of adjusting the turning effort, provided the ball joint is still good. (Fig. 14-26)

To check ball joint side clearance, you should:

1. Raise and support the vehicle on a hoist or jack stands.

2. Push the bottom of the tire in and out while watching for movement between the steering knuckle and the axle in the area next to the lower ball joint. (Fig. 14-27)

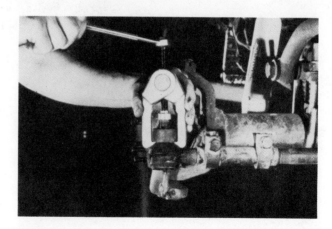

FIGURE 14-24.
When measuring the turning effort on a 4WD axle, the tie rod end should be disconnected; the tie rod can be removed from the steering arm by using a special puller as shown here. (Courtesy of Moog Automotive)

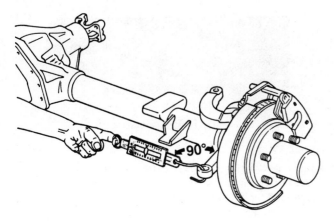

FIGURE 14-25.
Turning effort on a 4WD axle is measured using a spring scale attached to the end of the steering arm as shown here. Pull on the scale and note the amount of pull required to turn the steering knuckle. (Courtesy of Dana Corporation)

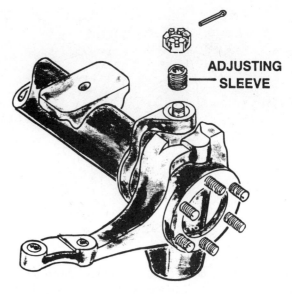

FIGURE 14-26.
Some 4WD axles have a provision for adjusting the turning effort; the adjusting sleeve is turned inward to increase the amount of turning effort. (Courtesy of Dana Corporation)

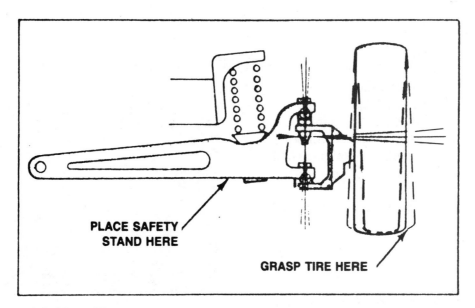

FIGURE 14-27.
The ball joints on a twin-I beam axle are checked in a manner similar to the checking of a friction-loaded ball joint; shake the tire in and out while looking for any visible side motion between the axle and the steering knuckle. (Courtesy of Ford Motor Company)

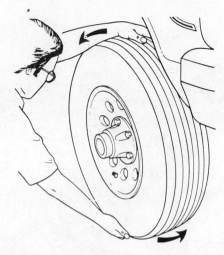

FIGURE 14-28.
When shaking the tire for ball joint checks, grasp the top and bottom of the tire, push inward with one hand while pulling outward with the other, and then reverse the pressure. (Courtesy of Ford Motor Company)

3. Repeat Step 2, pushing the top of the tire in and out while watching for side clearance at the upper ball joint.

4. If the amount of side motion exceeds the manufacturer's specifications, the ball joint should be replaced; if specifications are not available, a rule of thumb is 0.031 or 1/32 inch (0.8 mm). (Fig. 14-28)

14.3.6 Checking King Pin Clearance

Although king pins are not ball joints, they have been included with these checks because a king pin performs a similar function and the checking method is similar. King pins are used on solid axles and some twin I-beam axles; they are normally checked by measuring the **side shake** of the tire.

To check a king pin for excessive clearance, you should:

1. Raise and support the vehicle on a hoist or jack stands.

2. Eliminate wheel bearing clearance by installing a brake pedal jack to apply the brakes or by tightening the spindle nut. (Fig. 14-29)

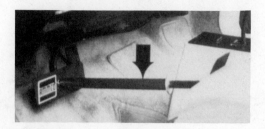

FIGURE 14-29.
A brake pedal jack (arrow) installed between the seat and brake pedal will hold the brakes in an applied position; application of the brakes will lock the brake drum/rotor to the steering knuckle and eliminate play between them. (Courtesy of Ammco)

3. Position a dial indicator at the lower part of the tire with the dial indicator stylus in a horizontal position. (Fig. 14-30)

4. Push in and out on the tire while observing the dial-indicator readings. Worn king pins are indicated if there is more than:

1/4 or 0.250 inch (6.35 mm) of side motion on 16 inch or smaller wheels.

3/8 or 0.375 inch (9.5 mm) for 17 to 18 inch wheels.

1/2 or 0.500 inch (12.7 mm) for wheels larger than 18 inches.

FIGURE 14-30.
King pin clearance is measured by mounting a dial indicator as shown here and moving the bottom of the tire in and out while watching the amount of travel on the dial indicator. (Courtesy of Moog Automotive)

14.4   CONTROL ARM BUSHING CHECKS

As previously mentioned, most control arm bushings are made from rubber and are of a torsilastic style. The outer metal sleeve is pressed into the control arm, and the tube-like inner sleeve is clamped tightly into the frame mount. As the control arm pivots up and down, the rubber portion twists and stretches to allow for suspension travel. This action transfers a certain amount of energy into the rubber bushing material, which generates heat. Heat from any source tends to harden rubber; as the rubber bushings become harder and stiffer, they tend to crack, break, and then disintegrate. The life of rubber bushings is determined somewhat by their temperature; rough, bumpy roads will cause more suspension action, more heat, and a shorter life. This bushing should not be lubricated; besides doing no good, a petroleum product will attack and shorten the life of the rubber.

Worn suspension bushings allow the control arm to move inward and outward or forward and backward as well as up and down. This results in an alignment change of the tires which, in turn, will cause tire wear and handling difficulties. This looseness often causes suspension noises, usually "clunks," when driving over rough roads or when the brakes are applied. Faulty rubber control arm bushings can usually be seen during a visual inspection. In locations where the bushings are difficult to see, faulty bushings are identified by excessive control arm motion either through an in and out or a sideways direction.

To check rubber control arm bushings, you should:

1. If possible, check the upper control arm bushings from under the hood. Use a light so you can get a good look at the rubber parts of the bushing. Ignore small, light cracks as long as the rubber is still solid and resilient. Look for heavy cracks, rubber material breaking out, or rubber distortion, which allows the control arm to change position. The pivot bolt should be centered in the bushing. Bushings that are

FIGURE 14–31.
Two faulty rubber bushings: the upper one is breaking up and losing some of the rubber material; the lower one shows signs of excessive cracking and distortion of the rubber. (Courtesy of SKF Automotive Products)

distorted, breaking up, or getting ready to break up should be replaced. (Fig. 14–31)

2. Raise and support the car on a hoist or jack stands.

3. Visually check the bushings on the lower control arm, looking for the same sort of problems. Also, check the sides of the control arm and the frame metal next to it for signs of metal contact, which indicate bushing failure.

4. Swing the tire rapidly back and forth while forcing it to bump at the steering stops; also, force the tire in and out. While doing this, watch the control arm for any motions that would indicate bushing failure.

NOTE: *The bellows at the end of the rack can be distorted if the checking motions are made too quickly on some cars with manual rack and steering gears.*

5. On single lower control arms, try prying the inner end of the control arm sideways using a pry bar or large screwdriver; a slight motion is acceptable. Larger motions indicate weak bushings.

Metal control arm bushing faults are more difficult to see. These bushings consist of a pair of large nuts that are locked into the control arm; the inner control arm shaft is bolted to the frame and is threaded on the ends to accept the large nut-like bushings. Suspension action turns the bushings on the shaft much like a nut turning on a bolt. These bushings must be lubricated to prevent wear, and they must be sealed to prevent dirt or water from entering the bushing. When this bushing runs out of lubricant it will begin to wear and squeak; noise is usually the first indication of metal bushing failure.

To check metal control arm bushings, you should:

1. Bounce the suspension while listening for squeaks or other bushing-related abnormal noises. If possible, place your finger lightly on the bushing while bouncing the front end; a noisy bushing will often have a rough, harsh feel. Noisy bushings can sometimes be cured by greasing them, but if they have squeaked for very long, they are probably worn and should be replaced.

2. Raise and support the car on a hoist or jack stands.

3. Swing the tire back and forth rapidly so the turning stops strike rather hard, and watch the control arm bushings. A very slight amount of side motion is acceptable, but a definite motion or jumping of the control arm on the shaft indicates a faulty bushing.

14.5 STRUT ROD BUSHING CHECKS

Strut rod bushings are rubber bushings that are compressed tightly against each side of an opening in the frame bracket. If they become weak, the outer end of the lower control arm will have an excessive amount of travel in a forward and backward direction. (Fig. 14-32)

Strut rod bushing failure is often indicated by a "thump" or "clunk" as the brakes are applied. These bushings are checked visually.

FIGURE 14-32.
A badly worn metal control arm bushing and shaft; note the worn areas on the control arm shaft (arrows).

To check strut rod bushings, you should:

1. Raise and support the car on a hoist or jack stands.

2. Grip the bushing end of the strut rod and shake it up and down; any free play indicates a faulty bushing.

3. Inspect the bushing for hard cracks, rubber breaking, and severe distortion of the rubber; also, check for signs of contact between the metal back up washer and the bushing bracket. Any of these indicate a faulty bushing. (Fig. 14-33)

14.6 STRUT CHECKS

Occasionally the tire on a strut suspension shows excessive camber wear, which is an indicator that the strut body or strut piston rod might be bent. There are several checks that can be made to determine if this has occurred.

One check is to compare the included angle or the camber angle and steering axis inclination angles to the specifications. This procedure requires wheel alignment equipment and is described in Section 17.18.

A quick check for a bent strut body is to measure the distance between the strut body and the brake rotor on both sides of the car and compare the measurements. If they differ by

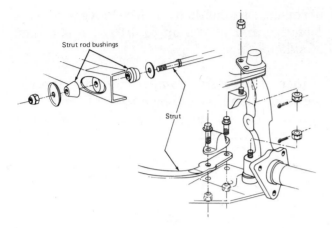

FIGURE 14-33.
Strut rod bushings allow the strut rod to swing up and down without moving forward or backward. (Courtesy of Ford Motor Company)

WORN BUSHING

NEW BUSHING

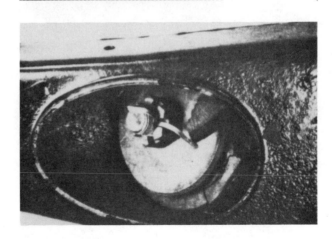

FIGURE 14-34.
Two worn (top and bottom) and one new (center) strut rod bushings. (Courtesy of SKF Automotive Products)

more than a few thousandths of an inch (mm), one of the struts is bent. (Fig. 14-34)

To check for a bent piston rod, loosen the piston rod at the upper mount and then rotate the piston rod while you watch for side motion at the top of the tire or strut body. A good strut rod should rotate evenly with no side motion of the tire or strut body.

14.7 STRUT DAMPER/INSULATOR CHECKS

A faulty strut damper assembly can cause noises; it will also allow the upper end of the strut to sag to an improper position, changing the alignment of the strut as well as the tire and wheel. (Fig. 14-35)

To check a strut damper, you should:

1. With the car's weight on the tires, note the relative position of the upper strut mount to the fender opening.

2. Raise the car by the frame, and note any change in the position of the mount assembly. A slight downward movement is normal. The manufacturer's specifications should be checked to determine the allowable clearance for a particular car. One manufacturer allows up to 1/2 inch (12.7 mm). (Fig. 14-36)

3. With the tires off the ground, reach around the tire and grip the strut's spring as close to the upper mount as possible. Alternately push inward and pull outward on the strut and spring while you watch the upper end of the strut piston rod. A slight movement is

FIGURE 14-35.
A ruler (arrow) is being used to measure the rotor to strut distance, checking for a bent spindle or strut. The measurement should be the same on the other side of the car.

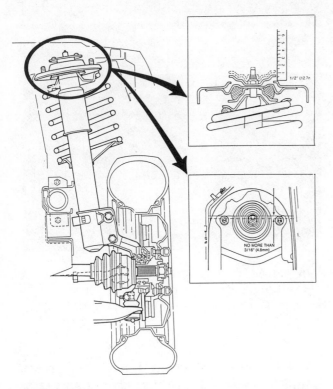

FIGURE 14-36.
A faulty strut damper will allow excessive movement of the upper strut mount as a car is lifted and the load is removed from the tires; there can also be an excessive amount of side play as the strut is pushed in and out.

normal; there should be no free movement. One manufacturer allows up to 3/16 inch (4.8 mm) of side motion.

If there is excessive damper motion or if a visual inspection reveals separation of the rubber portions, the damper portion of the strut mount should be replaced.

14.8 STEERING LINKAGE CHECKS

The checks for steering linkage wear are commonly done at the same time as the suspension checks. It is easier for us to study the suspension system separately from the steering system, but the average front-end technician does not separate them when it comes to repairing them or making a wheel alignment. These two systems are usually serviced together.

There are two major styles of steering systems: standard steering gears with parallelogram linkage and rack and pinion steering gears and linkage. Each system has some different components to check. The most common steering system fault is looseness; any clearance in any portion of the steering linkage will cause loose, sloppy steering or the tires to change toe angle as they go down the road. (Fig. 14-37)

The various portions of the steering system can be checked individually, as described in the following sections, or together by performing a **"dry-park"** test. Dry-park is the steering action some people make while parking a car—turning the steering wheel while the car is standing still. This places a much higher load on the steering components than turning with the tires rolling.

To perform a dry-park test, you should:

1. Park the car on a level surface where there is access to the steering linkage. An alignment rack is ideal because the car's weight should remain on the tires; if an alignment rack is used, the turntables should stay locked so they don't turn.

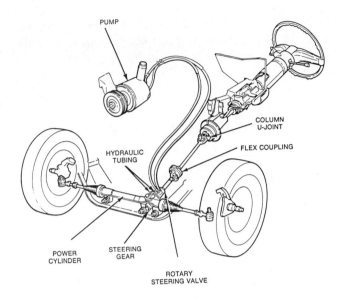

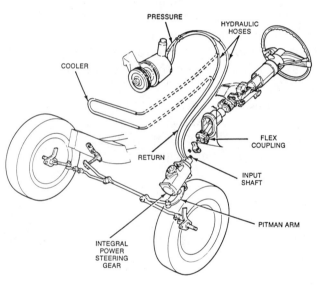

FIGURE 14–37.
Two general types of steering systems are used on cars; each system has its own type of inspection checks. (Courtesy of Ford Motor Company)

2. Have a helper turn the steering wheel in a back and forth manner, keeping a constant, alternating motion.

3. Observe the entire steering system for signs of play, being sure to check for:
    a. loose or worn steering shaft couplers
    b. loose steering gear mounts and bushings
    c. loose steering gear, internal
    d. worn Pitman-arm-to-center-link connection

    e. worn idler-arm-to-center-link connection
    f. worn idler arm bushing
    g. worn outer and inner tie rod ends.
A rotational or reciprocating motion of these parts is normal; free or sloppy motion indicates worn parts.

14.8.1 Tie Rod End Checks

Tie rod ends are ball and socket joints that are preloaded to have zero clearance. Don't forget that the tie rod ends must allow a pivoting and swiveling action between the tie rod and the steering arm.

Tie rod ends must be lubricated to prevent wear, and they must be sealed to keep dirt and water out. If the sealing boot fails, the tie rod end will soon fail. Traditionally, tie rod ends have had either a grease/zerk fitting or a plug for the purpose of installing a grease fitting so grease could be added to the joint. Many newer tie rod ends are lubricated and sealed during manufacture or use a rubber bushing. These joints require no lubrication; they have no provision for adding grease.

Tie rod ends that use rubber bushings will have no free play at all; resistance should be felt in the joint in a rotary as well as an up and down or a sideways direction.

Worn tie rod ends allow loose, sloppy steering action with steering wheel free play; they will also sometimes cause a **pop** or **snap** noise while turning.

To check tie rod ends, you should:

1. Raise and support the car on a hoist or jack stands.

2. Grip the tie rod next to the end and firmly push up and pull down while watching for excessive movement. Any free play is excessive. About 1/16 to 1/8 inch (1.5 to 3 mm) of motion, resisted by a spring, is allowable; travel greater than this is excessive and indicates a faulty tie rod end. (Fig. 14–38)

OR One manufacturer recommends measuring the distance from the side of the tie rod end to the end of the stud using vernier or dial

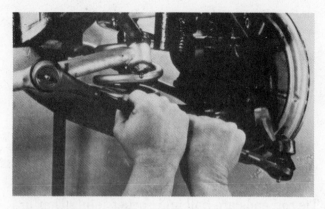

FIGURE 14-38.
A quick check of a tie rod end is made by pushing and pulling on the tie rod in the direction of the stud; a good tie rod end will have no visible looseness. (Courtesy of SKF Automotive Products)

calipers. This distance should be measured with the tie rod end pushed together and then pulled apart; the difference between these two measurements is the amount of clearance in the tie rod end. This manufacturer has a specification for the maximum allowable clearance.

3. Grip both front tires and firmly try to pull them together and then push them apart. Feel and watch for free play while doing this; free play indicates a faulty pivot point. (Fig. 14-39)

4. Rotate the tie rod; it should turn smoothly with no catches or binding for about 10 to 30°.

5. Inspect the boots for breaks and tears to ensure they are in good condition; a tie rod end with a bad boot should be replaced.

14.8.2 Parallelogram Steering Linkage and Idler Arm Checks

Parallelogram steering linkage, which is used with standard steering gears, uses a centerlink that is supported on one end by the Pitman arm from the steering gear and from the other end by an idler arm. The idler arm pivots from a bracket that is bolted to the frame. The idler arm is the same length and in the same relative position as the Pitman arm. The centerlink travels sideways across the car during turning maneuvers, and the two tie rods connect to the center link through a tie rod end. There is also a pivot joint, similar to a tie rod end, at each connection between the centerlink and the idler or Pitman arm.

Wear checks for the centerlink pivot points are made in the same manner as tie rod end checks. Worn Pitman arm and centerlink joints will often show up when the tires are pulled together and then spread apart. At this same time, a worn idler arm bushing will let the center link end of the idler arm raise and lower. Worn idler arm or Pitman arm bushings will also cause loose, sloppy steering.

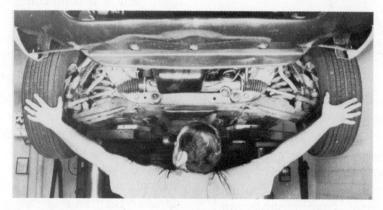

FIGURE 14-39.
To check the tie rod ends and the other portions of the steering linkage, push the tires apart and then pull them together while you watch for any play or free movement in the steering system. Any free movement indicates a fault. (Courtesy of Moog Automotive)

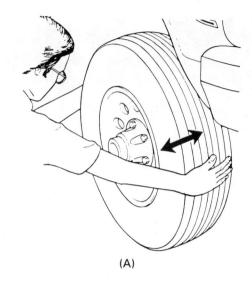

(A)

CHECKING STEERING LINKAGE WEAR
AS VIEWED FROM ABOVE

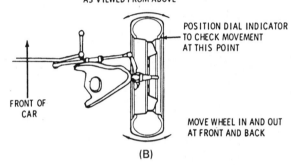

POSITION DIAL INDICATOR
TO CHECK MOVEMENT
AT THIS POINT

FRONT OF
CAR

MOVE WHEEL IN AND OUT
AT FRONT AND BACK

(B)

FIGURE 14–40.
Steering linkage wear can be measured by lifting one wheel off the ground and rocking the tire sideways (A); the amount of wear can be measured by mounting a dial indicator in the position shown at B. (A is courtesy of Ford Motor Company; B is courtesy of Chevrolet)

To check an idler arm bushing, you should:

1. Raise and support the car on a hoist or jack stands.

2. Grip the center link near the end of the idler arm, and firmly push up and pull down while watching the vertical motion. 1/8 inch (3 mm) or so is normal. The manufacturer's specifications should be used to determine the allowance for each car; some manufacturers allow up to 1/4 inch (6 mm). Movement greater than this indicates a faulty idler arm. (Figs. 14–40 to 14–42)

3. Inspect the frame area surrounding the idler arm mounting bolts for cracks and other signs of breakage. Cracks or breaks should be repaired by someone competent in frame repair.

14.8.3 Rack and Pinion Steering Linkage Checks

Rack and pinion steering systems are much simpler than the standard style. A tie rod connects to each end of the steering rack, which attaches to the frame through rubber bushings. The inner tie rod end, where the tie rod connects to the rack, is a ball and socket joint that is enclosed in the flexible, rubber bellows/boot.

All outer tie rod ends are checked in the same manner as described in Section 14.8.1.

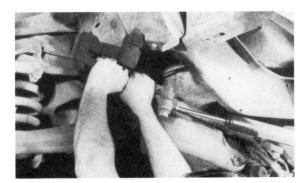

FIGURE 14–41.
An idler arm can be quickly checked by pushing up and pulling down on the center link close to the end of the idler arm. A very slight movement is normal on some cars. (Courtesy of Moog Automotive)

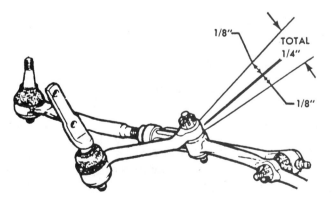

1/8″

TOTAL
1/4″

1/8″

FIGURE 14–42.
This manufacturer allows a total movement of 1/4 inch at the end of the idler arm; the idler arm should be replaced if the wear is greater than this. (Courtesy of Chevrolet)

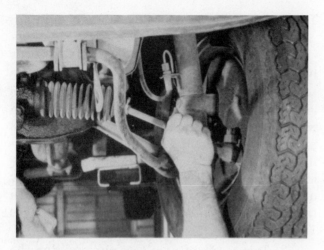

FIGURE 14–43.

The outer tie rod ends on rack and pinion steering are checked in the same manner as with standard steering; push up and down on the tie rod. The end should be replaced if there is any vertical looseness. (Courtesy of Moog Automotive)

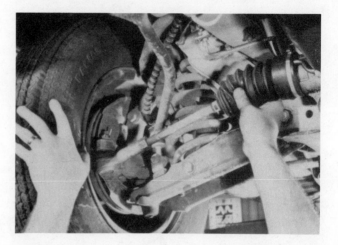

FIGURE 14–44.

The inner tie rod socket is checked by squeezing the bellows so you can feel the socket and tie rod while you push and pull on the tire; if you can feel movement between the tie rod and socket, the tie rod end should be replaced. The bellows should also be checked to ensure that there are no cracks or splits. (Courtesy of Moog Automotive)

The inner tie rod ends are a little more difficult to check because they are enclosed in the boot. (Fig. 14–43)

To check inner tie rod ends, bellows, and rack bushings, you should:

1. Raise and support the car on a hoist or jack stands.

2. Turn the steering to one of the stops and carefully check the bellows, which has been stretched out, for excessive cracks in the rubber, tears, breaks, or distorted pleats. Turn the steering to the other stop, and inspect the remaining bellows. Collapsed bellows can often be straightened by loosening the clamps and turning one end of the bellows. Torn or broken bellows should be replaced. Power steering fluid leaks from a bellows indicate a faulty steering gear.

3. Squeeze the bellows so you can feel the inner tie rod end socket and tie rod, and push and pull on the side of the tire. If you can feel any free play between the tie rod and the socket, the inner tie rod end should be replaced. If you have difficulty feeling the inner joint, loosen the large bellows clamp and slide the bellows aside so you can see the joint. (Fig. 14–44)

OR Some manufacturers recommend disconnecting the outer tie rod end from the steering arm and measuring the amount of force that is needed to swing the tie rod back and forth. If the tie rod swings too easily, the inner end should be replaced. (Fig. 14–45)

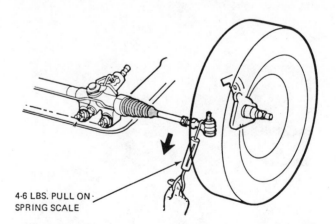

4-6 LBS. PULL ON
SPRING SCALE

FIGURE 14–45.

Inner tie rod socket preload can be measured by checking the articulation effort with a spring scale as shown here; if the tie rod swings too easily, the tie rod socket should be adjusted (some models) or the tie rod end should be replaced. (Courtesy of Ford Motor Company)

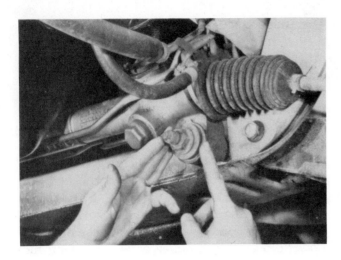

FIGURE 14-46.
The rack mounting bushings should be checked for wear or deterioration; faulty bushings should be replaced. (Courtesy of Moog Automotive)

4. Carefully inspect the rack mounting bushings for signs of wear, deterioration, or excessive rack movement. Faulty or oil-soaked bushings should be replaced. (Fig. 14-46)

14.9   STEERING GEAR CHECKS

Steering gear looseness can be checked at two separate points: from the steering wheel at the beginning of the inspection procedure and from under the car during the procedure. The first point gives an indication of whether or not play is present; the second check helps pinpoint the exact location of the free play. Steering gear clearance checks should be made with the tires in a straightahead position. A standard/conventional steering gear normally has a slight play in the turning positions.

Rock the steering wheel back and forth using a very light pressure, and feel for free movement. The engine should be running on power steering equipped cars. No clearance is acceptable; a very slight amount of play is tolerable. If it is difficult to feel any free play, watch the left front tire—a perceptible turning motion should be seen as soon as the steering wheel is turned.

On most cars, it is possible to reach up from under the car and turn the steering shaft while watching the linkage. Again, no clearance is acceptable, and a very slight clearance can be tolerated. If clearance is felt, observe the various parts to determine where the free play occurs—that part should be adjusted or repaired. On standard steering gears, you should also grip the Pitman arm and try to move it up and down as well as sideways in a steering direction; again, only a barely perceptible amount of play can be tolerated. (Fig. 14-47)

Also from underneath the car, check the security and tightness of the steering gear mounting bolts; any faults should be noted or repaired.

Conventional steering gears of the manual type should be checked to ensure they have an adequate supply of lubricant. Lubricant level is checked at the filler plug at the top surface of the gear housing or by removing one of the sector gear cover retaining bolts. The correct level is normally even with the bottom of the opening. (Fig. 14-48)

Adjustments and service of steering gears are described in Chapter 16.

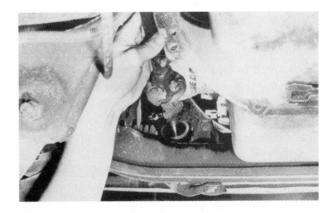

FIGURE 14-47.
The flexible steering coupler should be checked for looseness and wear; a worn coupler should be replaced. Also rotate the coupler and steering shaft and try to move the steering shaft in and out of the steering gear; the steering gear should be serviced if there is excess internal play. (Courtesy of Moog Automotive)

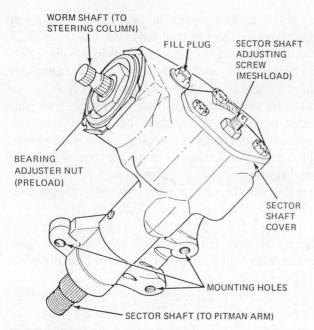

FIGURE 14-48.
The lubricant level of a manual steering gear should be even with the bottom of the fill plug opening; if the steering gear does not have a filler plug, check the level at the upper-most sector shaft cover bolt. (Courtesy of Ford Motor Company)

14.10   POWER STEERING CHECKS

These additional checks should be made on cars that are equipped with power steering: hose leaks and condition, pump and gear leaks, drive belt condition and tension, and fluid level and condition.

Check both power steering hoses for leaks, signs of deterioration, excessive cracking, sponginess, and rubbing or chafing, which might lead to failure. Faulty hoses should be replaced. Hose replacement is discussed in Chapter 16.

From underneath the car, check for signs of power steering fluid leakage. Look for the reddish (if ATF is used) or yellow-green (if power steering fluid is used) colored fluid. If fluid is found, follow the leak upward and forward to the faulty component. Also, run your hand along the hoses, pump, or gear as you feel for wetness to determine the exact location of the leak. (Figs. 14-49 & 14-50)

Inspect the pump drive belt for cracks, frayed fabric, very shiny or glazed sides, or other signs of deterioration; doubtful belts should be replaced. Some experts recommend replacing all the drive belts on an engine every four years.

Potential belt failure is not easily recognized on newer belts, and it is much less expensive to replace a belt too early than too late. The belt tension should also be checked and readjusted, if necessary. (Figs. 14-51 & 14-52)

Power steering fluid level is checked by removing the filler cap and checking the dipstick. Low fluid level can cause erratic steering motions as well as humming/buzzing noises during turning maneuvers. This should be done with the fluid hot (above the temperature where you can comfortably hold the end of the dipstick); cold fluid will check lower than normal. While checking the fluid level, inspect the condition of the fluid. Dirty fluid that has lost its bright, new appearance or that has begun to smell like varnish should be changed and replaced with new fluid. (Figs. 14-53 & 14-54)

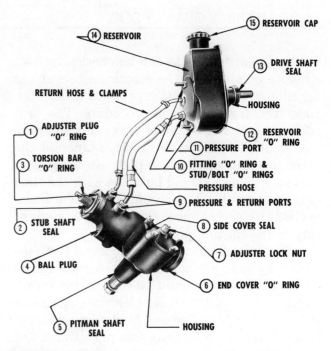

FIGURE 14-49.
Most power steering leaks occur at these locations. (Courtesy of Saginaw Division, General Motors Corporation)

power steering

QUICK CHECK PROCEDURES

THE POWER STEERING SYSTEM OF YOUR VEHICLE OPERATES AS A UNIT AND THERE ARE MANY COMPONENTS BETWEEN THE TIRES AND THE STEERING WHEEL THAT CAN AFFECT THE WAY A VEHICLE HANDLES. HERE ARE A FEW COMMON PROBLEMS THAT YOU CAN LOOK FOR AND POSSIBLY CORRECT.

FLUID
SHOULD BE CHECKED PERIODICALLY FOR PROPER LEVEL AND CONDITION. LOOK FOR METAL PARTICLES, BURNT CONDITION AND OTHER SIGNS OF CONTAMINATION OF FLUID.

COUPLINGS
SPLIT OR DENTED
STRIPPED THREADS
MUST BE TIGHT. CHECK FOR EXTERNAL LEAKAGE, DAMAGED THREADS AND DAMAGE TO TUBING.

BELTS
MUST BE IN GOOD CONDITION WITH CORRECT TENSION. CHECK FOR CRACKS, PEELING, EXCESSIVE LOOSENESS AND GLAZING.

PUMP
LOOK FOR SIGNS OF LEAKAGE AT PULLEY SHAFT AND CASING SEAMS, NOISE DURING OPERATION AND EXCESSIVE LOOSENESS. CHECK BRACKETS.

HOSES
DETERIORATION
WEAR FROM RUBBING
CHECKING   LOOSE COUPLINGS
LOOK FOR SWELLING, BRITTLENESS, FRAYED OR OTHER EXTERNAL DAMAGE. CHECK FOR FLUID SEEPAGE OR LEAKS.

GEAR BOX
CHECK FOR FLUID LEAKAGE AT SEAMS, NOISE DURING OPERATION, EXCESSIVE PLAY AND LOOSE MOUNTS.

STEERING
LOOK FOR EXCESSIVE PLAY IN STEERING WHEEL, ROUGH FEELING, HIGH EFFORT REQUIRED TO TURN. CHECK ENTIRE STEERING AND SUSPENSION SYSTEM.

FIGURE 14–50.
An inspection of a power steering system should include each of these points.
(Courtesy of Moog Automotive)

ADJUSTING POWER STEERING BELT

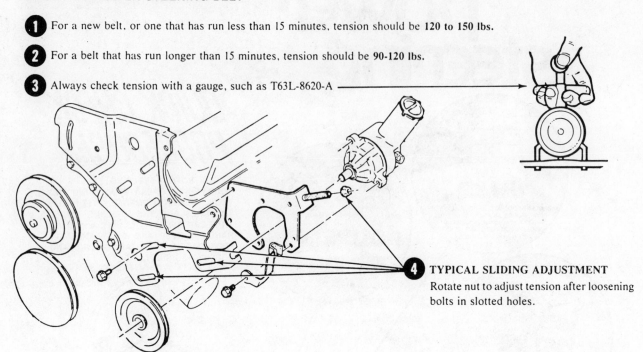

1 For a new belt, or one that has run less than 15 minutes, tension should be **120 to 150 lbs.**

2 For a belt that has run longer than 15 minutes, tension should be **90-120 lbs.**

3 Always check tension with a gauge, such as T63L-8620-A

4 TYPICAL SLIDING ADJUSTMENT
Rotate nut to adjust tension after loosening bolts in slotted holes.

FIGURE 14-51.
Belt tension should be measured using a tension gauge placed over the belt; the power steering pump mounting brackets are also shown. (Courtesy of Ford Motor Company)

FIGURE 14-52.
This chart shows the recommended tension for power steering pump drive belts. (Courtesy of Mitchell Information Services, Inc.)

TENSION IN LBS. (KG.) USING STRAND TENSION GAUGE

Application	New Belt	Used Belt
American Motors		
"V" Belts	140 (64)	100 (45)
Serpentine	190 (86)	150 (68)
Chrysler Corp.	120 (54)	70 (32)
Ford Motor Co.		
1.6L	50-90 (23-40)	40-60 (18-27)
2.3L HSC	150-190 (68-86)	140-160 (64-73)
Diesels	120-160 (54-73)	110-130 (50-59)
2.3 OHC, V6, V8	120-160 (54-74)	90-120 (41-54)
Ribbed V-Belts		
4 Rib	110-150 (50-68)	100-130 (45-59)
5 Rib	130-170 (59-77)	120-150 (54-68)
6 Rib		
Fixed	140-180 (64-82)	130-160 (59-73)
Tensioner	85-140 (39-64)	80-140 (36-64)
General Motors		
1.6L & 2.0L	145 (66)	70 (32)
1.8L & 2.5L	165 (75)	100 (45)
V6 Gas	135 (61)	65-80 (29-36)
V8 Gas		
Exc. Corvette	150 (68)	80 (36)
Corvette	130 (59)	
All Diesels	170 (77)	90 (41)

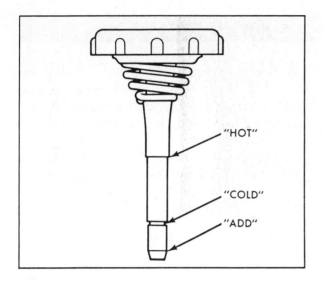

FIGURE 14-53.
A power steering filler cap and dipstick; note the markings for three different fluid levels. (Courtesy of Chevrolet)

14.11 CONCLUSION

After completing the inspection, the technician has the information needed to make a decision on what to do for the car. If no worn parts are found and there are no handling or driving problems, nothing needs to be done. If no worn parts are found but there is abnormal tire or driving problems, a wheel alignment should be performed. This will be described in Chapter 18. If worn parts are located, they should be replaced or adjusted. This will be described in Chapters 15 and 16.

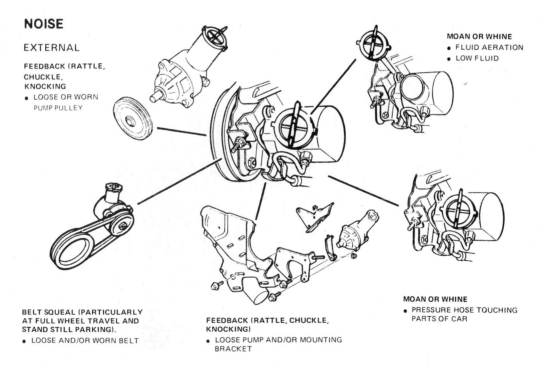

FIGURE 14-54.
Unusual power steering pump noises can be caused by these problems. (Courtesy of Ford Motor Company)

REVIEW QUESTIONS

1. A car has badly worn ball joints. Technician A says that this can allow the camber angle of the front tires to become more negative. Technician B says that a badly worn ball joint might come apart. Who is right?

   a. A only
   b. B only
   c. both A and B
   d. neither A nor B

2. A wear indicator ball joint should be checked with the joint

   a. unloaded by a jack under the control arm.
   b. unloaded with a jack under the frame.
   c. loaded and the car's weight on the tires.
   d. loaded and the tires in the air.

3. The ball joint boot
   A. is still all right if it has a few tiny tears in it.
   B. should be filled partially with grease.

   a. A only
   b. B only
   c. both A and B
   d. neither A nor B

4. A load-carrying ball joint mounted in the lower control arm is being checked. Technician A says the jack should be placed directly under the spring to unload the joint. Technician B says the allowable radial clearance for all ball joints is 0.250 inch. Who is right?

   a. A only
   b. B only
   c. both A and B
   d. neither A nor B

5. Ball joint clearance is
   A. usually measured using a dial indicator.
   B. can be measured with a caliper.

   a. A only
   b. B only
   c. both A and B
   d. neither A nor B

6. A friction-loaded ball joint is being inspected. Technician A says there should be no perceptible play in the joint as the steering knuckle is being shaken. Technician B says that if the joint were disconnected, it should take 0 to 5 inch pounds of torque to rotate the ball stud of a good joint. Who is right?

   a. A only
   b. B only
   c. both A and B
   d. neither A nor B

7. King pin bushing wear is checked by lifting the axle and measuring the
   A. side motion at the tire tread.
   B. vertical motion of the steering knuckle.

   a. A only
   b. B only
   c. both A and B
   d. neither A nor B

8. A faulty rubber control arm bushing will

   a. have rubber portions breaking out of it.
   b. have a center sleeve sagged off center.
   c. show signs of metal contact between the control arm and the frame.
   d. any of these.

9. Technician A says a squeaky metal control arm bushing is most likely bad and should be replaced. Technician B says that with metal bushings, a side motion of up to 0.125 inch is acceptable. Who is right?

   a. A only
   b. B only
   c. both A and B
   d. neither A nor B

10. A motorist is complaining about a loud, clunking noise when the brakes are applied. Technician A says this can be caused by bad control arm bushings. Technician B says this can be caused by bad strut rod bushings. Who is right?

    a. A only
    b. B only
    c. either A or B
    d. neither A nor B

11. Which of the following is not an indicator of a worn front end?

    a. front end vibration at 50 MPH
    b. scalloped tire wear
    c. sloppy steering with some degree of wander
    d. popping and clunking from the front end during different driving maneuvers.

12. A faulty strut rod bushing will show signs of
    A. rubber breakup.
    B. metal-to-metal contact between the backup washer and the frame.

    a. A only   **c.** either A or B
    b. B only   **d.** neither A nor B

13. A strut damper is being discussed. Technician A says you can check it by trying to shake the top of the strut in and out while the tires are in the air. Technician B says that a vertical motion of 0.250 inch at the top of the strut while lifting the tires off the ground is acceptable. Who is right?

    a. A only   **c.** both A and B
    b. B only   **d.** neither A nor B

14. Idler arm wear checks are being discussed. Technician A says a worn idler arm will allow excessive vertical motion of the right end of the center link. Technician B says a worn idler arm will usually show up when the backs of the tires are pushed apart and pulled together. Who is right?

    a. A only   **c.** both A and B
    b. B only   **d.** neither A nor B

15. An outer tie rod end can be checked by watching for play in the joint as it is

    a. squeezed together with a pair of pump pliers.
    b. squeezed together using dust cap pliers.
    c. pushed and pulled by hand.
    d. pried apart using a screwdriver.

16. Rack and pinion steering checks are being discussed. Technician A says that the steering should be turned outward to extend the bellows as it is being checked for cracks and breaks. Technician B says that the inner tie rod end can be checked by gripping the end of the rack and the end of the tie rod through the bellows and feeling for play as the tire is moved back and forth. Who is right?

    a. A only   **c.** both A and B
    b. B only   **d.** neither A nor B

17. Checks of a manual rack and pinion steering gear normally do not include

    a. lubricant level
    b. mounting bushing condition
    c. amount of internal play
    d. amount of internal drag

18. While inspecting a power steering system, you find a hard, glazed, cracking pump drive belt; this is probably caused by
    A. a leaky steering control valve.
    B. a leaky pump seal.

    a. A only   **c.** both A and B
    b. B only   **d.** neither A nor B

19. Power steering problems are being discussed. Technician A says that a steering wheel that "kicks" in one direction as the engine is started is probably caused by excessive pump pressure. Technician B says that a buzzing or humming noise as the wheels are turned can be caused by air in the system. Who is right?

    a. A only   **c.** both A and B
    b. B only   **d.** neither A nor B

20. Technician A says that a dry-park test is made by turning the steering wheel back and forth with the tires on the ground. Technician B says that a free movement of the steering wheel while making this check indicates wear or looseness in the steering gear or linkage. Who is right?

    a. A only   **c.** both A and B
    b. B only   **d.** neither A nor B

Chapter 15

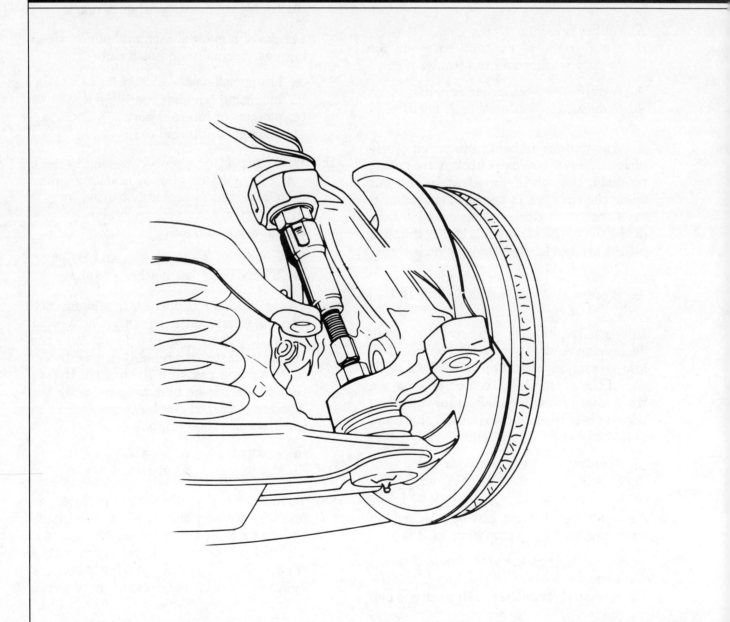

SUSPENSION COMPONENT SERVICE

After completing this chapter, you should:

- Be able to remove and replace upper and lower ball joints.

- Be able to remove and replace upper and lower control arms.

- Be able to remove and replace rubber and metal control arm bushings.

- Be able to remove and replace strut rod bushings.

- Be able to remove and replace king pins.

15.1   INTRODUCTION

After determining that a suspension component is faulty, the next operation is to remove and replace that component. This chapter concentrates on the replacement of ball joints, king pins, and suspension bushings. Service of the components of the steering system is described in Chapter 16. As in other portions of this text, the repair methods given are very general. A service manual should be consulted for operations specific to certain car models.

The operations described in this chapter can be done separately or together in various combinations. One school of thought prefers to rebuild the front end, replacing all bushings and ball joints at one time; another school of thought prefers to replace all bushings and any suspicious ball joints at one time; and a third school of thought prefers to replace a part only when that part absolutely needs replacing. There are pros and cons with each of these approaches. A few of the major considerations when making a decision are the mileage and value of the car, how long the car will be kept by that owner, how often the front end or other under-car parts are checked, how often the owner is willing to tie the car up, or how secure the owner needs to be about the car's condition. With front ends, there can be real labor savings when several operations are completed at one time; for example, once the ball joint is disconnected for replacement, the spring becomes very easy to replace. A properly rebuilt front end should give many thousands of miles and many years of safe, smooth, trouble-free, new car suspension driving with a minimum of tire wear.

SAFETY NOTE

Front end work can be hazardous; there are heavy parts and strong springs to work with. To prevent injury to yourself and to produce a safely operating car, the following practices should be followed:

- The car should be properly and securely supported when working under it.
- Follow the repair procedure that is recommended by the vehicle manufacturer.
- Determine what the result will be before removing each bolt or nut. Do not allow a part to fall or a spring to expand in an uncontrolled fashion.
- Use the proper tool for the job and use that tool in the correct manner.
- Spring compressors should be installed to keep a spring compressed to a safe length.
- Do not allow a strain to be placed on brake hoses.
- Do not let grease, oil, solvent, or brake fluid get on the brake lining or the braking surfaces of the rotor or drum.
- When using an air hammer, be careful; they operate rather violently and very noisily. Eye and ear protection should be worn.
- All damaged, worn, and bent parts must be replaced.
- Replacement parts should be checked against the old parts to ensure an exact or better quality replacement.
- Suspension and steering parts should not be heated.
- All replacement bolts must be of the same size, type, and grade as the originals.
- All bolts and nuts must be tightened to the correct torque and locked in place by the correct method.
- Parts using rubber bushings should be tightened while the car is on the ground and with the steering straight ahead.
- Check and adjust the wheel alignment whenever there is a possibility that the replacement parts have changed the alignment angles.
- Make sure the brake pedal operates with a normal feel before moving the car.
- Carefully road test the car to ensure that it will operate in a safe and correct manner.

After replacing front end, suspension, or steering system parts, a wheel alignment should be done to make sure that the tires are in correct alignment.

15.2   TAPER BREAKING

One of the difficult operations encountered during front suspension disassembly is breaking loose the tapers on ball joints or tie rod ends. The stud of a ball joint is locked into the steering knuckle, and the stud of a tie rod end is locked into the steering arm. If these parts are not locked tightly together, the strong side-loads they encounter will cause looseness. Not only will any looseness in these parts cause loose, sloppy steering and suspension action, but the movement will cause wear, which increases to the point of eventual failure. (Fig. 15-1)

To eliminate play and wear, these parts are connected using locking tapers. The taper used on the stud and hole is about 2 inches per foot

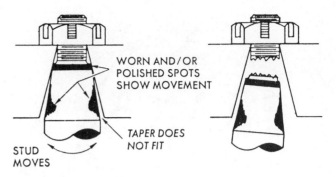

FIGURE 15-2.
If the ball joint stud taper does not fit and lock into the steering knuckle, movement of the stud can cause stud breakage. Always check for worn or polished spots on the ball joint stud and in the boss, which indicate that this problem has occurred. (Courtesy of Moog Automotive)

(51 cm per 305 mm) or about 9.5°. This taper can vary among car models or even, in a few cases, redesigned parts for the same car model. When the tapered stud is inserted into the tapered hole, the two tapers will interfere and connect; at that point, they should lock together over the entire length of the taper. Even after the nut is loosened, the tapers will stay locked together. Locking/Morse tapers are also used with some machine tools to lock drill bits in drill presses and centers or drill chucks in lathe tail stocks. (Fig. 15-2)

Several methods are commonly used to break these tapers loose to allow parts replacement; **shock, separator tool,** and pressure from a **puller** or **spreader.** Shock, usually from one or two hammers or an air hammer, is fast and often effective; it is usually the first method to try. **Separator tools,** commonly called **pickle forks** because of their shape, are fast and usually effective, but they usually destroy the ball joint or tie rod end boot and possibly the joints themselves. Separator tools should only be used when the ball joint or tie rod end is going to be replaced with a new one. **Pullers** and **spreaders** are special tools for ball joints or tie rod ends; they are fast and effective. They also break the tapers in a professional way so as to not damage any of the parts. (Fig. 15-3)

When breaking the taper on ball joint studs, it is a good practice to let the spring help you or keep it from hindering you. If the ball

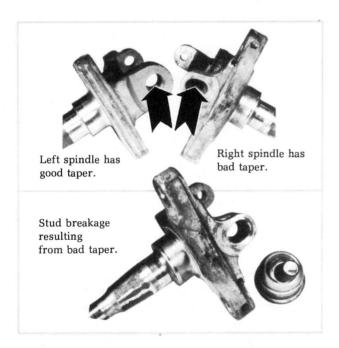

Left spindle has good taper.

Right spindle has bad taper.

Stud breakage resulting from bad taper.

FIGURE 15-1.
This ball joint stud (bottom right) was not tightened correctly; as a result, it worked loose, elongated the hole in the steering knuckle, and then broke the ball joint stud. (© 1976, courtesy of Replacement Parts Division, TRW, Inc.)

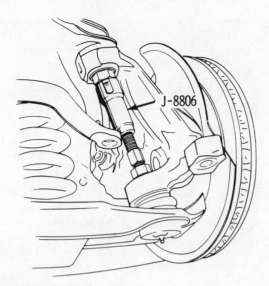

FIGURE 15-3.
The stud taper of tension-loaded ball joints can usually be broken loose using a ball joint taper breaker (tool # J-8806); the taper breaker is extended to exert an outward pressure on the ball joint studs. Note that the nuts on the ball joint studs must be loosened several turns. (Courtesy of Oldsmobile)

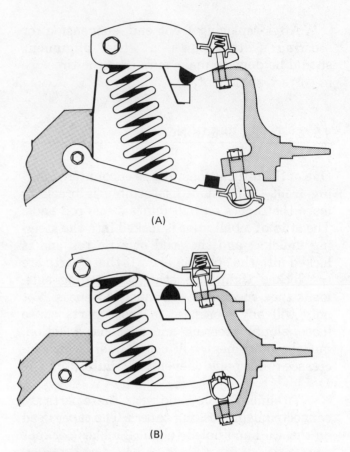

FIGURE 15-4.
When breaking the taper on tension-loaded ball joints (A), spring pressure, trying to separate the ball joint stud, will aid you. When breaking the taper on compression-loaded joints (B), the spring pressure will hold the tapers together; the spring should be compressed.

joint is tension-loaded, the spring on an S-L A suspension is trying to separate the ball joint stud; on compression-loaded ball joints, it is trying to keep the stud and the steering knuckle together. Before breaking the stud taper on tension-loaded ball joints, the car should be raised and supported by the frame, and the ball joint stud nut should be loosened a few turns. Loosen it just enough to let the taper break loose. With compression-loaded ball joints, a jack should be placed under the lower control arm or a spring compressor should be installed to remove the spring pressure from the ball joint stud. (Fig. 15-4)

If both ball joints on one side are to be serviced, it is often faster if both tapers are broken at the same time before disconnecting either one. If the follower ball joint is tension-loaded, it is often easier to break the taper while the spring pressure is helping you, as described earlier.

15.2.1 Breaking a Taper Using Shock

Once the nut has been loosened, a locking taper can often be broken loose by shock or vibration. The side of the ball joint boss in the steering knuckle or steering arm should be struck by a quick and firm blow, being careful not to bend or damage the part being struck or any of the surrounding parts. When breaking loose a tie rod end, a backup hammer is necessary to prevent bending the steering arm and to increase the amount of shock. Front end technicians often use an air hammer with a hammer-like punch for this operation. (Figs. 15-5 & 15-6)

FIGURE 15–5.
The taper on a tie rod end and sometimes a ball joint stud can often be broken loose by a sharp, quick blow of a hammer. When striking the side of the steering arm end, it is a good practice to use a second hammer for a backup. (Courtesy of Bear Automotive)

FIGURE 15–6.
After loosening the ball joint stud nut a few turns, the stud taper can often be broken loose by vibrating the steering knuckle boss using an air hammer and punch.

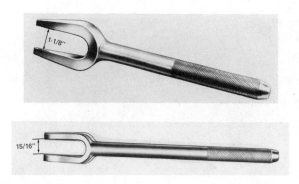

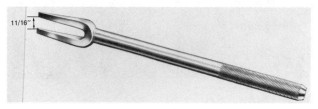

FIGURE 15–7.
Three different separators or wedges, commonly called "pickle forks"; the different sized openings allow them to be used for Pitman arms (top), ball joints (center), and tie rod ends (bottom). (Courtesy of OTC)

15.2.2 Breaking a Taper Using a Separator Tool

Separator tools/pickle forks are available with different sized openings designed for tie rod end, ball joint, and steering arm removal. They are available in larger, longer sizes to be struck with standard hammers or in a shorter version as an attachment for an air hammer. The wedge-like separator is driven into the gap between the ball joint and the steering knuckle or be-tween the tie rod end and the steering arm. The wedging action of the separator plus the shock and vibration of the hammer blows, along with a prying action from the technician, are fairly effective in separating these parts. However, this action cuts and destroys the sealing boots and also tends to pull the stud and ball out of the socket of the joint; it can destroy the joint. (Figs. 15–7 to 15–9)

15.2.3 Breaking a Taper Using Pressure

Pressure applied by pulling or spreading tools often called **taper breakers** is a fast, quiet, neat, and professional way to separate tapers. The tools, which are designed for this special purpose, are placed in position between the ball joint studs and tightened to produce the necessary outward force to break loose the tapers. Sometimes, a slight blow on the side of the stud boss is used after the puller or spreader is tightened. Taper breakers cannot be used with compression-loaded ball joints. (Figs. 15–10 & 15–11)

Tie rod end pullers often resemble two-jaw pullers; some designs use a single fixed jaw. The

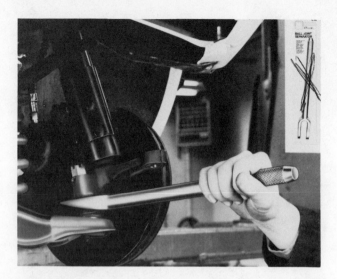

FIGURE 15-8.
A ball joint separator being used to break loose the ball joint stud taper on this modified strut suspension; the end of the tool will be struck with a hammer. Note that the stud nut is loosened several turns before starting. (Courtesy of OTC)

pressure from the puller screws pushes the tie rod stud into and through the steering arm. Most ball joint taper breakers resemble a long nut and bolt; one end is positioned at each ball joint stud. When the parts of the taper breaker are threaded apart, a spreading force is exerted on the ends of the ball joint studs. Taper break-

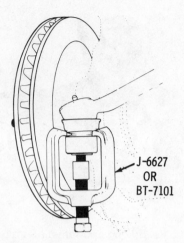

FIGURE 15-10.
Tool J-6627 or BT-7101 is being used to break loose the tie rod end stud taper. (Courtesy of Oldsmobile)

ers for some FWD cars use the ball joint nut to apply pressure onto the ball joint stud. (Fig. 15-12)

FIGURE 15-9.
Separator tools are also available for use in an air hammer; here one is being used to break loose the taper on a tie rod end stud. (Courtesy of Superior Pneumatic)

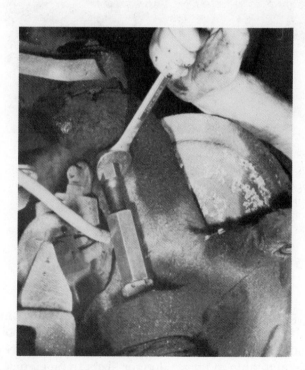

FIGURE 15-11.
This ball joint stud taper breaker has been positioned between the ends of the two ball joint studs; expanding it with a wrench places enough pressure on the ends of the studs to break them loose. (Courtesy of Moog Automotive)

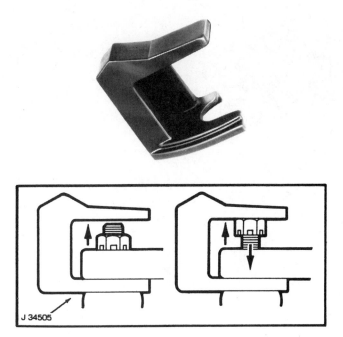

FIGURE 15–12.
This tie rod stud remover is used on FWD cars with tapered ball joint studs; the ball joint nut is removed, turned upside down, and then threaded against the tool to exert a downward pressure on the stud. (Courtesy of Kent-Moore)

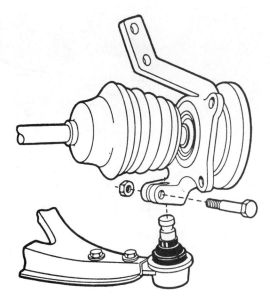

FIGURE 15–13.
Many FWD cars use a pinch-bolt to lock the ball joint stud into the steering knuckle. Note that removal of the bolt is necessary to remove the stud. (Courtesy of Moog Automotive)

15.2.4 Removing and Replacing Ball Joint Studs Locked by Pinch-Bolts

Some newer cars with strut suspensions do not use locking tapers on their ball joint studs; the stud is locked in place by a **clamp** or **pinch-bolt** in the steering knuckle boss. The steering knuckle boss is slotted so the pinch-bolt can clamp it tightly onto the ball joint stud. The pinch-bolt also passes through a notch in the stud to help ensure that the stud cannot come out of the boss, even if the pinch-bolt loosens. This type of connection is also used to secure the strut to the steering knuckle on some cars. (Fig. 15–13)

To separate this style of ball joint stud from the steering knuckle, you simply remove the pinch-bolt and slide the ball joint stud out of the steering knuckle boss. Occasionally it is necessary to place a screwdriver into the slot in the steering knuckle and pry it open to loosen its grip on the stud. To install it, you slide the stud into the boss, making sure that the notch

in the stud is aligned with the pinch-bolt hole, slide the pinch-bolt into position, and tighten the nut to the correct torque. (Fig. 15–14)

15.3 BALL JOINT REPLACEMENT

Ball joint replacement is necessary when the clearance of a vehicle-loaded ball joint exceeds specifications, a follower type joint shows clearance, or if either style of joint has a cut or torn boot or is too tight. If one ball joint of a set has excessive wear, many technicians prefer to replace the entire set of ball joints, using the theory that all of the ball joints have operated for the same length of time under the same operating conditions. The remaining ball joints will possibly fail in a relatively short period of time anyway. It is less expensive to replace all of the ball joints at one time than to replace one at a time. Also, doing a more complete repair should give the car owner more security for a longer period of time.

Ball joints can be replaced with the control arm attached to the car; many technicians pre-

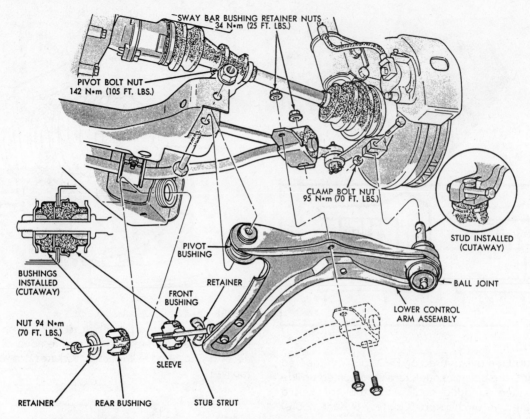

SWAY BAR BUSHING RETAINER NUTS
34 N•m (25 FT. LBS.)

PIVOT BOLT NUT
142 N•m (105 FT. LBS.)

CLAMP BOLT NUT
95 N•m (70 FT. LBS.)

STUD INSTALLED
(CUTAWAY)

BUSHINGS
INSTALLED
(CUTAWAY)

BALL JOINT

PIVOT
BUSHING

RETAINER

LOWER CONTROL
ARM ASSEMBLY

FRONT
BUSHING

NUT 94 N•m
(70 FT. LBS.)

SLEEVE

RETAINER

REAR BUSHING

STUB STRUT

FIGURE 15-14.
A pinch/clamp bolt locks this ball joint stud into the steering knuckle. Note that
the stud has a single notch that must be properly aligned in order to install the
clamp bolt. (Courtesy of Chrysler Corporation)

fer this because the arm is held securely during
replacement operations, and the time needed to
remove and replace the control arm is saved.
Or, ball joints can be replaced with the control
arm off the car; many technicians prefer this
because they have total access to the control
arm and ball joint. Whether the control arm
should be removed or not depends on the skill
and preference of the technician, the type of
equipment available, and what other jobs are to
be done. In either case, it is necessary to disconnect the ball joint stud from the steering
knuckle. Removing and replacing ball joints is
described in Section 15.5.

In a few cases it is necessary to replace the
control arm with the ball joint; the ball joint is
not available from the car manufacturer by itself. In some cases, individual parts are available from aftermarket suppliers, while O.E.M.
parts come as assemblies.

15.4 REMOVING A CONTROL ARM

This section describes the procedure to remove
a lower control arm on an S-L A suspension
with a spring on the lower arm. A similar procedure is used to remove an upper arm from this
same suspension or a lower arm from a strut
suspension; there is no need to compress or remove the spring for these other styles of control arms. A similar procedure to that described here is used on an upper control arm
with a spring mounted on it. (Fig. 15–15)

To remove a lower control arm, you should:

1. Raise and support the car on a hoist or
jack stands.

2. Remove the wheel.

3. Remove the shock absorber and the stabilizer bar end link.

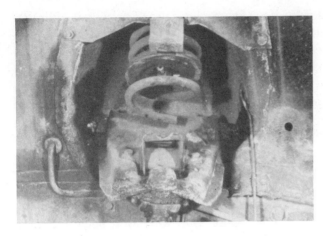

FIGURE 15-15.
Before removing an upper control arm with the spring mounted on it, the spring must be compressed. (Courtesy of Moog Automotive)

FIGURE 15-16.
When working on a lower control arm, the upper control arm and steering knuckle can be propped up out of the way after the ball joint has been disconnected. (Courtesy of Moog Automotive)

4. If the car has a tension-loaded lower ball joint, loosen the ball joint stud nut several turns, and using a spreader tool, separator, and/ or shock, break loose the ball joint stud taper.

5. Install a spring compresser and compress the spring enough to remove the spring load from the control arm. You can be sure the spring load is removed when the spring becomes free to move around or when the control arm can be moved upward. If the car has a compression-loaded ball joint, break loose the ball joint stud taper at this time. (Fig. 15-16)

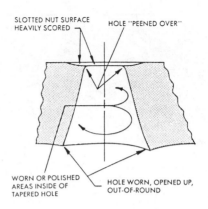

FIGURE 15-17.
After the ball joint stud has been removed, the tapered hole in the steering knuckle boss should be checked for any of these problems that are caused by a loose or improperly fitted ball joint stud. (Courtesy of Moog Automotive)

6. Disconnect the ball joint stud and separate the steering knuckle boss from the ball joint stud. It is often necessary to pry between the lower control arm and the brake-splash-shield-rotor-steering knuckle to keep the end of the control arm from catching on the splash shield. Inspect the tapered portion of the stud for any bright, worn metal, which would indicate a worn steering knuckle boss. The upper control arm, with the steering knuckle attached, can be raised and a block can be inserted between the control arm and the frame/body bracket to hold them up and out of the way. (Figs. 15-17 & 15-18)

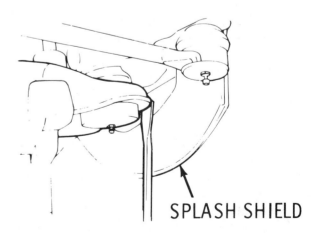

SPLASH SHIELD

FIGURE 15-18.
On some cars, the lower control arm tends to catch on the brake splash shield as the control arm is lowered downward; a large screwdriver or prybar can be used to prevent that. (Courtesy of Oldsmobile)

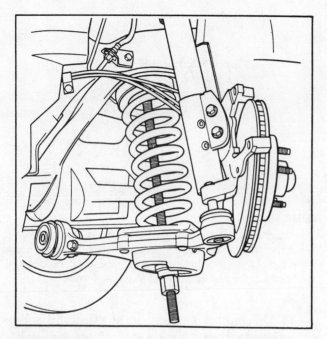

FIGURE 15-19.
This control arm is being removed by first removing the inner pivot bolts; the spring compressor was installed to remove the load on the bolts during their removal. (Courtesy of Ford Motor Company)

7. The lower control arm can be pivoted downward to remove the spring. If desired, the lower ball joint can be replaced. The lower control arm now can be removed by removing the inner bushing/pivot bolts. (Fig. 15-19)

The upper control arm can also be removed at this time by disconnecting the ball joint stud and the inner pivot bolts. If this is done before replacing the lower control arm, it will be necessary to support the steering knuckle with the brake rotor or drum and caliper or backing plate or disconnect the brake hose and completely remove the steering knuckle and brake assembly. Many front end technicians prefer to service the parts on the lower control arm, reattach the lower control arm to the steering knuckle and frame, tie the control arm in place (if necessary to keep it from falling and stressing the brake hose), and then service the parts on the upper control arm. The steering knuckle and brake assembly can be removed from the car, but it will require disassembling the tie rod end from the steering arm and the

brake hose from the caliper or wheel cylinder. If the brake hose is disconnected, it will be necessary to bleed the air from that brake after reassembly.

Many General Motors cars mount the upper control arm on a pivot shaft, which is bolted to a bracket on the frame; the pivot shaft is positioned inward of the bracket. This control arm must be slid inward and off the bolts to remove it. The steering shaft or exhaust manifold often prevents enough inward movement to allow the shaft to be slid off the mounting bolts. Use a socket, extension bar, and long socket handle or air impact wrench to turn the mounting bolts; this will loosen them. The bolts should work their way out of the frame bracket. It helps to pry outward on them as they are rotated. A special tool is available to press these bolts out of the frame bracket. With the bolts out of the way, the control arm can be lifted out. For replacement, the bolts are merely driven back in place after the control arm is repositioned. New bolts will be required if their serrations become too badly worn; if they do not fit tightly, the bolts may turn when replacing shims during a wheel alignment. (Fig. 15-20)

15.5 REMOVING AND REPLACING A BALL JOINT

In most cases, ball joints and bushings are available as individual parts and can be replaced as separate items. Ball joints are usually secured into the control arm by one of several methods: pressed into place, riveted or bolted into place, or threaded into place. If the O.E.M. ball joint is riveted into place, the replacement ball joint will be secured by bolts. The ball joint usually enters and leaves the control arm from the side opposite to the stud; in this position, normal operating pressures tend to seat the ball joint tighter, rather than try to pull it out of the control arm.

Whenever a ball joint is removed, the end of the control arm and the opening for the ball joint should be inspected for cracks or other signs of damage which would indicate a faulty control arm. (Fig. 15-21)

FIGURE 15–20.
The location of the steering shaft and the control arm mounting bolts block the removal of the upper control arm on this General Motors car; removal of the control arm mounting bolts will allow lifting the control arm out.

15.5.1 Removing and Replacing a Pressed-in Ball Joint

There are several methods of removing and replacing pressed-in ball joints. Most shops will purchase a specialized kit for this. Ball joint service kits usually include a series of pressing tools of different diameters to push on the ball joint and a series of various-sized support tools to fit around the ball joint to protect the control arm from distortion. Pressure to push the old joint out or the new joint in comes from either a hydraulic jack or a screw thread. It is not recommended to drive ball joints out or in using a hammer; there is too much danger in distorting the control arm or the new ball joint.

To remove a pressed-in ball joint, you should:

1. Clean the control arm in the area around the ball joint.

2. Select the smallest support or receiver adapter tube that will fit over the ball joint support boss. (Fig. 15–22)

3. Select the size of pressing adapter that fits the stud side of the ball joint best; this adapter must be able to pass through the control arm.

4. Assemble the ball joint press with the two adapters over the control arm, and press the ball joint out of the control arm. (Fig. 15–23)

To install a pressed-in ball joint, you should:

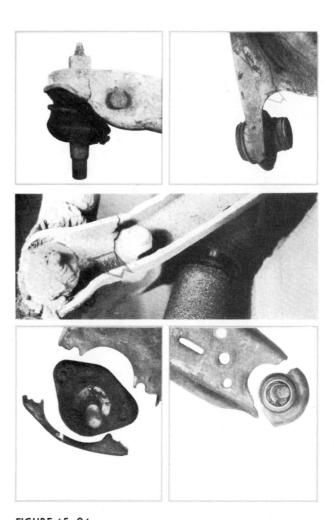

FIGURE 15–21.
During ball joint or bushing replacement, the control arm should be checked for cracks that might cause future failure. (© 1976, courtesy of Replacement Parts Div. TRW, Inc.)

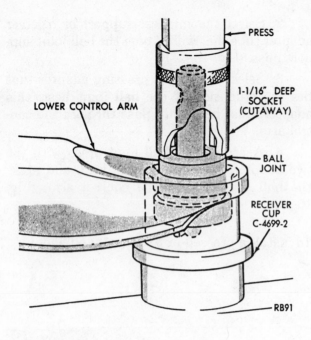

FIGURE 15-23.
This press tool is being used to remove a ball joint; note that the control arm is still attached to the vehicle. (Courtesy of Moog Automotive)

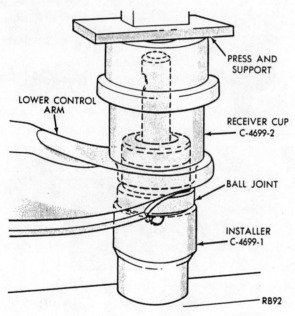

FIGURE 15-22.
A pressed-in ball joint being removed (top) and installed (bottom) in a control arm. Note the placement of the pressing and receiving tools in each operation. (Courtesy of Chrysler Corporation)

2. Position the new ball joint in the control arm and select a pressing adapter that fits the ball joint. (Fig. 15-25)

3. Place the ball joint press, support adapter, and pressing adapter over the control arm, and press the ball joint completely into the control arm. (Fig. 15-26)

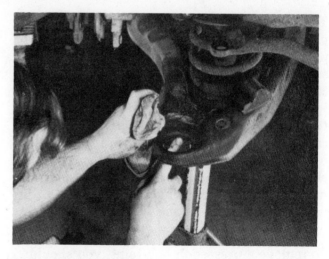

FIGURE 15-24.
After the ball joint has been removed, the control arm should be inspected for damage. (Courtesy of Moog Automotive)

1. SAFETY NOTE. Inspect the control arm to make sure there are no cracks and the opening is not distorted; a damaged control arm should be replaced. (Fig. 15-24)

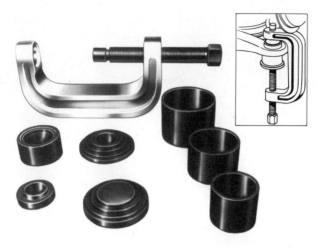

FIGURE 15–25.
This tool set can be used to remove and install pressed-in ball joints; note the various-sized adapters and receiver tubes to fit different sized ball joints. (Courtesy of OTC)

15.5.2 Removing and Replacing a Riveted/Bolted-in Ball Joint

Many ball joints are riveted in place at the factory; the replacement ball joint is bolted in place.

FIGURE 15–26.
This tool is being used to install a ball joint; note that tightening the two bolts will draw the ball joint into the control arm. (Courtesy of SKF Automotive Products)

FIGURE 15–27.
The rivets securing this ball joint to the control arm are being cut off using an air hammer and chisel. (Courtesy of Superior Pneaumatic)

To remove a riveted-in-place ball joint, you should:

1. Clean the control arm in the area around the ball joint.

2. Working from the ball joint side of the control arm, use a sharp chisel and an air hammer to slice off the head of the rivet next to the ball joint. Be careful not to damage the control arm. If you have difficulty cutting the rivets, drill a hole, about three-fourths of the diameter of the rivets, down through the center of the rivet. The hole should extend into the depth of the ball joint flange. After drilling, the rivet head should be easy to cut. (Figs. 15–27 & 15–28)

3. Use a punch to drive the body of the rivet out of the ball joint and the control arm. (Figs. 15–29 & 15–30)

4. Remove any bolts or nuts that may be used in addition to the rivets and remove the ball joint.

To install a riveted-in ball joint, you should:

1. **SAFETY NOTE.** Inspect the control arm to make sure there are no cracks and the opening is not distorted; a damaged control arm should be replaced.

2. Place the new ball joint in position, and assemble the bolts, nuts, washers, and any re-

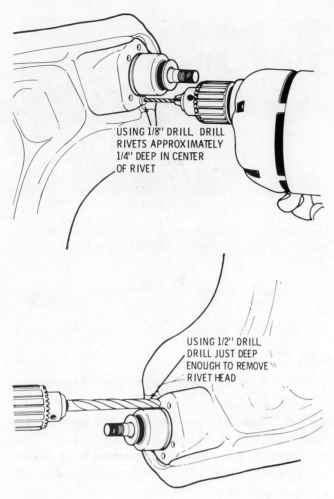

USING 1/8" DRILL, DRILL RIVETS APPROXIMATELY 1/4" DEEP IN CENTER OF RIVET

USING 1/2" DRILL, DRILL JUST DEEP ENOUGH TO REMOVE RIVET HEAD

FIGURE 15-28.
The heads of the rivets securing this ball joint are being removed with a drill; a small, pilot hole (top) helps align the larger drill that will remove the rivet head. (Courtesy of Oldsmobile)

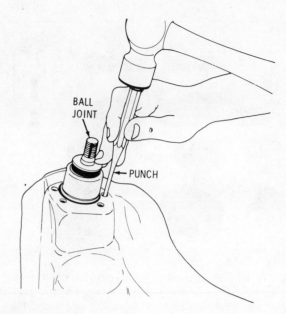

BALL JOINT

PUNCH

FIGURE 15-29.
After the rivet heads have been removed, the shank of the rivet can be driven out using a hammer and punch. (Courtesy of Oldsmobile)

To remove and replace a threaded ball joint, you should:

1. Clean the control arm in the area around the ball joint.

FIGURE 15-30.
An air hammer and punch are being used to drive the rivets out. (Courtesy of Superior Pneumatic)

tainers, following the directions provided with the ball joint. (Fig. 15-31)

3. Tighten the retaining bolts to the correct torque. (Fig. 15-32)

15.5.3 Removing and Replacing a Threaded Ball Joint

This style of ball joint can be identified by the wrench flats on the body of the ball joint. Several different sizes of special ball joint sockets are available to fit these flats.

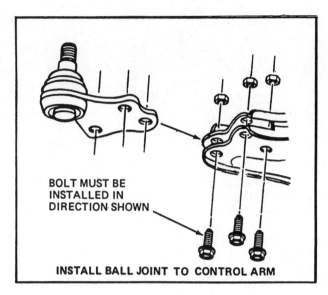

FIGURE 15-31.
The replacement ball joint for one that was riveted in is installed with bolts; the bolts and other parts should be installed in the proper manner. (Courtesy of Pontiac Division, GMC)

2. Secure the control arm, and using the correct socket and rather long socket handle for leverage, carefully unscrew the ball joint from the control arm. Penetrating oil can help ensure removal without damage to the control arm. (Fig. 15-33)

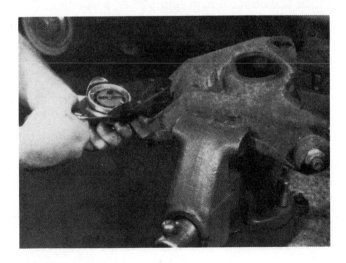

FIGURE 15-32.
After the ball joint retaining bolts have been installed, they should be tightened to the correct torque. (Courtesy of Moog Automotive)

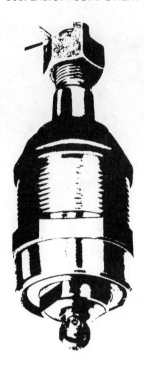

FIGURE 15-33.
A threaded type ball joint, note the wrench flats and very slight threads on the body of the ball joint. (Courtesy of Bear Automotive)

3. **SAFETY NOTE.** Inspect the control arm for cracks and damage to the ball joint threads.

4. When installing the new ball joint, make sure it does not cross-thread, and thread it into the control arm. Tighten the ball joint to the correct torque; if the correct torque cannot be obtained, the control arm must be replaced. (Fig. 15-34)

15.6 CONTROL ARM BUSHING REPLACEMENT

Most domestic cars use rubber control arm bushings; at one time, some used metal bushings. Most rubber control arm bushings used in domestic cars are of a three-piece (inner sleeve, rubber insert, and outer sleeve), torsilastic type; some import and a few domestic cars use a one- or two-piece (no outer sleeve) type. The outer

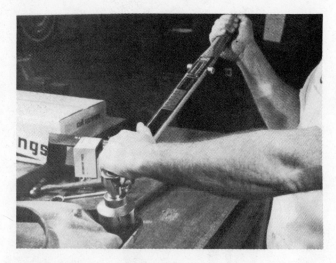

FIGURE 15-34.
A threaded ball joint is being installed into the control arm; it should be tightened to the correct torque. (Courtesy of Moog Automotive)

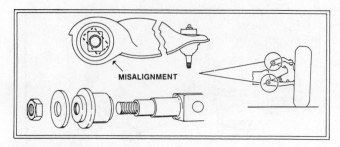

FIGURE 15-35.
When torsilastic bushings are installed, the retaining bolts/nuts must be tightened with the control arm in the correct position, usually with the suspension at ride height. If the bushings are tightened in the wrong position, the bushings and control arms will be placed in a stressed condition. (Courtesy of Moog Automotive)

sleeve of the three-piece type makes them relatively easy to install.

With torsilastic bushings, the control arm pivot bolts squeeze the serrations of the inner sleeve of the bushing against the shoulder of the control arm shaft or frame/body bracket; the inner sleeve becomes locked, and the control arm will be preloaded in this position. The rubber in the bushings must now twist if the control arm is raised or lowered. These bolts should be tightened in the normal ride-height position to prevent overstressing the rubber in the bushings. If the pivot bolts were tightened with the control arm in the rebound position, the bushings would be twisted at ride height and twisted an excessive amount at jounce. These bolts are usually left slightly loose until the front end is assembled and the car's weight is on the tires; then they are tightened to the correct torque. In cases where it is very difficult to tighten these bolts after assembly, the position of the control arm shaft to the control arm is carefully noted and marked during disassembly before the retaining bolts are loosened; then, after bushing replacement, the shaft is put back to this same, exact position, and the bolts are tightened. (Fig. 15-35)

Special bushing pullers and installers are available that allow replacement of the bush-ings with the control arm still in the car without disconnecting the ball joint. In some cases, there is adequate room to remove and replace the bushings under the hood. If under-hood space is restricted, the control arm can be pivoted on the ball joint so the bushing end of the control arm is in the fender well to gain working space. (Figs. 15-36 & 15-37)

In some cases, the inner sleeve of the bushing rusts onto the control arm shaft; this makes removal of the bushing very difficult. It is sometimes necessary to remove the bushing in pieces; removal of the outer sleeve and rubber insert allow access to a slit in the inner sleeve. Place the control arm shaft over a vise so you can strike the sleeve with a hammer or spread

FIGURE 15-36.
This upper control arm bushing is being pulled out of the control arm; the work is being done in the engine compartment with the control arm still attached to the ball joint.

FIGURE 15–37.
This upper control arm bushing is being pulled out of the control arm; the work is being done in the fender well with the control arm still attached to the ball joint.

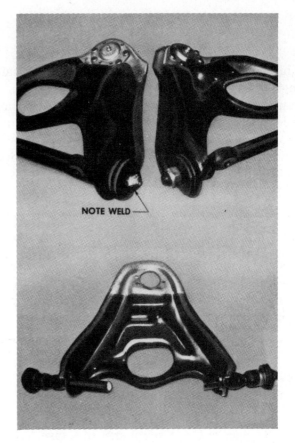

FIGURE 15–39.
The retaining nuts on some GM upper shafts were welded in place to prevent their working loose. The shaft has to be cut in order to replace the bushings; it must be replaced with a new shaft. (Courtesy of Moog Automotive)

the slit open with a chisel. This will allow it to be easily slid off of the control arm shaft. The rust-covered surfaces of the control arm shaft can be cleaned by wire brushing to ease the installation of the new bushings. (Fig. 15–38)

After the bushings have been removed, the control arm shaft is free to fall out of the control arm. Most control arm shafts are symmetrical so their position is not important, but it is a good practice to mark the position of the shaft before bushing removal to ensure its installation in the correct position.

Some General Motors vehicles have a spot weld on one or both of the retaining nuts on the control arm shaft to prevent the nuts from unthreading and falling off. When replacing the control arm bushings on these cars, it is necessary to cut the control arm shaft in half and remove the two halves of the shaft with the bushings. During bushing replacement, a new control arm shaft will be required along with the new bushings. If you encounter one of these models that is not spot welded and the nuts can be removed, the old shaft can be reused. It is a good practice to install a self-locking nut or lock washer to ensure retention of the nut. (Fig. 15–39)

FIGURE 15–38.
These control arm bushings were removed with an air hammer and chisel; note the chisel marks. The one at the left came out easy; the bushing next to it was rusted onto the control arm pivot shaft and had to be removed in pieces; and the other two required a little work to remove.

15.6.1 Removing Rubber Bushings Using an Air Hammer and Chisel

This popular way to remove control arm bushings is normally fairly fast. With experience, it can be done with the control arm on the car, but it is best to remove the control arm and secure the arm in a vise. The vibration of the air hammer probably has as much effect in removing the bushing as the force of the hammer blows.

> **SAFETY NOTE**
> Care should be taken not to damage the control arm with the chisel; personal caution should also be exercised because of the noise and violent manner of operation.

To remove a control arm bushing using an air hammer, you should:

1. Secure the control arm in a vise.
2. Clean the control arm in the area around the bushing.
3. Using an air hammer and chisel against the bushing flange or biting into the body of the bushing, force the bushing out of the con-

FIGURE 15-40.
An air hammer and chisel are being used to vibrate this control arm bushing out of the control arm. (Courtesy of Superior Pneumatic)

trol arm. On control arms where the bushing passes through two surfaces of the control arm, it is a good practice to vibrate the control arm right next to the bushing (with the air hammer and chisel or punch) and loosen each bushing-to-control-arm contact surface before starting the bushing removal. (Fig. 15-40)

15.6.2 Removing Rubber Bushings Using a Puller Tool

Pullers have the advantage that they can usually be used with the control arm on the car. The ball joint can be left attached to the steering knuckle, and the control arm pivoted to an area where there is access to the bushing with a puller. Most puller sets also include adapters to press the new bushing into the control arm. Pullers are quiet to use and normally will cause no damage to the control arm.

To remove a rubber bushing using a puller, you should:

1. Clean the area of the control arm around the bushing.
2. Select a sleeve large enough for the bushing to be pulled into. (Fig. 15-41)
3. Select an adapter that fits over the control arm shaft that will push against the end of the bushing.
4. Position the two adapters and the puller over the control arm and bushing, and tighten the puller to force the bushing out of the control arm. (Fig. 15-42)

OR Some bushing pullers are designed to be used with a press or special puller bolts; these units are often made specifically for one brand or model of car. (Fig. 15-43)

15.6.3 Installing a Rubber Bushing Using a Driver

This popular way to install bushings is normally quite fast, and bushing drivers, often called knockers, are fairly inexpensive. Care should be taken that the bushing is driven in

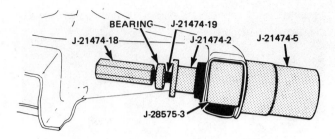

REMOVE LOWER CONTROL ARM BUSHING

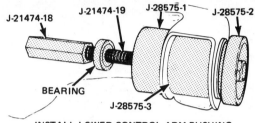

INSTALL LOWER CONTROL ARM BUSHING

FIGURE 15-41.
Tightening the nut (J-21474-18) will cause adapter J-21474-2 to push the bushing out of the control arm and into the receiver adapter, J-21474-5 (upper illustration); part J-28575-3 keeps the end of the control arm from collapsing. A new bushing has been installed using the setup in the lower illustration. Note that both of these operations can be done with the control arm connected to the steering knuckle. (Courtesy of Oldsmobile)

straight to ensure that the control arm is not damaged.

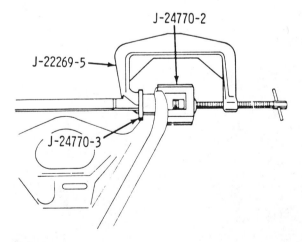

FIGURE 15-42.
This puller is set up to remove a control arm bushing; tool adapter J-24770-2 is large enough for the bushing to enter it. (Courtesy of Chevrolet)

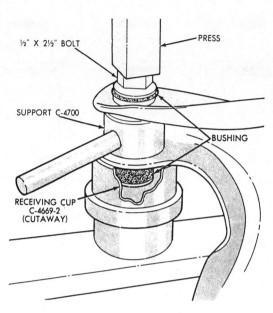

FIGURE 15-43.
This control arm is in a press, and the adapters are positioned so the bushing will be removed. (Courtesy of Chrysler Corporation)

To install a rubber bushing using a driver, you should:

1. Inspect the control arm to make sure it is not damaged.

2. Select a bushing driver that fits just around the rubber portion so it seats solidly onto the bushing flange. (Fig. 15-44)

3. Position the control arm with the bushing bore over the jaws of a vise; adjust the vise opening so it is slightly wider than the bushing.

4. If necessary, place the control arm shaft in place. Start the bushing into the control arm by hand and drive it into position using a hammer and bushing driver. (Fig. 15-45)

15.6.4   Installing a Rubber Bushing Using a Pressing Tool

Like the removal step, bushing installation can be done on the car, thus saving the time and trouble needed to remove and replace the ball joint stud. Pressing the bushing into place is fairly fast and usually cannot damage the control arm.

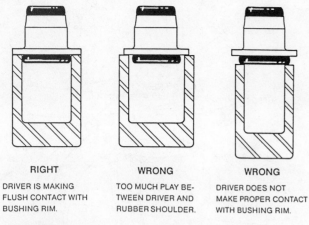

RIGHT

DRIVER IS MAKING
FLUSH CONTACT WITH
BUSHING RIM.

WRONG

TOO MUCH PLAY BE-
TWEEN DRIVER AND
RUBBER SHOULDER.

WRONG

DRIVER DOES NOT
MAKE PROPER CONTACT
WITH BUSHING RIM.

FIGURE 15–44.
A bushing driver should be selected so it fits flush against
the rim of the bushing. A driver that is too large will bend
the bushing rim; one too small will cut the bushing rubber.
(Courtesy of Moog Automotive)

FIGURE 15–45.
Installing a control arm bushing using a bushing driver and
hammer. The control arm should be supported by the jaws
of a vise while driving the bushing. (Courtesy of Moog Automotive)

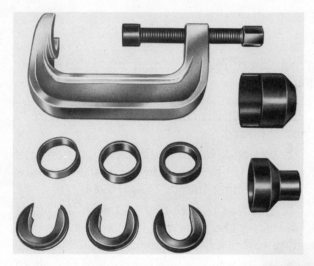

**To press a rubber bushing into a control arm,
you should:**

　1. Inspect the control arm to make sure it
is not damaged.

　2. Select a pushing tool that fits solidly
against the flange of the bushing. (Fig. 15–46)

　3. Select a support adapter that fits
around the control arm opening.

FIGURE 15–46.
This tool is being used to press a new bushing into the control arm (bottom). Note the various adapters used to remove and install bushings of different diameters. (Courtesy
of OTC)

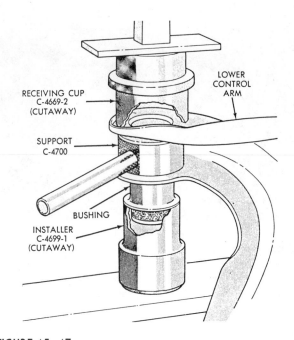

FIGURE 15–47.
The control arm is being pressed onto this bushing; note the relationship of the different adapters being used. (Courtesy of Chrysler Corporation)

4. If necessary, place the control arm shaft in position and start the bushings into the control arm by hand.

5. Assemble the bushing press and the two adapters over the control arm and press the bushing completely into the control arm. (Fig. 15–47)

15.6.5 Removing and Replacing Rubber Bushings Without Outer Sleeves

These bushings are used on some import cars as well as a few domestic cars; they are removed and replaced in a manner similar to the metal-sleeved bushings. The tight fit into the control arm causes them to work like a sleeved torsilastic bushing; this tight fit can cause difficulty on installation. Bushing removal and installing tools are available to make this job fast and fairly easy. The bushing and opening should be lubricated with tire rubber lube or a vegetable-based cooking oil before installing; do

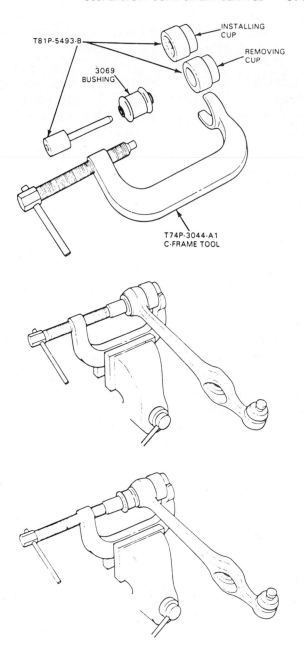

FIGURE 15–48.
The control arm pivot bushing tool set (top) is used to remove and replace rubber control arm bushings that do not have a metal outer sleeve. The setup for removing a bushing is shown at the center, and the installation is shown at the bottom. The bushing should be lubricated with a vegetable oil or tire lube during installation. (Courtesy of Ford Motor Company)

not use petroleum/motor oil or brake fluid as a lubricant, as these will cause deterioration of the rubber. (Fig. 15–48)

15.6.6 Removing and Replacing Metal Bushings

Metal bushings are removed by unscrewing them from the control arm; the control arm shaft is removed after removal of the bushings. The pivot shaft is normally replaced with a new one as the bushings are replaced; a worn shaft must be replaced during bushing replacement. A kit with rubber bushings and pivot shaft to fit them is available from aftermarket suppliers for those people who prefer the advantages of the rubber bushing.

During replacement, the shaft is placed in the control arm in the correct position, and the new bushings with seals are threaded into the control arm and onto the shaft. The bushings should be tightened alternately to ensure that the shaft is centered in the control arm, and they should be tightened to the correct torque. These bushings must be lubricated; in some cases, it is necessary to grease them before installing the control arm in the car because there is not enough room to grease them after installation. (Figs. 15–49 & 15–50)

FIGURE 15–50.
Installation of metal control arm bushings is the reverse procedure from removal; the control arm shaft should be centered in the control arm during installation. (Courtesy of Moog Automotive)

15.7 INSTALLING A CONTROL ARM

This section describes the procedure to replace a lower control arm and spring on an S-L A suspension. A similar procedure is used on other control arms.

To install a lower control arm with spring on an S-L A suspension, you should:

1. Place the control arm in position in the frame/body brackets and install the inner pivot bolts. Replace the nuts and tighten them finger tight.

2. Turn the ball joint stud to align the cotter pin hole lengthwise to the car so the cotter pin will be easy to install. Clean and inspect the ball joint stud hole in the steering knuckle boss.

3. Replace the compressed spring and align it to the correct position, as described in Section 12.4.1. Swing the control arm upward so the ball joint stud enters the steering knuckle boss, and thread the nut onto the ball joint stud.

4. Tighten the ball joint stud nut to the correct torque and install the cotter pin. If a prevailing torque nut is used, it is sometimes

FIGURE 15–49.
Metal control arm bushings are removed by unscrewing them from the control arm and pivot shaft. (Courtesy of Moog Automotive)

FIGURE 15-51.
The tapered hole in the steering knuckle boss should be cleaned before installing the ball joint stud. (Courtesy of Moog Automotive)

SAFETY NOTE

When ball joint studs are replaced, there are a few cautions to observe:

- Do not overtighten the retaining nut. After the tapers meet, there is not much effective bolt length to accept any further stretching; overtightening can cause enough stress to break off the threaded end of the stud.

- Do not back off the nut to align the cotter pin slots. Backing off or loosening of the nut can cause a slight clearance between the nut and the stud boss; this can allow the tapers to come loose, which will cause play and wear.

- Tighten the nut, first to the minimum torque specification, and then, if necessary, continue tightening to the nearest cotter pin slot.

- Make sure that the stud and the hole are clean to ensure a good locking action of the tapers. Some technicians run a shop towel in and out of the stud hole to ensure its cleanliness. (Figs. 15-51 & 15-52)

- On cars that use self-locking nuts, first use a plain nut of the correct thread to lock the stud taper, and then install the self-locking nut to the correct torque.

- Tighten the bolts that secure rubber pivot bushings when the suspension and the bushing are in the normal, ride height position. Rubber bushings in the steering linkage are tightened with the steering in the straightahead position.

5. Remove the spring compressor, making sure the spring seats correctly.

6. Install the shock absorber and stabilizer bar end links. Install the tires and wheels and lower the car onto the tires.

7. Tighten the inner pivot bolts to the correct torque.

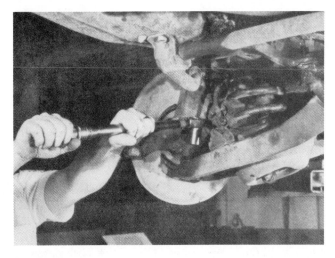

FIGURE 15-52.
The ball joint stud nut should be tightened to the lowest torque specification so it can be tightened farther if necessary to install the cotter pin. Never back the nut off to install the cotter pin. (Courtesy of Moog Automotive)

difficult to tighten the nut because the stud tends to rotate. In this case, it is usually necessary to either pretighten the stud using a plain nut, use a stud holder, or tap the control arm upward to try to seat the stud taper. (Figs. 15-53 & 15-54)

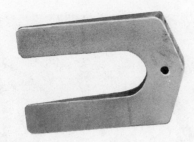

FIGURE 15–53.
This simple tool can be used to keep the ball joint stud from turning when self-locking nuts are used. It is also possible to lock the taper using a plain nut before installing the self-locking nut. (Courtesy of Moog Automotive)

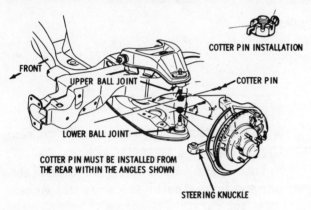

FIGURE 15–54.
Before installing the ball joint studs, it is best to position the cotter pin hole so the cotter pin can be easily installed. (Courtesy of Oldsmobile)

15.8 REMOVING AND REPLACING STRUT ROD BUSHINGS

Most strut rod bushings can be replaced while the control arm is off during a bushing or ball joint replacement or as a separate operation with the control arm connected to the steering arm and frame. On many cars, the strut rod is connected to the control arm with one or two bolts and can be easily removed after these bolts and the nut at the strut rod bushing are removed. On some cars, the strut rod passes through the lower control arm and is retained by a nut; strut rod bushing replacement on these cars is more involved. The lower control arm will need to be removed or moved away

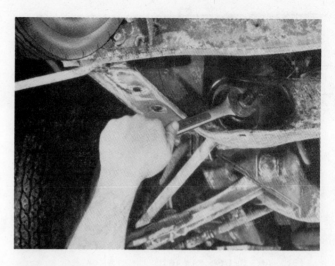

FIGURE 15–55.
Strut rod bushing replacement usually begins with the removal of the nut at the end of the strut rod. (Courtesy of Moog Automotive)

from the strut rod to make strut rod removal possible.

To replace strut rod bushings, you should:

1. Remove the strut rod bushing nut; if two nuts are used at the bushings, leave the inner nut in position so the caster adjustment will not be disturbed. Also, remove the strut rod to control arm bolts. (Fig. 15–55)

2. Lift the strut rod from the control arm and slide it out of the frame bracket. Remove the old bushings and washers. (Fig. 15–56)

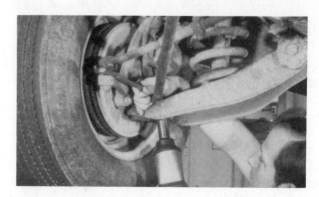

FIGURE 15–56.
Removing the strut rod from the lower control arm is usually a matter of removing the bolts and nuts that connect them. (Courtesy of Moog Automotive)

FIGURE 15–57.
With the strut rod disconnected, it can be removed from the frame bracket and old bushings. (Courtesy of Moog Automotive)

3. Following the directions for the bushing set, assemble the new inner bushing and retaining washer. Slide the strut rod into the bracket, assemble the remaining bushing parts, and install the retaining nut. (Fig. 15–57)

4. Hold the free end of the strut rod in its normal ride height position and tighten the retaining nut to snug-up and align the bushing parts. (Fig. 15–58)

FIGURE 15–58.
The new bushings should be placed in the correct position as the strut rod is slid back into the frame bracket. (Courtesy of Moog Automotive)

FIGURE 15–59.
Strut rod bushing replacement is completed by tightening the strut rod nut, connecting the strut rod to the control arm, and then tightening all of the nuts and bolts to the correct torque. (Courtesy of Moog Automotive)

5. Install the strut rod to control arm bolts and nuts and tighten all nuts and bolts to the correct torque. (Fig. 15–59)

15.9 REMOVING AND REPLACING KING PINS

Most trucks and pickups using a solid axle or twin I-beam axles use a reversed Elliott king pin design, which locks the king pin in the end of the axle and has a pair of king pin bushings in the steering knuckle. Two types of bushing materials are used: **bimetal** (bronze bushing material in a steel shell) or **synthetic** (plastic or nylon). Synthetic bushings tend to reduce steering effort and reduce road shock; another plus is that they can be installed without machining. They also tend to wear faster, especially under dusty conditions. Bimetal bushings tend to resist contaminants better and usually last longer under adverse conditions. Metal bushings usually have to be machined (reamed or honed) after installation to bring them to the correct size. (Fig. 15–60)

To remove and replace king pins and bushings, you should:

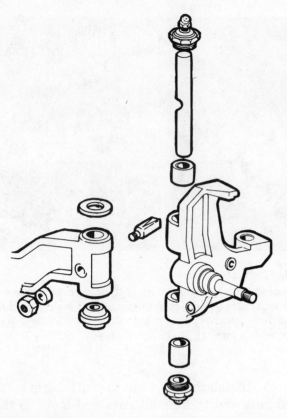

FIGURE 15-60.
An exploded view of an axle end and steering knuckle showing the king pin and the other parts of a king pin set. (Courtesy of Moog Automotive)

1. Raise and support the vehicle on a hoist or jack stands.

2. Remove the wheel and the brake caliper and rotor or brake drum and backing plate. Using a wire, hang the caliper or backing plate inside the fender.

3. Remove the tie rod end from the steering arm; this operation is described in Chapter 16.

4. Remove the retaining nut and drive the king pin lock bolt out of the axle.

5. Remove the king pin end caps. If they have wrench flats, they can be unscrewed. If they are soft plugs/metal discs, drive a punch into them to loosen them and pry them out using a screwdriver.

6. Using a punch and hammer, drive the king pin out of the axle and steering knuckle. (Fig. 15-61)

FIGURE 15-61.
After removing king pin end caps, the nut from the locking bolt, and driving the locking bolt out of the axle, the king pin can be driven out of axle and steering knuckle. (Courtesy of Moog Automotive)

7. Remove the steering knuckle, bearing, and shims; save the shims for possible reinstallation. (Fig. 15-62)

8. Drive the old bushings out of the steering knuckle using the right size bushing driver and a hammer. (Fig. 15-63)

9. If using synthetic bushings, slide the new bushings into the steering knuckle bosses.

OR If using bimetal bushings, carefully drive or press the new bushings into the steering knuckle making sure the grease holes in the

FIGURE 15-62.
With the king pin removed, the steering knuckle, thrust bearing, and shims can be removed; at this time, the end of the axle beam should be checked for damage. (Courtesy of Moog Automotive)

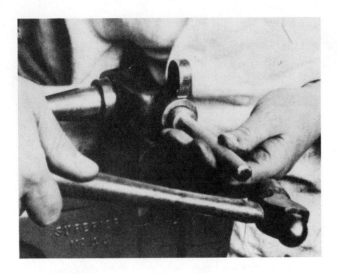

FIGURE 15-63.
The old bushings can be driven out of the steering knuckle using a punch or better yet, a bushing driver. (Courtesy of SKF Automotive Products)

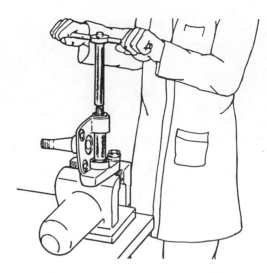

FIGURE 15-65.
Metal king pin bushings must be reamed or honed to the correct size; it is important that the two bushing bores are kept in straight alignment during this operation. (Courtesy of Dana Corportaion)

bushings line up with the zerk fitting holes in the steering knuckle bosses. A correctly sized bushing driver must be used to ensure that the bushings are not damaged during installation. The bushing bores should be honed (preferred)

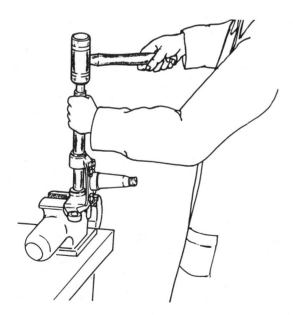

FIGURE 15-64.
New metal bushings are pressed or driven into the steering knuckle using a properly sized bushing driver. Nylon bushings can usually be slid into place. (Courtesy of Dana Corporation)

or reamed to size the bores to the king pin diameter and ensure that the two bores are in exact alignment. Honing or reaming of the bushing bores can be done by most automotive machine shops. (Figs. 15-64 & 15-65)

10. If used, place the sealing "O" rings in the recesses in the steering knuckle bores, lubricate the "O" rings and bushings, place the steering knuckle (correct side up) over the axle, pack the thrust bearing with grease, and place the bearing between the bottom of the axle and the lower steering knuckle bore. (Fig. 15-66)

11. Slide a new set or the original shims into the gap between the top of the axle and the upper steering knuckle boss. Now, try to slide a 0.005 inch (13 mm) feeler gauge alongside of the shims; if it enters, larger shims should be installed so the gap is 0.005 inch or less. (Fig. 15-67)

12. Slide the new king pin into the steering knuckle and axle bores making sure that the notch or flat in the king pin aligns with the lock pin hole. (Fig. 15-68)

13. Install the lock pin, and tighten the nut to the correct torque. (Fig. 15-69)

FIGURE 15–66.
An "O" ring seal is used with some king pins; it should be installed and lubricated before the steering knuckle is replaced. (Courtesy of Moog Automotive)

FIGURE 15–68.
The new king pin should be lubricated and slid in place so the notch in the king pin aligns with the locking pin hole. (Courtesy of Moog Automotive)

15.10 REAR SUSPENSION SERVICE

The wearing points of most rear suspensions in most cases are the rubber bushings, which are almost the same as those already discussed. If these bushings become excessively worn, rear wheel misalignment or noises will result. Rear wheel control arm bushing service is done following a procedure that is very similar to front suspension bushing replacement; many of the same tools are used.

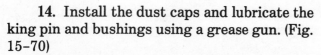

14. Install the dust caps and lubricate the king pin and bushings using a grease gun. (Fig. 15–70)

15. Replace the brake assemblies and the tires and wheels.

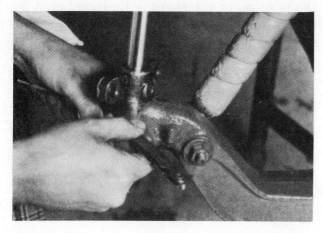

FIGURE 15–67.
The thrust bearing and the bushings should be lubricated before replacing the steering knuckle onto the axle making sure the steering knuckle is not upside down. Shims should be added at the top of the axle until a 0.005 inch feeler gauge can not be slid alongside of the shim pack. (Courtesy of SKF Automotive Products)

FIGURE 15–69.
The lock pin bolt should be tapped into place and then tightened to the correct torque. (Courtesy of SKF Automotive Products)

FIGURE 15-70.
King pin replacement is completed when the end caps have been replaced and the parts have been given a final greasing. (Courtesy of Moog Automotive)

Rear suspension service procedures are described in manufacturer's and technician's shop manuals, if there is doubt as to the exact procedure for these operations.

15.11 COMPLETION

Several checks should always be made after completing any suspension system service operation. These are:

- Ensure proper tightening of all nuts and bolts with all locking devices properly installed.

- Ensure no incorrect rubbing or contact between parts or brake hoses.

- Ensure proper feel of the brake pedal.

- Check and readjust the wheel alignment if there is a possibility that the replacement parts have changed the alignment.

- Perform a careful road test to ensure correct and safe vehicle operation.

REVIEW QUESTIONS

1. A ball joint stud taper can be broken loose using a

    a. sharp hammer blow against the stud boss.
    b. stud separator.
    c. pickle fork.
    d. any of these.

2. Technician A says the pressure from the car's spring can help you break the stud taper of a compression-loaded ball joint. Technician B says that in all cases, the spring should be compressed before breaking the ball joint studs loose. Who is right?

    a. A only   c. both A and B
    b. B only   d. neither A nor B

3. If a ball joint has a torn boot, the

    a. hole in the boot should be glued shut.
    b. boot should be replaced.
    c. ball joint should be replaced.
    d. any of these.

4. A pickle fork is a quick way of breaking ball joint studs loose, but it
    A. will ruin the boot.
    B. can distort the ball joint socket.

    a. A only   c. both A and B
    b. B only   d. neither A nor B

5. Technician A says that many ball joints are secured to the control arm by a press fit. Technician B says that some ball joints are threaded into the control arm. Who is right?

    a. A only
    b. B only
    c. both A and B
    d. neither A nor B

6. Rivets used to secure the ball joint to the control arm can be removed using a
    A. cutting torch.
    B. drill and chisel.

    a. A only
    b. B only
    c. both A and B
    d. neither A nor B

7. A torsilastic control arm bushing's
    A. inner sleeve is locked to the control arm shaft when the mounting bolt is tightened.
    B. rubber portion should be lubricated with a special grease.

    a. A only
    b. B only
    c. both A and B
    d. neither A nor B

8. Technician A says that metal control arm bushings must be pressed into the control arm an equal amount on each side. Technician B says that metal control arm bushings must be lubricated during or after installation. Who is right?

    a. A only
    b. B only
    c. both A and B
    d. neither A nor B

9. Technician A says that the installing tool for rubber control arm bushings should push against the outer sleeve. Technician B says that the bolts retaining rubber bushings should be tightened only with the car at correct ride height. Who is right?

    a. A only
    b. B only
    c. both A and B
    d. neither A nor B

10. Common Rubber control arm bushings are pressed in place
    A. by a force applied on the rubber portion.
    B. until they seat solidly in the control arm.

    a. A only
    b. B only
    c. both A and B
    d. neither A nor B

11. Metal control arm bushings are installed
    A. by threading them into the control arm and onto the control arm shaft.
    B. an equal amount on each bushing.

    a. A only
    b. B only
    c. both A and B
    d. neither A nor B

12. Sleeveless rubber bushings can be installed easier if they are lubricated with

    a. lithium grease
    b. brake fluid
    c. vegetable oil
    d. any of these

13. Technician A says that a worn ball stud hole can be caused by a mechanic installing the stud into a dirty hole. Technician B says that a broken ball joint stud can be the result of a mechanic backing off the nut to align the cotter pin. Who is right?

    a. A only
    b. B only
    c. both A and B
    d. neither A nor B

14. Technician A says that the ball joint stud can be held from turning with a special tool when installing a self-locking nut. Technician B says that the stud can be locked in place with a standard nut before installing the self-locking nut. Who is right?

    a. A only
    b. B only
    c. both A and B
    d. neither A nor B

15. When removing or replacing the ball joint stud on a car with S-L A suspension, a spring compressor should be installed to make the job
    A. easier.
    B. safer.

    a. A only
    b. B only
    c. both A and B
    d. neither A nor B

16. When replacing strut rod bushings, the
    A. control arm must always be removed.
    B. retaining nut should be tightened with the strut rod in the normal, ride height position.

    a. A only
    b. B only
    c. both A and B
    d. neither A nor B

17. Technician A says that ball joint studs that are locked in place using a pinch-bolt are more difficult to remove than the other style. Technician B says that during replacement, the notch in the ball joint stud must be carefully aligned with the pinch-bolt. Who is right?

    a. A only
    b. B only
    c. both A and B
    d. neither A nor B

18. Technician A says that bimetal king pin bushings are presized and do not require machining after installation. Technician B says that shims should be added if the gap between the axle and the thrust bearing is greater than 0.005 inch. Who is right?

    a. A only
    b. B only
    c. both A and B
    d. neither A nor B

19. Plastic/nylon king pin bushings should be
    A. lubricated after installation.
    B. reamed to the correct size after installation.

    a. A only
    b. B only
    c. both A and B
    d. neither A nor B

20. After new ball joints or control arm bushings have been installed, the
    A. alignment should be checked and adjusted if necessary.
    B. car should be road tested.

    a. A only
    b. B only
    c. both A and B
    d. neither A nor B

Chapter **16**

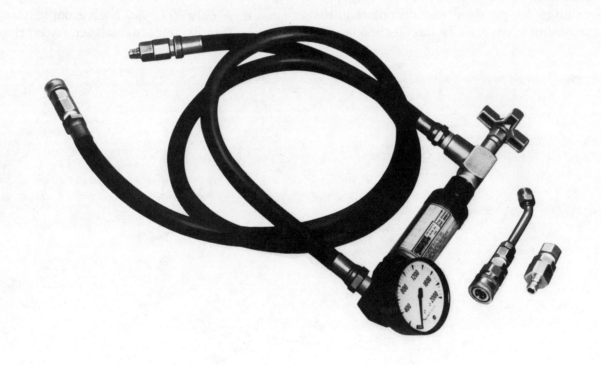

STEERING SYSTEM SERVICE

After completing this chapter, you should:

- Be able to remove and replace tie rods and tie rod ends on conventional and rack and pinion steering systems.

- Be able to remove and replace idler arms, center links, and Pitman arms.

- Be able to remove and replace conventional and rack and pinion steering gears and mounting bushings.

- Be able to remove and replace steering wheels, steering columns, and steering shaft couplers.

- Be able to make the necessary adjustments on a conventional or rack and pinion steering gear.

- Be able to overhaul a conventional or rack and pinion steering gear.

- Be able to repair fluid leaks and replace the fluid in a power steering system.

- Be able to test power steering system pressures and determine whether or not they are correct.

- Be able to adjust power steering control valves.

- Be able to overhaul a power steering pump.

16.1   INTRODUCTION

After determining that a steering system component is faulty, the next step is to repair, adjust, or remove and replace that component. This chapter focuses on the replacement, adjustment, and repair of tie rod ends, idler arms, manual and power steering gears, and power steering hoses, pumps, and belts.

Steering system service should be done with great care so that there is no possibility of failure after the work is done. Failure of the steering system can have devastating effects on the operation of the car. Safety first must be the attitude of a professional technician while working on steering systems.

SAFETY NOTE

The precautions listed in the Safety Notes in Sections 15.1 and 15.7 for suspension system repair apply also to steering system repair operations.

Most manual steering gears can be easily disassembled, inspected, reassembled, and adjusted in a short period of time, and the service operations involved require only a few, if any, special tools. Many technicians have successfully rebuilt manual steering gears. Power steering gears can also be rebuilt in garages or shops. Special tools are usually required to perform some of the service operations, and special equipment is needed to test the effectiveness of the repair. Removal and replacement of a power rack and pinion steering gear on some late model FWD cars can be a difficult and time-consuming process. It is normally considered too difficult and time-consuming to install a unit in the car for testing with the possibility that it will have to be removed for further work. Several major companies are rebuilding or remanufacturing power rack and pinion steering gears to new or better-than-new condition. These units are thoroughly tested to ensure proper, leak-free operation. In the case of power

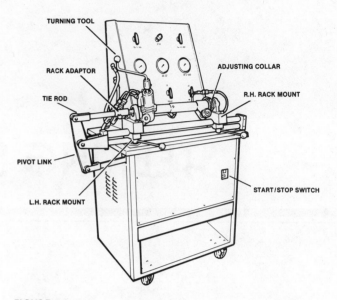

FIGURE 16-1.
A repaired or rebuilt power rack and pinion steering gear should be checked out on an analyzer like this one before replacing it on the car. This unit allows a thorough check of the steering gear. (Courtesy of Branick Industries)

steering units, many technicians will install a rebuilt unit to complete the job quicker with fewer potential problems. (Fig. 16-1)

Faulty steering linkage components are removed and replaced with new parts. Other steering linkage services are lubrication of the pivot points (if required) and adjustment of the tie rods to correct toe. Toe and other front end adjustments are described in Chapter 18.

The repair methods described in this chapter are presented in a general manner, so it is highly recommended to check a service manual when doing specific service operations on most car models.

16.2   STEERING LINKAGE REPLACEMENT

Like ball joints, most steering linkage components are connected using locking tapers. Tie rod ends and centerlink connections at the Pitman arm and idler arm require taper breaking methods much like ball joint taper breaking. Hammer blows, separators or pickle forks, or pullers are the common methods of breaking loose tapers; of these, pullers are the most professional and will cause the least amount of damage to the parts.

After replacement of any part of steering linkage, toe must be checked and readjusted, if necessary, to prevent tire wear or handling difficulties.

16.2.1 Removing and Replacing an Outer Tie Rod End

Tie rod ends can be replaced individually or together as part of a replacement tie rod. Toe should always be checked and readjusted if necessary after a tie rod end replacement. When replacing tie rods or tie rod ends, time can be saved by preadjusting the tie rod length before installing the tie rod studs. Measurements of tie rod length are made before tie rod disassembly, and the new parts are adjusted to the same length during replacement. It is also possible, and sometimes quicker, to count the number of turns used to remove the tie rod end and then turn the new end the same amount during installation. When using this second method, check to ensure that the threaded portions of the old and new tie rod ends are the same length. (Fig. 16-2)

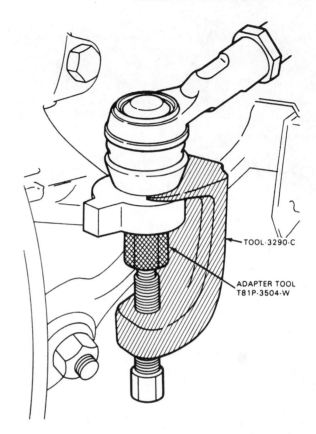

FIGURE 16-3.
A tie rod end puller can be used to separate the tie rod end from the steering arm or centerlink in an easy and quick manner. (Courtesy of Ford Motor Company)

FIGURE 16-2.
It is a good practice to measure the length of a tie rod, usually from the sleeve to zerk fitting or plug, before removing a tie rod end; this allows a quick, fairly accurate presetting of the toe adjustment. (Courtesy of Moog Automotive)

To remove an outer tie rod end, you should:

1. Raise and support the car on a hoist or jack stands.

2. Loosen the clamp bolt at the tie rod adjuster sleeve or the jam nut at the tie rod end.

3. Remove the cotter pin and nut from the tie rod stud.

4. Install a tie rod end puller and tighten it to break the taper so the stud can be removed from the steering arm. (Fig. 16-3)

OR Drive a separator tool/pickle fork between the steering arm and the tie rod end to break the taper and allow stud removal. (Fig. 16-4) Note that this tool will tear the boot and also possibly damage the socket.

OR Hold the head of hammer securely against one side of the steering arm boss while striking the other side of the boss with a sharp,

FIGURE 16-4.
Driving the tie rod wedge/pickle fork between the center link or steering arm and the tie rod end will break loose the tie rod stud. (Courtesy of SKF Automotive Products)

FIGURE 16-5.
A sharp hammer blow will often break loose a tie rod stud taper. Note the pry bar giving a separating force. (Courtesy of Bear Automotive)

quick blow from another hammer. When the taper is broken loose, lift the stud out of the boss. (Fig. 16-5)

5. Measure the distance from some point on the tie rod end, usually the center of the stud while it is in a centered position, to some point on the tie rod, usually the adjuster sleeve or count the number of turns as you remove the tie rod end from the tie rod.

To replace a tie rod end, you should:

1. Thread the new tie rod end into the tie rod sleeve/tie rod the number of turns equal to the amount used for removal or until the tie rod

length is correct. On three-piece tie rods, each tie rod end should have about the same amount of thread contact with the adjusting sleeve. (Fig. 16-6)

2. Inspect and clean the tie rod stud hole in the steering arm. (Fig. 16-7)

3. Position the dust boot onto the tie rod end, if separate, and install the tie rod stud into the steering boss. Replace the nut on the tie rod stud, tighten it to the correct torque, and install a new cotter pin.

NOTE: If the tie rod end is of the rubber type, the front wheels must be in a straightahead position when the stud is tightened into the steering arm. A memory steer or pull will develop if the studs are installed in a wheels-turned position; this will also cause an overstretching of the bushing material when a turn is made in the opposite direction. (Fig. 16-8)

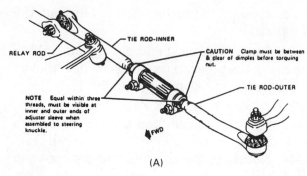

(A)

FIGURE 16-6.
Each tie rod end should be threaded into the adjusting sleeve an equal amount. (A is courtesy of Chevrolet; B is courtesy of Pontiac Division, GMC)

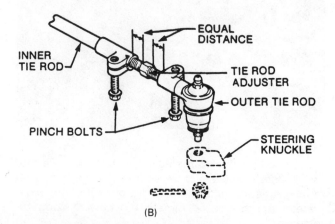

(B)

FIGURE 16–7.
Before replacing a tie rod end, make sure that the tapered hole is clean and not damaged. (Courtesy of SKF Automotive Products)

4. Check and adjust toe, center both the tie rod ends to their studs, and tighten the clamp/jam nut on the tie rod. This step is described in more detail in Chapter 18.

16.2.2   Removing and Replacing an Inner Tie Rod End, Standard Steering Gear

This operation is nearly identical to the procedure just described. The only real difference is that the inner end is removed from the center-link while the outer end is removed from the steering arm when it is replaced.

If both tie rod ends are removed from the adjuster sleeve at the same time, care should be taken to ensure that both ends are threaded into the sleeve the same number of turns, and the sleeve is centered on the two ends. The concern is to ensure secure connections between the adjuster sleeve and the tie rod ends.

16.2.3   Removing and Replacing an Inner Tie Rod End, Rack and Pinion Steering

This procedure varies slightly depending on the method used to lock the inner tie rod end to the steering rack gear. At this time there are six different methods used to make sure that the inner tie rod socket/pivot does not come loose from the rack. Most tie rods connect to the ends of the rack; one style connects to the center. In some cases, it is possible to unlock the tie rod end socket with the steering gear mounted in the car; it is then possible to remove and replace the inner tie rod end without removing the steering gear. If it is not possible to remove

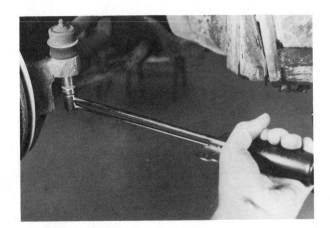

FIGURE 16–8.
When a tie rod stud is replaced, the nut should be tightened to the correct torque and the cotter pin, if used, should be replaced.

the lock pin, it will be necessary to remove the steering gear in order to change the tie rod end.

On most cars, the inner tie rod socket is a complete assembly that is adjusted to the correct preload. On some cars the tie rod end is several parts. A preload spring and bushing is part of the socket; the tie rod and tie rod end housing form the rest of the socket. It is necessary to adjust the preload tension on the tie rod end as this second style is being replaced onto the rack.

When the tie rod end sockets are being unscrewed from the rack gear (after removal of the locking device), the turning force on the socket and rack gear can damage the teeth on the pinion or rack gear; the rack should be held from turning by placing an adjustable open-end wrench over the tooth portion of the rack. The contact area between the wrench jaws and the rack teeth is large enough to keep the rack from turning without damage. (Fig. 16–9)

It is necessary to remove the boot/bellows when replacing the inner tie rod ends. It is normal for some gear oil to be in the boots of manual steering gears. On power steering units, a large amount of power steering fluid in the boots indicates faulty seals inside the steering

FIGURE 16–10.
The boot retainers are usually cut off using side cutting pliers ("dikes"); the boot can then be slid or cut off. (Courtesy of Saginaw Division, General Motors Corporation)

FIGURE 16–9.
When replacing the inner tie rod ends on a rack and pinion gear, the rack is held from turning by gripping it with an adjustable wrench; tool T81P-3504-G is engaging the tie rod socket so it can be unscrewed. (Courtesy of Ford Motor Company)

gear. New boots and clamps should be installed during tie rod replacement.

To remove an inner tie rod end on a rack and pinion steering gear, you should:

1. Raise and support the car on a hoist or jack stands.

2. Remove the outer tie rod end from the steering arm.

3. Remove both bellows clamps and slide the bellows down the tie rod. Note that on power steering units, a vent/breather tube is connected into each bellows. (Fig. 16–10)

4. Check for the locking method used on the tie rod socket. If it is a plain jam nut, a swaged connector, or an accessible roll pin, solid pin, or set screw, the tie rod end should be able to be replaced on the car. If a solid pin, roll pin, or set screw is in a nonaccessible location, the steering gear should be removed. Removal of a rack and pinion steering gear will be described later in this chapter. If removal of the steering gear is necessary, it should be mounted securely in a holding fixture or vise during the repair steps. (Fig. 16–11)

FIGURE 16-11.
When working on rack and pinion steering gear out of the car, it should be secured in a manner that will not damage the gear or gear housing; this gear is held in a special rack and pinion gear and strut vise. (Courtesy of Moog Automotive)

5. If a plain jam nut is used, place one wrench on the tie rod socket and a second wrench on the jam nut and loosen the jam nut. Next, place an adjustable wrench over the rack teeth, adjust the jaws to fit tightly onto the gear teeth, and unscrew the tie rod socket with the tie rod from the rack.

OR If a swaged end is used, place an adjustable wrench over the teeth on the rack, tighten the wrench jaws onto the rack, and using a second wrench, unscrew the tie rod socket from the rack. The swaged portion of the socket will bend outward as it is unscrewed. (Fig. 16-12)

OR If a roll pin is used, install a special roll pin puller into the roll pin and remove it from the tie rod socket. A roll pin can also be removed by using a screw extractor/easy out to rotate the roll pin until it works its way out of the hole. After the roll pin is removed, hold the rack gear from turning with one wrench and unscrew the tie rod socket with a second wrench. (Figs. 16-13 & 16-14)

NOTE: There are two styles of roll pins. A normal roll pin is a tubular shaped strip of metal that is rolled into almost one revolution; it is soft enough to be drilled out if necessary. A spiral lock pin is also tube shaped, but it has several revolutions of hardened steel. A spiral lock

FIGURE 16-12.
In most cases, the tie rod end can be unscrewed from the rack after the locking device has been loosened. This particular end is staked/swaged, and the swaged portion is expanded as the end is unscrewed. Note that the rack is being held in a vise (with soft jaws) to keep it from turning. (Courtesy of Saginaw Division, General Motors Corporation)

pin cannot be drilled out; it is too hard. (Fig. 16-15)

OR If a solid pin is used, install a drill guide so it is centered over the roll pin, set the drill depth limiter so the drill depth will just touch the rack, and drill the old lock pin out.

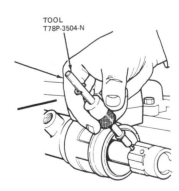

FIGURE 16-13.
If the tie rod end is locked by a roll pin, the roll pin can be removed using a special puller tool. This tool threads into the roll pin, and the slide hammer is used to provide an outward force. (Courtesy of Ford Motor Company)

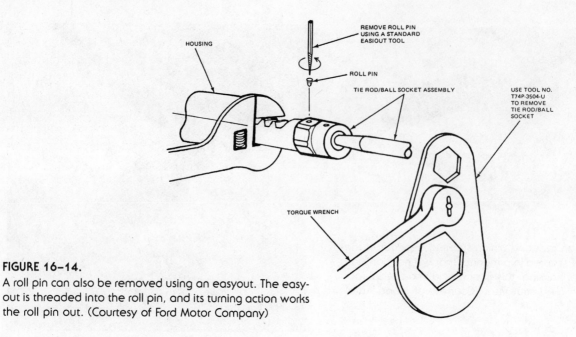

FIGURE 16-14.
A roll pin can also be removed using an easyout. The easy-out is threaded into the roll pin, and its turning action works the roll pin out. (Courtesy of Ford Motor Company)

Since the tie rod socket and jam nut are going to be replaced, it is possible to simply use an oversize drill bit to drill out the pin; be very careful not to drill any farther into the rack than absolutely necessary. After the pin is removed, place one wrench on the tie rod socket and one wrench on the jam nut and loosen the jam nut. Next, grip the rack teeth with an adjustable wrench, and unscrew the tie rod socket with a second wrench. (Fig. 16-16)

OR If a set screw is used, remove the set screw using an allen wrench, grip the rack teeth with an adjustable wrench, install a special wrench on the tie rod socket, and unscrew the tie rod socket from the rack. (Fig. 16-17)

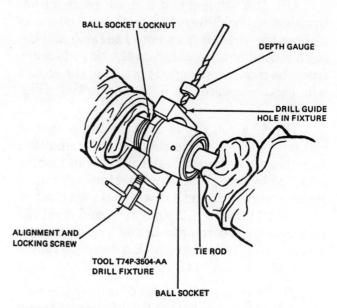

FIGURE 16-16.
If the tie rod end is locked by a solid pin, the pin must be removed with a drill; it is recommended to use a guide and drill depth gauge so the drill hole will be in the right place and not too deep. (Courtesy of Ford Motor Company)

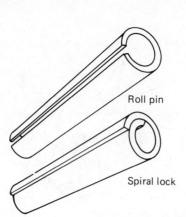

FIGURE 16-15.
A roll pin and a spiral lock pin, note how far each of them turns. A roll pin is soft enough to drill out if necessary; a spiral lock pin is too hard.

FIGURE 16-17.
If the tie rod end is locked by a set screw, it should be loosened before unscrewing the end. (Courtesy of Saginaw Division, General Motors Corporation)

SAFETY NOTE
The rack should be checked for other lock pin drillings; each end of the rack can have a maximum of two lock pin holes. The steering gear should be replaced if one of the inner tie rod ends has been replaced before.

NOTE: I have tried to describe each style of locking device used on domestic cars; some import cars use other methods to secure the tie rod socket onto the end of the rack. Whenever in doubt, check the service manual for that particular car.

To replace an inner tie rod end on a rack and pinion steering gear, you should:

1. Thread the new tie rod end socket with tie rod onto the rack, and while holding the rack from turning with an adjustable wrench, tighten the tie rod socket to the correct torque. (Fig. 16-18)

OR In a few cases, the replacement tie rod end will require an adjustment. A spring scale is used to measure the force required to swing/articulate the tie rod in the socket; tightening the tie rod end will increase the tightness of the joint and the amount of force needed to swing the tie rod. The tie rod socket should be tightened enough to obtain the correct articulation force. (Fig. 16-19)

2. Lock the tie rod end in place. If a jam nut is used, use two wrenches and tighten the jam nut against the tie rod socket.

OR Stake or swage the new inner tie rod socket onto the flats at the ends of the rack; a special tool is available to crimp the socket if the steering gear is on the car. If the unit is off the car, rest the tie rod socket over a vise and strike the socket using a hammer and punch to

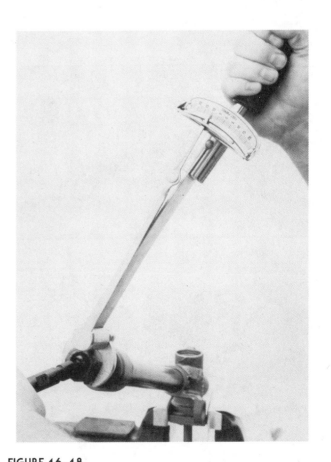

FIGURE 16-18.
The new tie rod end should be tightened onto the rack using a torque wrench; make sure that the tie rods swing freely after tightening. (Courtesy of Saginaw Division, General Motors Corporation)

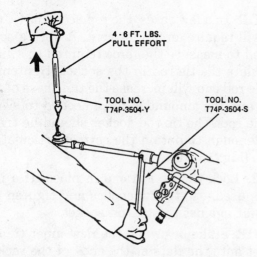

FIGURE 16-19.
Some tie rod end sockets are assembled on the end of the rack. These sockets are tightened the right amount to produce the correct tie rod socket preload; this is measured by measuring the pull effort needed to swing the tie rod. (Courtesy of Ford Motor Company)

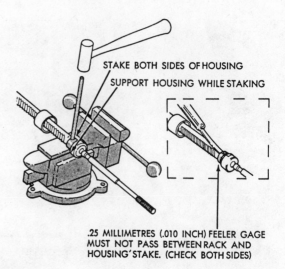

FIGURE 16-21.
An inner tie rod end can be staked/swaged by placing the rack so it is solidly supported and striking the socket with a punch and hammer. It is recommended that the swaged portions should be solidly against the rack or no further away than the dimensions shown here. (Courtesy of Chrysler Corporation)

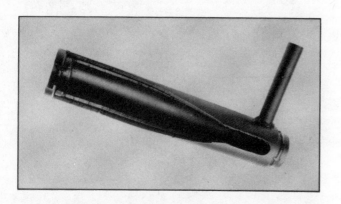

FIGURE 16-20.
The tool at the top allows removal of a swaged inner tie rod end with the steering gear mounted in the car; the bottom tool is used to crimp/swage an inner tie rod when it is replaced on the car. (Courtesy of Moog Automotive)

crimp the socket onto the flats. Check for proper swaging by trying to slide a 0.010 inch (0.25 mm) feeler gauge under the staked portion; if it does not enter, the stakes/crimps are adequate. (Figs. 16-20 & 16-21)

OR Drive a new roll pin into the tie rod socket. (Fig. 16-22)

OR Drill a new hole into the junction of the socket and the tight jam nut; the hole should be the correct depth so the pin will just enter into the rack. A drill guide is available to ensure a properly drilled hole. Drive the new lock pin into the hole and punch-mark around the pin to lock it in place. (Fig. 16-23)

OR Tighten the set screw to the correct torque. (Fig. 16-24)

3. Slide the new bellows/boot into position and tighten the large clamp. The small clamp will be tightened after the toe check. A special crimping tool is often required to secure the clamps. (Fig. 16-25)

4. Replace the outer tie rod end and check and adjust toe, if necessary. After adjusting toe, tighten the small clamp on the bellows.

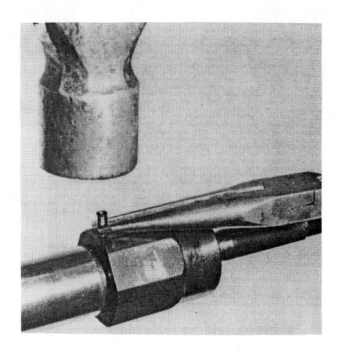

FIGURE 16-22.
After the tie rod end has been tightened to the correct torque, a spiral pin is driven into the hole in the tie rod socket. (Courtesy of Ford Motor Company)

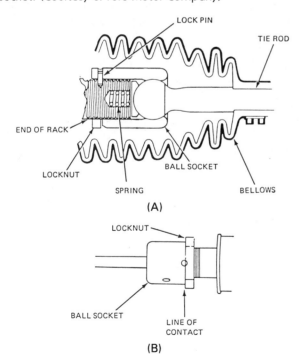

(A)

(B)

FIGURE 16-23.
If the tie rod socket is locked by a solid pin, a hole must be drilled to the correct depth in the rack (A); a new lock pin is driven into the hole and staked in place using tool T74P-3504-X or a center punch. (Courtesy of Ford Motor Company)

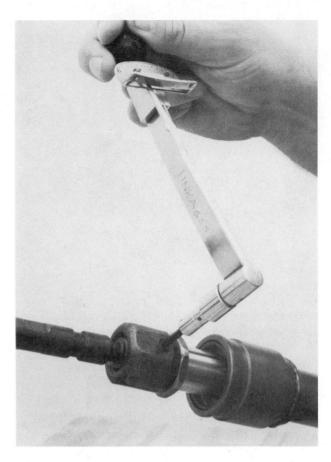

FIGURE 16-24.
If the tie rod socket is locked with a set screw, the set screw should be tightened to the correct torque. (Courtesy of Saginaw Division, General Motors Corporation)

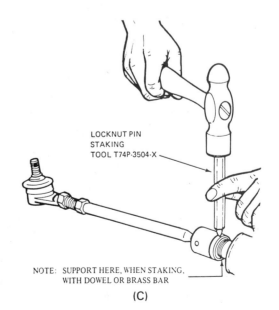

NOTE: SUPPORT HERE, WHEN STAKING, WITH DOWEL OR BRASS BAR

(C)

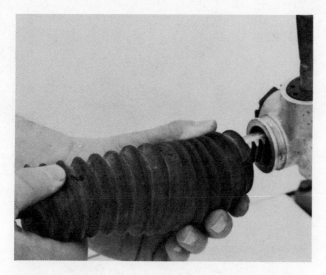

FIGURE 16-25.
After the inner tie rod socket has been installed, a new boot/bellows should be slid into place and secured with a clamp. Many clamps require a special pliers to crimp them. (Courtesy of Saginaw Division, General Motors Corporation)

16.2.4 Removing and Replacing Inner Tie Rod End Pivot Bushings, Rack and Pinion Steering Gear with Center-Mounted Tie Rods

This tie rod design uses a bushing instead of a ball and socket type joint; it is currently used on some late-model General Motors FWD cars. The tie rods are secured to the steering gear by a pair of bolts that pass through a slot in the gear housing and are threaded into the rack gear.

If both tie rods are removed, care should be taken to hold the boot and other parts in alignment. It is recommended that you remove the first tie rod and replace the bolt before removing the second bolt and tie rod. (Figs. 16-26 & 16-27)

To remove and replace an inner tie rod bushing, you should:

1. Raise and support the car on a hoist or jack stands.

2. Remove the outer tie rod end from the steering arm.

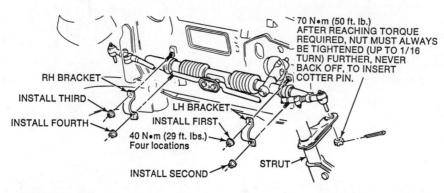

FIGURE 16-26.
This rack and pinion steering gear has the inner tie rod ends mounted at the center of the rack; note also that the steering arms are attached in the strut. (Courtesy of Pontiac Division, GMC)

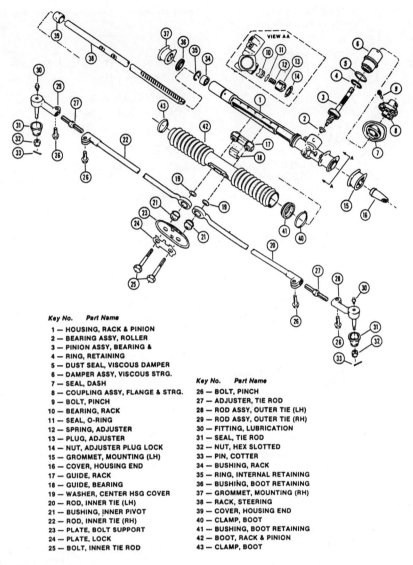

Key No.	Part Name
1 —	HOUSING, RACK & PINION
2 —	BEARING ASSY, ROLLER
3 —	PINION ASSY, BEARING &
4 —	RING, RETAINING
5 —	DUST SEAL, VISCOUS DAMPER
6 —	DAMPER ASSY, VISCOUS STRG.
7 —	SEAL, DASH
8 —	COUPLING ASSY, FLANGE & STRG.
9 —	BOLT, PINCH
10 —	BEARING, RACK
11 —	SEAL, O-RING
12 —	SPRING, ADJUSTER
13 —	PLUG, ADJUSTER
14 —	NUT, ADJUSTER PLUG LOCK
15 —	GROMMET, MOUNTING (LH)
16 —	COVER, HOUSING END
17 —	GUIDE, RACK
18 —	GUIDE, BEARING
19 —	WASHER, CENTER HSG COVER
20 —	ROD, INNER TIE (LH)
21 —	BUSHING, INNER PIVOT
22 —	ROD, INNER TIE (RH)
23 —	PLATE, BOLT SUPPORT
24 —	PLATE, LOCK
25 —	BOLT, INNER TIE ROD

Key No.	Part Name
26 —	BOLT, PINCH
27 —	ADJUSTER, TIE ROD
28 —	ROD ASSY, OUTER TIE (LH)
29 —	ROD ASSY, OUTER TIE (RH)
30 —	FITTING, LUBRICATION
31 —	SEAL, TIE ROD
32 —	NUT, HEX SLOTTED
33 —	PIN, COTTER
34 —	BUSHING, RACK
35 —	RING, INTERNAL RETAINING
36 —	BUSHING, BOOT RETAINING
37 —	GROMMET, MOUNTING (RH)
38 —	RACK, STEERING
39 —	COVER, HOUSING END
40 —	CLAMP, BOOT
41 —	BUSHING, BOOT RETAINING
42 —	BOOT, RACK & PINION
43 —	CLAMP, BOOT

FIGURE 16–27.
An exploded view of a center take-off rack and pinion steering gear. Note that the guide (17) can slip out of position if both inner tie rod bolts (25) are removed at the same time. (Courtesy of Pontiac Division, GMC)

3. Bend tabs of the lock plate away from the bolt head and remove the inner tie rod bolt. Remove the tie rod. (Fig. 16–28)

4. Using the correct tool, remove the old bushing from the tie rod end. (Fig. 16–29)

5. Coat the new bushing with a light film of grease and, using the special tool, press the bushing into the tie rod end.

6. Make sure the center housing cover washers are fitted into the rack and pinion boot, place the tie rod end in position, and install the inner tie rod bolt.

7. Tighten the tie rod bolt to the correct torque and bend the flats of the lock plate against the bolt head to secure the bolt in position.

8. Replace the outer tie rod end.

16.2.5 Removing and Replacing an Idler Arm

Several styles of idler arms and brackets are used with parallelogram steering. The most

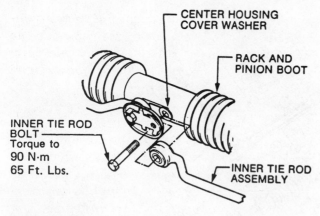

FIGURE 16–28.
The inner tie rod end of a center take-off type is removed by removing the inner tie rod bolt; note that this bolt is secured by a lock plate. (Courtesy of Pontiac Division, GMC)

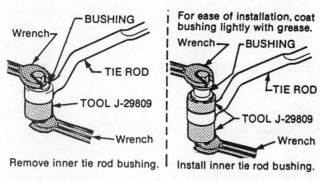

FIGURE 16–29.
The rubber bushing at the inner end of a center take-off tie rod can be removed and replaced using the method shown here. (Courtesy of Pontiac Division, GMC)

common is an assembly with the idler arm integral with the mounting bracket. A worn idler arm is usually replaced as an idler arm and bracket assembly; in a few cases, only the idler arm or idler arm bushing is replaced.

In cases where rubber idler arm bushings are used, the idler arm should be tightened onto the bracket with the tires in a straightahead position. Like with rubber tie rod ends, tightening the bushings in a turned position can cause a steering pull in the direction that the arm was when it was tightened and early failure of the bushing.

To remove and replace an idler arm, you should:

1. Raise and support the car on a hoist or jack stands.

SAFETY NOTE

Some car models have changed the length of their idler arms during model change-over, and in some cases, the two different arms are almost interchangeable. Too short an idler arm can allow the idler arm to go overcenter and lock the steering at the extreme position of a turn. *Always check the new idler arm against the old one to ensure an exact duplicate.*

2. Remove the cotter pin and nut from the idler arm or centerlink stud. Install a puller to break the stud taper and separate the idler arm from the centerlink. (Fig. 16–30)

3. Remove the bolts securing the idler arm bracket to the frame and remove the idler arm.

4. Clean and inspect the frame mounting location and the stud taper and mounting hole to make sure they are in good condition. (Fig. 16–31)

FIGURE 16–30.
The first step in removing an idler arm is to disconnect the idler arm from the centerlink; this usually requires the use of a puller to break loose the stud taper. (Courtesy of Moog Automotive)

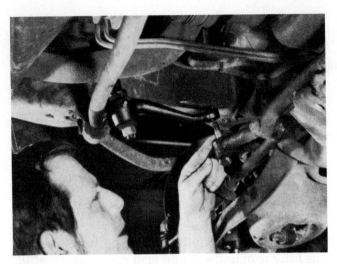

FIGURE 16–31.
Be sure the tapers in the hole and the stud are clean and undamaged before making the idler arm to centerlink connection. (Courtesy of Moog Automotive)

5. Place the new idler arm bracket in position, install the mounting bolts, and tighten the bolts and nuts to the correct torque.

NOTE: Some idler arms have adjustable slots at the mounting bolts. On these cars, **steering linkage parallelism** *should be adjusted. With the steering exactly centered, measure the distance from the center of the inner tie rod stud*

to the center of the upper control arm mounting/pivot bolt on each side of the car. Slide the idler arm up or down so these two measurements are the same or within 0.060 inch (1.5 mm) of each other. (Fig. 16–32)

NOTE: Some idler arms use a threaded style bushing; the idler arm can be unthreaded off the bracket. When installing this style of idler arm, the arm should be threaded as far as possible onto the bracket, but with enough room to rotate for normal operation. Thread the idler arm completely onto the bracket until it stops, and then back it off one-half to one turn. Some idler arm assemblies have a setting dimension for this same purpose. (Fig. 16–33)

6. Connect the idler arm to the centerlink, install the nut, tighten the nut to the correct torque, and install the cotter pin. (Fig. 16–34)

7. Check toe and adjust if necessary.

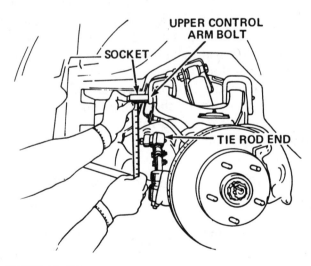

FIGURE 16–32.
One manufacturer recommends a steering parallelism adjustment. The distance between the tie rod stud and a socket on the pivot bolt is measured on each side, and the threaded portion of the idler arm pivot is adjusted to make both measurements equal. (Courtesy of Oldsmobile)

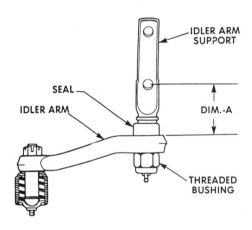

DIVISION	BODY	DIMENSION A	
		MILLIMETRES	INCHES
Buick	A-B-C-E	59.5 ± 1.6	$2^{11}/_{32} \pm ^{1}/_{16}$
Olds	A-B-C	59.5 ± 1.6	$2^{11}/_{32} \pm ^{1}/_{16}$
Pontiac	F-X	75.4 ± 1.6	$2^{31}/_{32} \pm ^{1}/_{16}$
	A-B	59.5 ± 1.6	$2^{11}/_{32} \pm ^{1}/_{16}$
Cadillac	C-D	59.5 ± 1.6	$2^{11}/_{32} \pm ^{1}/_{16}$
	K	75.4 ± 1.6	$2^{31}/_{32} \pm ^{1}/_{16}$

FIGURE 16–33.
When installing an idler arm with a threaded bushing, dimension A should be adjusted to the correct specification. (Courtesy of Saginaw Division, General Motors Corporation)

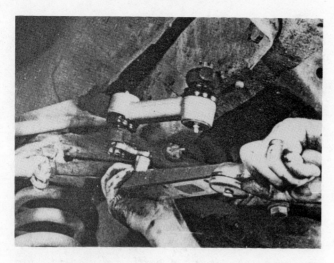

FIGURE 16-34.
The bolts securing the idler arm bracket to the frame should be tightened to the correct torque. (Courtesy of SKF Automotive Products)

16.2.6 Removing and Replacing a Centerlink

Two styles of centerlinks are commonly used with parallelogram steering linkage: those with wearing sockets and those without. Centerlinks without wearing sockets seldom require any service; the wear points are in the idler arm and Pitman arm. If the sockets are in the centerlink and they wear out, the centerlink will need to be replaced. (Fig. 16-35)

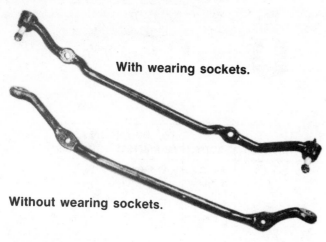

With wearing sockets.

Without wearing sockets.

FIGURE 16-35.
Two styles of centerlinks, the one without wearing sockets requires no service; if the sockets wear out, the centerlink with wearing sockets must be replaced. (Courtesy of SKF Automotive Products)

To remove and replace a centerlink, you should:

1. Raise and support the car on a hoist or jack stands.

2. Remove the cotter pins and retaining nuts from the studs of the inner tie rod ends and the idler arm and Pitman arm to centerlink joints.

3. Using a puller, break the taper of the four studs and remove the centerlink.

4. Clean and inspect the four studs and holes.

5. Place the centerlink in position, install the idler arm and Pitman arm studs, and install the two nuts.

6. Place the inner tie rod end studs in place, install the two nuts, tighten all four nuts to the correct torque, and install new cotter pins.

7. Check toe and adjust if necessary.

16.2.7 Removing and Replacing a Pitman Arm

The Pitman arm used with nonwearing centerlinks has a socket in it that can wear. Pitman arms are secured to the Pitman shaft with a splined and tapered connection; a puller is required to separate them. One or more **missing** or **master splines** are usually used at this connection to ensure that the arm is replaced onto the Pitman shaft in the proper position. It is still a good practice to mark the Pitman-arm-to-shaft position to ensure correct mounting. (Fig. 16-36)

To remove and replace a Pitman arm, you should:

1. Raise and support the car on a hoist or jack stands.

2. Remove the cotter pin and nut from the Pitman arm/centerlink stud. Break the taper on this stud, preferably with a puller, and separate the Pitman arm from the centerlink. (Fig. 16-37)

FIGURE 16-36.
Two styles of Pitman arms, the lower one requires no service; the upper one with the stud and socket must be replaced if the socket wears out. (Courtesy of SKF Automotive Products)

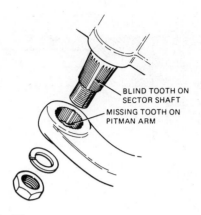

BLIND TOOTH ON SECTOR SHAFT
MISSING TOOTH ON PITMAN ARM

FIGURE 16-38.
The blind tooth/master spline is used to help ensure that the Pitman arm is replaced in the correct position. (Courtesy of Ford Motor Company)

3. Mark the arm-to-Pitman-shaft position and loosen the retaining nut several turns or until it is even with the end of the threads.

NOTE: The nut should be left on the Pitman shaft to help protect the threads and to prevent the Pitman arm from popping off.

4. Install a puller and tighten it to break loose the taper between the Pitman arm and shaft. Remove the puller, the retaining nut, and the Pitman arm.

5. Clean and inspect the splines on the Pitman shaft, the splined hole in the Pitman arm,

and the hole and stud at the centerlink connection.

6. Connect the Pitman arm to the centerlink, and place the front tires in a straight-ahead position.

7. Align the marks between the Pitman arm and shaft or turn the steering wheel to the center-most position of the steering gear with the steering wheel centered, and slide the Pitman arm onto the Pitman shaft. (Fig. 16-38)

8. Install the retaining nut and lock washer on the Pitman shaft, and tighten the nut to the correct torque.

9. Tighten the Pitman arm or centerlink stud retaining nut to the correct torque and install a new cotter pin.

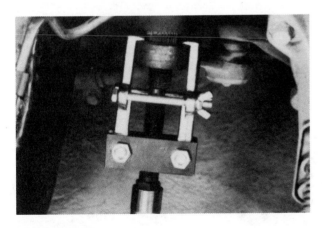

FIGURE 16-37.
A Pitman arm puller is being used to remove this Pitman arm; some mechanics prefer to leave the loosened nut on the end of the Pitman shaft threads to help protect the end of the Pitman shaft. (Courtesy of SKF Automotive Products)

16.2.8 Removing and Replacing an Adjustable Drag Link Socket

Drag links are used on some pickups, 4WDs, and trucks to connect the Pitman arm to the third arm on the left steering knuckle. Most drag link sockets resemble a tie rod end; some are disassembled for removal and need to be adjusted after reassembly. (Fig. 16-39)

To disconnect an adjustable drag link socket, you should:

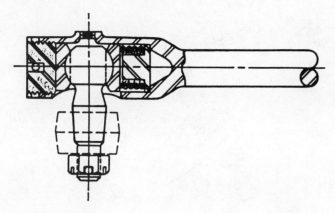

FIGURE 16-39.
A drag link socket, note that unscrewing the screw at the left will provide enough room to lift the drag link off of the ball stud. (© 1965, courtesy of Replacement Parts Div., TRW, Inc.)

1. Raise and support the car on a hoist or jack stands and remove the left tire and wheel.

2. Remove the cotter pin from the drag link and unscrew the adjuster plug enough to allow the drag link socket to be lifted off the ball stud.

To connect a drag link socket, you should:

1. Clean and inspect the ball stud to ensure that it is not worn out of round.

2. Place the drag link socket over the ball stud and turn the adjuster plug to tighten the socket against the ball stud.

3. Turn the steering to the right and left turn stops to check for binding; loosen the adjuster screw if necessary.

4. Install a new cotter pin and lubricate the drag link socket.

16.3   STEERING GEAR SERVICE

Most standard, manual steering gears are relatively easy to service. Normal service operations are worm shaft bearing preload and sector gear lash adjustments and steering gear rebuilding. Rebuilding steps include disassembly, cleaning and inspection, and reassembly. Service procedures for rack and pinion steering

gears are similar to standard units; however, the adjustment methods on different makes of gear units are not as standardized.

Service of standard, power steering gears follows the same general procedures as that for manual gears with these additions: a control valve with seals, outer and, sometimes, inner piston seals, and high pressure seals at the shafts and any other openings in the gear housing. Additional care must be taken to ensure that the gear housing and internal pressure areas are completely sealed during assembly to prevent leaks. Seal kits for most popular steering gears are available from the vehicle manufacturer and aftermarket sources. (Fig. 16-40)

The **gear lash** and sometimes **worm shaft bearing preload** adjustments can usually be made in the car on most standard steering gears. When doing this, one end of the Pitman arm must be disconnected to allow a better feel of the steering gear. Gear adjustments are measured by the small amount of turning effort it takes to rotate the worm shaft. Some technicians prefer to remove, disassemble, and inspect the internal parts of a steering gear before adjustment. They feel that if the gears and bearings have worn enough to require an adjustment, they might be worn far enough to cause a failure. They also point out that in the case of many manual steering gears, after the Pitman arm has been disconnected, there are

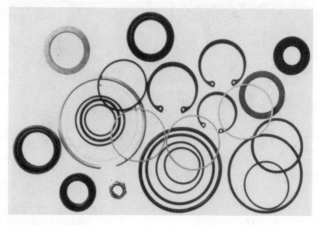

FIGURE 16-40.
A power steering gear sealing kit contains those "O" rings and seals needed to prevent fluid leaks. (Courtesy of Moog Automotive)

only the shaft coupler and a few bolts holding the steering gear in the car; steering gear removal is fairly quick from this point. If the steering gear adjustments are made with the unit in the car, care should be taken to feel for any suspicious rough or erratic turning motions that might indicate faulty bearings or worn gears.

In a few cases, rack and pinion steering gears can be adjusted in the car; in most cases, they must be removed. Adjustment procedures for rack and pinion steering gears vary too much to be described in a text of this type; the manufacturer's or technician's service manual should be consulted for the exact procedure.

16.3.1   Steering Gear Removal and Replacement, Standard Steering Gear

The removal and replacement for power steering gears is similar to that used for a manual gear with the addition of the two fluid lines. When removing and replacing a power steering gear, check the condition of the fluid and hoses. Dirty, contaminated fluid should be changed; faulty hoses or ones with internal deterioration should also be changed. As a power steering unit is replaced, the fluid level should be adjusted, any air should be bled out, and the fluid level should be rechecked. Filling and bleeding procedures are described later in this chapter.

The steering coupler should be marked before disassembly to ensure replacement in the same position; steering wheel position will be changed if the coupler is replaced improperly. Some coupler and steering shaft splines have a flat or missing master spline to ensure proper alignment. The condition of the coupler should always be checked; damaged couplers must be replaced or rebuilt. (Fig. 16–41)

To remove a standard steering gear, you should:

1. If equipped with power steering, disconnect and cap the two fluid lines to reduce fluid loss and contamination. (Fig. 16–42)

2. Mark and disconnect the steering shaft coupler. Usually the coupler can not be slid off

FIGURE 16–41.
Most steering couplers such as this one can be replaced in only one position; others, without the flat or master spline, must be marked to ensure their replacement onto the steering shaft in the proper position. (Courtesy of Oldsmobile)

the shaft until the gear is removed from the car. (Fig. 16–43)

3. Raise and support the car on a hoist or jack stands.

4. Mark the Pitman-arm-to-shaft position and remove the Pitman arm from the shaft.

5. Remove the bolts and nuts securing the gear housing to the frame and remove the steering gear from the car. The steering shaft coupler should be slid off the shaft as the steering gear is removed. (Fig. 16–44)

To replace a standard steering gear, you should:

1. Place the gear housing in the proper position in the frame and replace the mounting bolts. It is often necessary to slide the coupler

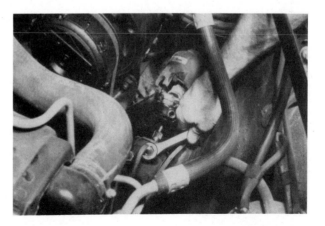

FIGURE 16–42.
When removing a power steering gear, the fluid lines are removed from the gear and capped to reduce fluid loss and contamination. (Courtesy of Moog Automotive)

FIGURE 16-43.
Before disconnecting the steering shaft coupler, check to see if there is a flat spot or master spline to provide proper positioning during replacement; if there is none, mark the shaft and coupler. (Courtesy of Moog Automotive)

onto the steering shaft at this time. Tighten the mounting bolts to the correct torque.

2. Align the Pitman arm to the shaft and replace the Pitman arm. Tighten the retaining nut to the correct torque.

3. Align the coupler to the steering shaft and slide it onto the steering shaft. Tighten the pinch-bolt to the correct torque. Note that

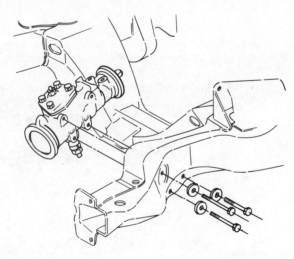

FIGURE 16-44.
Most steering gears are secured to the frame with several bolts; after disconnecting the hoses, Pitman arm, and steering shaft, these bolts are removed to allow removal of the gear. (Courtesy of Chevrolet)

aligning and installing the coupler onto the steering shaft are often done during Step 1.

4. On power steering cars, reconnect the two fluid lines, fill the system with fluid, and bleed the air from the system. After road testing, check the system for possible leaks and correct the fluid level if necessary.

16.3.2   Steering Gear Removal and Replacement, Rack and Pinion Steering Gear

Removal and replacement of a rack and pinion steering gear follows a procedure that is similar to a standard gear. Access to some rack and pinion units is very difficult, especially on some FWD cars with the rack mounted on the engine compartment bulkhead. You should follow the specific procedure outlined in the manufacturer's or technician's service manual. The description given here is very general.

To remove a rack and pinion steering gear, you should:

1. Raise and support the car on a hoist or jack stands and remove the front tires and wheels. On some cars, it is easier to do Steps 2 and 3 from under the hood before raising the car.

2. If equipped with power steering, disconnect the two fluid lines from the power steering pump and cap the lines to reduce the fluid loss. (Fig. 16-45)

3. Mark and disconnect the steering shaft coupler; on some cars, this is easier after Step 5. (Fig. 16-46)

4. Disconnect both outer tie rod ends. (Fig. 16-47)

5. Remove the bolts and nuts securing the gear housing to the frame and remove the steering gear from the car. (Fig. 16-48)

To replace a rack and pinion steering gear, you should:

1. Check the mounting bushings in the

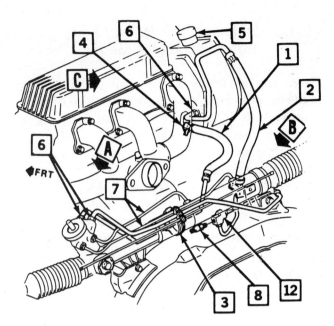

FIGURE 16-45.
Before removing a power rack and pinion steering gear, the pressure (1) and return (2) hoses should be disconnected and plugged to reduce fluid loss and contamination. (Courtesy of Moog Automotive)

gear housing and, if necessary, replace any faulty, deteriorated bushings. (Fig. 16-49)

2. Slide the steering gear into position, install the mounting bolts and nuts, and tighten them to the correct torque. On many cars, it is necessary to do Steps 2 and 3 at the same time.

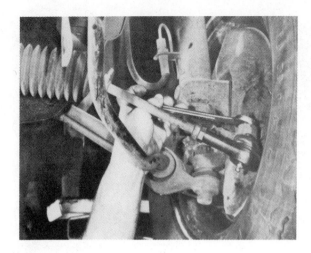

FIGURE 16-47.
The tie rods are normally disconnected from the steering arms during removal of a rack and pinion steering gear. (Courtesy of Moog Automotive)

3. Align the coupler and the steering shaft and replace the coupler. Tighten the pinch-bolt to the correct torque.

4. Reconnect the outer tie rod ends.

5. On power steering cars, reconnect the two fluid lines, fill the system with fluid, and bleed the air from the system. After road testing, check the system for any possible fluid leaks and correct the fluid level if necessary.

6. Check and adjust toe if necessary.

FIGURE 16-46.
When removing a rack and pinion steering gear, mark the steering shaft to coupler location to ensure proper positioning during replacement. (Courtesy of Moog Automotive)

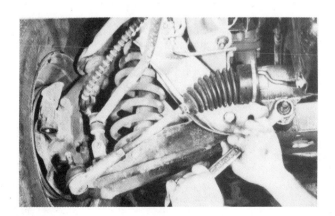

FIGURE 16-48.
After disconnecting the power steering hoses, steering shaft, and both tie rod ends, remove the mounting bolts so the steering gear can be removed. At this time check the condition of bushings at the rack mounts. (Courtesy of Moog Automotive)

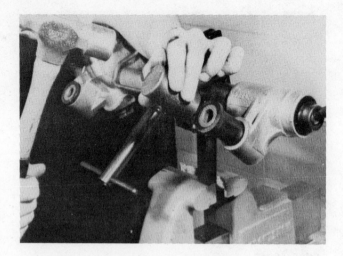

FIGURE 16-49.
Faulty rack mounting bushings can be removed and replaced with new ones. (Courtesy of Moog Automotive)

16.3.3 Steering Gear Adjustments, Standard Steering Gear

There are normally three adjustments to be made on standard steering gears. Two of these (**worm shaft bearing preload** and **sector gear lash**) are made from the outside of the gear housing, and one (**sector gear/shaft end play**) can only be checked with the sector gear removed from the unit. (Fig. 16-50)

The worm shaft bearing preload adjustment provides for easy, play-free turning of the steering shaft; it is also called **worm bearing preload**. The bearings are adjusted so there is a light pressure or preload on them at all times; this pressure is just enough to eliminate bearing and worm shaft end play. Steering shaft

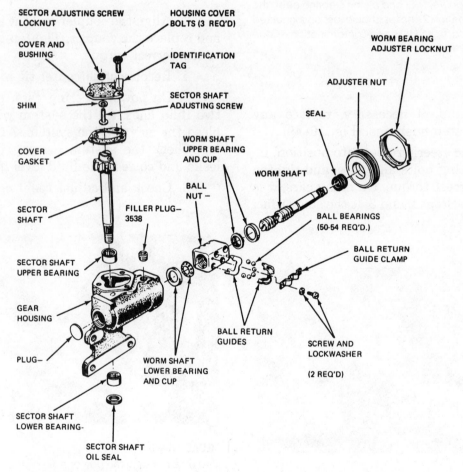

FIGURE 16-50.
An exploded view of a recirculating ball nut steering gear; all of these parts are normally removed from the gear housing during a steering gear overhaul. (Courtesy of Ford Motor Company)

FIGURE 16-51.
Steering gear preload/turning effort can be measured by attaching a spring scale to the rim of the steering wheel; pulling on the scale will show how much "rim pull" it takes to turn the steering shaft. (Courtesy of Ford Motor Company)

preload can be measured with a torque wrench and socket on the worm shaft or on the steering wheel retaining nut; it can also be measured with a spring scale, measuring the pull at the rim of the steering wheel. (Figs. 16-51 & 16-52)

The sector gear lash adjustment is also called the **over center adjustment**. This adjustment ensures that there is no lash between the worm gear (worm and sector gear set) or ball nut (recirculating ball nut gear set) and the sector gear when the steering is straight ahead. There is actually a preload, as gear lash at this time would cause sloppy steering and wander. As the steering is turned from the center position, the teeth of the sector gear swing away from the worm gear or ball nut, and lash will usually occur between the two gears. Gear lash during a turn is not critical because the wheel alignment angles are trying to push the tires back to a straightahead position; this self aligning force from the tires keeps pressure on the gears to eliminate lash. Sector gear lash, actually preload at center position, is also measured with a torque wrench. (Fig. 16-53)

Sector gear endplay is important because the sector gear tends to drop downward into the worm gear or ball nut as the steering is turned off center. If it drops too far, the gears might bind and wedge when the steering is turned back to center.

These adjustments are normally made in the following order when rebuilding a steering

FIGURE 16-52.
Steering gear preload/turning effort can also be measured with a torque wrench attached to the steering wheel retaining nut. (Courtesy of Saginaw Division, General Motors Corporation)

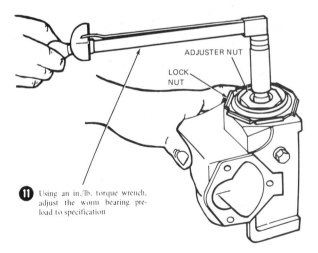

⑪ Using an in./lb. torque wrench, adjust the worm bearing preload to specification

FIGURE 16-53.
A torque wrench and socket can be attached directly to the steering shaft to measure steering shaft preload. Note that the sector gear is not in the gear box allowing the worm shaft bearings to be adjusted with no interference. Turning the adjuster nut inward will increase the amount of preload. (Courtesy of Ford Motor Company)

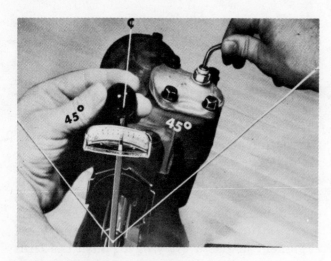

FIGURE 16-54.
When adjusting sector gear preload/gear lash, the steering shaft preload must be measured in a 90° portion of the center of the steering shaft travel; turning the adjuster screw (at Allen wrench) inward will increase the amount of gear preload. (Courtesy of Saginaw Division, General Motors Corporation)

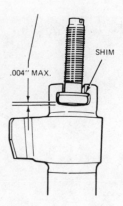

FIGURE 16-55.
If the clearance between the sector gear and the adjusting screw is excessive, the sector gear will have too much end play, and this can cause binding in the gear box. (Courtesy of Ford Motor Company)

gear: (1) **sector gear end play**—as the sector gear is removed, (2) **worm shaft bearing preload**—as the worm shaft and bearings are being installed into the gear housing, and (3) **sector gear lash**—after the gear cover has been replaced. If these adjustments are made with the gear in the car, the Pitman arm must be disconnected. A careful check of the turning feel of the gear set should also be made to ensure that the internal gears and bearings are in good shape. Check a service manual for the adjustment procedure and the specifications for a particular car. (Fig. 16-54)

Measuring Sector Shaft End Play

To measure sector gear end play, you should:

1. Remove the sector gear/shaft and adjuster screw from the cover by turning the sector gear adjusting screw all the way inward.

2. Try to slide a 0.002 inch (0.05 mm) feeler gauge between the head of the adjuster screw and the bottom of the slot; it should not enter. If it does not enter and the screw rotates freely, no change is required. If the clearance is greater than 0.002 inch, a thicker shim is required. A

thinner shim is required if the screw does not rotate freely. (Fig. 16-55)

Adjusting Worm Shaft Bearing Preload

To measure and adjust worm shaft bearing preload, you should:

1. Select the smallest 12-point socket that will slide over the splines of the worm shaft, and with a low reading torque wrench, measure the torque required to turn the worm shaft. Note that if the steering gear is assembled, the sector gear lash adjuster should be turned completely out or up to eliminate any possible drag between the two gears. If an excessive amount of slippage occurs between the socket and the shaft splines, a tighter fit can be obtained by wrapping a strip of cardboard around the shaft and forcing the socket over the shaft and cardboard.

2. Obtain the worm shaft bearing preload specification and compare it to your readings. If specifications are not available, a rule of thumb is 5 inch-pounds (0.5 N·M) of torque.

NOTE: As a general rule, the average technician can turn a worm shaft with a thumb and forefinger and notice a definite turning resistance.

3. If an adjustment is required, loosen the lock nut for the adjuster plug and rotate the

adjuster plug inward to increase preload or outward to reduce it. When the effort to turn the worm shaft matches the specifications, tighten the adjuster lock nut.

Adjusting Sector Gear Lash

To measure and adjust sector gear lash, you should:

1. Rotate the steering shaft one direction until it stops. Next, rotate it the opposite way, counting the number of turns it takes to reach the other stop; divide this distance by two to determine the center of travel for the steering gear.

2. Turn the shaft one turn from either end stop and, using a low-reading torque wrench, measure the steering shaft preload. Record this measurement.

3. Turn the steering shaft to the centermost position, and remeasure the steering shaft preload. The difference between these two measurements is the sector gear lash measurement.

4. Obtain the sector gear lash specifications and compare them to your measurements. If specifications are not available, a rule of thumb is 10 inch-pounds (1 N•M) total or 5 inch-pounds (0.5 N•M) greater than the amount of steering shaft bearing preload.

NOTE: *As a general rule, the average technician must use both thumbs and forefingers to exert 10 inch-pounds of torque and rotate the steering shaft as it passes through the center of travel.*

5. If an adjustment is required, loosen the adjuster lock nut and turn the adjuster screw inward to increase the turning effort and outward to decrease it. Remember that the turning effort should only be measured at the centermost position; if you check it, you will notice that the turning effort drops off when the steering shaft is turned about one-half turn off center.

6. When lash is adjusted, tighten the adjuster lock to the correct torque.

16.3.4 Steering Gear Overhaul, Standard Steering Gear

If the steering shaft turns with a rough or erratic motion or there is free play that cannot be adjusted out, the steering gear should be replaced or overhauled. Most of the internal steering gear parts are available from the O.E.M. manufacturer or aftermarket sources. Many steering gears have a metal identification tag under one of the bolts or a number stamped into the gear housing or cover. This identification number should be used when ordering replacement parts. The procedure and specifications recommended by the manufacturer should be followed while doing the operations.

To disassemble a standard steering gear, you should:

1. Clamp a mounting ear of the gear housing in the jaws of a vise.

2. Remove the bolts securing the sector gear cover to the gear housing and remove the gear cover with the sector gear/Pitman shaft from the gear housing. The Pitman arm must be off the Pitman shaft. (Fig. 16–56)

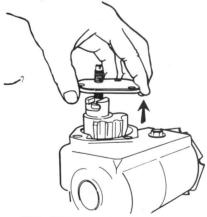

NOTE: GEAR WILL NORMALLY BE FILLED WITH GREASE.

FIGURE 16–56.
The first step in disassembling a standard steering gear is to remove the sector cover retaining bolts and lift the sector cover and gear out of the gear housing; turning the adjuster all the way inward will allow removal of the gear from the cover. (Courtesy of Ford Motor Company)

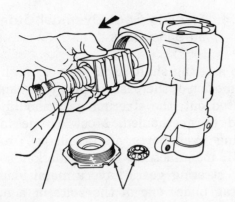

FIGURE 16–57.
After removing the worm shaft bearing adjuster, the worm shaft and ball nut assembly can be removed from the housing. Note the angle of the teeth on the ball nut as it is removed. (Courtesy of Ford Motor Company)

3. Loosen the lock nut, and thread the sector gear adjustment screw into and through the gear cover to allow removal of the sector gear from the cover. Measure the sector shaft end play.

4. Loosen the lock nut and remove the worm shaft bearing adjuster plug. The worm shaft, with ball nut and worm shaft bearings, is removed along with the adjuster plug. (Fig. 16–57)

5. Remove the ball guides from the ball nut and remove the balls to allow separation of the ball nut from the worm shaft. It is recommended that you do this operation over a shop rag or carpet section to help prevent loss of the balls. (Fig. 16–58)

6. Clean the parts in solvent and dry them with air to allow inspection. Inspection of the parts should include:

  a. worm shaft and Pitman shaft bearings,

  b. worm shaft and Pitman shaft bearing surfaces,

  c. worm shaft and Pitman shaft seals (should be replaced),

  d. ball nut and steering shaft grooves.
All the damaged parts should be replaced.

To assemble a standard steering gear:

1. As the steering gear is assembled, the parts should be lubricated with the same lubri-

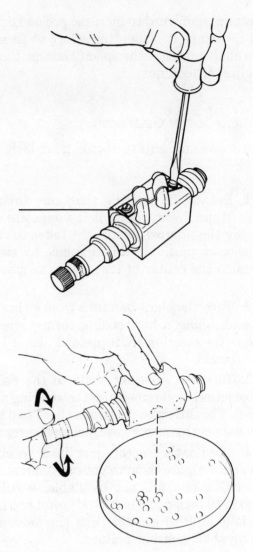

FIGURE 16–58.
To remove the ball nut from the worm shaft, first remove the clamp for the ball return guides and the return guides. Turn the assembly upside down over a container or shop cloth as you rotate the worm shaft to work all of the balls out of the grooves. (Courtesy of Ford Motor Company)

cant that will be used in the unit; this will usually be gear oil, power steering fluid, or ATF. Place the worm shaft into the ball nut making sure that the ball nut is in the correct position (note the teeth on the ball nut). Hold the steering shaft centered in the ball nut and insert the balls, one at a time, into the openings of the ball grooves. It is recommended that the balls be installed at only one end of each groove. Push the balls inward with a heavy wire or small screwdriver or rotate the shaft slightly to help

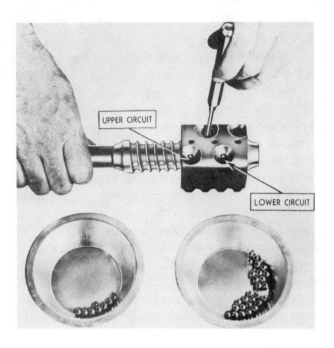

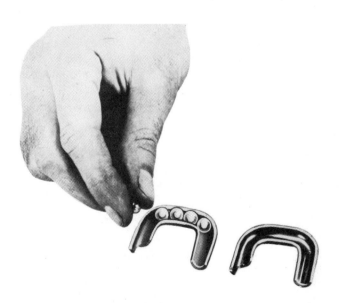

the balls enter. When the grooves are almost full (push a ball in one end, and the ball at the other end should lift), divide the remaining balls into two groups. Some ball guides have loading slots; in this case, place the ball guides in position and drop the remaining balls into the slots in the guides. If the ball guides do not have loading slots, use petroleum jelly to stick a group of balls into each of the ball guides and place the ball guides, with balls, into the ball nut. Secure the ball guides to the ball nut. (Fig. 16–59)

2. Test the ball-nut-to-worm-shaft fit by holding the ball nut so the worm shaft is vertical; gravity should cause the worm shaft to rotate smoothly downward. Reverse the nut position, and, again, the steering shaft should rotate downward.

3. Place the worm shaft with one of the bearings into the gear housing, making sure the bearing and races are completely seated and the ball nut is positioned correctly. Position the remaining bearing and install the adjuster plug. Adjust the worm shaft bearing preload.

4. Connect the sector gear to the gear cover by threading the adjuster screw through the gear cover until the gear is almost against the cover.

5. Place the cover gasket in position, and place the sector gear/shaft into the housing so the two gears are meshed correctly. Install and tighten the cover retaining bolts to the correct torque.

6. With the steering shaft in the centermost position, thread the sector gear adjusting screw into the housing until you feel the sector

FIGURE 16–59.

When assembling a worm shaft and ball nut, first center the worm shaft in the ball nut, and working from one end of each circuit, fill both circuits with balls. A punch is being used to help work the balls into the grooves. When the circuits are almost full, the remaining balls can be dropped into the opening in the ball guides or, if there is no opening, loaded in the guides; petroleum jelly can be used to hold the balls in the ball guides. (Courtesy of Saginaw Division, General Motors Corporation)

gear contact the ball nut; from this point, complete the sector gear lash adjustment.

7. After installation of the steering gear in the car, fill the housing with the correct lubricant; this is usually a gear oil or a semifluid gear grease.

16.3.5 Steering Gear Adjustments, Rack and Pinion Steering Gear

As mentioned earlier, service on rack and pinion steering gears varies among manufacturers. Some manual units are nonserviceable and nonadjustable; if they become faulty, they should be replaced with a new or rebuilt unit. Most rack and pinion steering gears can be serviced, and the overhaul procedure of the various units is similar. It is described in the next section. Adjustment of the different units varies; most units have a rack support bearing adjustment; some have a pinion bearing adjustment. A manufacturer's or technician's service manual should always be followed when adjusting rack and pinion steering gears. These

adjustments are usually made with the unit off the car. (Fig. 16-60)

Most modern rack and pinion steering gears use an adjustable rack support opposite to the pinion gear; the purpose is to hold the two gears together to eliminate gear lash. The **rack support** is also called a **support yoke, rack bearing, rack plunger, friction plunger,** or **pressure pad** by the different manufacturers. A spring is usually used in the rack support to maintain a constant pressure. A common method of adjusting the rack support is to thread the adjuster plug inward to a specific torque or until it bottoms and then back it off a specified amount. The adjustment is then checked by measuring the rotating torque of the pinion shaft and comparing it to specifications; a torque wrench and socket or spring scale and adapter are used to measure the rotating torque. (Fig. 16-61)

A few makes of rack and pinion steering gears require an adjustment of the pinion bearings. When an adjustment is required, there will be a shim pack or threaded adjuster at the bearing retainer. (Fig. 16-62)

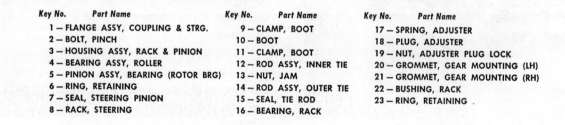

Key No.	Part Name	Key No.	Part Name	Key No.	Part Name
1 — FLANGE ASSY, COUPLING & STRG.		9 — CLAMP, BOOT		17 — SPRING, ADJUSTER	
2 — BOLT, PINCH		10 — BOOT		18 — PLUG, ADJUSTER	
3 — HOUSING ASSY, RACK & PINION		11 — CLAMP, BOOT		19 — NUT, ADJUSTER PLUG LOCK	
4 — BEARING ASSY, ROLLER		12 — ROD ASSY, INNER TIE		20 — GROMMET, GEAR MOUNTING (LH)	
5 — PINION ASSY, BEARING (ROTOR BRG)		13 — NUT, JAM		21 — GROMMET, GEAR MOUNTING (RH)	
6 — RING, RETAINING		14 — ROD ASSY, OUTER TIE		22 — BUSHING, RACK	
7 — SEAL, STEERING PINION		15 — SEAL, TIE ROD		23 — RING, RETAINING .	
8 — RACK, STEERING		16 — BEARING, RACK			

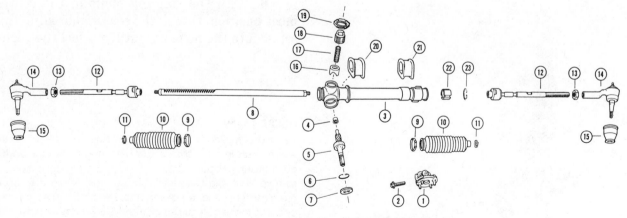

FIGURE 16-60.

An exploded view of a manual rack and pinion steering gear. (Courtesy of Saginaw Division, General Motors Corporation)

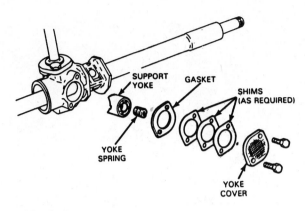

FIGURE 16–61.
Most rack and pinion gears have an adjustable support yoke/rack bearing; removing shims will increase the pressure between the support yoke and the rack on this one. (Courtesy of Ford Motor Company)

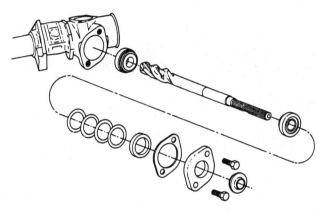

FIGURE 16–62.
Some rack and pinion steering gears have adjustable pinion shaft bearings; removing shims will decrease bearing preload on this one. (Courtesy of Ford Motor Company)

16.3.6 Steering Gear Overhaul, Rack and Pinion Steering Gear

Most rack and pinion steering gears can be overhauled rather easily; the serviceable portions are pinion bearings and seals, the rack support, and a rack housing bushing. Lubrication of these units is usually accomplished by packing the parts with grease during assembly or by pouring the correct amount of gear oil into the housing before installing the second bellows. When overhauling a particular unit, check the service manual for the exact repair, adjustment, and lubrication procedure. (Fig. 16–63)

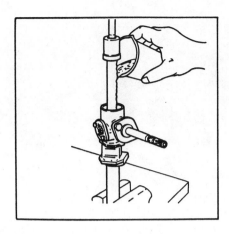

FIGURE 16–63.
Some manual rack and pinion steering gears are lubricated by pouring gear oil into the gear housing; make sure the boot is installed at the other end of the housing before pouring the oil in. (Courtesy of Ford Motor Company)

To disassemble a manual rack and pinion steering gear, you should:

1. Support the unit in a rack and pinion holding fixture or a standard shop vise by gripping a mounting ear. (Fig. 16–64)

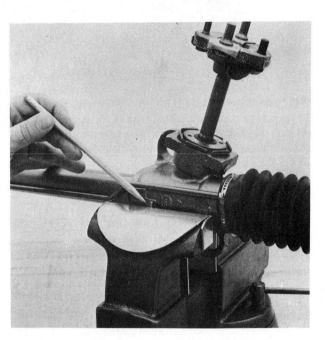

FIGURE 16–64.
The rack housing should be securely supported before beginning disassembly; this part of this particular gear housing is strong enough to be clamped is a shop vise. (Courtesy of Saginaw Division, General Motors Corporation)

FIGURE 16-65.
Disassembly of a rack and pinion steering gear often begins with the removal of the pinion shaft bearings and pinion gear. (Courtesy of Saginaw Division, General Motors Corporation)

2. Remove the bellows and tie rod ends from the ends of the rack gear and housing; on some units, the tie rod end at the end opposite to the gear teeth can be left in place.

3. Remove the retaining bolts or locking ring for the pinion bearing and remove the pinion retainer, the pinion shaft and gear, and the bearing and seal. (Fig. 16-65)

4. Remove the rack support retainer and the rack support assembly. (Fig. 16-66)

5. Slide the rack gear out of the housing.

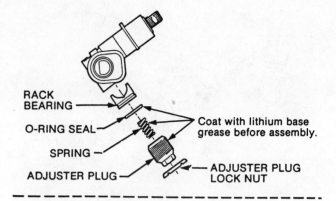

RACK BEARING
O-RING SEAL
SPRING
ADJUSTER PLUG
Coat with lithium base grease before assembly.
ADJUSTER PLUG LOCK NUT

NOTICE: Due to tolerances, some sockets will require a wrapping of card stock around the pinion serrations to make a tight enough fit.

Pinion torque — 0.9 to 2.3 N·m (8 In. Lbs. to 20 In. Lbs.)

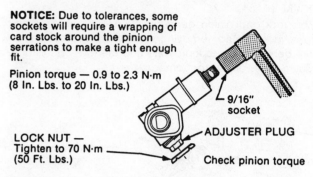

9/16" socket
ADJUSTER PLUG
Check pinion torque

LOCK NUT — Tighten to 70 N·m (50 Ft. Lbs.)

FIGURE 16-66.
After the rack support and the pinion gear have been removed, the rack can be slid out of the gear housing. Note the method of adjusting the rack bearing during reassembly. (Courtesy of Pontiac Division, GMC)

6. Wash the parts in solvent and dry them using shop air. Inspection of the parts should include:
   a. pinion bearings and seals,
   b. rack and pinion gear teeth,
   c. rack support,
   d. rack housing bushing.

All the damaged parts, seals, and rack bellows should be replaced.

To assemble a manual rack and pinion steering gear, you should:

1. Slide the rack gear into the gear housing with the rack teeth positioned to the pinion gear location. Don't forget to lubricate the parts, if required.

2. Install the pinion gear and shaft assembly and tighten the retaining bolts to the correct torque. A pinion bearing adjustment is required on some cars.

3. Install and adjust the rack support assembly.

4. Check the pinion shaft turning effort and compare it to specifications. If specifications are not available, pinion shaft rotation should be smooth, not harsh or erratic, and require about 10 inch-pounds (1 N·M) of torque. If the effort is too high or too low, readjust the rack support; when it is correct, tighten the adjuster lock to the correct torque.

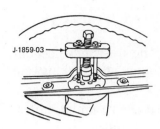

FIGURE 16–67.
Most steering wheels are drilled and tapped at the hub so a puller can be attached for pulling the steering wheel off of the steering shaft. (Courtesy of Pontiac Division, GMC)

SAFETY NOTE

A rack gear will tend to wear more in the center than at the ends; after making an adjustment, check the gear travel from lock to lock to ensure that there is no binding.

5. Replace the tie rod ends and bellows; do not tighten the small bellows clamp until toe is checked and adjusted.

16.4 REMOVING AND REPLACING A STEERING WHEEL

Most steering wheels are indexed to the steering shaft by a mark on the steering wheel hub and another mark on the end of the shaft. Look for these marks, often at the top of the hub when the steering wheel is in the centered position. It is a good practice to disconnect the wire at the horns or horn relay before beginning the removal procedure.

To remove a steering wheel, you should:

1. Remove the horn ring to provide access to the steering wheel retaining nut. The ring is often retained by two or three screws entering from the back side or a spring clip.

2. Locate the steering wheel index marks; if you cannot find any, use a scratch awl or scribe to make your own.

3. Loosen and remove the steering wheel retaining nut.

4. Install the bolts for a puller into the threaded holes in the steering wheel hub and tighten the puller to pull the steering wheel off the shaft. (Fig. 16–67)

OR Tighten the steering wheel retaining nut one-fourth to one-half turn to unseat the splines and then remove the nut. Grip the steering wheel with one hand at each side or at the top and bottom, and pull firmly on the wheel rim in an alternating fashion, first on one side, then the other. The steering wheel should creep off the shaft. If not, rotate the steering wheel one-fourth turn and repeat the alternating pulling action. (Fig. 16–68)

FIGURE 16–68.
A steering wheel can often be removed by gripping the rim at each side and then pulling in an alternating manner; first though, the nut should be tightened slightly to break the splines loose before removing it.

To install a steering wheel, you should:

1. Slide the steering wheel onto the splines of the steering shaft making sure the index marks are aligned.

2. Install the retaining nut and tighten it to the correct torque.

3. Install the horn ring and reconnect the wire for the horns, if necessary.

16.5 POWER STEERING SERVICE OPERATIONS

Normal power steering maintenance operations include fluid level and condition checks, drive belt tension and condition checks, and inspection of the rest of the system to ensure that there are no leaks and the pump mounts are good. In addition to this, there are also diagnostic, problem-solving operations when something goes wrong with the system and repair operations when the cause of the problem is determined.

The most commonly occurring complaints with power steering systems are hard steering, loose or rough steering, noise, and leaks. Some of these problems, such as leaks, are often easy to diagnose; some are more difficult and require a systematic trouble-shooting procedure. For example, hard steering can be caused by a number of things; the most probable faults are a faulty pump or steering gear. Most experienced front end technicians will still follow a diagnostic chart such as the one in Table 16-1 to help them locate the cause of the problem and ensure that they do not overlook or forget some possibilities. Don't forget that some of these problems can be caused by components outside of the steering gear and pump; they might be caused by steering linkage or suspension components.

16.5.1 Power Steering System Checking/Problem Diagnosis

Power steering problem diagnosis begins with a thorough inspection of the system. It is a

TABLE 16-1
The Most Probable Causes of Poor Power Steering System Performance

Symptom	Possible Cause(s)
Hard steering	Low power steering fluid level Low tire pressure Slipping drive belt or pulley Binding in steering linkage Restricted hose Low pump output pressure Overly tight steering gear adjustments Sticking flow control valve in pump Internal leak in steering gear Binding in steering column
Hard steering —parked	All of the above Low engine idle speed
Poor steering wheel return	Low tire pressure Binding in steering linkage Tight steering gear Incorrect wheel alignment Binding in steering column
Hard steering in one direction	Defective seals in gear valve assembly Defective seal in rack housing
Excessive free play	Worn or loose wheel bearings Worn steering linkage Improper gear adjustments Worn gear
Pulls to one direction	Unequal tire pressures Incorrect wheel alignment Worn control valve in gear Improperly adjusted gear control valve Brake drag or pull
Noise	Air in hydraulic system Restricted hose Defective pump mount Improper pump control valve operation Slipping drive belt

good practice not only to cure the immediate problem, but also to make sure the system will operate trouble-free for a reasonable period of time. Begin at the pump and then, depending on the nature of the problem, check out the entire system.

To check out a power steering system, you should:

1. Check the condition of the drive belt; if it is worn, badly cracked or frayed, or has very shiny or glazed sides, the belt should be replaced. Some sources recommend a replacement of this belt every four years.

2. Check the tension of the belt; if it is too loose, readjust it. The best way to measure belt tension is with a tension gauge, comparing the reading to the specifications. Most power steering belts will use a tension of about 70 pounds (32 Kg). If a tension gauge is not available, push downward at the center of the belt with a pressure of about 5 pounds (2.25 Kg); the belt should deflect or move about 1/16 inch (1.5 mm) if it is new or 1/8 inch (3 mm) if it is used. (Fig. 16–69)

3. Readjust the drive belt, if necessary, by loosening the pivot and adjusting bolts and swinging the pump on the pivot bolt. Many pump brackets and mounts have provision for

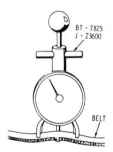

BELT TENSION

	5/16 WIDE	3/8 WIDE	15/32 WIDE
NEW BELT	350 N Max. 80 Lbs. Max.	620 N Max. 140 Lbs. Max.	750 N Max. 165 Lbs. Max.
USED BELT	200 N Min. 50 Lbs. Min.	300 N Min. 70 Lbs. Min.	400 N Min. 90 Lbs. Min.
USED COGGED BELT		250 N Min. 60 Lbs. Min.	

FIGURE 16–69.
Power steering belt tension should be checked with a belt tension gauge; if the tension is different from the specifications, the belt should be readjusted. (Courtesy of Chevrolet)

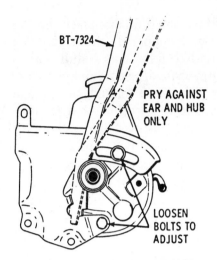

PULLEY REMOVED FOR PHOTO PURPOSE

FIGURE 16–70.
This power steering pump bracket is constructed so a pry bar can be inserted and used to pry outward when adjusting belt tension. Some brackets are made with wrench flats or square openings to allow a wrench to be used for the same purpose. Never pry against the pump reservoir when adjusting belt tension. (Courtesy of Chevrolet)

attaching a wrench or socket handle to aid in moving the pump; do not pry on the pump reservoir. (Fig. 16–70)

4. Before checking the fluid level, start the engine and turn the steering almost to the right and left stops several times. Shut off the engine and remove the dipstick. Don't forget the full level is for hot fluid, about 170° F (77° C). If necessary, add fluid and repeat this step. (Fig. 16–71)

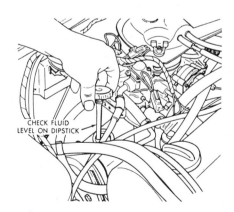

CHECK FLUID LEVEL ON DIPSTICK

FIGURE 16–71.
Power steering fluid level is usually checked at the dipstick attached to the reservoir cap. (Courtesy of Chrysler Corporation)

5. Smell the fluid and place some on a piece of clean, white paper. If the fluid has a burned or varnish-like odor, has a dull, very dark, or brown appearance, or contains particles of hose or other debris, it should be changed. Also check for fluid aeration that causes foamy, milky fluid. The procedure to change the fluid and bleed air out is described in Section 16.5.2.

6. Check the system for leaks at the pump, hoses, and gear. A leak in the high-pressure areas can be made to show up better by starting the engine and holding the steering wheel briefly against the right or left turn stops. This will raise the pump output pressure to the setting of the relief valve. (Fig. 16–72)

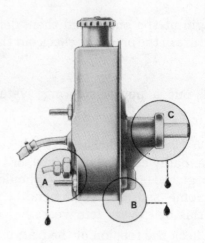

SAFETY NOTE

DO NOT hold the steering wheel against the stops longer than 4 or 5 seconds; fluid aeration and overheating or pump damage can occur.

---

7. Check the hoses for heavy rubbing or abrasion, excessive rubber cracking, excessive swelling, soft, spongy areas, damaged metal ends, or other signs of deterioration.

8. Replace any damaged hoses. After replacement, fill the system with fluid, bleed the air out, and check for leaks. New "O" rings should be used on hose ends that use "O" ring seals. If leaks occur at the hose connections sealed by a double-flare fitting, loosen the tube nut, rotate the tubing slightly to reseat it with the brass seat, and retighten the tube nut to the correct torque. (Figs. 16–73 & 16–74)

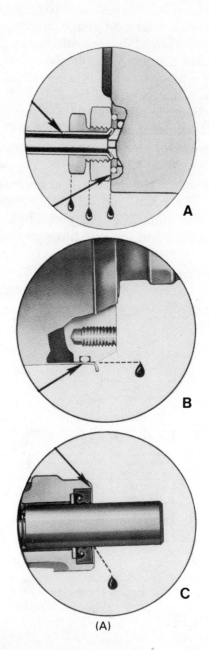

FIGURE 16–72.
The normally encountered power steering leakage areas of a standard steering gear, a pump, and a rack and pinion steering gear. (A and B are courtesy of Saginaw Division, General Motors Corporation; C [see page 438] is courtesy of Ford Motor Company)

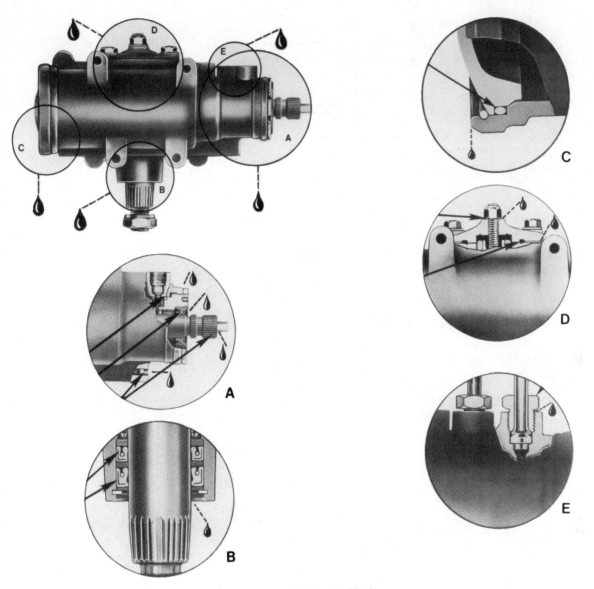

FIGURE 16-72 (B)

9. If the complaint is hard steering and if the problem has not been corrected at this point, check the pump and system operating pressures. Power steering pressure gauge sets are available for testing purposes. The pressure checking procedure is described in Section 16.5.3. (Fig. 16-75)

16.5.2 Power Steering Fluid Change and Air Bleeding

Contaminated power steering fluid should be changed. During a fluid change or when fluid, hoses, or any other part of the system is replaced, air can enter the system. Air causes a moaning or groaning noise or an erratic steering feel. Normally, air in the system will be caught by the fluid flow and be brought to the reservoir where it will escape; occasionally air will get trapped inside the power steering gear.

EXTERNAL LEAKAGE

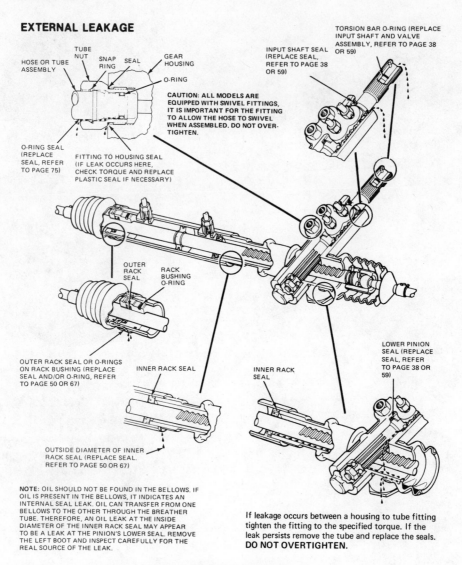

CAUTION: ALL MODELS ARE EQUIPPED WITH SWIVEL FITTINGS, IT IS IMPORTANT FOR THE FITTING TO ALLOW THE HOSE TO SWIVEL WHEN ASSEMBLED. DO NOT OVER-TIGHTEN.

HOSE OR TUBE ASSEMBLY
TUBE NUT
SNAP RING
SEAL
GEAR HOUSING
O-RING
O-RING SEAL (REPLACE SEAL, REFER TO PAGE 75)
FITTING TO HOUSING SEAL (IF LEAK OCCURS HERE, CHECK TORQUE AND REPLACE PLASTIC SEAL IF NECESSARY)

TORSION BAR O-RING (REPLACE INPUT SHAFT AND VALVE ASSEMBLY, REFER TO PAGE 38 OR 59)
INPUT SHAFT SEAL (REPLACE SEAL, REFER TO PAGE 38 OR 59)

OUTER RACK SEAL
RACK BUSHING O-RING

LOWER PINION SEAL (REPLACE SEAL, REFER TO PAGE 38 OR 59)

OUTER RACK SEAL OR O-RINGS ON RACK BUSHING (REPLACE SEAL AND/OR O-RING, REFER TO PAGE 50 OR 67)

INNER RACK SEAL

INNER RACK SEAL

OUTSIDE DIAMETER OF INNER RACK SEAL (REPLACE SEAL. REFER TO PAGE 50 OR 67)

NOTE: OIL SHOULD NOT BE FOUND IN THE BELLOWS. IF OIL IS PRESENT IN THE BELLOWS, IT INDICATES AN INTERNAL SEAL LEAK. OIL CAN TRANSFER FROM ONE BELLOWS TO THE OTHER THROUGH THE BREATHER TUBE. THEREFORE, AN OIL LEAK AT THE INSIDE DIAMETER OF THE INNER RACK SEAL MAY APPEAR TO BE A LEAK AT THE PINION'S LOWER SEAL. REMOVE THE LEFT BOOT AND INSPECT CAREFULLY FOR THE REAL SOURCE OF THE LEAK.

If leakage occurs between a housing to tube fitting tighten the fitting to the specified torque. If the leak persists remove the tube and replace the seals. **DO NOT OVERTIGHTEN.**

FIGURE 16-72 (C)

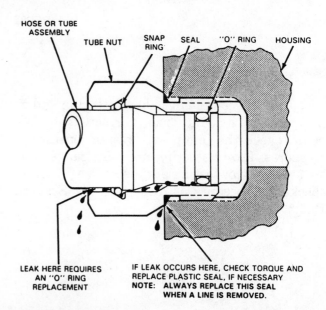

HOSE OR TUBE ASSEMBLY
TUBE NUT
SNAP RING
SEAL
"O" RING
HOUSING

LEAK HERE REQUIRES AN "O" RING REPLACEMENT

IF LEAK OCCURS HERE, CHECK TORQUE AND REPLACE PLASTIC SEAL, IF NECESSARY
NOTE: ALWAYS REPLACE THIS SEAL WHEN A LINE IS REMOVED.

FIGURE 16-73.
A quick connect type of hose fitting uses an "O" ring and a plastic seal to prevent fluid leakage. (Courtesy of Ford Motor Company)

FIGURE 16–74.
Most power steering pressure hoses are sealed by either a flare at the end of the tubing or an "O" ring at the connections. (Courtesy of Moog Automotive)

To change power steering fluid, you should:

1. Remove the return hose from the pump and place it in a container. Be ready for a fluid loss from the hose and reservoir as the hose is removed.

2. After the fluid has drained from the reservoir opening, cap the opening.

3. Refill the reservoir, start the engine, observe the flow from the hose, and be ready to add additional new fluid into the reservoir. New fluid should be added at the same rate that it is pumped out; the reservoir should not be allowed to become empty.

NOTE: It is better to change fluid by cranking the engine rather than running it. This requires a good battery, disconnecting the ignition coil, and cranking the engine in 10-second bursts to prevent overheating the starter.

4. When all of the old fluid is gone and the fluid flow appears new, stop the engine, reconnect the return hose, and fill the reservoir to the correct level with new fluid.

To bleed air from a system, you should:

1. Start the engine and slowly turn the steering wheel to the right and left turn stops; do not hold the steering wheel at the stops. Some technicians prefer to lift the front tires partially off the ground while doing this to remove the steering load and strain from the

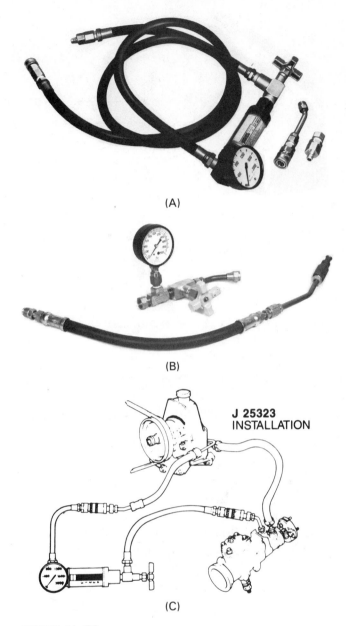

FIGURE 16–75.
A and B are two different styles of power steering testers; unit A can measure flow in addition to pressure; either unit is used by installing the gauge in the pressure line between the pump and the gear. (Courtesy of Kent-Moore)

steering linkage and gear and to reduce scrubbing of the tires.

2. If the car is equipped with a hydro-boost power brake unit, depress the brake pedal several times while the steering wheel is being turned.

3. Check the fluid condition; if it is excessively foamy or aerated, let the system sit with

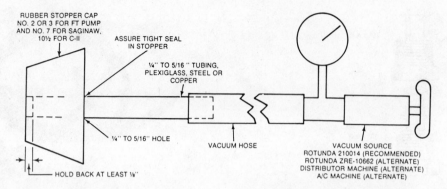

FIGURE 16-76.
A device such as this can be constructed so a vacuum pump can be attached to the power steering pump reservoir; running the pump with a vacuum will help remove trapped air. Air bubbles can cause a moan or whine. (Courtesy of Ford Motor Company)

the engine off for several minutes to allow the foam to dissipate. In extreme cases, it should sit for several hours.

4. One manufacturer recommends using a vacuum pump to remove air bubbles in extreme cases where the moaning noises persist. A stopper is connected to the pump reservoir filler neck, and a vacuum pump is connected to the stopper. A vacuum of 15 inches (51 kPa) is then applied to the pump reservoir while the engine is running at idle speeds for two or three minutes. The system is then stopped, the fluid level checked, and fluid is added, if necessary. Next, a 15 inch vacuum is reapplied, and the system is operated at idle speed for about five minutes while the steering wheel is slowly turned to the right and left turn stops. (Fig. 16-76)

16.5.3   Pressure Testing a Power Steering System

Power steering system pressure gauge sets usually include a 0–2,000 psi (0–13,790 kPa) pressure gauge, a shut-off valve, and a series of adapters to connect to the fluid ports and hose ends. Some testers include a gauge for measuring the amount of fluid flow. Pressure specifications are available for most systems in the manufacturer's or a technician's service manual. These specifications are for fluid temperatures of about 150 to 170° F (65 to 77° C); if

the system being tested is colder than this, warm the fluid by operating it a few minutes with the valve on the tester partially closed to provide a pressure of about 350 psi (2,400 kPa).

To pressure test a power steering system, you should:

1. Connect the test unit into the pressure line between the pump and the steering gear. Usually the most accessible end of the pressure hose is disconnected, the tester is connected to the port where the hose was, and the hose is connected to the tester. (Fig. 16-77)

2. With the tester valve open, start the engine and let it run at idle speed. Observe the system pressure; with the valve open, it should be about 30 to 80 psi (200 to 550 kPa). If the

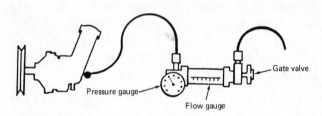

FIGURE 16-77.
Four pressure checks are made using a flow/pressure type power steering analyzer: Test 1 is made at idle speed with the gate valve fully open and flow and pressure are noted. Test 2 is made at idle speed with the valve closed far enough to generate 750 psi and flow is noted. Test 3 is made at idle speed with the valve closed and opened three times and the pressure noted each time. Test 4 is made at 1,500 rpm with the valve fully open and the flow is noted. The pressure and flow from these tests are compared with specifications to determine if there is a problem in the pressure relief valve, the steering gear, or the pump. (Courtesy of Ford Motor Company)

pressure is above 200 psi (1,380 kPa), check for restrictions in the hoses or steering gear valve. Many systems do not specify a pressure for this part of the test.

3. Close the tester valve completely, observe the pressure, and reopen the valve. *Do not keep the valve closed longer than four or five seconds.* The pressure should raise to within 50 psi (345 kPa) of the specifications; specified pressures are normally between 600 to 1,300 psi (4,100 to 8,950 kPa). If the pressure is low, a faulty pump or pump relief valve is indicated. A faulty relief valve is indicated if the pressure is too high. (Fig. 16–77)

4. Repeat Step 3 two more times, and compare the readings. A faulty flow control valve is indicated if the three pressures are not within 50 psi (345 kPa) of each other.

5. If the readings in Steps 3 and 4 indicate problems, remove and clean the flow control valve; if the fluid contains debris, change the fluid. Repeat Steps 3 and 4.

6. Open the valve completely and observe the pressure while you turn the steering wheel to each stop. The pressure while the steering wheel is at the stop should be at the same maximum pressure as in Step 3 or 4; if it is lower, the steering gear has an internal leak.

If the system pressures are correct with the steering wheel turned to each stop, the pump, flow and pressure control valve, and the internal seals of the steering gear and valve are good.

16.5.4 Power Steering Pump Service

There is a variety of power steering pumps made by the various manufacturers; these units are similar in design. The commonly performed power steering pump service operations are control valve cleaning, leak repair, and pump rebuilding. Most of these service operations are done with the pump off the car.

Cleaning a Control Valve

On some pump designs and installations, the control valve can be removed and replaced with the pump mounted on the engine. Most pumps mount the valve under the fluid outlet fitting where the high-pressure line connects.

To remove a control valve assembly, you should:

1. Remove the high pressure hose from the pump.

2. Remove the outlet fitting from the pump.

3. Use a narrow magnet to lift the control valve out of the bore, and also the control valve spring. (Fig. 16–78)

4. Some control valves can be disassembled for cleaning. Carefully clamp the valve in a soft-jawed vise and unscrew the hex-headed relief ball seat. Remove the shims and other internal valve parts and clean them. The shims are used to adjust pump relief pressure. After cleaning, reassemble the control valve assembly. (Fig. 16–79)

To replace a control valve assembly, you should:

1. Slide the spring for the control valve into the bore.

2. Lubricate the control valve with power steering fluid or ATF and slide it into the bore. Make sure the valve is not cocked by pushing inward on the valve; you should be able to feel it move against the spring pressure.

3. Replace the outlet fitting and tighten it to the correct torque. Replace the hose.

Removing and Replacing Pump Seals

The most commonly encountered pump leak locations are at the hose connections, at the reservoir to pump body connection, around the reservoir retaining bolts, at the filler cap, and at the front shaft seal. Seal and "O" ring kits are available from the vehicle manufacturer and aftermarket suppliers. (Fig. 16–80)

To repair power steering pump leaks, you should:

1. Disconnect the hoses and remove the pump from the engine.

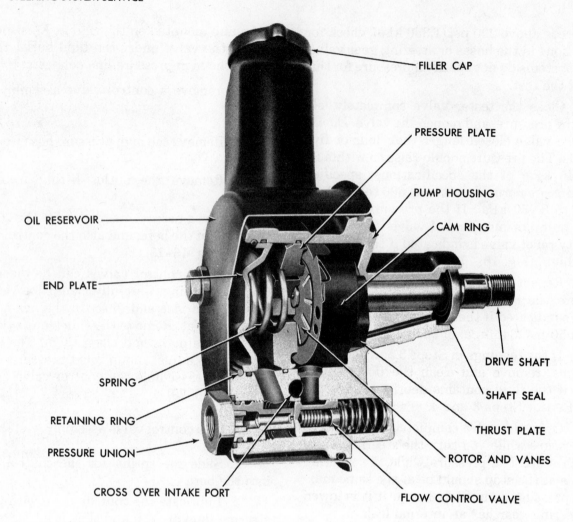

FILLER CAP

PRESSURE PLATE

PUMP HOUSING

CAM RING

OIL RESERVOIR

END PLATE

SPRING

RETAINING RING

PRESSURE UNION

CROSS OVER INTAKE PORT

DRIVE SHAFT

SHAFT SEAL

THRUST PLATE

ROTOR AND VANES

FLOW CONTROL VALVE

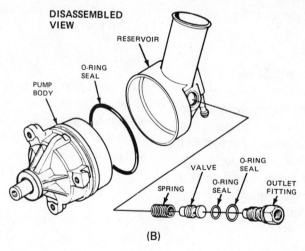

DISASSEMBLED
VIEW

RESERVOIR

PUMP
BODY

O-RING
SEAL

SPRING

VALVE

O-RING
SEAL

O-RING
SEAL

OUTLET
FITTING

(B)

FIGURE 16-78.
The control valve is normally held in place by the pump's outlet fitting/pressure
union. (A is courtesy of Chevrolet; B is courtesy of Ford Motor Company)

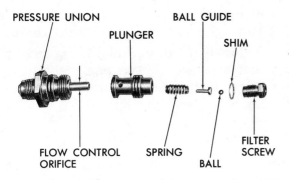

PRESSURE UNION

PLUNGER

BALL GUIDE

SHIM

FLOW CONTROL ORIFICE

SPRING

BALL

FILTER SCREW

FIGURE 16–79.
Removing the filter screw from the valve plunger allows the pressure relief ball, guide, and spring to be removed for cleaning; a thicker shim will lower pump relief pressures. (Courtesy of Chevrolet)

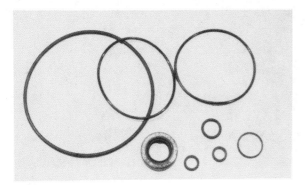

FIGURE 16–80.
A power steering pump seal kit usually includes a shaft seal, a reservoir "O" ring, and "O" rings for the reservoir bolts. (Courtesy of Moog Automotive)

STEERING PUMP PULLEY REMOVER J-25034

FIGURE 16–81.
This puller slides into a groove in the pulley and puts pressure on the shaft to pull the pulley off of the pump shaft. (Courtesy of Kent-Moore)

2. Remove the front pulley. A special alternator-power steering pump pulley remover is available to do a quick and neat job; a two- or three-jaw puller can also be used. (Figs. 16–81 & 16–82)

3. Clean the dirt and rust from the drive shaft. Using a sharp chisel or punch, cut a hole in the old seal, and pry it out of the pump housing. (Fig. 16–83)

4. Lubricate the lip of the new seal with power steering fluid or ATF and, using a driver that engages the whole face of the seal cartridge, drive the seal into the housing bore. (Fig. 16–84)

5. Remove the reservoir-retaining bolts and the outlet fitting. Tap and rock the reservoir to loosen it, and work the reservoir off the

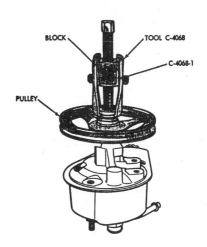

BLOCK

TOOL C-4068

C-4068-1

PULLEY

FIGURE 16–82.
A power steering pump pulley being removed using a special puller. (Courtesy of Chrysler Corporation)

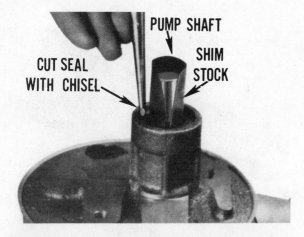

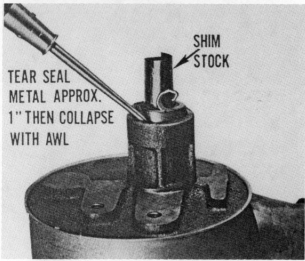

FIGURE 16–83.
To remove a pump shaft seal, first cut into the body of the seal with a small chisel; next collapse the body of the seal and pry it out using an awl. The shim stock is being used to protect the shaft. (Courtesy of Saginaw Division, General Motors Corporation)

8. Install the retaining bolts and outlet fitting and tighten them to the correct torque.

9. Install the drive pulley using an installing tool attached to the pump shaft; some pumps can be damaged by pressing the pulley directly onto the shaft. (Fig. 16–87)

Power Steering Pump Overhaul

Pump overhaul should be done in a clean work area; it is a process of disassembling, inspection, replacement of worn parts, and reassem-

FIGURE 16–84.
To install a pump shaft seal, first clean the shaft with a shop rag and, if necessary, crocus cloth; next drive the seal in using a driver tool that fits flat against the outer portion of the seal body. (Courtesy of Saginaw Division, General Motors Corporation)

pump body. Discard the old "O" ring seals. (Figs. 16–85 & 16–86)

6. Place the new "O" rings in position; if necessary, stick them in place using petroleum jelly. Note that there should be an "O" ring around each opening in the reservoir.

7. Place a film of petroleum jelly on the large "O" ring around the pump body and reservoir sealing surface, and slide the reservoir over the pump body, making sure the bolt holes are aligned.

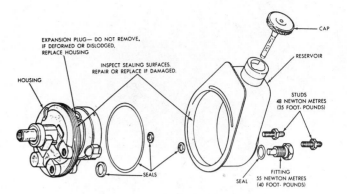

FIGURE 16–85.

Pump reservoir leaks are cured by removing the reservoir, making sure the sealing surfaces are in good condition, installing new sealing rings, and replacing the fitting and bolts to the correct torque. (Courtesy of Chrysler Corporation)

POWER STEERING PUMP ASSEMBLY
(N SERIES-REMOTE RESERVOIR)
(P SERIES-SUBMERGED)

1-SHAFT, DRIVE
2-SEAL, DRIVE SHAFT
3-SEAL, O-RING (HOUSING)
4-HOUSING ASM., PUMP
5-SPRING, FLOW CONTROL
6-VALVE ASM., CONTROL
7-SEAL, O-RING (HOUSING)
8-SEAL, O-RING (PRESSURE & END PLATE)
9-PIN, DOWEL
10-PLATE, THRUST
11-ROTOR, PUMP
12-RING, SHAFT RETAINING

13-VANE, PUMP
14-RING, PUMP
15-PLATE, PRESSURE
16-SPRING, PRESSURE PLATE
17-PLATE, END
18-RING, END PLATE RETAINING
19-SEAL, O-RING (HOUSING TO STUD)
20-RESERVOIR ASM.
21-CAP ASM., RESERVOIR
22-STUD OR BOLT, PUMP MOUNTING
23-SEAL, O-RING (FITTING ASM.)
24-FITTING ASM. (CONNECTOR &)
25-MAGNET
26-RESERVOIR ASM.
27-HOUSING ASM., PUMP
28-TUBE, RETURN

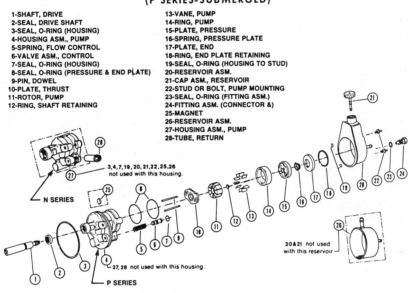

POWER STEERING PUMP ASSEMBLY
(TC SERIES-REMOTE RESERVOIR)

1-HOUSING ASM., HYD. PUMP
2-SLEEVE, ASM.
3-PIN, DOWEL
4-SEAL, O-RING
5-SPRING, PRESSURE PLATE
6-SEAL, O-RING
7-PLATE, PRESSURE
8-PIN, PUMP RING DOWEL (2)
9-VANE (10)
10-ROTOR, PUMP
11-RING, PUMP
12-SEAL, O-RING
13-PLATE ASM., THRUST
14-RING, THRUST PLATE RETAINING

15-TUBE, RETURN
16-SEAL, DRIVE SHAFT
17-SHAFT, DRIVE
18-BEARING ASM., BALL
19-RING, RETAINING
20-SPRING, FLOW CONTROL
21-VALVE ASM., CONTROL
22-SEAL, O-RING
23-FITTING, O-RING UNION

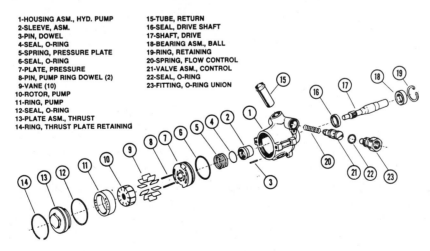

FIGURE 16–86.

Exploded views of two different styles of power steering pumps; note that the internal parts are essentially the same. (Courtesy of Chevrolet)

FIGURE 16–87.
This tool threads into the power steering pump shaft and is used to push the pump pulley onto the shaft. (Courtesy of Kent-Moore)

bly. After assembly, the pressure plate must fit tight against the cam ring so the pump can work. On some pumps, the drag of the pressure plate "O" ring is greater than the strength of the pressure plate spring. It is a good practice to lightly press the pressure plate into position. Disassembly and reassembly procedures will vary slightly among different makes of pumps; it is a good practice to follow a procedure for the specific pump you are working on as outlined in the manufacturer's or technician's service manual.

To overhaul a power steering pump, you should:

1. Remove the pump pulley and reservoir.

2. Remove the end plate retaining ring and tap the pump body to remove the end plate.

3. Carefully tap inward on the front of the drive shaft to remove the pump assembly, pressure plate, thrust plate, cam ring, rotor, shaft, alignment dowels, and the vanes, rollers, or slippers. Note the relationship of the various parts as they come apart; some of them have slight differences on each side. (Fig. 16–88)

4. Most of the pump wear will occur on the inside surfaces of the cam ring, the vanes, and the shaft bushing; inspect them for damage.

5. Remove the old "O" rings and replace them with new ones. Apply a film of petroleum jelly on each "O" ring.

6. Install the pump assembly parts into the pump body, making sure the parts are correctly aligned. Make sure the pressure plate is completely seated.

7. Pour a small amount of power steering fluid or ATF into the pump assembly, place the pressure plate spring in position, and install the end plate and retaining ring.

8. Replace the pump pulley and reservoir.

16.5.5 Power Steering Gear Service

The most commonly performed power steering gear service operations are leak repair and gear assembly overhaul. Some seals (the outer steering shaft seal and Pitman shaft seal) are relatively easy to remove and replace; steering gear disassembly is necessary to replace the remaining seals.

Disassembly, inspection, reassembly, and adjustment of a standard type of gear requires a procedure similar to that of a manual gear. This operation is done in many dealerships, garages, and repair shops. A few special tools are required when working on some gear sets. (Fig. 16–89)

Disassembly, inspection, reassembly, and adjustment of a power rack and pinion gear is also similar to a manual rack and pinion unit. However, quite a few additional steps are required because of the internal and external fluid seals and lines. Additional special tools are also required. Many technicians have found it ad-

1. REMOVE AND INSTALL DRIVE SHAFT SEAL WITHOUT DISASSEMBLING THE PUMP.

REMOVE

1. Protect drive shaft with shim stock.
2. Use chisel to cut seal and remove.

INSTALL

1. Coat drive shaft seal with hydraulic pump fluid, refer to inset for drive shaft seal installation.

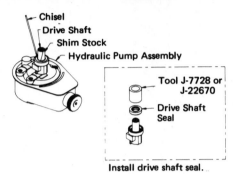

- Chisel
- Drive Shaft
- Shim Stock
- Hydraulic Pump Assembly

Tool J-7728 or J-22670

Drive Shaft Seal

Install drive shaft seal.

2. REMOVE AND INSTALL PUMP RESERVOIR ASSEMBLY.

REMOVE

1. Drain oil from reservoir assembly before removal.
2. Remove parts as shown.

INSTALL

1. Use all new seals and lubricate with power steering fluid before installation.
2. Install parts as shown.

Welch Plug
Do not remove.
If deformed or dislodged, replace housing assembly.

Magnet
clean before reassembly.

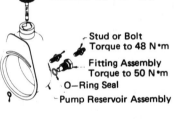

- Reservoir Cap Assembly
- Stud or Bolt Torque to 48 N•m
- Fitting Assembly Torque to 50 N•m
- O—Ring Seal
- Pump Reservoir Assembly
- O—Ring Seals
- Housing Assembly

3. REMOVE AND INSTALL END PLATE.

REMOVE

1. Refer to inset for retaining ring removal

INSTALL

1. Lubricate end plate and retaining ring, install parts as shown. Refer to inset for positioning of retaining ring in housing.

- Screwdriver
- Punch

Remove retaining ring

Press — Retaining Ring Locate ring gap at position shown.

Positioning of retaining ring.

- End Plate
- Pressure Plate Spring
- End Plate Retaining Ring
- Control Valve Assembly
- Flow Control Spring
- Housing Assembly

NOTICE: Before proceeding, examine this part of the drive shaft. If it is corroded, clean with crocus cloth before removing. This will prevent damage to the shaft bushing which might require replacement of the entire housing.

4. REMOVE AND INSTALL ROTATING GROUP.

REMOVE

1. Using a rubber mallet, tap lightly on drive shaft until pressure plate is free.
2. Remove retaining ring from drive shaft and discard. Remove parts as shown.

INSTALL

1. Install parts as shown on drive shaft. Install new retaining ring on drive shaft and install in pump housing.
2. Refer to inset for positioning of pump ring in housing.

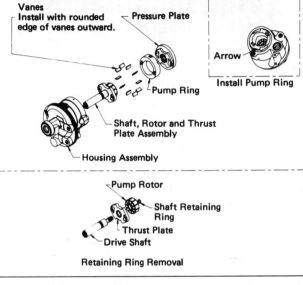

Vanes
Install with rounded edge of vanes outward.

- Pressure Plate
- Pump Ring
- Shaft, Rotor and Thrust Plate Assembly
- Housing Assembly

Arrow

Install Pump Ring

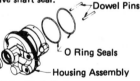

- Pump Rotor
- Shaft Retaining Ring
- Thrust Plate
- Drive Shaft

Retaining Ring Removal

5. REMOVE AND INSTALL DRIVE SHAFT AND O—RING SEALS.

REMOVE

1. Remove parts as shown.

INSTALL

1. Refer to the inset for drive shaft seal installation. Use all new seals and lubricate seals with power steering fluid before installation.
2. Install parts as shown.

Tool J-7728 or J-22670

Install drive shaft seal.

- Dowel Pins
- O Ring Seals
- Housing Assembly

Drive Shaft Seal

NOTE: OMIT STEP 2 WHEN SERVICING N SERIES — REMOTE RESERVOIR PUMP.

FIGURE 16-88.

This is the procedure used to overhaul a power steering pump. (Courtesy of Oldsmobile)

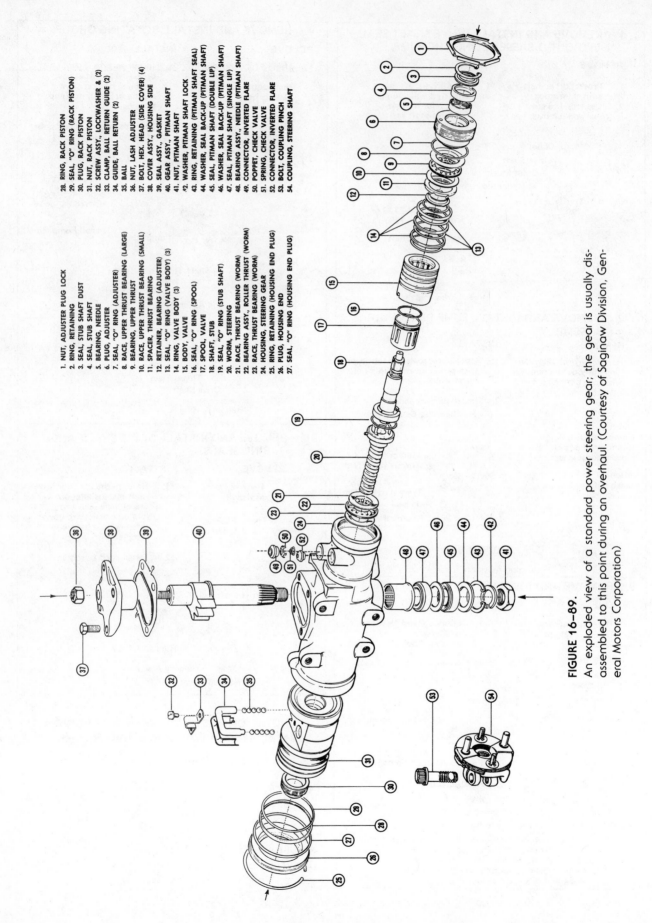

1. NUT, ADJUSTER PLUG LOCK
2. RING, RETAINING
3. SEAL, STUB SHAFT DUST
4. SEAL, STUB SHAFT
5. BEARING, NEEDLE
6. PLUG, ADJUSTER
7. SEAL, "O" RING (ADJUSTER)
8. RACE, UPPER THRUST BEARING (LARGE)
9. BEARING, UPPER THRUST
10. RACE, UPPER THRUST BEARING (SMALL)
11. SPACER, THRUST BEARING
12. RETAINER, BEARING (ADJUSTER)
13. SEAL, "O" RING (VALVE BODY) (3)
14. RING, VALVE BODY (3)
15. BODY, VALVE
16. SEAL, "O" RING (SPOOL)
17. SPOOL, VALVE
18. SHAFT, STUB
19. SEAL, "O" RING (STUB SHAFT)
20. WORM, STEERING
21. RACE, THRUST BEARING (WORM)
22. BEARING ASSY., ROLLER THRUST
23. RACE, THRUST BEARING (WORM)
24. HOUSING, STEERING GEAR
25. RING, RETAINING (HOUSING END PLUG)
26. PLUG, HOUSING END
27. SEAL, "O" RING (HOUSING END PLUG)

28. RING, RACK PISTON
29. SEAL, "O" RING (RACK PISTON)
30. PLUG, RACK PISTON
31. NUT, RACK PISTON
32. SCREW ASSY., LOCKWASHER & (2)
33. CLAMP, BALL RETURN GUIDE (2)
34. GUIDE, BALL RETURN (2)
35. BALL
36. NUT, LASH ADJUSTER
37. BOLT, HEX. HEAD (SIDE COVER) (4)
38. COVER ASSY., HOUSING SIDE
39. SEAL ASSY., GASKET
40. GEAR ASSY., PITMAN SHAFT
41. NUT, PITMAN SHAFT
42. WASHER, PITMAN SHAFT LOCK
43. RING, RETAINING (PITMAN SHAFT)
44. WASHER, SEAL BACK-UP (PITMAN SHAFT)
45. WASHER, PITMAN SHAFT (DOUBLE LIP)
46. WASHER, SEAL BACK-UP (PITMAN SHAFT)
47. SEAL, PITMAN SHAFT (SINGLE LIP)
48. BEARING ASSY., NEEDLE (PITMAN SHAFT)
49. CONNECTOR, INVERTED FLARE
50. POPPET, CHECK VALVE
51. SPRING, CHECK VALVE
52. CONNECTOR, INVERTED FLARE
53. BOLT, COUPLING PINCH
54. COUPLING, STEERING SHAFT

FIGURE 16–89.

An exploded view of a standard power steering gear; the gear is usually disassembled to this point during an overhaul. (Courtesy of Saginaw Division, General Motors Corporation)

448

vantageous to replace a faulty power rack and pinion unit; it saves them time and provides the customer with a professionally rebuilt or re-manufactured unit. It is often less expensive than a garage repair because of the shop time that is saved. Many rebuilt units come complete with tie rods and bellows or in a short rack version without tie rods and bellows. (Fig. 16-90)

Power steering gear service is too varied to try to describe in a text of this type. Manufacturer's and technician's service manuals provide a thorough explanation of repair procedures for specific units. It is a good practice to follow these procedures if you decide to rebuild a power steering gear.

16.6 COMPLETION

Several checks should always be made after completing any spring or shock absorber service operation:

1. Ensure proper tightening of all nuts and bolts, with all locking devices properly installed.

2. Ensure that there is no incorrect rubbing or contact between parts or brake hoses.

3. Ensure that there are no fluid leaks.

4. Ensure an even and smooth movement of the steering from one lock to the other.

5. Ensure proper feel of the brake pedal.

6. Check and readjust the wheel alignment if there is a possibility that the replacement parts have changed the alignment.

7. Perform a careful road test to ensure correct and safe vehicle operation.

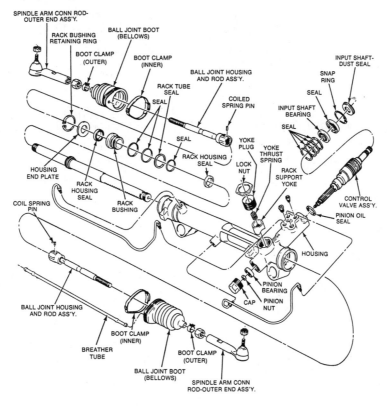

FIGURE 16-90.
An exploded view of a power rack and pinion steering gear; the gear is usually disassembled to this point during an overhaul. (Courtesy of Ford Motor Company)

REVIEW QUESTIONS

1. Technician A says that the tapered stud of a tie rod end can be loosened using tools and methods similar to those used on ball joints. Technician B says that a pickle fork can damage the boot and the socket of the tie rod end. Who is right?

   a. A only   **c.** both A and B
   b. B only   **d.** neither A nor B

2. Before disassembling, the _____ of a tie rod should be measured to allow a preadjustment of the toe.
   A. diameter
   B. length

   a. A only   **c.** both A and B
   b. B only   **d.** neither A nor B

3. Technician A says that idler arms and tie rod ends that use rubber bushings can be tightened in place with the steering turned to any position. Technician B says that these joints should be lubricated with chassis grease during installation. Who is right?

   a. A only   **c.** both A and B
   b. B only   **d.** neither A nor B

4. When threading inner tie rod ends off or on a rack gear, the rack can be held from turning with
   A. an adjustable wrench tightened over the rack teeth.
   B. a pipe wrench.

   a. A only   **c.** either A or B
   b. B only   **d.** neither A nor B

5. An inner tie rod end can be locked onto the rack gear by a

   a. steel pin.
   b. set screw.
   c. swaged metal collar.
   d. any of these.

6. When replacing a multipiece inner tie rod end socket, the socket should be
   A. adjusted to obtain the correct articulation effort.
   B. locked onto the rack gear.

   a. A only   **c.** both A and B
   b. B only   **d.** neither A nor B

7. Technician A says that some cars mount the inner tie rod ends to the center of the rack gear. Technician B says that rubber bushings are used on these tie rod ends. Who is right?

   a. A only   **c.** both A and B
   b. B only   **d.** neither A nor B

8. When idler arms with _____ are replaced, the installation height/length should be adjusted.
   A. slotted mounting holes
   B. threaded bushings

   a. A only   **c.** both A and B
   b. B only   **d.** neither A nor B

9. When replacing idler arms,
   A. the frame brackets and stud hole should be checked for damage.
   B. all bolts and nuts should be tightened to the correct torque.

   a. A only   **c.** both A and B
   b. B only   **d.** neither A nor B

10. Which of the following is *not* one of the steps in removing a steering gear?

    a. disconnect shaft coupler
    b. disconnect idler arm
    c. disconnect Pitman arm
    d. remove mounting bolts

11. Steering gear adjustments are being discussed. Technician A says that steering shaft preload and sector gear lash can be

adjusted with the steering gear either on or off the car. Technician B says that sector gear end play can only be adjusted with the steering gear assembled. Who is right?

a. A only   c. both A and B
b. B only   d. neither A nor B

12. A properly adjusted steering gear will allow steering shaft rotation with
A. a slight drag.
B. a slight end play.

a. A only   c. either A or B
b. B only   d. neither A nor B

13. Technician A says that a standard steering gear should turn with the same amount of preload from one lock to the other. Technician B says that this gear should have a slight amount of lash between the ball nut and sector gear all the way through the gear travel. Who is right?

a. A only   c. both A and B
b. B only   d. neither A nor B

14. Sector gear end play is adjusted by
A. turning an adjustment screw.
B. changing a shim.

a. A only   c. both A and B
b. B only   d. neither A nor B

15. A properly adjusted manual rack and pinion steering gear will rotate
A. from one lock to the other with the same amount of preload.
B. smoothly from one lock to the other.

a. A only   c. both A and B
b. B only   d. neither A nor B

16. Technician A says that tightening the rack support bushing will increase the amount of preload in a rack and pinion steering gear. Technician B say that this gear is lubricated through an oil level plug at the top of the gear housing. Who is right?

a. A only   c. both A and B
b. B only   d. neither A nor B

17. Power steering drive belt tension should be checked using
A. the belt deflection method.
B. a belt tension gauge.

a. A only   c. either A or B
b. B only   d. neither A nor B

18. Normal power steering service includes

a. checking the fluid level and condition.
b. checking the belt tension and condition.
c. inspecting and curing leaks.
d. all of these.

19. A hard steering complaint is being discussed. Technician A says that the cause can be determined by using a special pressure gauge set. Technician B says that this gauge should be installed between the pump outlet port and the steering gear inlet port. Who is right?

a. A only   c. both A and B
b. B only   d. neither A nor B

20. Technician A says that a power steering system should have about 500 to 750 psi at idle speed. Technician B says that pump pressure should increase when the valve on the tester is closed. Who is right?

a. A only   c. both A and B
b. B only   d. neither A nor B

Chapter 17

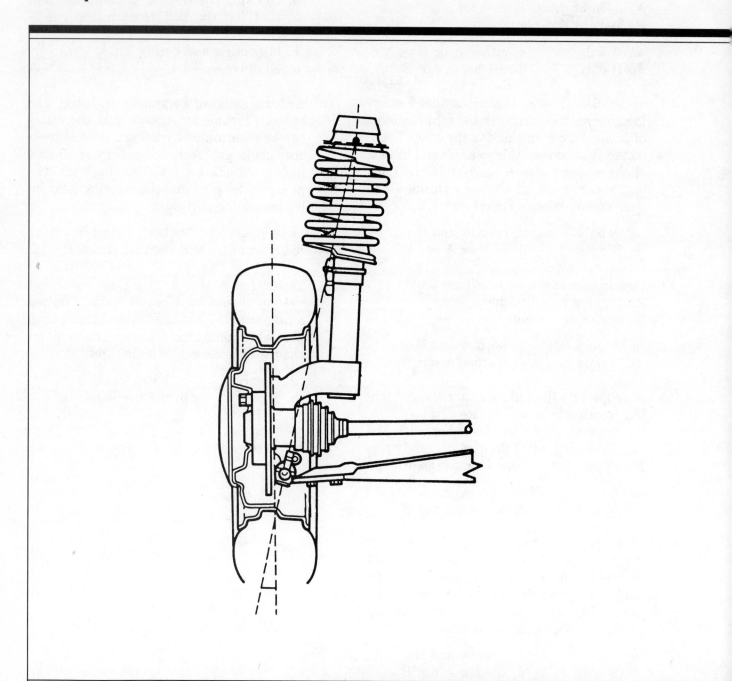

WHEEL ALIGNMENT— FRONT AND REAR

After completing this chapter, you should:

- Be familiar with the terms commonly used with wheel alignments.

- Understand the purpose for the various front and rear wheel alignment angles.

- Understand the effect that improper alignment angles have on vehicle operation.

17.1   REQUIREMENTS

For a tire to roll down the road in an efficient and controlled fashion, it must be aligned correctly. Usually the tire will be vertical or straight up and down (zero camber) and the center of the tire will point straight down the road (zero toe). **Camber** and **toe** are **tire position angles.** If they are not correct, they will cause tire wear; they can also cause direction control problems. If we did not steer the car, this is all that wheel alignment would be, but the car would travel only in a straight line. (Fig. 17-1)

Since we do steer the car so that we can turn corners, we also have to think about the position of the **steering axis.** The steering axis is an imaginary line that runs through the two ball joints of a car with S-L A suspension, the lower ball joint and the steering pivot of a car with strut suspension, or the king pin of most solid or swing axles. **Caster** and **steering axis inclination (SAI)** are thought of as two different angles of the steering axis; they ensure that the front wheels are easily and safely steerable. They are two different planes of the steering axis. We normally think of them as two separate and distinctly different angles. They are called **directional control angles.** Since they do not directly control the position of the tire, they

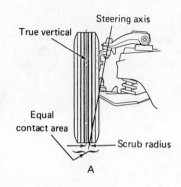

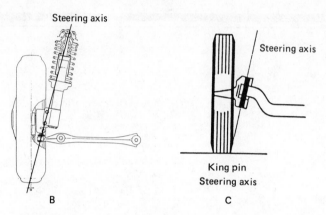

FIGURE 17-2.

The steering axis is the point where the steering knuckle pivots on turns. It is in line with the ball joints in an S-L A suspension (A), in line with the ball joint and upper pivot in a strut suspension (B), and in line with king pin on a solid axle. (A is courtesy of Ford Motor Company; B is courtesy of Volkswagen of America; C is courtesy of Ammco)

cannot cause tire wear. They do affect the driveability and the handling of the vehicle. (Fig. 17-2)

The fifth alignment angle, **toe out on turns,** also called **turning angle** or **turning radius,** ensures that the tires are turned to the correct angle while the vehicle is turning. Tire wear will occur during turns if this angle is not correct.

Before we begin a closer study of front end alignment, we must remember that the rear wheels must also be aligned to the car. Camber and toe must be correct in order to have the maximum traction for vehicle control plus maximum tire life. We should also understand that the car's direction, while traveling in a straight line, is mainly controlled by the rear wheels. It can be said that the front tires turn the car, but the rear tires steer, at least in the straight-ahead direction. The straightahead direction of

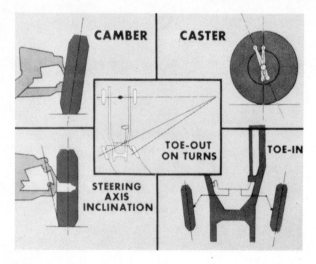

FIGURE 17-1.

There are 5 different angles involved in front end alignment. (Courtesy of Chrysler Corporation)

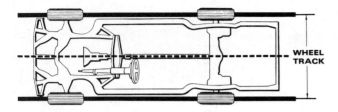

FIGURE 17–3.
On most cars, the rear wheels do not steer so we only need to maintain correct camber, toe, and track; if the track is not correct, the car will go down the road slightly crosswise. (Courtesy of Hunter Engineering)

a vehicle is on a path that is parallel to the rear wheels. This is often called the **thrust line**. It is also called **track**; the rear tires should make tracks directly behind the front tires. (Fig. 17–3)

17.2 MEASURING ANGLES

Most wheel alignment specifications are printed in degrees or parts of degrees. A degree is a single slice out of a circle in which 360 equally sized slices have been made. A degree is an international measurement; all countries use the same system. The symbol "°" is used to indicate degree.

There are some differences when measuring parts of a degree. Traditionally, in the United States, a portion of a degree has been referred to as its fractional or decimal part; for example, 1/2° or .5°. Our scientists and engineers plus most of the rest of the world have measured portions of a degree in **minutes (')** and **seconds (")**; much like the division of an hour, there are 60' in a degree and 60" in a minute. 1/2° or .5° would equal 30'°. An equivalency chart is included in the appendix of this book to help you convert fractional or decimal portions of a degree into minutes and seconds.

Those of you who are familiar with trigonometry know that if you know the length of two sides of a right triangle, it is fairly easy to compute the angle between the hypotenuse and the long side. For example, if I know that if a tire 28 inches in diameter is leaning 1/8 inch from vertical, I can compute that the tire has

a camber angle of .256 or about 1/4°. This is described in Figure 17-4 and also in a portion of Section 19.12.

17.3 CAMBER

Camber is a term used to describe the position of the tire as seen from the front or back. If the tire is exactly vertical, it has a camber angle of zero. If the top of the tire leans outward, away from the car's center, the tire has a **positive camber** angle; an inward leaning tire has **negative camber**. Camber is directly controlled by the spindle; the center line of the tire is at a right angle, 90° from the spindle (assuming there is no looseness in the wheel bearings). A spindle that drops 1° at the outer end will cause 1° of positive camber. FWD cars do not have a front spindle, but the hub and bearings in the steering knuckle serve the same purpose as the front spindle of an RWD car. (Figs. 17–5 & 17–6)

Most modern cars will use about 1/4 to 1/2° of positive camber on their front wheels. This is usually just enough to cause near-to-zero camber when driving on the road. (Fig. 17–7)

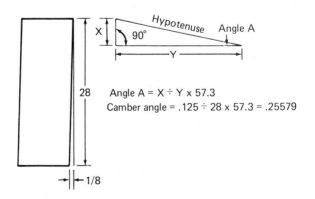

$$\text{Angle A} = X \div Y \times 57.3$$
$$\text{Camber angle} = .125 \div 28 \times 57.3 = .25579$$

FIGURE 17–4.
If we know the vertical distance from the ground to the top of the tire (side Y) and the amount the tire is from being exactly vertical (side X), we can determine the camber angle (angle A) by using the above formula.

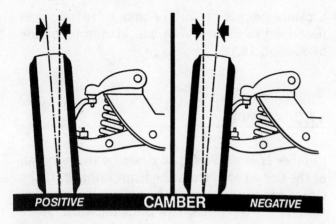

FIGURE 17–5.
A tire with positive camber will lean outward at the top; it will lean inward if it has negative camber. (Courtesy of Hunter Engineering)

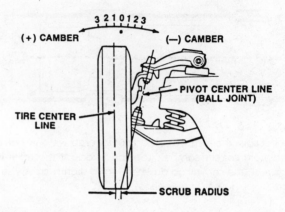

FIGURE 17–7.
The camber angle is measured in degrees of tilt from true vertical; if the tire was straight up and down, it would have 0 camber. (Courtesy of Ford Motor Company)

17.4 CAMBER AND SCRUB RADIUS

At one time, when tires were taller and narrower, cars had more positive camber to help reduce the scrub radius. **Scrub radius** is a term referring to the distance between the steering

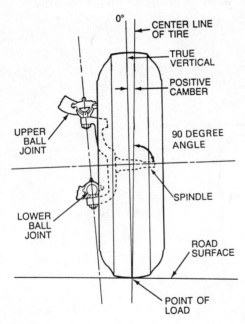

FIGURE 17–6.
The tire's centerline is at a 90° angle to the centerline of the spindle; if the spindle tilts downward, the tire will have positive camber. (Courtesy of Chrysler Corporation)

axis (at ground level) and the center of the tire. A **positive scrub radius**, where the steering axis is inboard of the tire center, will tend to cause the tire to turn or toe outward as the car is driven. This effect is especially true if the tire is flat or it meets a bump or pothole in the road; a definite pull or dart will occur. The drag of the tire pushes it toward the rear of the car while the car is pushing forward on the steering axis. Front wheel drive cars will usually have a **negative scrub radius**; the steering axis meets the road outboard of the tire centerline. This offsets the effect of the tires trying to toe inward as they pull the car down the road. It also produces an effect that cancels front end pull if a front tire goes flat while the car is being driven. (Fig. 17–8)

A slight scrub radius helps give the front end directional stability. There is always a slight amount of compliance in the moving parts of the steering linkage, and this can cause the front tires to pivot in and out slightly. Compliance results from the slight flexibility of the ball joints and rubber bushings; it allows the tire to change position slightly. With positive scrub radius, the tire, trying to turn outward, will load the steering parts in one direction and compress the compliance to a stable, loaded position. (Fig. 17–9)

Besides reducing scrub radius, positive camber was said to be used to help strengthen the wooden spokes of wheels used in the past.

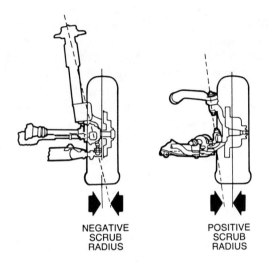

FIGURE 17-8.
The distance between the road contact point of the steering axis and the centerline of the tire is called "scrub radius." If the tire center is inward of the steering axis, it is called negative, and if it is outward, it is called positive. (Courtesy of Chrysler Corporation)

This reason, of course, is now obsolete, and scrub radius is now controlled more by the steering axis inclination and wheel offset than by camber. There is no longer a valid reason for

having very much positive camber on a rolling tire. We might set a tire to a slightly positive camber angle on the alignment rack, but it should have zero camber when it rolls down the road.

17.5 CAMBER-CAUSED TIRE WEAR

The major theory with modern tires, which are lower and wider, is that the tread should be flat on the road so it has full tread contact. Camber tends to lift one side of the tire upward. Reducing the pressure between the tread and the road can reduce tire traction, causing a loss of vehicle handling and stopping ability.

It is easy to see that if we lift part of the tread, the remaining tread portion will carry a higher load, more than its share. Because of this, incorrect camber will cause abnormally fast tire wear. Camber that is too positive will cause rapid wear on the outside of the tread, and camber that is too negative will cause rapid wear on the inner side of the tread. Wear will be greatest on one side of the tread, and the

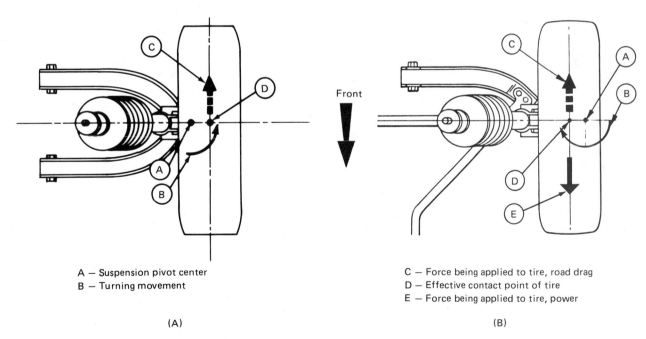

A — Suspension pivot center
B — Turning movement

C — Force being applied to tire, road drag
D — Effective contact point of tire
E — Force being applied to tire, power

(A)

(B)

FIGURE 17-9.
Road drag on the tire will cause it to turn outward on cars with positive scrub radius (A). The tire will tend to turn inward from negative scrub radius, but front wheel drive forces will offset this effect with an outward turning action (B). (Courtesy of Ford Motor Company)

FIGURE 17–10.
Incorrect camber angles will cause a tapered wear pattern across the tire tread. (Courtesy of Dana Corporation)

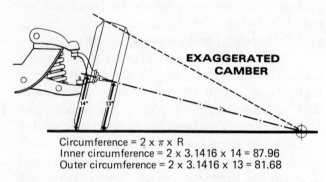

EXAGGERATED CAMBER

Circumference = 2 x π x R
Inner circumference = 2 x 3.1416 x 14 = 87.96
Outer circumference = 2 x 3.1416 x 13 = 81.68

FIGURE 17–11.
In this exaggerated example, the circumference difference between the inside of the tread (88″) and the outside (82″) will cause 6 inches of tread slippage on each tire revolution. (Courtesy of Hunter Engineering)

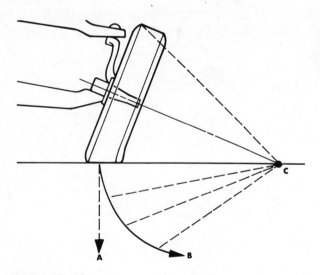

FIGURE 17–12.
A cambered tire will try to roll in a circle just as if it were the end of a cone. (Courtesy of Ford Motor Company)

amount of wear will steadily decrease across the tread. (Fig. 17–10)

Some people feel that camber wear is a result of tire scrub caused by a shorter tire radius on one side of the tire than on the other. The effective tire radius is the distance between the ground and the horizontal center of the tire. The circumference of the tire, which is the distance the tire will roll down the road in one revolution, is the result of multiplying the radius (r) by 2 and then by Pi (π) or 3.1416 (circumference $= 2\pi r$). A tire that is cambered will have two different tire radii, a different radius at each side of the tread. This causes the tire to have two different circumferences, which results in a tire trying to travel two different distances on each revolution. This, of course, is impossible, so portions of the tread will slip or scrub as the tire rolls. (Fig. 17–11) In any event, regardless of which theory seems most correct, incorrect camber will cause tire wear. That is why it is referred to as a tire-wearing angle.

Another negative effect of incorrect camber is tire pull. A cambered tire rolls as if it were the end of a cone. A cone will not roll straight; it rolls in a circle, with the point/apex of the cone being the center of the circle. We see this effect when we are riding a bicycle or motorcycle and turn corners by leaning toward the direction we want to turn. Excessively positive camber will cause the car to pull or turn toward that side. Excessively negative camber will cause a pull toward the opposite side. A rule of camber: the car will pull toward the most positive side. (Fig. 17–12)

17.6   CAMBER SPREAD AND ROAD CROWN

Roads are usually crowned or raised in the middle so that rain water will run off. Like a ball, a tire rolling across an inclined surface will tend to turn and roll down hill. This effect would cause the car to pull to the right on most crowned roads. Some manufacturers will specify slightly more positive camber (about 1/4°) on the left wheel than on the right; this is called **camber spread**. This spread will generate a

FIGURE 17–13.
If a car is driven on a hillside or high crowned road, the car will have a tendency to turn downhill. (Courtesy of Hunter Engineering)

camber pull to the left; if the camber pull equals the road crown pull, the car will go straight. (Fig. 17–13)

Because camber is a tire-wearing angle, many alignment technicians prefer to adjust camber to the least tire-wearing angle and use a caster spread to compensate for road crown (see Section 17.15). Caster can also cause a slight pull to offset road crown, and it will not cause tire wear.

17.7   CAMBER CHANGE

Vertical wheel travel on cars with independent suspension will cause the camber to change.

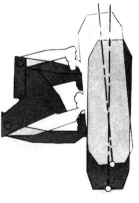

NEGATIVE CAMBER IN JOUNCE

FIGURE 17–14.
On cars with independent suspension, a change in the camber angle will occur as the tires move up or down. (Courtesy of Chrysler Corporation)

Depending on the suspension type, this can be an undesirable change, as seen on swing axles, or a carefully planned change that is designed into some S-L A suspensions. The important thing to remember is that vehicle height is important. If the car is too high or too low, the camber angle will probably be out of specifications. (Fig. 17–14)

The procedure for measuring and adjusting camber is described in Chapter 18.

17.8   TOE IN

Toe in, like camber, is an angle of the tire, and it will cause tire wear if it is incorrect. We commonly use the term toe in, but we probably should use the term **toe** because many FWD cars use **toe out**. Toe is a comparison of the distance between the front of the tires and the rear of the tires. If the two tires are parallel, these two distances will be the same, and there will be zero toe. If the front measurement is 1/8 inch (3.17 mm) shorter than the rear, there will be 1/8 inch (3.17 mm) toe in. If the rear measurement is 1/8 inch (3.17 mm) shorter, there will be 1/8 inch (3.17 mm) of toe out. Toe is usually measured at the height of the center of the hub to ensure measuring at the most forward and rearward portions of the tire. Toe is controlled by the length of the tie rods. A longer tie rod (mounted to the rear of the tire centerline) will cause an increase in toe in. (Figs. 17–15 & 17–16)

Traditionally, toe specifications have been printed in inches for domestic cars and millimeters for imported cars. Many modern cars are now specifying **toe angles**. This is the angle between the two tires (combined) or between a single tire and straight-ahead (individual). A toe specification can be converted from a distance to an angle by using this formula:

$$\text{Toe angle in degrees} = \frac{\text{toe distance}}{\text{tire diameter}} \times 57.3$$

A conversion chart for toe distance and degrees is included in the appendix of this book.

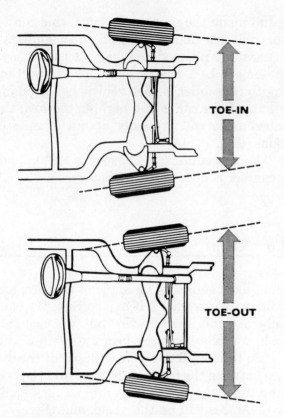

FIGURE 17–15.
Toe in refers to the tires being closer together at the front; if they are closer at the rear, it is toe out. (Courtesy of Hunter Engineering)

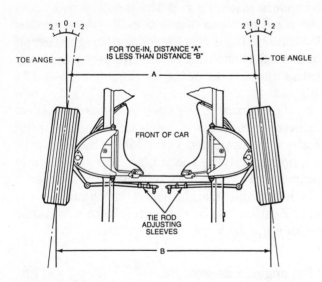

FIGURE 17–16.
Toe has traditionally been measured as the difference in distance between the tires at front and rear; toe angle dimensions and methods of measuring are becoming more commonplace. (Courtesy of Chrysler Corporation)

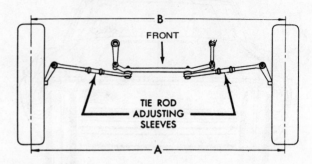

FIGURE 17–17.
If these tie rod adjusting sleeves were made longer, toe in would increase. (Courtesy of Chevrolet)

A small amount of toe in, about 1/16 inch (1.58 mm), is usually used on the front tires of RWD cars. This amount of toe in is used to offset the tendency of most front tires to toe outward as they are pushed down the road. Each tie rod end has a small amount of compliance that allows the tie rod to lengthen or shorten slightly under pressure; the ideal amount of toe in exactly matches the change in length of the tie rods so the tires will be exactly parallel with zero toe when the car is going down the road. (Fig. 17–17)

A small amount of toe out is often used on FWD cars; the front tires of FWD cars tend to toe inward as the front tires pull the car down the road. This tendency will produce zero toe if the tires are adjusted to the correct amount of toe out. (Fig. 17–18)

17.9   THINGS THAT AFFECT TOE

Toe is affected by the rolling resistance of the tires and the scrub radius. Tires with more road drag (lower air pressure, wider tread, bias ply) will tend to toe out more than tires with less road drag (higher air pressure, narrower tread, radial ply). If the scrub radius is increased by the use of wheels with a greater negative offset than stock, the toe out tendency will also be increased. Worn bushings, idler arms, and/or tie rod ends will also affect the amount that the tires will toe out on the road because of the increased compliance. These factors are usually

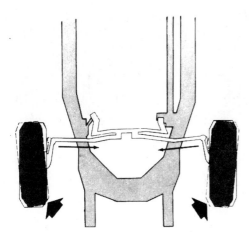

TOE-IN OFFSETS DEFLECTIONS

FIGURE 17-18.
Road forces on the tires of an RWD car will cause the tires to tend to toe outward. Toe in would try to occur if the front tires were pulling the car like in FWD cars. (Courtesy of Chrysler Corporation)

considered when the skilled alignment technician adjusts toe. (Fig. 17-19)

17.10 TOE-CAUSED TIRE WEAR

If toe is not correct, the tire will have to scuff or scrub sideways as the car travels down the road. This side-scuffing causes an easily recognizable wear pattern on the tread called a

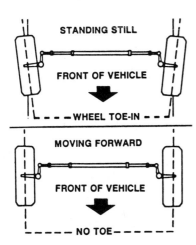

FIGURE 17-19.
On an RWD car, road forces will cause the toe to change from toe in, standing still, to zero toe when moving forward. (Courtesy of Ford Motor Company)

FIGURE 17-20.
Incorrect toe angles will cause an easily recognizable, abnormal wear pattern of the tire tread. Note the "feather edge" toward the left; this tire was scuffing toward the right. (Courtesy of Ford Motor Company)

feather edge or **saw tooth.** It is most noticeable if you run your hand inward across the tread and then back out. If the roughness is felt as your hand is passed inward, excessive toe out is indicated; if the roughness is felt on the way out, excess toe in is indicated. (Fig. 17-20)

A toe error of 1/16 inch (1.58 mm) causes the tire to scuff sideways about 11 feet (3.35 m) for each mile (1.61 km) the car is driven. This is because the car is traveling one way while the tire goes in a different direction. The resulting sidescuffing not only wears tires; it also causes a reduction in gas mileage. It takes a lot more power to push a tire sideways than it does to roll it. For an example of this, try pushing a car forward or backward (in neutral with the brakes released); then try pushing it sideways. (Fig. 17-21)

17.11 TOE CHANGE

On most cars, toe will change slightly as the front wheels travel up and down over bumps. This is caused by the steering knuckle and spindle moving in an arc as it rises and falls. The tie rod end also moves in an arc, with its center being the inner tie rod end. If these two arcs do not match exactly, the steering knuckle will be

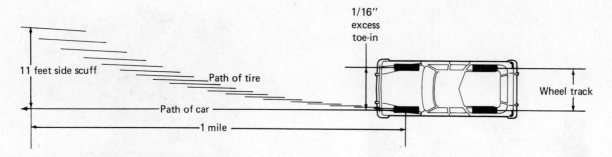

FIGURE 17–21.
A tire that is toed 1/16 inch (total) more or less than it should will have to scrub sideways about 11 feet during every mile it travels.

turned slightly, and toe will change. It is hoped that this change is very slight and only occurs for a brief period of time. However, a change in the position or length of the steering linkage components or suspension components can cause more severe toe change and, therefore, tire wear and handling problems. (Figs. 17–22 & 17–23)

Some toe gauges allow a fairly easy check to see if toe change is occurring; these are discussed in later sections. Exact measurement of toe change is often done when setting up a competition car; this is discussed in Chapter 19.

Because of toe change, toe is correct only at one vehicle height—curb height. Like the other wheel alignment angles, toe is always checked at ride height. The procedure for measuring and adjusting toe along with centering of the steering wheel is described in Chapter 18. (Fig. 17–24)

17.12 CASTER

Caster is a directional control angle that helps the driver keep the front tires turned straight down the road and return the steering to straight ahead after turning a corner. It is an angle of the steering axis when viewed from the side; the steering axis leans toward the front or rear of the car. If the steering axis were exactly vertical, when viewed from the side there would be zero caster. If the upper end of the steering

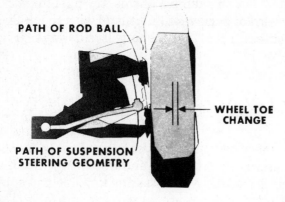

FIGURE 17–22.
A toe change will occur during bump travel if the suspension geometry does not match the geometry of the tie rod. (Courtesy of Chrysler Corporation)

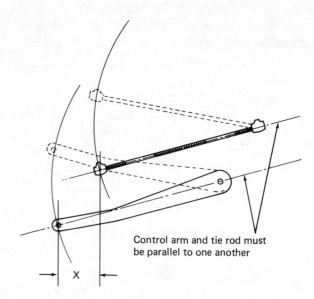

FIGURE 17–23.
The tie rod must be parallel to the control arm or else toe change will result from the differences in the arcs that the ends of the arm and tie rod travel. Toe change will occur if distance "x" changes.

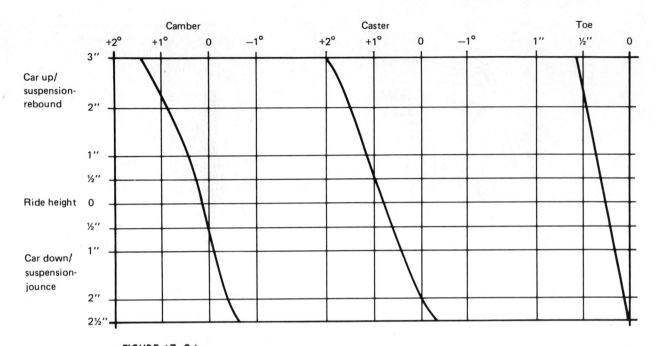

FIGURE 17–24.
These camber, caster, and toe readings were taken from a 1986 domestic FWD car; note the rather consistent change in readings taken from various heights.

axis is toward the rear of the car, caster is **positive**; it is **negative** if the upper end of the steering axis is toward the front. (Fig. 17–25)

Caster can be understood more easily if we consider a furniture caster or caster on a bicycle (even though there is a slight difference in operation). A furniture caster has a vertical pivot axis that can be compared to the steering axis of a car. The wheel axle, and with it the road contact of the wheel, is positioned to one

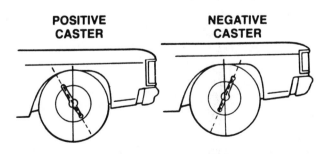

FIGURE 17–25.
Caster is defined as the forward or rearward tilt of the steering axis. If the top of the steering axis is toward the rear, the caster angle is positive; if the top is toward the front, the angle is negative. (Courtesy of Ford Motor Company)

side of the pivot axis. As this distance gets greater, caster effect gets stronger. When the furniture is moved, wheel drag will cause the wheel to fall back and follow the vertical pivot. This action brings the wheel in alignment with the direction of travel. To get caster effect on a bicycle, the steering axis (front fork pivots) is inclined so the steering axis ends up ahead of the road contact of the tire. More inclination of the steering will increase the caster effect, and this will cause the vehicle to be more stable while riding straight. (Fig. 17–26)

If the steering axis is vertical (in a sideways plane), there will be 0° caster. The actual amount of caster equals the amount of lean of the steering axis; 5° of caster tells us the steering axis is leaning 5°. Negative or positive tells us which way the steering axis is leaning. Forward at the top is negative caster, and rearward is positive caster. Positive caster angles are much more common because this is a more stable condition.

Caster is measured in degrees between the steering axis, viewed from the side, and true vertical. Most domestic cars will specify a caster angle of somewhere between 0° and 3°

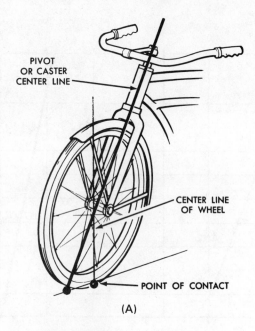

PIVOT OR CASTER CENTER LINE

CENTER LINE OF WHEEL

POINT OF CONTACT

(A)

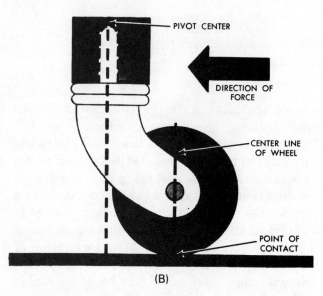

PIVOT CENTER

DIRECTION OF FORCE

CENTER LINE OF WHEEL

POINT OF CONTACT

(B)

FIGURE 17-26.
Caster effect can be demonstrated using a furniture caster or a bicycle fork. In each case the pivot center (comparable to the steering axis), leads the contact point of the tire; road drag will cause the tire to follow the pivot axis. (Courtesy of Ford Motor Company)

or 4° positive; some imported cars use as much as 10° of caster. Unless they are specified as being negative, most wheel alignment angles are considered to be positive. A few cars will use 1° or 2° of negative caster. (Fig. 17-27)

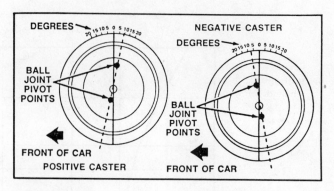

FIGURE 17-27.
Caster is measured in degrees; if the steering axis is exactly vertical, there will be zero degrees of caster. (Courtesy of Ford Motor Company)

17.13 CASTER EFFECTS

Positive caster places the ground level point of the steering axis toward the forward end of the road contact patch of the tire. The drag of the tire on the road relative to the steering axis has the same effect that the drag of the wheel has on the pivot axis of the furniture caster. The ground level point of the steering axis is also affected by SAI, which we will study further on. (Fig. 17-28)

Caster causes a tire's camber angle to change as the tires are turned to the right or left. A right turn will cause the camber on the right tire to become more positive and the left

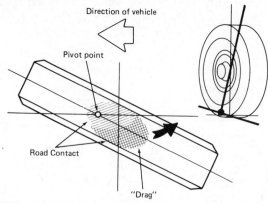

Direction of vehicle

Pivot point

Road Contact

"Drag"

FIGURE 17-28.
With positive caster, the road contact point of the steering axis leads the contact patch of the tire much like that of a furniture caster. Road drag will try to bring the tire back to a straightahead position. (Courtesy of Ford Motor Company)

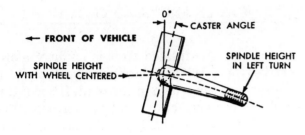

FIGURE 17-29.
Because of the steering axis angle, positive caster will cause camber to change during a turn; looking at the left steering knuckle during a left turn, camber has become more positive. (Courtesy of Ford Motor Company)

tire camber to become more negative (with positive caster); the camber changes will be just the opposite with negative caster. This camber change is considered when selecting caster angles for competition cars raced on road circuits. (Fig. 17-29)

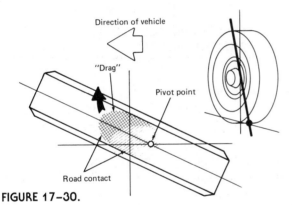

FIGURE 17-30.
With negative caster, the road contact patch is ahead of the steering axis; as the tire is turned, road drag will try to turn the tire farther from straight ahead. (Courtesy of Ford Motor Company)

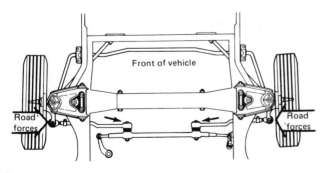

FIGURE 17-31.
Caster places a load on the tie rod, pushing or pulling, across the car from each tire; if the loads are equal, there will be no side pull. (Courtesy of Chrysler Corporation)

Negative caster places the ground level point of the steering axis toward the rear of the tire contact patch. Placing the steering axis behind the center of the contact makes the car easier to steer. This effect can be illustrated by comparing the effort required to steer while the car is going forward and backward. Some manufacturers specify negative caster on their cars with manual steering gears to get easier steering. Positive caster is nearly always used on cars equipped with power steering. (Fig. 17-30)

On an automobile, the caster effect of one wheel acts in opposition to the caster effect of the other wheel. These opposing forces, like those of toe, work against each other through the tie rod. If the forces are even, the tie rod will be centered, and the car will go straight ahead. If the forces are not equal, the car will pull or turn away from the side of the car having the most positive caster. (Fig. 17-31)

17.14   THINGS THAT AFFECT CASTER

Like toe, caster is affected by the road drag of the tire. Tires with a greater rolling resistance generate a greater caster effect than tires with less rolling resistance. An example of this can be seen by comparing the caster specifications for 1975–1980 model cars with those made before 1970. The older cars, with less caster, used bias ply tires while the newer ones are designed for radial tires and usually specify more caster.

Caster is also affected by the frame angle of the vehicle. Adding a heavy load to the trunk of the car will usually lower the back of the car. This will change the frame/body angle and, therefore, the angle of the steering knuckle and caster. Sagging rear springs will also increase caster, making it more positive. Raising the back of the vehicle or lowering the front will cause caster to become more negative. The amount of caster change relative to the change in height is determined by the car's wheel base. If you want to know exactly how much, use this formula to compute the actual amount of change: Wheelbase $\times$ 2 $\times$ π $\div$ 360 = change in height that causes a 1° change in caster. (Fig. 17-32)

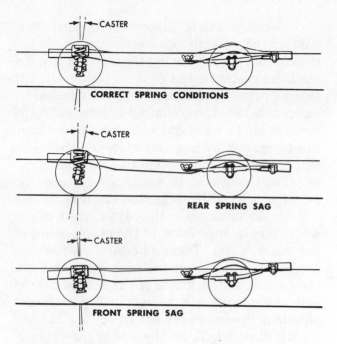

FIGURE 17-32.
If the fore-aft angle of the vehicle changes, the amount of caster angle will change with it; this type of change often occurs because of spring sag or when weight is added or removed. (Courtesy of Ford Motor Company)

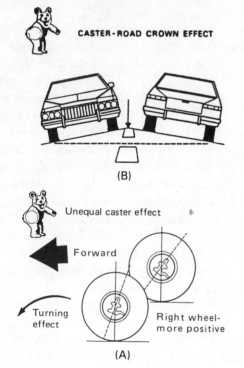

FIGURE 17-33.
The road crown (A) will often cause the car to steer toward the side of the road; more positive caster on the right tire can be used to offset the effects of road crown. (Courtesy of Bear Automotive)

17.15 CASTER AND ROAD CROWN

Caster is often adjusted so it is slightly more positive (1/4 to 1/2°) on the right wheel to offset the effects of the crown in the road. If the right wheel has the correct amount of more positive caster than the left wheel (called **caster spread**), the car's caster pull (toward the left) will equal the road crown pull (toward the right), and the car will go straight down the road. Many alignment technicians prefer to use caster rather than camber to correct for road crown because caster is a non-tire-wearing angle. (Fig. 17-33)

The amount of caster spread used on a car varies with the crown of the road, the weight of the car, and the road drag of the tire. This is often a trial-and-error adjustment until experience teaches you to judge the vehicle and where it is driven.

17.16 EFFECTS OF TOO LITTLE OR TOO MUCH CASTER

The correct amount of caster gives us a car that drives straight down the road, requires only a little or occasional correction to go straight, and steers easily. Too little caster gives us a car that **wanders**; it goes right or left and requires continous correction to travel straight. Too much caster gives us a car that travels straight with no wander, but is harder to steer. Too much caster might even cause **tramp**, also called **shimmy**, especially in vehicles with a solid axle. Tramp is a violent and continuous turning of the steering back and forth from right to left to right, continuing on and on. Tramp places the car in a potentially dangerous situation; the driver of a shimmying car should slow down immediately. Many alignment technicians vary the amount of caster to tailor the steering to the most stable point. (Fig. 17-34)

17.17 STEERING AXIS INCLINATION (SAI)

Like caster, **SAI** is an angle of the steering axis and is a directional control angle. It is a more

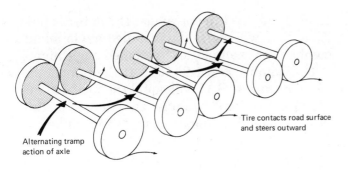

FIGURE 17-34.
Tramp, sometimes called shimmy, causes a violent oscillation of the axle along with a dangerous back and forth steering motion.

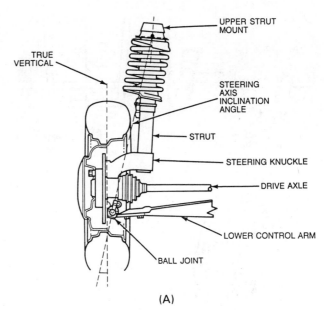

(A)

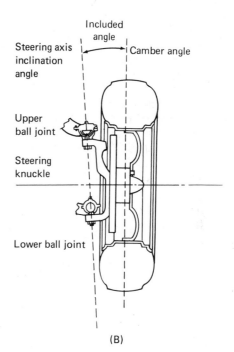

(B)

FIGURE 17-35.
Steering axis inclination on a strut (A) and S-L A suspension (B). Note that the strut is on an FWD car so there is a negative scrub radius. (Courtesy of Chrysler Corporation)

17.18   INCLUDED ANGLE

The SAI specification can be given in degrees by itself, relative to true vertical, or as part of the **included angle**. The included angle is the angle between SAI and camber; it includes

powerful directional control angle than caster. The steering axis is always angled outward at the bottom so the ground level end is somewhere under the contact patch of the tire. SAI, with camber, controls scrub radius to minimize the effects that bumps and chuckholes have on steering. (Fig. 17-35)   This angle is also called **steering axis angle (SAA), ball joint inclination or angle (BJI or BJA),** or **king pin inclination or angle (KPI or KPA).**

SAI works through our simplest and most reliable physical force, gravity. Because of the angle of the steering knuckle, the spindle lowers or drops (relative to the car) when the wheels are turned from straight ahead. A similar type of height change will also occur because of the caster angle. The end of the spindle cannot drop; it is supported by the tire and wheel. So the effect of SAI during a turn is to lift the car. This can be seen if you watch the front fender of a car rise and fall as the wheels are turned into a turn and back. Gravity is always trying to pull the car to the lowest position possible, which, because of SAI, is straight ahead. When you place a car's front tires on turntables, you can demonstrate SAI by turning the tires into a sharp turn and then releasing them. They will tend to turn back toward a straightahead position. (Figs. 17-36 & 17-37)   Most cars have an SAI angle of about 7° to 12°; a few have as much as 16°.

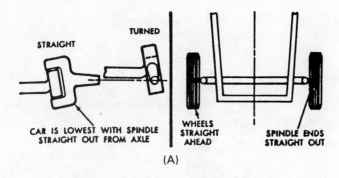

(A)

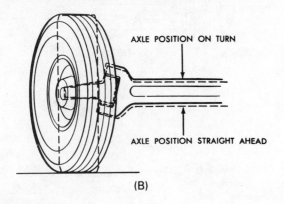

(B)

FIGURE 17-36.

Because of SAI, a vehicle is lifted whenever the front tires are turned from straightahead; gravity acting on the car will always try to pull the car to its lowest position with the tires straightahead. (Courtesy of Ford Motor Company)

camber and SAI in one angle. If camber is positive, SAI will equal the included angle minus camber; for example, a 7° included angle minus +1/2° camber = 6 1/2° SAI. If the camber angle is negative, SAI will equal the included angle plus the camber angle; an example of this is a 7° included angle plus −1/2° camber = 7 1/2° SAI. We go to the trouble of manipulating camber from the included angle because our measuring equipment measures both camber and SAI from true vertical, and different manufacturers print their specifications in different manners. It should be noted that all alignment measuring equipment will not measure SAI. Some manufacturers publish separate camber and SAI specifications (this makes it easy); others publish camber and included angle specifications (making you compute the SAI angle. (Figs. 17-38 & 17-39)

SAI is not always found with the other alignment specifications because, traditionally, this angle is not always checked during a front end alignment. A change in SAI, without a

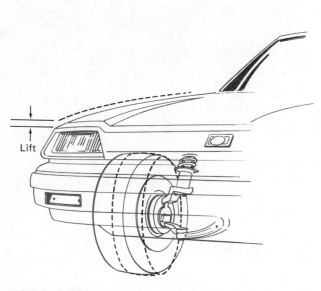

FIGURE 17-37.

As the steering wheel is turned from a straightahead position, the angle of the steering axis will cause the car to be lifted; the weight of the car will try to turn the wheels back to a straightahead position.

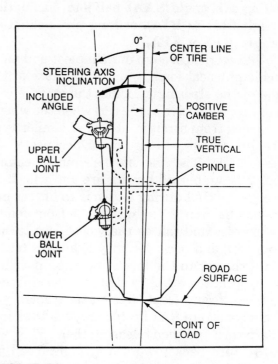

FIGURE 17-38.

The included angle is the amount of angle between the steering axis and the centerline of the tire. (Courtesy of Chrysler Corporation)

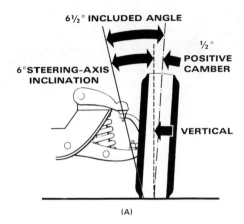

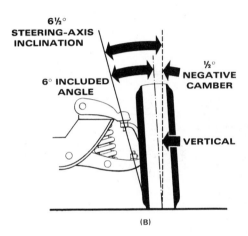

FIGURE 17-39.
If the camber angle is positive, it is added to the SAI angle to get the included angle; if camber is negative, it is subtracted from SAI to get the included angle. (Courtesy of Hunter Engineering)

change in camber, would require that the motorist must bend the steering knuckle or spindle without damaging any other suspension component. This is not very probable, except for some strut suspensions, so many alignment technicians are not in the habit of checking SAI. The competent technician does check SAI when there is a problem such as wander, pull, or excess road shock, which is not cured by the normal alignment steps.

The measurement of SAI can be very helpful when the alignment technician is trying to correct a camber problem on a car with a nonadjustable strut suspension. If camber is not adjustable, the correct way to obtain the right camber angle is to replace the parts that are worn or bent. There is sometimes a problem in

determining exactly which parts are or are not bent. If the included angle is correct and the camber angle is wrong, the strut or steering knuckle is not bent, and it will not do any good to replace it. If the included angle is not correct, the strut, or at least the spindle, is bent and should be replaced or corrected.

17.19   TOE OUT ON TURNS

Like its name implies, **toe out on turns** causes the front wheels to toe outward when a car turns a corner. This is necessary because the inner wheel has to make a sharper turn than the outer wheel; the inner wheel has to travel around a smaller circle. This angle is also called **turning angle** or **turning radius**. It is sometimes referred to as **Ackerman angle** because the linkage that causes toe out on turns was developed by Rudolph Ackerman (sometime between 1818 and 1820). (Fig. 17-40)

The **Ackerman linkage**, which causes toe out on turns, consists of a pivoting steering axis

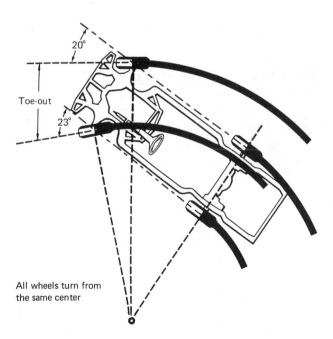

FIGURE 17-40.
To keep from scuffing, a tire must be at a right, 90° angle to the center point of the turn; this makes it necessary for the front tires to toe out on a turn. (Courtesy of Hunter Engineering)

with nonparallel steering arms. If these steering arms were parallel, the two front wheels would turn exactly the same amount during a turn, and one or both tires would have to scrub or scuff when turning corners. To understand toe out on turns, we have to look at what happens when a lever moves. We can accurately think of the steering arm as a lever with the steering axis as the fulcrum. When a tie rod moves the end of the steering arm across the car, the end of the steering arm also moves closer to or farther away from the axle. On the outside wheel, the end of the steering arm moves slightly away from the axle as it moves toward the outside of the car; it will pass through an arc where most of the motion is across the car (parallel to the axle). On the other side of the car, the steering arm is passing through an arc toward the axle, as well as across the car. This added motion—toward the axle—produces a greater turn on the inside wheel. (Figs. 17-41 & 17-42)

As a car is driven in a circle, the center of a turn is almost in line with the rear axle. Each tire must be at a right angle (90°) to a line from the center of the circle to the center of the tire or else the tire will scrub or scuff.

To produce these turning angles, the steering arms are mounted onto the steering knuckles so they follow a line that points approximately toward the center of the rear axle. If the steering arms are mounted in front of the axle, they must form this same angle. This is sometimes difficult because the tie rods will usually interfere with the tires or brakes, a common problem on many 4WD vehicles. (Fig. 17-43)

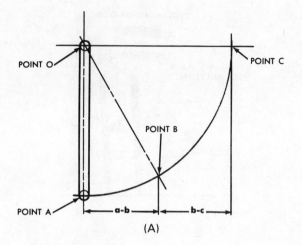

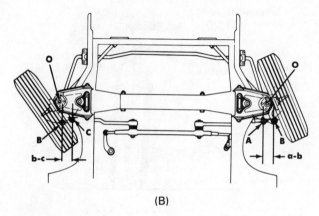

FIGURE 17-42.
Note how distance a-b is equal to b-c but the distance from point A to point B is much less than the distance from point B to point C; the lever is moving upward as well as sideways. Distances a-b and b-c compare to the sideways movement of the tie rods while the angles compare to the turning angles of the tires in the lower picture. (Courtesy of Ford Motor Company)

The amount of turning difference in the angles of the front tires is usually about 1 or 2° when one of the tires turns 20°. The actual amount of toe out on turns depends on the track width and the wheelbase of the vehicle.

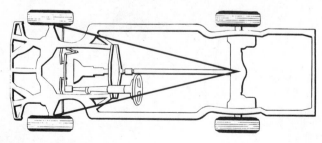

FIGURE 17-41.
Toe out on turns is a result of Ackerman angle, which is an angle of the steering arms pointing toward the center of the rear axle. (Courtesy of Hunter Engineering)

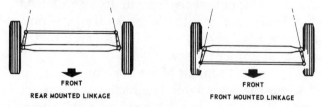

FIGURE 17-43.
When the tie rods are mounted behind the axle, the steering arms angle inward; when the tie rods are in front of the axle, they should angle outward to get the correct Ackerman angle. (Courtesy of Ammco)

17.20   TOE OUT ON TURNS—PROBLEMS

If the front wheels do not turn to the correct angles, one or both of them will have to scrub as they turn a corner. Sometimes this scrubbing can be heard as tire squeal as the car makes a slow turn on a very smooth, slick surface; rubber scuff marks are also evidence of this. This tire scrub will tend to roll rubber off the edge of the tire tread, giving a wear pattern that is similar to camber wear but with a more rounded outside edge. (Figs. 17–44 & 17–45)

The procedures for measuring and adjusting (when possible) toe out on turns are described in Chapter 18.

17.21   SET BACK

Set back should be considered during any discussion of wheel alignment even though it is not a normal wheel alignment angle. Most people believe that a car is made geometrically square or rectangular, with each of the four tires in perfect placement at each corner. This is true with some cars, but not true with many. Some cars have a wider track at the front or rear. Some

FIGURE 17–44.
This vehicle is making a sharp, right turn; the faint scrub marks behind the outer edge of the tire are a result of incorrect toe out on turns.

FIGURE 17–45.
This abnormal tire wear was possibly the result of incorrect toe out on turns; note how the shoulder of the tire has worn.

cars, through damage, wear, or design, have one of the axles at an angle other than 90° to the car's center line. On rear axles, this is called "track" or "thrust" and is discussed in Section 17.23. When one of the front tires is behind the other, it is called **set back**. Set back can be the result of worn suspension or accident damage; some new cars are built with as much as 1 1/2 inch (38 mm) of set back of the right front wheel. This set back is intended to divert the car sideways during a severe accident to help protect the occupants from severe injury or reduce the effects of torque steer. (Fig. 17–46)

Inproper set back on a driving axle can cause **torque steer**: the vehicle turns to one side during a hard acceleration. It can also cause odd tire wear or handling difficulties. Set back can also cause difficulties in adjusting toe to get a straight steering wheel; this is discussed more completely in Chapter 18.

17.22   REAR WHEEL ALIGNMENT

Because the rear wheels do not turn/steer, we only worry about two alignment angles: camber and toe. These are the tire-wearing angles,

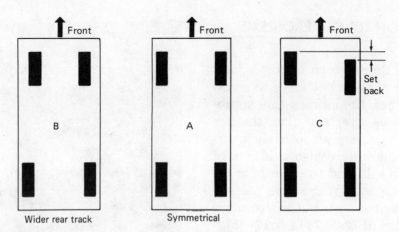

FIGURE 17–46.
Many cars are symmetrical with the tires placed in a rectangular position (viewed from above) (A); some cars have a wider axle at the front or rear (B); and some cars have set back with one of the tires of an axle behind the other.

and they can cause rear wheel tire wear as easily as front tire wear.

With most RWD cars, rear wheel alignment is controlled by the rear axle housing; its size and strength usually keeps the wheels aligned with no problem. RWD cars with independent suspension (IRS) usually have provision for camber and toe adjustments. Most FWD cars also have methods in which rear wheel camber or toe can be adjusted. **Four**

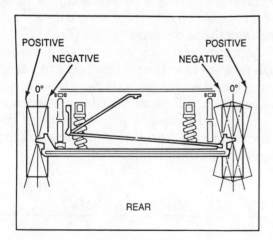

FIGURE 17–47.
Rear wheel camber is the same as front wheel camber; toe is the same also. (Courtesy of Chrysler Corporation)

wheel alignment has become very common with the increased popularity of FWD cars. (Figs. 17–47 to 17–49)

17.23 TRACK/THRUST LINE

In addition to camber and toe, rear wheel alignment should be concerned with **track** or **thrust**. If the rear wheels are parallel to each other and to the center line of the car, they will follow (track) directly behind the front wheels, and the car will be straight as it goes down the road. If the rear tires are parallel to each other but not parallel to the car's center line, the car will go down the road slightly sideways, with the rear wheel tracks to the right or left of the front wheel tracks. The car follows a line that is parallel to the average toe of the rear wheels, called the thrust line. If this line does not run down the center of the car, as it should, the driver turns the steering wheel slightly to compensate, and the car **dogtracks** (goes slightly sideways) down the road. Some cars are adversely affected by thrust angles as small as 0.06°. (Fig. 17–50)

The procedures for measuring and adjusting (when possible) rear wheel alignment are described in Chapter 18.

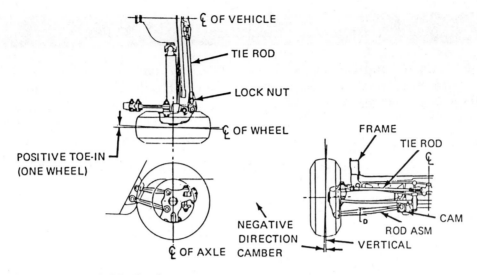

FIGURE 17-48.
Camber on this Corvette rear suspension can be adjusted by turning cam; toe can be adjusted by changing the length of the tie rod. (Courtesy of Chevrolet)

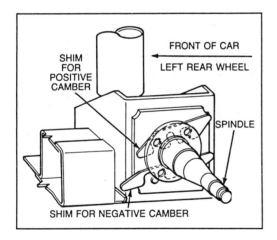

FIGURE 17-49.
Many FWD cars use shims to adjust rear wheel camber and toe. The shim can be placed at the rear to increase toe in or at the front to decrease toe in. (Courtesy of Chrysler Corporation)

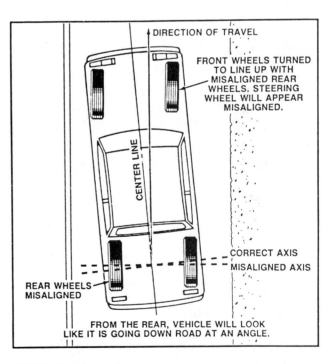

FIGURE 17-50.
If the rear wheels are not aligned so their average toe angle does not run down the center of the car, the car will go down the road slightly sideways; the "direction of travel" is the same as the rear wheel thrust line. (Courtesy of Ford Motor Company)

REVIEW QUESTIONS

1. Statement A: Caster and camber are angles of the tire. Statement B: Caster and SAI are angles of the steering axis. Which statement is correct?

    a. A only   **c.** both A and B
    b. B only   **d.** neither A nor B

2. If the front spindle is at a slight downward angle at the outer end,
    A. camber is negative.
    B. caster is positive.

    a. A only   **c.** both A and B
    b. B only   **d.** neither A nor B

3. Slightly positive camber is used on most front tires to

    a. produce scrub radius.
    b. place more of the car's load on the outer wheel bearing.
    c. offset road crown effects.
    d. none of these.

4. Technician A says that excessively positive camber on the left front tire will cause a pull to the left. Technician B says that this will cause the tire to wear at the inner edge of the tread. Who is right?

    a. A only   **c.** both A and B
    b. B only   **d.** neither A nor B

5. The distance between the center of the tires is 54 1/2 inches at the rear and 54 3/8 inches at the front. This car has

    a. 1/16 inch of toe in.
    b. 1/8 inch of toe out.
    c. 1/8 inch of toe in.
    d. none of these.

6. Technician A says that toe in will increase if the tie rods (rear mounted) are made longer. Technician B says that the tie rod sleeves are used to adjust toe out on turns. Who is right?

    a. A only   **c.** both A and B
    b. B only   **d.** neither A nor B

7. Incorrect toe settings will cause a characteristic tread wear pattern that

    a. has a feather edge at one side of each tread bar.
    b. is worn heavy at one side of the tread.
    c. is worn heavy at both sides of the tread.
    d. none of these.

8. The tires of an RWD car are normally toed in slightly so
    A. a slight amount of toe out will occur as road forces push on the tire.
    B. camber angle will not cause a pull.

    a. A only   **c.** both A and B
    b. B only   **d.** neither A nor B

9. Statement A: Caster is a directional control angle that helps keep the car going straight down the road. Statement B: Caster is caused by the forward or rearward tilt of the steering axis. Which statement is correct?

    a. A only   **c.** both A and B
    b. B only   **d.** neither A nor B

10. Technician A says that negative caster is usually used on power steering cars. Technician B says that negative caster places the upper ball joint to the rear of the lower one. Who is right?

    a. A only   **c.** both A and B
    b. B only   **d.** neither A nor B

11. If a lot of weight is placed in the trunk of a car causing the back end to lower a few inches, the

    a. caster angle will become more negative.
    b. caster angle will become more positive.

c. camber angle will become more positive.

d. none of these will occur.

12. Technician A says that road crown pull can be offset by adjusting the left wheel's caster to slightly more positive. Technician B says that this is called caster spread. Who is right?

   a. A only
   b. B only
   c. both A and B
   d. neither A nor B

13. Too much caster can cause
   A. vehicle wander.
   B. hard steering.

   a. A only
   b. B only
   c. both A and B
   d. neither A nor B

14. The angle of the steering axis, leaning inward at the top, is called

   a. SAI
   b. BJA
   c. KPI
   d. all of these

15. Statement A: SAI causes the front wheels to seek a straightahead position. Statement B: SAI is a directional control angle. Which statement is correct?

   a. A only
   b. B only
   c. both A and B
   d. neither A nor B

16. The included angle equals
   A. the SAI angle plus the camber angle (if positive).
   B. the SAI angle minus the camber angle (if negative).

   a. A only
   b. B only
   c. both A and B
   d. neither A nor B

17. The included angle is controlled by the

   a. position of the ball joints.
   b. angle of the spindle.
   c. shape and dimensions of the steering knuckle.
   d. none of these.

18. The angle created by the steering arms pointing inward at the rear is called

   a. Ackerman angle.
   b. turning radius.
   c. toe out on turns.
   d. all of these.

19. Technician A says that the front tires need to toe out on turns so they will steer easier. Technician B says that the outside tires turn sharper than the inside tires. Who is right?

   a. A only
   b. B only
   c. both A and B
   d. neither A nor B

20. The thrust line is
   A. a line that is between and parallel to the rear tires.
   B. the direction that the car goes while traveling straightahead.

   a. A only
   b. B only
   c. both A and B
   d. neither A nor B

Chapter 18

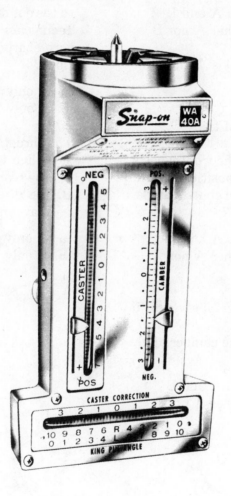

WHEEL ALIGNMENT— MEASURING AND ADJUSTING

After completing this chapter, you should:

- Be able to determine the cause of abnormal vehicle handling such as pulling, wandering, hard steering, or poor return and recommend the needed repairs.

- Be able to measure camber, caster, SAI, toe out on turns, and toe on the front wheels of an automobile.

- Be able to measure camber and toe on the rear wheels of an automobile.

- Be able to adjust those front and rear alignment angles which are adjustable on a given car.

- Be able to check a rear axle assembly for bending or misalignment.

- Be able to adjust a steering wheel to a centered position.

18.1   INTRODUCTION

Of the five different alignment angles, **SAI** is never adjustable, **toe** is always adjustable, **caster** and **camber** are often adjustable, and **toe out on turns** can sometimes be adjusted. Three of these—**camber, toe,** and **toe out on turns**—are angles of the tire; two of them—**caster** and **SAI**—are angles of the steering axis. The three that are angles of the tire will cause tire wear if they are not correct and are called **tire-wearing** angles. Caster and SAI are referred to as **directional control** angles. All five of these angles can affect vehicle handling under various circumstances. Since rear tires do not turn for steering, only two of these angles, camber and toe, apply to them.

The operation of performing a wheel alignment normally means to measure these angles (although SAI and toe out on turns are often skipped), compare the measurements with the specifications for the particular vehicle, and then adjust any of the angles that do not meet the specifications. The procedure for measuring most of the angles is fairly easy, especially with some of the more modern, computerized equipment. The procedure for correcting alignment angles can be fairly easy or extremely difficult depending on whether a means of adjustment is provided by the manufacturer. At one time, most cars provided for adjusting caster, camber, and toe; with many modern cars, there is a provision for adjusting toe only.

Manufacturers are building many cars so that caster, camber, SAI, and toe out on turns angles are correct when the car leaves the assembly line. With modern assembly methods using precise fixtures, it is faster and less expensive to build a car with the front tires in correct alignment. Adjustable parts are more expensive to build, and adjustments take time. On cars with nonadjustable alignment angles, if any of these angles become incorrect at a later date, something must be wrong (bent, worn, or sagged); in this case, the alignment is corrected by repairing the problem.

All alignment checks are made with the car on a level surface. This is because most mea-suring instruments or gauges measure the angles relative to true vertical or true horizontal. The vehicle should also be at the correct curb height; a lower or higher than normal height will change some of the alignment angles. It is also important that the tires, at least on each axle, be the same size and have the correct inflation pressure.

In all cases, the suspension and steering systems must be in good shape before performing a wheel alignment. For example, it won't do much good to measure and adjust toe if a worn tie rod end or idler arm allows the toe to change as the car goes down the road; tire wear or handling difficulties will still occur after the alignment is completed.

While doing an alignment, special wrenches and tools are available from various sources to make adjustments easier and faster. The mounting or adjusting bolts on some cars are very difficult to get to and to turn using ordinary wrenches, especially if the car has optional accessories. In some cases it is difficult to see or touch these bolts, much less turn them. The special wrenches are often bent to a configuration that will fit one or two particular bolts on one particular model of car. The special tools are designed to speed up an adjustment or make that adjustment more accurate. Special tools can be of a general nature, such as a tie rod adjusting sleeve wrench, or of a very specific nature for one particular model of car. (Fig. 18–1)

18.2   MEASURING ALIGNMENT ANGLES

Because some alignment angles change each time the tires are turned from a straightahead position and also change with vertical suspension travel, care should always be taken that the tires are pointed straight forward and the car is at the correct ride height. Some alignment angle checks require turning the tires a certain number of degrees from straight ahead; the tires are placed on turntables with degree scales to allow these measurements. If these precautions are not followed, there will be an

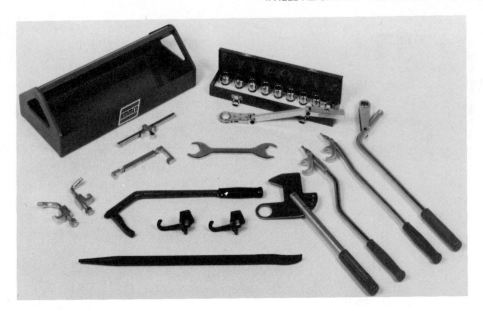

FIGURE 18–1.
A special set of alignment tools; in some cases, they are used to make the adjustments easier and faster; in some cases, the job cannot be done without them. (Courtesy of Ammco)

error in the measurements and corresponding adjustments.

Caster-camber gauges are commonly available that attach to the front hubs using a strong magnet. These **magnetic gauges** use a simple bubble level with a scale so the camber angle can be read. The gauge is usually used along with a pair of portable **turntables** or **slip plates**; these ensure that the tires can move/ scuff sideways when they are lowered and that the suspension moves to the correct ride height. The turntables can be placed directly on the floor, but some shops use two **portable stands** for the turntables and two more stands for the rear tires to provide a level surface for the car while the measurements are being made. The stands also allow access to the suspension and steering components when the adjustments are made. (Figs. 18-2 & 18-3)

In cases where the center of the wheel does not allow attachment of the gauge or there is damage to the end of the hub, adapters can be used to attach the gauge onto the wheel rim. Bubble level gauges are also available in a hand-held type; to read camber or caster, this gauge unit is simply held against the side of the wheel. (Figs. 18-4 to 18-6)

SAFETY NOTE

All wheel alignments should result in a car that will drive and handle in a safe and efficient manner. To ensure this, it is necessary to observe these precautions:

- The car must be given a prealignment inspection to ensure there are no worn or unsafe parts.

- The alignment angles should be adjusted to the specifications recommended by the vehicle manufacturer using the recommended methods.

- All nuts and bolts should be tightened to the correct torque rating.

- The car should be road tested to ensure proper driveability.

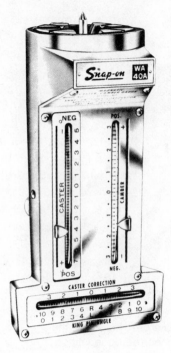

FIGURE 18-2.

A magnetic caster-camber gauge; note that this particular gauge also has a king pin angle/caster correction scale. (Courtesy of Snap-On)

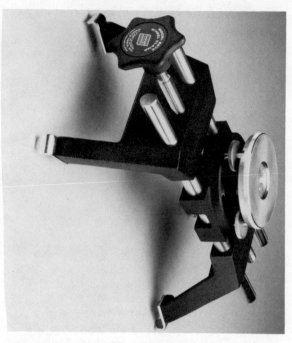

FIGURE 18-4.

A wheel clamp allows mounting a magnetic gauge to the wheel instead of the hub; use of this tool will often result in more accurate gauge attachment because compensation can be made for mounting errors. (Courtesy of Ammco)

FIGURE 18-3.

A tire being lowered onto a turntable. At this time the tire should be centered to the turntable and pointed straight ahead; the turntable scale should be at zero.

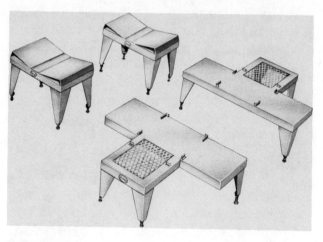

FIGURE 18-5.

A set of floor stands. The turntables are placed on the larger pair of stands, and these two stands are placed under the front tires. The rear tires are placed onto the other pair of stands. This places the car and the suspension at a better working height with access under the car for adjustments. (Courtesy of Snap-On)

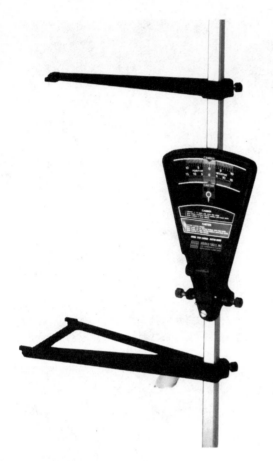

FIGURE 18-6.
This caster-camber gauge is hand-held and placed with the projections at the left against the wheel rim while making measurements. (Courtesy of Ammco)

More sophisticated electronic gauges are available. These are often called **alignment machines** or **systems**. They also attach to the wheel, and they measure the angles electronically or project a light beam onto a graduated screen. They usually measure camber, caster, toe, and sometimes SAI. Most of the newer computerized machines electronically compensate for mounting error or wheel runout. Most are **four-wheel systems** that can measure the location and angle of all four wheels at the same time, and some give a printed readout comparing the vehicle measurements with the specifications. Electronic gauges tend to be more accurate, more sensitive, more fragile, and much more expensive. These gauges are often quicker and easier to use than the magnetic gauges. One manufacturer claims the ability to measure front wheel camber, caster, and toe and rear wheel camber and toe in 5 1/2 minutes from beginning to attach the wheel clamps to getting a readout of the measurements.

Electronic alignment systems use three basic ways of displaying alignment measurements: a light beam, a group of meters (**analog type**), and a digital readout; in addition, some of the newer systems will print out the readings on a sheet of paper. All alignment systems give two-wheel readouts; some will measure and display the angles of all four wheels. Most systems use wheel rim mounting for the **sensing** or **projector units**; these units determine the position of the tire so the measurements can be taken. They usually must be **compensated** so they are perfectly aligned with the tire to eliminate mounting runout. Some systems require a manual runout compensation of the sending units. Some newer computerized systems have electronic compensation; all the operator needs to do is push one or two switches while rotating the tire. (Fig. 18-7)

Alignment systems sometimes take a little longer to mount and set up on a car than magnetic gauges, but they measure and read out more angles and the display is much larger so they can be read accurately from a distance, even from under the car. With a little experience, a wheel alignment can be done much faster and more accurately using these systems than with portable equipment.

Each style of alignment system has some means of checking and **calibrating** itself to maintain accuracy. Some computerized systems run their own self-checks.

Alignment systems are used in conjunction with **alignment racks**. Racks provide a convenient, level surface so the front tires (usually) are on turntables and all four tires are at the same, level height while the measurements are being made. They also provide access to the suspension and steering components so that checks or adjustments can be made quickly and easily. Most racks have built-in, air-powered jacks so the car or suspension parts can be easily raised for service.

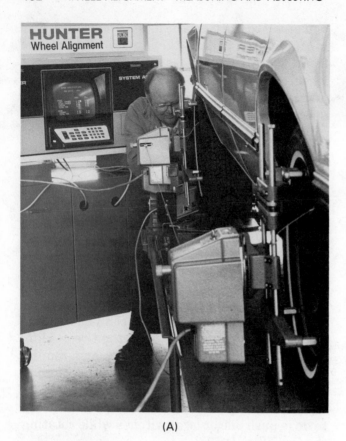

(A)

(B)

FIGURE 18–7.
Three different styles of wheel alignment systems; system A is computerized with a CRT readout, system B is computerized with analog meter readout, and system C uses a light projection readout. All three are four-wheel systems capable of measuring the angles on all four wheels. (Courtesy of Hunter Engineering)

The following different styles of racks are in common use:

1. **Pit racks,** also called **half** or **stub racks.** A pit is constructed in the floor of the shop with an island or peninsula to support the turntables; the pit provides access to the suspension and steering parts. The car is driven straight onto the pit with the wheels to be aligned over the turntables.

2. **Fixed/power rack.** Two metal runways with the turntables built into the front end are secured to the shop floor. The rear of these runways can be lowered so the car can be driven on and then raised to a level position. The height of the runways provides a convenient working height and access to the suspension and steering parts.

(C)

3. **Hoist rack.** Two metal runways with turntables are attached to a standard, single-post shop hoist. Raising the car on the hoist provides access for repairs or adjustments. The rack is lowered onto four stands (built into the rack) to provide a level surface while the measurements are made. (Fig. 18–8)

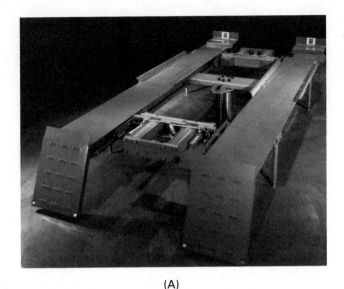

(A)

(B)

(C)

FIGURE 18–8.
Three different styles of alignment racks; system A is attached to a single post shop hoist; it allows raising the car for repairs and lowering onto supports while the alignment is being made. System B is a power rack; the rails are lowered so a car can be driven on and raised for a level alignment surface. System C is a pit rack with the top set at floor level; the vehicle is merely driven onto the alignment stands. (Courtesy of Hunter Engineering)

18.3   WHEEL ALIGNMENT SEQUENCE

Most technicians follow a particular sequence of operations while doing a wheel alignment. Part of the sequence is personal preference, part is not. Toe is always the last angle to check and adjust because it is affected by the others. For example, a more positive camber adjustment will move the top and also the center of the tire outward; this will produce less toe in on cars with rear-mounted tie rods. When using modern computerized equipment, the measurement sequence is programmed into the machine.

Camber and caster are always measured first and adjusted at the same time or in an order that is determined by the technician's preference or the car. For example, on a car with S-L A suspension and a sliding adjustment,

they are made together. On cars with shim or eccentric adjustments, they can be made in any order.

Toe out on turns is measured along with caster because this is the easiest and quickest time to do it. Measurement of SAI is often skipped unless there is a problem with the car. These two angles are normally not adjusted.

18.4   MEASURING CAMBER

Since it is compared to true vertical, camber could be measured with a carpenter's level or a plumb bob. However, the bulge on the lower side of the tire would get in the way and affect the measurement. Each gauge measures camber from true vertical. The greatest error in measuring camber probably results from the attachment of the gauge to the measuring surface, either the hub or the wheel rim. If these surfaces are not true or straight, the measurement will be wrong, and the resulting adjustment will also be wrong. If a hub mount gauge is used, the face of the hub should be checked to ensure that it is true; all wheel rim mount systems should be adjusted to compensate for rim runout or mounting errors.

18.4.1 Measuring Camber Using a Magnetic Gauge

Many small shops prefer **portable alignment gauges** because they are less expensive and are relatively durable. They also take only a small amount of skill and are fairly fast to use.

Magnetic gauges can be checked for accuracy by one of two different and fairly easy methods. If a vertical, iron surface is available—some door frames for example—attach the gauge to the metal surface; if the surface is truely vertical, the camber gauge should read zero. The gauging surface can be checked with a carpenter's level if there is any question. If a flat piece of metal that has two smooth and parallel sides is available, clamp the metal in a vise and measure the camber on each side of the metal. If both sides read zero or if one side shows a negative reading that is equal to a positive reading on the other side, the gauge is accurate. Most magnetic gauges have a provision for calibrating or correcting the camber scale. (Fig. 18–9)

To measure camber using a magnetic gauge, you should:

1. Park the car on a level surface.
2. Remove the hub cap and the front hub dust cap. Clean the vertical, machined surface at the end of the hub. Check this surface for any bumps or burrs that might prevent the gauge from seating completely.
3. Raise the tire off the ground, grip the tire at the top and bottom, and rock the tire to check for loose wheel bearings. Readjust the wheel bearings if necessary.

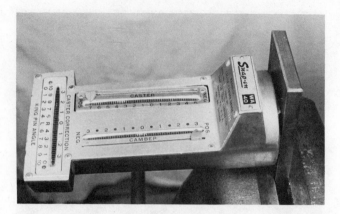

FIGURE 18–9.
The accuracy of a magnetic caster-camber gauge can be checked by placing the gauge on both sides of a flat metal plate; if one reading is negative and the other one equally positive or if both readings are zero, the camber scale is accurate. If not, the gauge needs to be calibrated.

4. Center the magnetic gauge with the spindle, and carefully, because of the strong magnet, attach the gauge to the hub. (Fig. 18–10)

5. With the tire still in the air, cross-level (level in a horizontal or front to back direction) the gauge, and read the amount of camber in-

FIGURE 18–10.
After the hub surface is cleaned, the magnetic gauge is centered and then attached to the hub. (Courtesy of Snap-On)

dicated on the gauge (this reading is only to check for possible errors). Rotate the tire 1/2 turn or 180°, rotate the gauge 1/2 turn, reread the gauge, and compare the two readings; if they are different, there is a defect in the gauge mounting that is causing an error in one or both readings. If there is an error or the gauge cannot be mounted directly onto the hub, a rim mount adapter should be used or the hub surface should be trued. This unit is attached to the wheel rim and has a provision for calibrating or compensating for errors. Calibration procedures will vary between different manufacturers.

To use a rim mount adapter, you should:

1. Attach the rim mount adapter securely to the wheel rim and attach the magnetic gauge to the adapter. (Fig. 18-11)

2. With the adapter straight up and down, cross-level the gauge and read the camber, hold the gauge still while you rotate the tire 1/2 turn, and reread the gauge.

NOTE: Some rim mount adapters have three compensating screws. When using

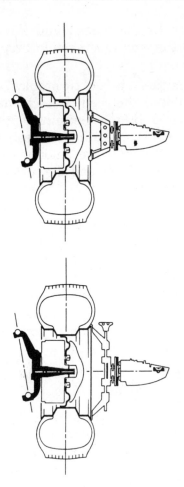

FIGURE 18-12.
Runout of the wheel clamp is corrected by measuring the camber with the wheel rotated to three different positions; the adjustment thumb wheels are then turned so the gauge reads the average reading. After compensating for runout, all three readings should be the same. (Courtesy of Ammco)

one of these units, a camber reading should be taken every 1/3 turn or 120°. A reading should be taken with each compensating screw straight up. (Fig. 18-12)

3. If the two (three) readings are not the same, determine the average reading and adjust the compensating screw(s) until the camber gauge reads the average reading. For example, if the readings are +1/4° and +3/4° degrees, the compensating screw would be adjusted to a reading of +1/2°.

NOTE: The compensating screw must be in a vertical position during this adjustment.

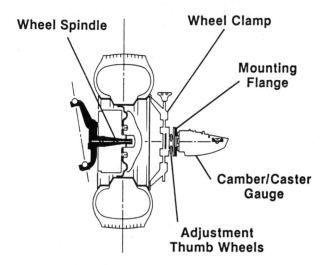

Wheel Spindle · Wheel Clamp · Mounting Flange · Camber/Caster Gauge · Adjustment Thumb Wheels

FIGURE 18-11.
A wheel clamp attaches to the outside or inside edge of the bead flange and provides a mounting flange for a magnetic caster/camber gauge; note the adjustment/compensation thumb wheels to correct gauge runout. (Courtesy of Ammco)

4. Repeat Steps 2 and, if necessary, 3 until the readings taken in Step 2 are the same.

6. Position a pair of turntables so they are centered under the road contact patch of the tire. If caster or toe measurements are also going to be made, the front wheels should be straight ahead with the amount of toe split equally at each wheel, and the turntable scales should be at zero.

To **split the toe** you should:

1. Sight past the inner edge of the right front tire and note where your line of sight meets the right rear tire.

2. Sight past the inner edge of the left front tire and note where your line of sight meets the left rear tire.

NOTE: On cars with toe out, it is necessary to sight along the outside of the tires. (Fig. 18–13)

3. If the lines of sight do not meet the rear tires in the same place, strike the appropriate sidewall of one of the front tires so the tires are turned slightly to the right or left.

4. Repeat Steps 1, 2, and 3 until the sight lines are equal.

NOTE: If caster is going to be measured, the rear tires should be placed on blocks that are equal to the height of the turntables; the car must be level in a front-to-back as well as a side-to-side direction. (Fig. 18–14)

FIGURE 18–14.
When doing alignment checks with the turntables placed on the floor, the rear tires should be placed on blocks to keep the car level.

7. Pull the locking pins from the turntables and lower the car onto the turntables. Bounce the end of the car up and down; you should be able to see the turntables and the tires move sideways during the bouncing action.

8. Cross-level the magnetic gauge and read the camber scale on the gauge. The amount of camber is indicated where the center of the bubble aligns with the gauge scale. (Fig. 18–15)

OR On one popular gauge type, the camber scale is rotated until the bubble is centered, and then the amount of camber is read on the scale where the pointer indicates. (Fig. 18–16)

18.4.2 Measuring Camber Using an Alignment System

At this time, there are several different types of alignment systems. The specific operation of each system varies among different models and manufacturers and is too varied to describe in a text of this type. The following description is very general.

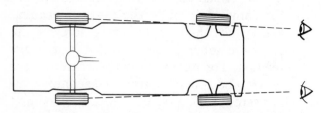

FIGURE 18–13.
Splitting the toe is done by sighting down the edges of both front tires. The tires are turned until the line of sight meets each rear tire at the same location (relative to the center of the car).

FIGURE 18-15.
The camber angle is read by locating the center of the bubble on the camber scale and reading the scale at that point. This gauge is showing about −1/4°.

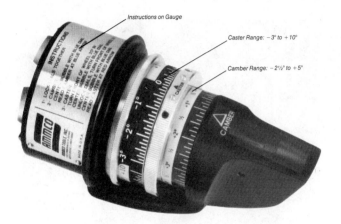

(A)

(B)

FIGURE 18-16.
This camber gauge shows about +1°. It is read by turning the camber scale (B) until the bubble is centered and then reading the scale at the arrow. (Courtesy of Ammco)

FIGURE 18-17.
This sensor head of a wheel alignment system is placed in position at the wheel bead flange and is being locked in place. (Courtesy of Hunter Engineering)

To measure camber using an alignment system, you should:

1. Drive the car onto the alignment rack and position it with the tires over the turntables.

2. Attach the sensors/projectors securely onto the wheel. (Fig. 18-17)

3. Raise the tire off the turntable and, using the correct procedure, compensate for any sensor or projector mounting errors. (Fig. 18-18)

4. With the tires in a straightahead position, lower them onto the turntables, making

FIGURE 18-18.
One of runout compensating screws being turned to correct mounting errors. Readings are taken with the wheel turned to three different positions, and the screws are adjusted to an average position. Newer computer units will self-compensate for mounting error in a quick and easy step. (Courtesy of Hunter Engineering)

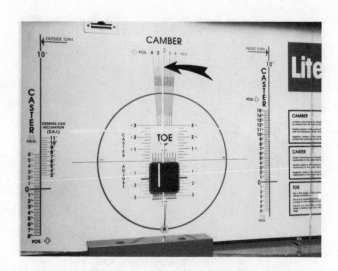

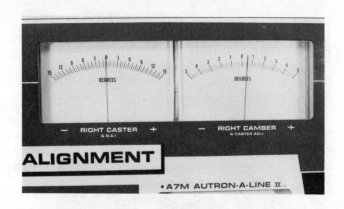

sure that the tires are centered on the turntables.

    5. Bounce the car up and down to ensure that the suspension is settled to ride height.

    6. Read the camber display. (Fig. 18–19)

18.5   CAMBER SPECIFICATIONS

All car manufacturers publish wheel alignment specifications that include camber. A typical specification would be 0° to 3/4° with a maximum variation or spread between the two tires of 1/8°. This means that a camber measurement of 0° or positive 3/4° or anywhere in between would be correct, as long as both readings are within 1/8° of each other. All specifications are assumed to be positive unless negative is specified. Another way of publishing a specification is 1/2° ±1/4. With this specification, +1/2° − 1/4 = +1/4° and +1/2° + 1/4 = +3/4°; anywhere in between +1/4° and +3/4° is acceptable. In this case, +1/2° would be the most ideal setting. (Fig. 18–20)

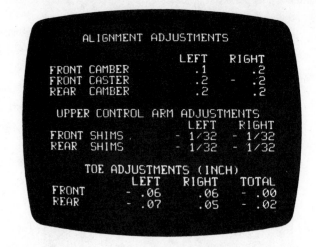

FIGURE 18-19.
The camber readout from three types of alignment systems. The light projection unit (A) is reading +1/2° (at arrow). The analog meter unit (B) is also reading +1/2°. The screen on the computer unit is displaying all of the alignment readings plus indicating what shim changes are needed to adjust the settings. (Courtesy of Hunter Engineering)

SPECIFICATIONS

FRONT WHEEL ALIGNMENT	Acceptable Alignment Range	Preferred Setting
CAMBER—All Models	−0.2° to +0.8° (−1/4° to +3/4°)	+0.3° (+5/16°)
TOE—All Models		
Specified in Inches	7/32″ OUT to 1/8″ IN	1/16″ OUT ±1/16″
Specified in Degrees	.4° OUT to .2° IN	0.1° OUT ±.1°

FIGURE 18–20.
Wheel alignment specifications. Note that this manufacturer prints their specifications in both decimals and fractions and that the readings are given in an acceptable range and a preferred setting. (Courtesy of Chrysler)

Specifications for domestic cars are usually given in degrees or fractions of a degree (2° or 1/4°). Decimals are occasionally used for portions of a degree, for example, .25° instead of 1/4°. Imported car specifications are also given in degrees, but angles less than 1° are measured in minutes and sometimes even seconds.

Some alignment technicians feel that the most ideal camber angle as far as tire wear and driveability is concerned, for the front wheels of an RWD car, is 1/4° to 1/2° positive; they will try to make their final settings (within range of the specifications) to be within this 1/4° range.

18.6   ADJUSTING CAMBER

If the camber readings fall within the specifications and no camber wear is present on the tire, there is no need for an adjustment. If the measurements are outside of the specification range or they are within the range and the tire is showing camber wear, camber should be adjusted. The camber range on some cars is sometimes quite wide, sometimes as much as 2°, and some cars will show tire wear when they are as little as 1/4 to 1/2° from ideal. It is very possible for a tire to be at the positive end of the camber range, and be within specifications, and show positive camber wear on the tire. It should be adjusted to a more negative position. (Fig. 18–21)

The amount of camber change that will occur at the tire while an adjustment is being made varies from car to car depending on the height of the steering axis (between the two steering pivots). Because we are working with degrees, the amount of change adjustment becomes an arc of a circle with the radius of the circle being the length of the steering axis. For example, if an adjustment is being made on a car with S-L A suspension and there is a span on the steering knuckle of 14 inches (35.6 cm) between the ball joints, we will be working with a radius of 14 inches. A sideways movement of 1/8 inch (3mm) will cause a camber change of .5 or 1/2°. The circumference of a circle having a radius of 14 inches is almost 88 inches (223 cm); if we divide 88 by 360, the result is .24 inches (6.1 mm). If an arc of .24 inches is equal to 1°, then an arc of .125 inch (1/8 inch) would equal about .5°. If we were adjusting a car equipped with a strut and there was a span of 36 inches (91 cm) between the ball joint and the upper strut mount, the radius would be 36 inches, the circumference would be 226 inches (574 cm), a

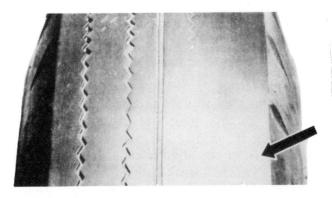

FIGURE 18–21.
This tire is showing camber wear. If the outside was on the same side as the arrow, camber was too positive. Camber should be then adjusted to a more negative position. (Courtesy of Snap-On)

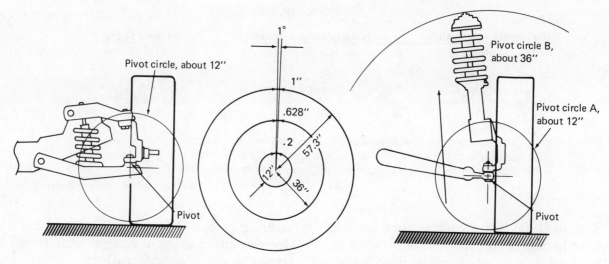

FIGURE 18–22.
As the distance between the pivot and the adjustment increases, larger and larger changes must be made to get the same amount of angle change. A 1° change would require a 1-inch shim if the adjustment distance was 57.3 inches, a 0.628-inch shim at 36 inches, or a 0.2-inch shim at a 12-inch distance.

degree would become a span of .63 inches (1.6 cm), and the 1/8 inch adjustment would now cause a camber change of only .2°. (Fig. 18–22)

Many cars provide for adjusting camber. These adjustments are made by moving the top or bottom of the steering knuckle and tire inward or outward. The usual adjustment methods are to add or remove shims between the control arm mounts and the frame; turn eccentrics where the control arm attaches to the frame or ball joint or where the strut connects to the steering knuckle; or slide the control arm pivot shaft or the top of the strut inward or outward or the strut-to-steering-knuckle connection, along slotted mounting holes. (Figs. 18–23 & 18–24)

Some front suspensions are manufactured with no provision for adjustment. In these cases, if camber goes out of specifications, something is worn or bent. It is now the responsibility of the alignment technician to determine whether a weak spring, worn bushings, or bent suspension arms, spindle and/or frame is at fault. A suspicion of a bent steering knuckle or strut can be confirmed by comparing the camber reading to the steering axis inclination specification or included angle specification.

Remember that the included angle is the angle between the camber and SAI angles. If the steering knuckle or strut is bent so the spindle changes position, the included angle will change. For example, let's imagine a car that comes in with a camber reading of −3/4°, a camber specification range of 0 to +1/2°, and

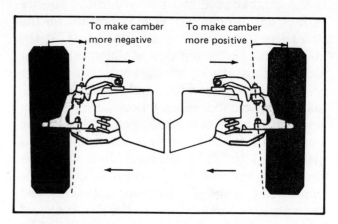

FIGURE 18–23.
Camber is adjusted to a more positive position by moving the top of the tire outward; this is done by moving the upper ball joint outward or the lower ball joint inward. Camber is made more negative by adjusting in the opposite direction. (Courtesy of Ford Motor Company)

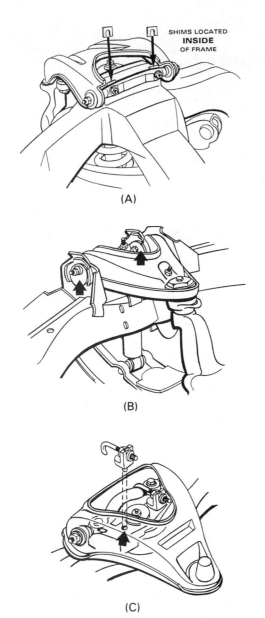

FIGURE 18-24.

S-L A suspensions are usually adjusted by adding or removing shims (A), turning one or two eccentric cams (B), or sliding the control arm shaft inward or outward along slotted holes in the frame (C). (Courtesy of Hunter Engineering)

an included angle specification of 9°. There is no specification for SAI, but an included angle of 9° minus a camber angle of +1/2° (ideal) gives us an SAI angle of 8.5°. If you were to measure the SAI on this car, and it measured 8.5° or close to it, you would know that the strut was bent and should be replaced. If the SAI measured about 10°, you would know that the included angle was almost correct (9 + 3/4) so the steering knuckle or strut is not bent,

but damage to the car has moved the upper strut mount or the lower control arm mounting locations, which has changed the SAI angle and, in turn, the camber angle. If a front end technician were to bend the strut to correct the camber (in the second example), camber might end up correct, but the car would probably have handling difficulties because of the incorrect SAI angle. Remember that a quick and simple check for a bent strut is to compare the distances between the brake rotor and the strut on each side of the car. The faulty parts should be corrected or replaced. (Fig. 18-25)

Camber adjustments will always change the track width slightly and, therefore, will change toe; it is always a good practice to check toe after adjusting camber or caster.

Bending tools are available from aftermarket suppliers to bend axles or steering knuckles into alignment. Bending often saves the technician time and the customer money, but it should be used with discretion as the strain that is placed on the parts could cause damage or failure of the part later. Since it does not correct the cause of the misalignment, many technicians feel that bending only hides a problem instead of curing it. **No car manufacturer recommends the bending of a suspension component.** (Fig. 18-26)

18.6.1 Adjusting Camber, Shims

Shim adjustments are the favorite of many alignment technicians. Shims are fairly easy to change, and it is easy to make very small or large adjustments and to keep track of how much adjustment was made. The shims can be located in pairs at the upper or lower control arm or in a single group at the lower, single control arm. The upper control arm shaft can be mounted inboard or outboard of the frame mounting bracket. Adding shims on a control arm shaft that is inboard of the mounting bracket will cause a more negative camber change; if the control arm shaft was outboard, a more positive change would occur. If the shims are used in pairs, and two equal-sized shims are added or removed, front and back, for camber adjustments; a caster change will

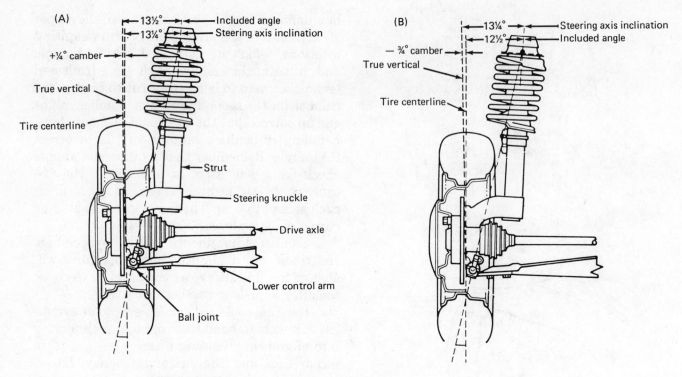

FIGURE 18-25.
This car has a camber angle of +1/4°, an SAI angle of 13 1/4°, and an included angle of 13 1/2° (A). The strut at B with −3/4° camber, an SAI angle of 13 1/4°, and an included angle of 12 1/2° must be bent.

occur if the number of shims changed at one end is different from the number changed at the other end. (Fig. 18-27)

Shims are available in various thicknesses between 1/64 and 3/16 inch (.4 and 4.7 mm). Most technicians use the largest shims they can in order to reduce the total number of shims. It is a good practice to use a maximum of five shims at any one location; the chance of control arm loosening and shim loss increases as the number of shims increases. Camber and caster are usually adjusted at the same time, using the same shims. Shim style caster adjustments are described in Section 18.8.1. (Figs. 18-28 & 18-29)

To adjust camber using shims, you should:

1. Measure camber, record the readings, and determine how much change is needed and in what direction the change should be made.

2. Locate the shims and determine whether shims will need to be added or removed.

3. Loosen the control arm mounting bolts, pry the control arm away from the bracket, and change the shims by the desired amount. (Fig. 18-30)

4. Snug down the control arm mounting bolts and remeasure camber.

5. If the camber angle is correct, tighten the mounting bolts to the correct torque. If the camber angle is still wrong, repeat Steps 3 and 4. At this time it should be fairly easy to determine how much of an additional shim change is needed by how much of a change occurred during the last adjustment.

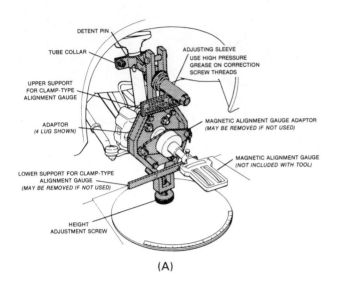

(A)

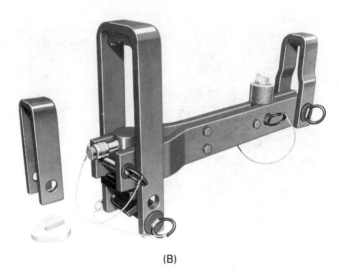

(B)

(C)

FIGURE 18–26.

When no adjustment is provided, camber can be adjusted by bending the parts; caution should be used when bending suspension parts because they can be severely weakened by the bending action. Tool A is set up to bend a strut, tool B is designed to bend a twin-I Beam axle, and the tool set (C) is designed to bend large truck axles. (A and B are courtesy of Branick Industries; C is courtesy of Bee Line)

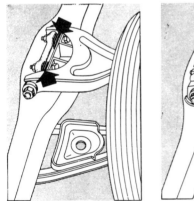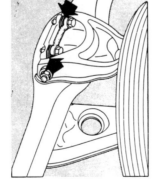

FIGURE 18–27.

Increasing the shim pack at the left will make camber more negative while increasing the shim pack at the right will make camber more positive. (Courtesy of Snap-On)

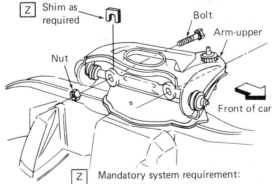

FIGURE 18–28.

Some of the practices that should be followed when changing shims to ensure a good job. (Courtesy of Buick)

Z Mandatory system requirement:

Shim as required to make caster and camber adjustment. A maximum of one each of two thinner shims (.030 or .060 in.) may be used per bolt.
Shims must be clean and free of foreign substance. No foreign substance may be added to aide assembly.

A. Maintain a minimum of .10 in. thread length beyond nut after assembly (not including chamfer).

B. Difference between front and rear shim pack must not exceed .40 in.

FIGURE 18–29.
Alignment shims are available in different widths to fit different bolt sizes. Thicknesses of 1/64 and 1/8 inch are also available. (Courtesy of Ammco)

18.6.2 Adjusting Camber, Eccentric Cams

Eccentric cams look like off-center washers that are attached to the ends of the inner control arm mounting bolt. The frame bracket for these cam bolts have slotted openings through which the bolts are positioned and a pair of raised shoulders that the eccentric cams act against. Turning the cam bolt will cause the bolt and the control arm bushing to move sideways, in and out, in the frame bracket. When the nut is tightened on the cam bolt, the bolt and the control arm are locked in that position. (Fig. 18–31)

Eccentric cams are used at the upper or lower control arms on S-L A suspensions, on some strut suspensions where the strut is connected to the steering knuckle, and on some IRS suspensions. The eccentric cams allow very fine or rather large adjustments. Like shims, eccentric cams can be used in pairs on control arms or singly on single bushing control arms. Also like shims, when cams are used in pairs, the cams, front and back, should be turned equally when adjusting camber; turning only one cam will change caster as well as camber.

Occasionally, on very rusty cars, cam bolts rust in place and become very hard to turn. Penetrating type oils or solvents help. They should be applied early in the alignment sequence, during the prealignment inspection, so they will get a chance to soak in by the time the adjustment is made. Sometimes it helps to

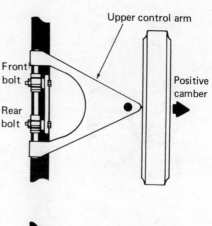

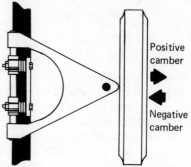

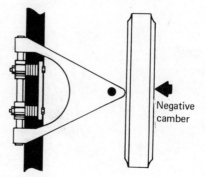

FIGURE 18–30.
When adjusting camber, an equal change should be made at the front and rear shim pack.

use an air hammer with a punch against the head of the bolt to vibrate the bolt loose or vibrate the penetrating oil in and around the bolt. Damaged or broken cams or cam bolts should be replaced with new ones.

To adjust camber using eccentric cams, you should:

1. Measure the camber, record the readings, and determine the amount of camber change desired.

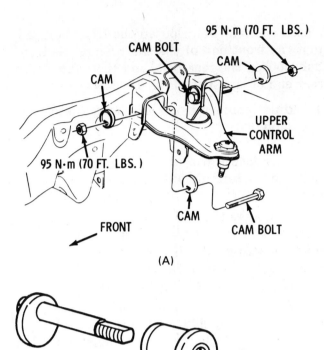

FIGURE 18-31.
This upper control arm (A) is equipped with eccentric cams to adjust camber and caster. Note the frame brackets on each side of the cam, which the cam pushes against, and the slotted hole in the frame for the cam bolt. A replacement cam bolt kit plus a control arm bushing (B). Note that this cam has two flats so it will have a stronger fit to the cam bolt. (A is courtesy of Oldsmobile; B is courtesy of Moog Automotive)

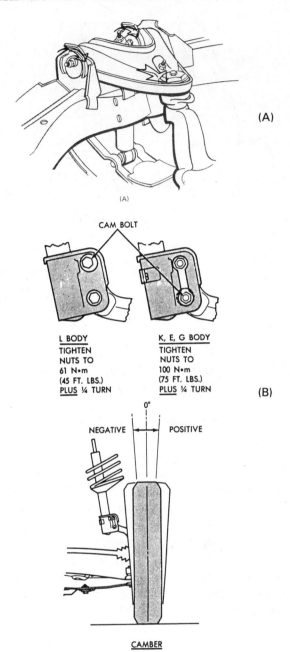

FIGURE 18-32.
Turning each eccentric cam the same direction an equal amount will change camber without changing caster (A). Some strut suspensions provide an eccentric cam where the strut attaches to the steering knuckle for a camber adjustment. (A is courtesy of Hunter Engineering; B is courtesy of Chrysler Corporation)

2. Locate the cams and determine which way they need to be turned. It is a good practice for beginners to mark the position of the cam in case it becomes necessary to return to the starting point.

3. Loosen the retaining nut on the cam bolt, and using a wrench on the head of the bolt, turn the cam bolt enough to move the control arm the desired amount. If a pair of cam bolts are used, turn the other cam bolt an equal amount in the same direction. (Fig. 18-32)

4. Remeasure camber and if necessary repeat Step 3. It should be fairly easy at this time to determine how far each cam needs to be turned to complete the adjustment.

5. When camber is correct, tighten the retaining nut to the correct torque.

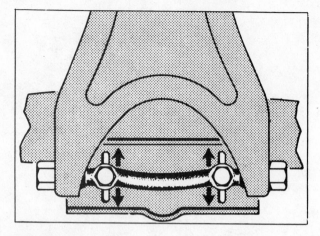

FIGURE 18-33.
To make a camber adjustment on some cars, the control arm inner shaft mounting bolts are loosened, the control arm is slid to the desired position, and the bolts are retightened. (Courtesy of Snap-On)

18.6.3   Adjusting Camber, Sliding Adjustment

Sliding adjustments are used at the upper control arm inner shaft on some S-L A suspensions, at the strut-to-steering-knuckle connection on some strut suspensions, and at the strut-to-fender-well mount on some cars. A slotted opening is provided in the frame bracket to allow the loosened control arm shaft mounting bolts to slide to different positions across the top of the bracket. The bottom of the control arm shaft is serrated to help lock it in position on the bracket when the mounting bolts are tightened. (Fig. 18-33)

Many technicians have difficulty making fine adjustments with sliding adjusters; when they loosen the mounting bolts to allow an adjustment, the control arm slides, moves too far, and loses the starting point. The mounting bolts should be loosened just enough to allow the adjustment and no more. Also, tools are available to help gradually move the control arm the desired amount and to hold it in that position. Some technicians use a jack under the frame to help them; the control arm tends to move inward when the car's weight is on the tires and outward when the frame is lifted. Some technicians prefer to move one end of the shaft at one time in a sequence like this: barely

loosen the rear bolt, loosen the front bolt, adjust the front end of the shaft, snug the front bolt, loosen the rear bolt, and then adjust the rear end of the shaft. (Fig. 18-34)

To adjust camber using sliding adjustments, you should:

1. Measure the camber, record the readings, and determine what camber changes are needed.

2. Locate the adjustment location and determine which way the adjustment should be made. It is a good practice for the beginner to mark the location of the sliding parts in case it is necessary to return to the starting point.

3. Loosen the mounting bolt(s), attach the correct tool, and slide the control arm shaft or strut mount bracket the desired amount. (Fig. 18-35)

4. Snug down the mounting bolt(s) and remeasure camber. Repeat Step 3, if necessary.

5. Tighten the mounting bolt(s) to the correct torque.

18.6.4   Adjusting Camber, Miscellaneous Styles

There are other methods of adjusting camber that are used on a rather limited basis. Most of these described here are built into the car by the manufacturer; a few are aftermarket modifications that provide an adjustment method on nonadjustable suspensions or provide additional adjustment on some suspensions that have a limited adjustment range. This text does not completely describe the adjustment procedure for each of these types; the procedure is usually described in manufacturers' or technicians' service manuals.

Camaros and Firebirds (RWD with front struts) and some General Motors FWD cars that use strut suspensions use a slotted/sliding adjustment at the upper strut mount. This allows the top of the strut to be moved inward or outward after the retaining nuts are loosened. A special tool is available to help move the upper strut mount to the exact position desired. (Fig. 18-36)

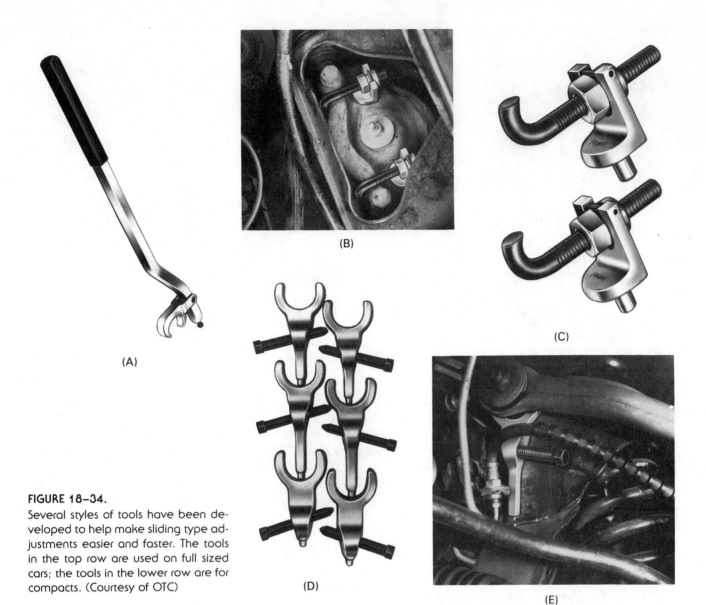

(A)

(B)

(C)

(D)

(E)

FIGURE 18–34.
Several styles of tools have been developed to help make sliding type adjustments easier and faster. The tools in the top row are used on full sized cars; the tools in the lower row are for compacts. (Courtesy of OTC)

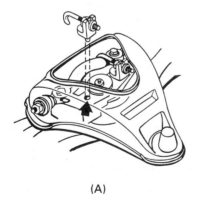

(A)

FIGURE 18–35.
If the upper control arm is slid outward (A), camber will become more positive; one of the special tools is extremely helpful with S-L A suspensions. Some cars with struts and sliding adjustments can be adjusted as shown in B. (A is courtesy of Hunter Engineering; B is courtesy of Oldsmobile)

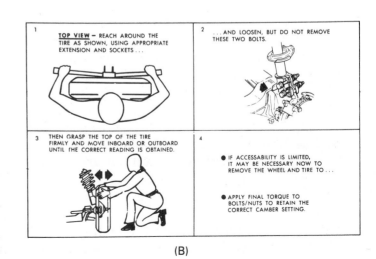

1. **TOP VIEW** – REACH AROUND THE TIRE AS SHOWN, USING APPROPRIATE EXTENSION AND SOCKETS . . .

2. . . . AND LOOSEN, BUT DO NOT REMOVE THESE TWO BOLTS.

3. THEN GRASP THE TOP OF THE TIRE FIRMLY AND MOVE INBOARD OR OUTBOARD UNTIL THE CORRECT READING IS OBTAINED.

4.
- IF ACCESSABILITY IS LIMITED, IT MAY BE NECESSARY NOW TO REMOVE THE WHEEL AND TIRE TO . . .
- APPLY FINAL TORQUE TO BOLTS/NUTS TO RETAIN THE CORRECT CAMBER SETTING.

(B)

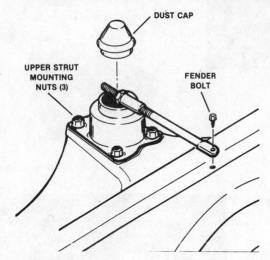

FIGURE 18-36.
This special tool is used to aid the camber adjustment on a late model Camaro or Firebird; the tool is attached to the upper strut mount as shown, the three mounting bolts are loosened, the tool is adjusted to obtain the correct camber setting, the mounting bolts are tightened, and the tool is then removed. (Courtesy of Branick Industries)

Some Ford products using a modified strut suspension are equipped with an upper strut mount that can pivot on one of the retaining bolts. After the retaining nuts are loosened and a pop rivet is removed (in some cases), the upper strut mount is rotated to obtain the correct camber, and the retaining nuts are retightened to lock the adjustment in place. A special tool is available to make this job quicker and easier. (Fig. 18-37)

Model year 1961 to 1976 Cadillacs used an eccentric bushing between the ball joint stud and the steering knuckle. Camber is adjusted by loosening the ball joint stud nut and rotating the bushing. (Fig. 18-38)

Late model Ford 4WDs and pickups (both 2WD and 4WD) use an eccentric sleeve between the axle and the upper ball joint stud. A similar, aftermarket bushing is available for

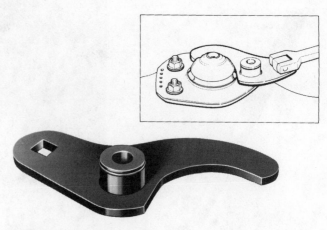

FIGURE 18-37.
This special tool is used to aid the camber adjustment on late model RWD Ford products. To make this adjustment, loosen the three mounting bolts, attach the tool to the upper strut mount as shown, attach a 1/2" drive socket handle to the tool, pry the strut to the correct camber position, and tighten the mounting bolts. (Courtesy of Branick Industries)

other makes of 4WDs. The sleeve can be replaced with one of a greater or lesser amount of offset, from 1/8° or 1/4° to 1 1/2° or 2° depending on the exact part. The bushing can also be rotated to reduce its effect or change caster, in some cases. (Fig. 18-39)

Tapered shims are available, aftermarket, to fit between the spindle and the steering knuckle on 4WD axles. They come in various tapers between 1/8° and 1 1/2°. The tire and

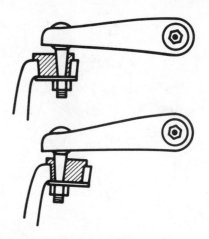

FIGURE 18-38.
An eccentric bushing connects the upper ball joint to the top of the steering knuckle on some Cadillacs; rotating this eccentric allows a camber adjustment. (Courtesy of Ammco)

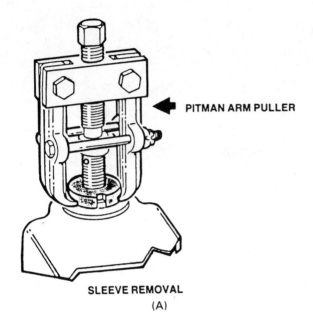

PITMAN ARM PULLER

SLEEVE REMOVAL
(A)

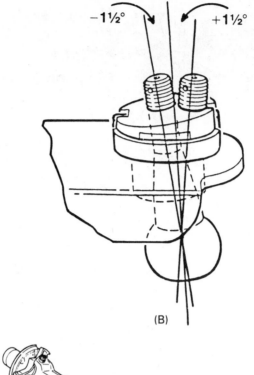

(B)

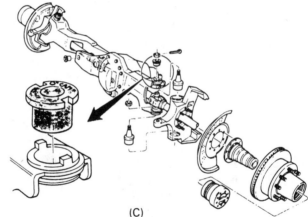

(C)

FIGURE 18-39.
An eccentric bushing is used on late model Ford pickups to connect the upper ball joint to the axle; this bushing can be replaced with one having a different angle or rotated to change the camber setting. A similar bushing will allow caster changes also. (Courtesy of Dana Corporation)

wheel, hub and wheel bearings, and spindle must be removed in order for the shim to be installed or changed. These shims should be used with discretion on some vehicles because they will cause a misalignment between the brake rotor and the caliper; this depends on how and where the caliper is mounted and how large a shim is required. (Fig. 18-40)

Solid beam axles must be bent to change camber; special fixtures or racks are required for this. Securing the outer end of the axle while jacking up just inboard will bend the axle so camber is more positive; jacking up the outer end with the axle secured just inboard will cause the camber to become more negative. (Fig. 18-41)

Many older cars with S-L A suspensions that used king pins provided an eccentric sleeve threaded onto the outer pivot for the upper control arm. Turning this sleeve adjusted the camber. (Fig. 18-42)

Aftermarket sources market special kits to provide a camber adjustment on some of the more popular imported cars with nonadjustable strut suspensions. These kits vary depending on the car model. On some cars, they are an offset inner control arm pivot bolt. On others, they are a way to convert a fixed inner control arm pivot bolt to one with an eccentric bushing; use of this style of kit usually requires some modification to the control arm mounting bracket. (Fig. 18-43)

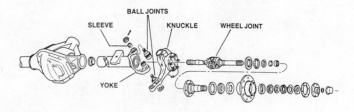

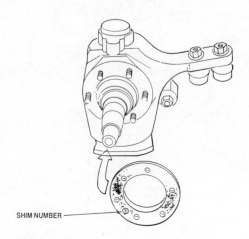

FIGURE 18-40.
Special, tapered shims are available to change the camber settings on 4WD vehicles; this shim, available in tapers of different degree amounts, is placed between the steering knuckle and spindle. (Courtesy of Dana Corporation)

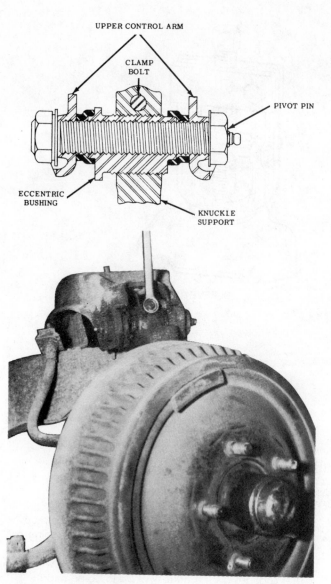

FIGURE 18-42.
Many older cars using king pins use an eccentric bushing to connect the top of the spindle support to the upper control arm; turning this bushing allows caster and camber to be adjusted. (Courtesy of Ammco)

FIGURE 18-41.
The fixtures on this truck axle are set up to bend the axle in such a way as to make camber more positive. (Courtesy of Bee Line)

Aftermarket kits are also available to provide additional camber adjustment for some domestic cars. Some of these models tend to sag at the front cross member and cause camber to become too negative. The car will come in with camber excessively negative and the adjustment as far as possible in the positive direction; the adjustment is used up. An increased amount of adjustment can be obtained by installing an offset inner control arm shaft, offset

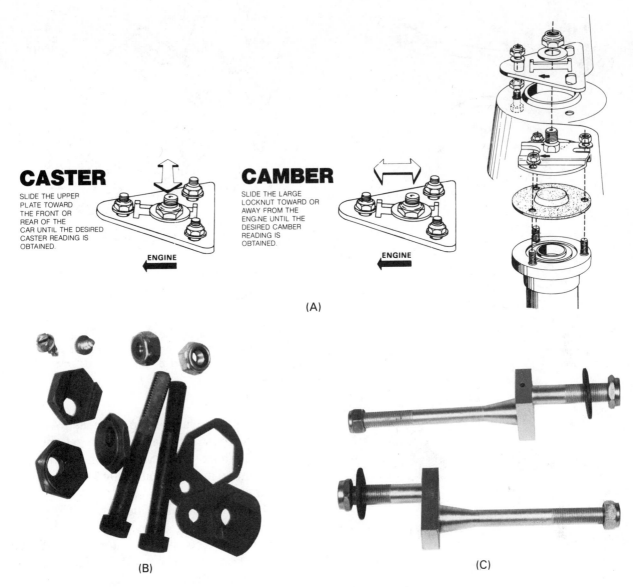

CASTER

SLIDE THE UPPER PLATE TOWARD THE FRONT OR REAR OF THE CAR UNTIL THE DESIRED CASTER READING IS OBTAINED.

← ENGINE

CAMBER

SLIDE THE LARGE LOCKNUT TOWARD OR AWAY FROM THE ENGINE UNTIL THE DESIRED CAMBER READING IS OBTAINED.

← ENGINE

(A)

(B)   (C)

FIGURE 18–43.
These aftermarket adapter kits are available to replace the upper strut mount or lower control arm inner pivot on some imported cars. A will allow adjustment of the top of the strut to provide camber and caster adjustments. B and C provide a camber adjustment; B requires a modification of the frame. (A is courtesy of Moog Automotive; B and C are courtesy of Arn-Wood)

control arm bushings, or ball joints with slotted mounting holes; the different styles depend on the exact car model and the aftermarket supplier. (Fig. 18–44)

18.7 MEASURING CASTER

If a steering knuckle were machined smoothly so a side of it was parallel to the steering axis, caster angle could be measured with a level and protractor. But it is not made that way, and it is usually too difficult to get clear access into the steering knuckle to measure it in that manner. Because caster inclines the steering knuckle, camber will change when the wheels are turned to the right or to the left. Positive caster will cause the camber to go more negative on the outer wheel and more positive on the inner wheel as a car goes around a corner. If you mea-

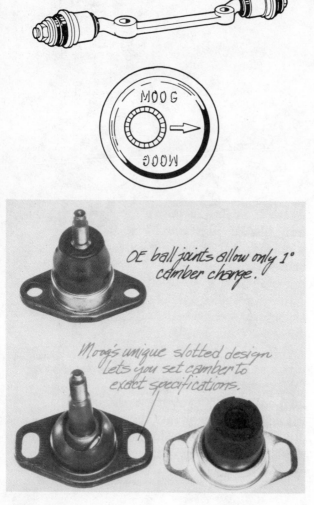

OE ball joints allow only 1° camber change.

Moog's unique slotted design lets you set camber to exact specifications.

FIGURE 18–44.
Offset inner control arm shafts, offset upper control arm bushings, and ball joints with slotted mounts are available for some domestic car models to provide additional camber adjustment. (Courtesy of Moog Automotive)

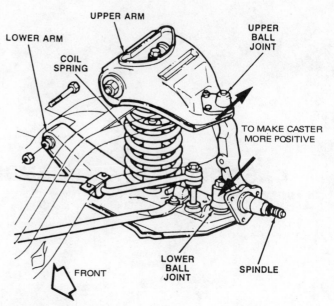

FIGURE 18–45.
Since caster is a forward/backward tilt of the steering knuckle, it is adjusted by moving either the top or bottom of the steering knuckle forward or backward. (Courtesy of Ford Motor Company)

sured the camber change as a wheel was turned from a 20° right turn to a 20° left turn and divided this change by 2, you would have the caster angle. All caster gauges measure the vertical wheel position (camber) at two different turning angles and convert that change into a caster reading. (Figs. 18–45 & 18–46)

When caster is measured, the front tires are placed on turntables that have degree scales built into them. This allows the wheels to be turned a specified amount. Most camber gauges have a caster scale built into them. Caster is also measured with alignment systems. After

the units have been set up to measure camber, a few additional steps are used to get the caster readings. These steps vary from one brand of equipment to another, but they generally follow the same routine.

Don't forget that caster readings vary with vehicle loading. Heavy objects in the trunk will compress the rear springs; this will cause caster to become more positive. The amount of caster change depends on the wheelbase of the car and the change in height. (Fig. 18–47)

Caster specifications, like camber specifications, are given in degrees and fractions of a degree for most domestic cars and in degrees and minutes for most imported cars. Caster specifications for many trucks and pickups are given with a relative frame angle; the angle of the frame is measured, and using this angle, the correct caster angle can be determined. (Fig. 18–48)

18.7.1 Measuring Caster Using a Magnetic Gauge

Caster is usually measured right after camber in the alignment sequence, using the same

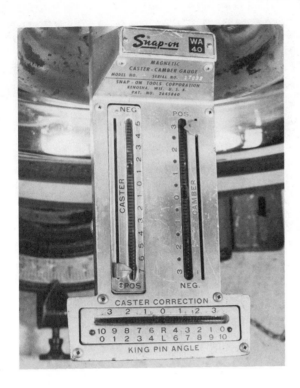

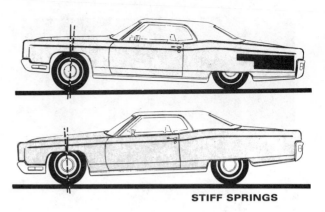

FIGURE 18-47.
Any height change at the back or front of a vehicle that changes the attitude of that vehicle will also cause a change in caster angle. (Courtesy of Hunter Engineering)

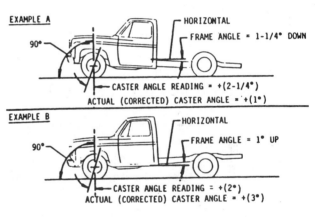

FIGURE 18-48.
Caster specifications for trucks and vans are often given with a reference to frame angle so compensation for load conditions can be made. (Courtesy of Chevrolet)

FIGURE 18-46
A magnetic caster-camber gauge at a straightahead and 20° turn position; note how the camber bubble changes position from +1/4° to +1 3/4°.

FIGURE 18-49.
Measuring caster requires two steps. The first step is to turn the tire in or out a specified amount as read on the turntable scale and to adjust the caster scale to zero (left). The second step is to turn the tire a specified amount and read the caster scale (right). (Courtesy of Snap-On)

gauge and turntables. Magnetic gauges have a caster scale built into them that is separate from the camber scale; the caster scale has a means to reset it. This scale is set to zero with the front wheels turned to a certain position determined by the gauge manufacturer; the usual position is a 20° turn. Turning the wheel the wrong direction at the beginning will reverse the readings; positive will become negative and vice versa. After zeroing the scale, the wheel is turned to a second, specified position, and the scale is read.

To measure caster using a magnetic gauge, you should:

1. Perform Steps 1 through 8 of Section 18.4.1.

2. Turn the front of the wheel inward or outward, as required by the gauge manufacturer, the correct amount, usually to a 20° left or right turn; the exact instructions for this are often printed on the gauge.

3. Cross-level the gauge and zero the caster bubble. Many technicians prefer to use

the end of the bubble, rather than the center, for zeroing and reading. (Fig. 18-49)

4. Turn the wheel the required amount in the opposite direction (usually enough to produce a 40° turn), cross-level the gauge, and read the caster scale. This is the amount of caster for that wheel.

18.7.2 Measuring Caster Using an Alignment System

Alignment systems use various methods of measuring caster. As with magnetic gauges, measuring caster is usually the next step after reading camber. Some machines display the directions to follow while measuring caster. In general, to measure caster, the unit is set up to measure camber, the wheels are turned to the right or left a certain amount, a button is pushed to perform the zeroing step, the wheels are turned in the opposite direction a certain amount, a second button is pushed, and the caster reading is displayed. (Fig. 18-50)

18.8 ADJUSTING CASTER

Caster is adjusted by moving either the top or bottom of the steering axis forward or backward. This is usually done by changing the length of the strut rod or repositioning the upper control arm (sometimes the lower) so the ball joint moves forward or backward. Moving the upper control arm in this manner is done by moving one end of the inner pivot toward the center of the car while moving the other end toward the outside of the car. This can be done by moving a shim from one end of the pivot shaft to the other, by turning one eccentric inward and one outward, or by sliding one end of the inner pivot shaft inward and the other end outward. Equal and opposite movement on each end of the pivot shaft will cause the caster to change with no change in camber.

Many technicians prefer to adjust camber first and then caster. When a front end technician really understands what is happening, both adjustments are made at the same time. For example, removing or adding a shim at one end of the control arm will change both camber and caster. In cases where it is impossible to get both adjustments exactly right, most technicians place a greater importance on camber; if it is off as little as 1/4 or 1/2°, tire wear will result. Caster can be off more than this, and as long as the spread is not too great, driveability problems will not occur. Incorrect caster will not cause tire wear.

As caster is being adjusted, it is rather tedious to keep remeasuring to see if you have completed the adjustment. For this reason, many caster gauges have a caster-correction scale built into them. A brake pedal jack is used to apply the brakes and lock the wheel to the steering knuckle. It is always a good practice to use a brake pedal jack while measuring caster to help keep the tire from walking off the turntable. With a brake pedal jack installed, as caster is adjusted and the steering knuckle angle is changed, the caster-correction scale will show this change through the slight rotation of the wheel. A caster change of 2° will cause a wheel rotation of 2°. The alignment technician can now measure caster, determine the amount

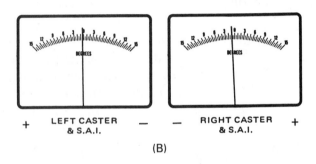

(A)

(B)

(C)

FIGURE 18-50.
Most alignment systems also require a two-step procedure to measure caster. The light projection unit is showing +2 1/2° of caster (A). The analog meter system is showing about +1/4° on the left wheel and +1° on the right wheel (B). The digital readout on the computer unit is very easy to read (C). (Courtesy of Hunter Engineering)

FIGURE 18-51.
Note the caster correction scale at the end of this gauge (arrow); it shares the level vial with the king pin angle scale. This scale can be set to show the amount of change that occurs while caster is being adjusted.

of desired change, and then adjust the caster until the correct change occurs on the caster-correction scale. (Figs. 18-51 & 18-52)

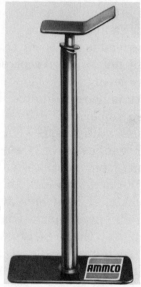

FIGURE 18-52.
A brake pedal jack, this tool is wedged between the brake pedal and the car seat to lock the brake rotor/drum to the steering knuckle; this causes the wheel to rotate with the steering knuckle so the caster correction scale can be used. (Courtesy of Ammco)

18.8.1 Adjusting Caster, Shims

Caster adjustment with shims is fairly easy as long as there are enough shims to work with. The adjustment is made by moving a shim from one of the shim packs to the other, from front to back or vice versa. This operation will cock the control arm slightly and cause the ball joint to move forward or backward. As the ball joint moves forward, caster becomes more negative; caster becomes more positive if the ball joint is moved toward the rear.

To adjust caster using shims, you should:

1. Measure caster, record the readings, and determine the change needed to meet the specifications and the desired road crown spread.

2. Adjust the caster-correction scale on the gauge so the amount of adjustment can be noted. There are several ways this can be done. Probably the simplest is to set the caster-correction scale to the existing reading, and then, when the correction scale reaches the desired reading, the caster is adjusted. (Fig. 18-53)

3. Install a brake pedal jack to lock the wheel and brake rotor or drum to the steering knuckle.

4. Locate the location of the shims, and determine which way a shim should be moved.

5. Loosen the control arm mounting bolts, remove a shim from one of the shim packs, and install it in the other.

6. Snug down the mounting bolts, and read the caster-correction scale.

7. After caster has changed the desired amount, tighten the mounting bolts to the correct torque. If the wrong amount of caster-correction has occurred, repeat Steps 5 and 6 until it is correct. If necessary, caster or camber can be remeasured.

18.8.2 Adjusting Caster, Eccentric Cams

Caster adjustment using eccentric cams follows a pattern similar to a shim-type adjustment. Moving one end of a control arm outward

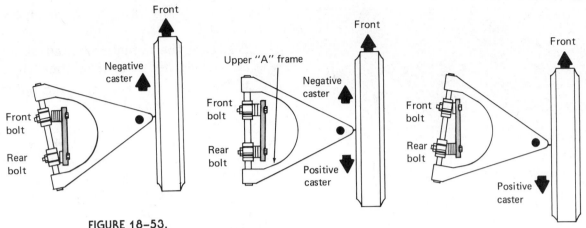

FIGURE 18-53.
Moving a shim from the front pack to the rear or vice-versa will cause caster to change without changing camber.

while moving the other end inward will move the ball joint forward or backward but not inward or outward.

To adjust caster using eccentric cams, you should:

1. Perform Steps 1, 2, and 3 of Section 18.8.1.

2. Locate the eccentrics and determine which way they should be turned.

3. Loosen the retaining nuts on the cam bolts and turn one of them outward and the

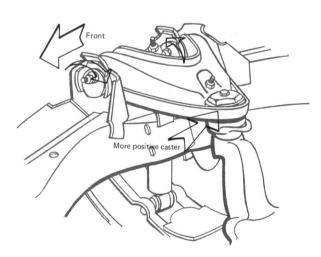

FIGURE 18-54.
If one eccentric cam is rotated inward while the other is rotated outward the same amount, caster will change without changing camber. (Courtesy of Hunter Engineering)

other one inward the same amount. Note the marks on the cams or the amount that the wrench moves to ensure that both cams turn the same amount. (Fig. 18-54)

4. Read the caster-correction scale; if it shows the desired amount of adjustment, tighten the retaining nuts to the correct torque. If the wrong amount of correction has occurred, repeat Step 3 until it is correct. If necessary, caster and camber can be remeasured.

18.8.3 Adjusting Caster, Sliding Adjustment

Caster adjustment using a sliding adjustment method follows a pattern similar to shim and eccentric cam adjustments. It can be a little more difficult to perform because of the difficulty in determining how far the control arm has been moved. Some of the special tools help greatly by allowing control arm movement in carefully controlled amounts.

To adjust caster using a sliding adjustment, you should:

1. Perform Steps 1, 2, and 3 of Section 18.8.1.

2. Locate the adjustment and determine which way the control arm should be moved.

3. Loosen the control arm mounting bolts, attach the proper alignment tool, if available,

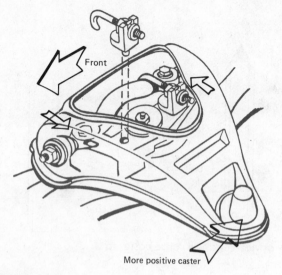

FIGURE 18-55.
If one end of the control arm mounting shaft is slid inward while the other end is slid outward the same amount, caster will change without changing camber. (Courtesy of Hunter Engineering)

and carefully move one end of the control arm inward and the other end outward the same amount. (Fig. 18-55)

4. Snug down the mounting bolts and read the caster-correction scale.

5. If caster has changed the desired amount, tighten the control arm mounting bolts to the correct torque. If the wrong amount of correction has occurred, repeat Steps 3 and 4 until it is correct. If necessary, caster or camber can be remeasured.

18.8.4 Adjusting Caster, Adjustable Strut Rod

Many cars that use a single, lower control arm use a strut rod that has an adjustable end mounting. Two nuts, one on each side of the strut rod bushings, are used to change the length of the strut rod. Making a trailing strut longer will move the lower ball joint toward the rear, which will make caster more negative.

To adjust caster using an adjustable strut rod, you should:

1. Perform Steps 1, 2, and 3 of Section 18.8.1.

2. Locate the adjustment and determine whether to lengthen or shorten the strut rod.

3. Loosen one of the strut rod adjusting nuts and tighten the other one.

4. Read the caster-correction scale; if it has changed the desired amount, tighten the adjusting nuts to the correct torque. If the wrong amount of correction has occurred, repeat Step 3 until it is correct. If necessary, caster can be remeasured; the camber angle should not have been disturbed. (Fig. 18-56)

18.8.5 Adjusting Caster, Miscellaneous Styles

There are a few additional methods used to adjust caster that are of limited use. For trucks and pickups using solid beam axles and leaf springs, adjust caster by using **tapered shims**. These shims, which are available in several widths, lengths, and various tapers between 1/2° and 6°, are placed between the axle and the spring; when the bolts are tightened, the

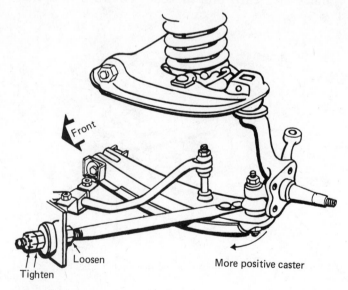

FIGURE 18-56.
If this strut rod is made shorter by loosening the inner nut and then tightening the outer nut, caster will become more positive. (Courtesy of Hunter Engineering)

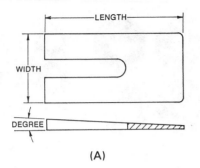

(A)

FIGURE 18-57.
Tapered shims are available in different widths and various degrees of taper; placing a shim between the solid axle and the spring will change the caster. (A is courtesy of Specialized Products; B is courtesy of Bear Automotive)

axle will tilt an amount equal to the shims. (Fig. 18-57)

Caster spread on a solid axle is changed by twisting the axle; special jacking fixtures are required for this operation. Twisting the axle will increase or reduce the caster angle on only one end of the axle. (Fig. 18-58)

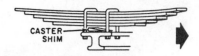

CASTER CORRECTION
(Straight Axle)

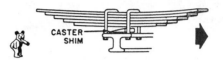

When thick part of shim is placed toward the rear, caster will increase. (Positive)

When thick part of shim is placed toward the front, caster will decrease. (Negative)

(B)

Special aftermarket eccentric bushings are available to adjust the caster on early twin I-beam axles. The radius arm bosses must be bored out to accept this bushing. Caster can also be adjusted on these axles by bending the radius arm. (Fig. 18-59)

FIGURE 18-58.
These attachments are set up to twist this truck axle; operating the jack will cause the caster to become more positive on the right wheel. (Courtesy of Bee Line)

FIGURE 18-59.
Eccentric bushings can be installed in the radius arm of this twin-I beam axle to provide a caster adjustment; the hole in the radius arm must be drilled oversize. (Courtesy of Arm-Wood)

Aftermarket kits are available for some nonadjustable imported cars. One kit converts a fixed-length strut rod into an adjustable one. (Fig. 18-60)

18.9   MEASURING SAI

SAI, like caster, could be measured with a level and protractor if there were an accurate place from which to take the measurements, but this place does not exist. So SAI, like caster, is measured by attaching a gauge to the wheel assembly, turning the wheel through a turn, and noting the change in gauge position. Many times this will be the same scale as that used for caster correction. SAI scales are found on some, not all, caster-camber gauges of both bubble type, magnetic, and electronic alignment machines. (Figs. 18-61 & 18-62)

Specifications are not always published for SAI, but it can be easily determined if the included angle is available. Remember that SAI equals the included angle minus the amount of

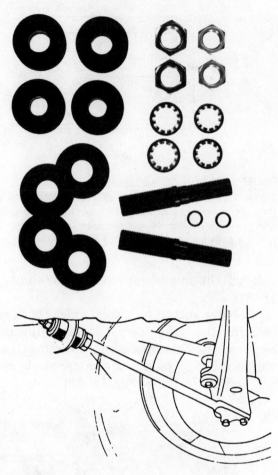

FIGURE 18-60.
This kit is designed to change the strut rod on an imported car from a fixed length strut rod to an adjustable rod to provide a caster adjustment. (Courtesy of Specialized Products)

positive camber; the amount of negative camber plus the included angle equals SAI.

18.9.1   Measuring SAI Using a Magnetic Gauge

Many magnetic gauges and some alignment systems include an SAI gauge in them. The procedure used to measure SAI is similar to a caster measurement.

To measure SAI using a magnetic gauge, you should:

1. Perform Steps 1 through 8 of Section 18.4.1.

FIGURE 18–61.
Measuring SAI is a two-step process much like measuring caster. Here the wheel is turned and the bubble (arrow) is adjusted to zero. Next the wheel will be turned the required amount, and the scale will be read.

2. Turn the front of the wheel the correct number of degrees, usually 20, in the correct left or right direction, as specified by the gauge manufacturer.

3. Zero the SAI bubble by rotating the gauge unit or turning the adjusting screw.

4. Turn the front of the wheel in the opposite direction the required amount and read the SAI scale. If the reading goes off the end

of the scale, repeat Steps 2 and 3, turn the wheel to straight ahead, read the scale, then reset the scale to zero, turn the wheel the rest of the way to the specified degree setting, reread the scale, and add the two readings together.

18.10 ADJUSTING SAI

SAI is considered a nonadjustable angle. This is because it is determined by the construction of the steering knuckle or strut and it cannot be changed without bending the spindle. Bending stresses might cause fracture and later failure of the spindle or steering knuckle or strut. The possibility of a broken steering knuckle or spindle presents a very high risk. If SAI is wrong and camber is correct, this tells us the included angle is wrong; either the spindle, steering knuckle, or strut must be bent or damaged. A competent alignment technician will locate and replace that bent part.

There is some argument as to whether a camber adjustment really adjusts camber or SAI and whether the steering knuckle really controls SAI or camber. These points are presented here as they are treated by the majority of professionals and manufacturers in the suspension and alignment industry. For most practical purposes, it makes sense to treat these views as being correct.

	LEFT	RIGHT
SAI	10.0	9.9
CAMBER	1.2	.7
INCLUDED ANGLE	11.2	10.6

FIGURE 18–62.
Some alignment systems will measure SAI; the display on the monitor of this computer system is very easy to read. (Courtesy of Hunter Engineering)

18.11   MEASURING TOE OUT ON TURNS

Specifications for toe out on turns are usually printed with one of the wheels at 20° and a specified number of degrees for the other wheel. If the specified angle is larger than 20°, it will be for the inside wheel; if it is smaller than 20° it will be for the outer wheel. If specifications are not available, one side of the car can be checked relative to the other side. If the readings come out different, something is wrong. If the readings are the same, toe out on turns is probably correct.

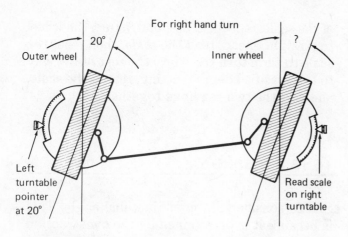

FIGURE 18-63.

Toe out on turns is measured by turning one of the tires 20° and then reading the amount of turn on the other turntable. The next step is to repeat this process in the other direction.

Toe out on turns is very easy to measure while caster measurements are being made. After one of the tires (depending on the specifications) has been set to a 20° position on the turntable, the technician can easily read the amount of turn from the scale on the other turntable. This step is then repeated as caster is measured on the second tire. (Fig. 18-63)

18.12   ADJUSTING TOE OUT ON TURNS

On most cars, toe out on turns is nonadjustable; if it is not correct, one or both steering arms is bent and should be replaced. A quick check for a bent steering arm is to measure the distance from the brake rotor to the steering arm on each side of the car and compare the two measurements. They should be very close to equal. In some cases the steering arm is bolted onto the steering knuckle, and it is an easy replacement. In others, the arm is cast as part of the steering knuckle, and the whole knuckle assembly has to be replaced. (Fig. 18-64)

Some steering arms are bolted onto the steering knuckle in such a way that a slight error in toe out on turns can be corrected by shimming between the arm and the steering knuckle. The amount of shim should be limited

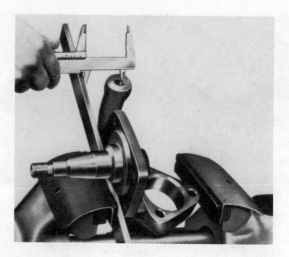

FIGURE 18-64.

A check is being made to determine if the steering arm is bent; if this dimension is not available, it should be compared to the same distance on the other steering arm. (Courtesy of Volkswagen of America)

to about 1/16 inch (1.5 mm) to ensure a secure tightening of the steering arm. The rear end of the steering arm should be moved toward the center of the car (increasing toe out on turns) if there is not enough difference measured between the two angles, or moved toward the outside of the car (decreasing toe out on turns) if there is too much difference. (Fig. 18-65)

TURNING RADIUS CORRECTION

(Increase)

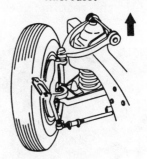

Place shim, not over 1/16" thick, at **rear** steering arm mounting. To correct beyond 1/16", replace steering arm.

FIGURE 18-65.

In cases where the steering arm is bolted to the steering knuckle, toeout on turns can be adjusted by shimming the steering arm. A shim at the rear will increase the difference in turning angles; a shim at the front will reduce the difference. (Courtesy of Bear Automotive)

18.13 MEASURING TOE

Since toe is a comparison of two distances, it can be measured with a ruler or tape measure. Measure the two distances and subtract one from the other. Front and rear wheel toe measurements are made in the same manner. Tires and wheels, however, are not perfectly true, and the tread, where the measurements should be taken, does not lend itself to accurate measuring. Also, you would have a great deal of difficulty running a tape measure from the back of one tire to the other. (Fig. 18-66)

To provide an accurate measuring point, a line can be scribed around the tire by pushing a sharp object against the tread while rotating the tire. **Tire scribes** are available to make this operation easy. Toe measurements are made from the scribed line on one tire to the line on the other tire. A trammel bar is a simple tool having two extended pointers to measure these distances accurately.

Another method of measuring toe is done with a **light beam**. This device projects a beam of light at a 90° (right) angle to the wheel. The amount of toe is read where the light beam

FIGURE 18-66.
Tire wear caused by incorrect toe. If we were looking toward the rear at the top of the right (driver's side) tire, the wear on the upper tire would be caused by excessive toe out and the lower tire by excessive toe in. (Courtesy of Snap-On)

meets the scale on the projector that is attached to the other wheel. This device is actually measuring toe angle, but the gauge is usually calibrated in fractions of an inch. This style of measurement is affected by the width of the car; a wider car would measure as having more toe than a narrower car. This discrepancy, however, is usually very slight.

An advantage to this style of measuring system is that it measures **individual toe**. Individual toe readings allow easier and faster centering of the steering wheel. Individual toe readings on the rear wheels of IRS cars also give a quick indication of the rear wheel thrust line.

Modern alignment systems measure toe individually, some in inches or millimeters, some by degrees, many by both measuring systems. The more modern equipment notes the position of the rear wheels so that toe and steering wheel position can be set relative to rear wheel thrust or front tire set back. If the rear wheel thrust line does not run down the center of the car, the rear wheels will steer to the right or left, and the driver will have to turn the steering off center, toward the side where the thrust line is, in order to go straight down the road. This positions the steering wheel off center and is a visual annoyance to the driver. The ideal cure for this is to align the rear wheels to correct the thrust line. On some cars this is fairly easy and relatively inexpensive; with others, it can be very expensive. The inexpensive cure is to readjust the tie rod lengths to center the steering wheel; this procedure is described in Section 18.16.3.

Set back will also affect toe settings and steering wheel position if the equipment being used measures toe angles. For example, if both tires were pointed straight ahead, zero toe angle, and the right tire was set back, the angle at the right tire would be greater than 90° and the angle of the left tire would be less than 90°. This difference would cause the equipment to read toe in on the right tire and toe out on the left tire. Some toe gauges can be adjusted to compensate for set back; some computerized equipment will make this compensation automatically. (Fig. 18-67)

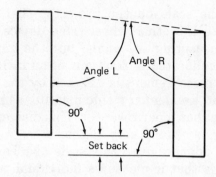

FIGURE 18-67.
When toe is measured using equipment that actually measures the angles, set back of one of the tires will cause an error in the instrument readings. In this case, there is zero toe, but the instruments will read toe out on the left tire and toe in on the right.

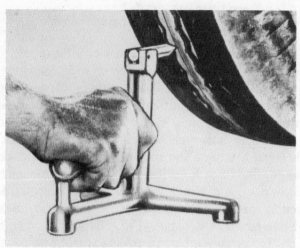

FIGURE 18-68.
When using a trammel bar to measure toe, the first step is to scribe a line around the tire; the tire scribe is pushed against the tire as the tire is rotated. (Courtesy of Snap-On)

The front tires must be in a straightahead position when measuring toe to prevent the effect of toe out on turns from changing the measurements. Also, the car should be at the correct ride height; the geometry of some suspension and steering systems causes a toe change, often called "bump steer," when the suspension moves up and down.

18.13.1 Measuring Toe Using a Trammel Bar

A **trammel bar** is simple, trouble-free, and inexpensive; it is merely a metal bar having a pair of stands and a pair of pointers. A scale is often built into one of the pointers to provide a means of measuring the amount of toe. Some technicians prefer to do the measuring with a scale or rule, rather than slide the pointer. A trammel bar is always used in conjunction with a tire scribe to obtain an accurate measuring point; if used correctly, this is an extremely accurate measuring method.

To measure toe using a trammel bar, you should:

1. Park the car on a smooth, level surface, preferably an alignment rack, and position the steering wheel in a straightahead position.

2. Raise the tires, place a tire scribe so the pointer presses against the most even portion of the tire tread, and rotate the tire so a fine line is scratched around the tire. If the line does not show up very well, mark a line around the tire using chalk or tire crayon, and scribe a scratch line in the marked portion. (Fig. 18-68)

3. Position a pair of turntables or slip plates under the tires, pull the locking pins, and with the tires in a straightahead position, lower the tires onto the turntables.

4. Bounce the car up and down to help the suspension settle to ride height.

5. Adjust the pointers so they are at hub height (even with the center of the hub), and slide the trammel bar under the car so the pointers are at the rear of the tires. (Fig. 18-69)

6. Adjust the width of both pointers so they are aligned with the scribed lines, and if you are planning on using the measuring scale on the bar, adjust it to zero. (Fig. 18-70)

7. Carefully, so you don't disturb the pointers, slide the trammel bar out from under the car and reposition it so the pointers are at the front of the tires.

8. Slide the trammel bar so the pointer without the measuring scale is aligned with its scribed line. The other pointer will probably not be aligned.

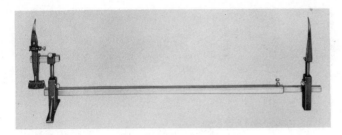

FIGURE 18-69.
A trammel bar. The pointers on this tool are adjusted to the width of the scribed lines at the back of the tires; the tool is then moved to the front of the tires to compare the pointer width to the scribed lines. (Courtesy of Ammco)

9. Measure the distance between the second pointer and the scribed line; this is the amount of toe. This can be done by repositioning the pointer and reading the scale or by holding a ruler or tape in line with the pointer and scribed line. (Fig. 18-71)

If a toe adjustment is made, it should be remembered that adjusting toe moves the rear and the front of the tire at the same time. When remeasuring toe with a trammel bar, it is necessary to repeat Steps 5 and 6 as well as Steps 7 and 8.

FIGURE 18-71.
After setting the pointers at the rear of the tires, the trammel bar pointers are placed at the front in order to measure the amount of toe.

FIGURE 18-70.
The left pointer of this trammel bar has been adjusted to align with the scribed line at the back of the tire; the right pointer is also aligned to its scribed line. Note that the pointer (arrow) is at hub height.

18.13.2 Measuring Toe Using a Light Gauge

Light gauges are more complex, fragile, and expensive than a trammel bar, but they are usually faster to use because they give continuous, individual readings. They consist of a pair of projectors that attach to each front wheel. Each unit has a light source, a means of focusing and calibrating the light beam, and a target or scale for the other light beam to shine onto. (Fig. 18-72)

It is a good practice to check the accuracy of a light gauge periodically and recalibrate it, if necessary, especially if it has been dropped or bumped. A calibration fixture is available that has machined ends of the correct size to accept the gauge units. When the units are attached, they should give a toe reading of zero. (Fig. 18-73)

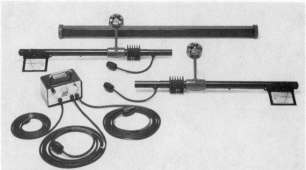

FIGURE 18-72.
A light projection/optical toe gauge set, this unit has a projector with scale that attaches to each front hub; the straight bar at the top is a calibration fixture. (Courtesy of Ammco)

FIGURE 18-74.
The light projection gauges are attached to each front hub or wheel clamps on the front wheels; toe can be adjusted while watching the light beam on the scales. (Courtesy of Ammco)

To measure toe using a portable light gauge, you should:

1. Place the car on a smooth, level surface, preferably an alignment rack, and turn the steering wheel to a straightahead position.

2. Remove the wheel covers or hub caps and the dust cap from the hub.

3. Attach the light gauge units to the vertical, machined surface of the front hubs. Wheel rim mounting adapters can be used on rear wheels or front wheels that do not allow mounting the units directly onto the hub. If a wheel rim mount adapter is used, it should be of the

type using three compensating screws so it has 360° of parallelism, not just up and down.

4. Level the gauge units, focus the light beam, and raise or lower the light beam as necessary to get a sharp image on the scales.

5. Read the center point of the light image on each scale. This is the amount of toe for each wheel; the readings must be added together to obtain the total amount of toe. (Fig. 18-74)

18.13.3 Measuring Toe Using an Alignment System

Alignment systems use various means of measuring toe. Depending on the system, a light beam is projected from one wheel to the other (in a manner similar to the portable units) or from the wheel units to a projection screen. Other systems connect a string between the two wheel units. The display for toe readings is usually a continuous reading for each wheel and is often calibrated for inches, millimeters, and degrees. Newer computer systems with digital readouts can be programmed to give readings in inches, millimeters, or angles. (Fig. 18-75)

Once an alignment system has been installed on a car so camber can be read, it is usu-

FIGURE 18-73.
The gauge units are positioned on the calibration bar (arrow) to check their accuracy; while on this fixture, the gauges should read zero toe. (Courtesy of Ammco)

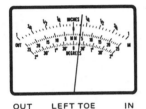
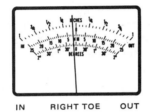

OUT   LEFT TOE   IN   IN   RIGHT TOE   OUT

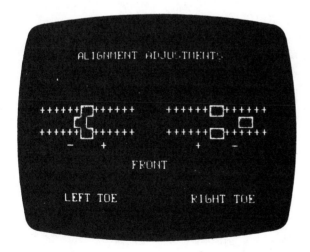

FIGURE 18–75.

Toe readings from three different styles of alignment machines. A is a light projection type showing between 3/16 and 1/4 inch or between 25 and 30 minutes of toe in. The analog meter unit is reading 1/16 inch, 2 mm, and about 5 minutes of toe in. The computer monitor is displaying a pair of bar graphs showing that each tire has too much toe in; as toe is adjusted the middle bar will move in relationship to the adjustment. (Courtesy of Hunter Engineering)

ally reading toe at the same time. It is sometimes necessary to connect a toe string or turn a switch.

18.14 TOE AND STEERING WHEEL POSITION

Equipment that measures individual toe is very convenient when an alignment technician is adjusting the position of the steering wheel. The steering wheel is indexed so that it is in the correct position when the steering gear is in the center of its travel (straightahead position). In this position, the amount of toe at each wheel should be equal and the Pitman and idler arms should be pointed straight forward or rearward. If they are not equal, the wheels are turned slightly, and the car will travel in a circle. What really happens is that the driver will turn the steering wheel slightly so there will be equal toe when the car is going down the road, but the steering wheel will be off center. Having the steering wheel off center is an annoying visual problem, but it also moves the steering gear off center. An off-center steering gear can have excessive clearance, which can allow vehicle wander. (Figs. 18–76 & 18–77)

The procedure to center a steering wheel is simply to shorten one tie rod and lengthen the other the same amount so as not to disturb toe. Centering a steering wheel is usually a part of a toe adjustment because both operations are performed by turning the tie rod sleeves or tie rods. A final centering adjustment is sometimes done after the wheel alignments and the road test, which should always follow a wheel alignment, shows an off-center steering wheel. Figure 18–78 illustrates the procedure to center a steering wheel; the procedure is described more completely in Section 18.16.3.

18.15 TOE SPECIFICATIONS

Toe specifications have traditionally been given in fractions of an inch for domestic cars and in millimeters for imported cars. This often re-

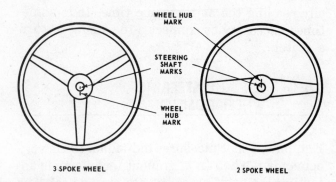

3 SPOKE WHEEL   2 SPOKE WHEEL

FIGURE 18-76.
When removing a steering wheel, make sure that there are index marks that can be used to replace it in the same position. A three-spoke wheel is normally centered with the middle spoke downward, and a two-spoke wheel with the spokes horizontal. (Courtesy of Ammco)

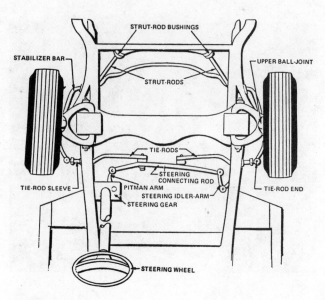

FIGURE 18-77.
When a steering wheel is centered, both tires should be straight ahead with the toe split equally. Note that if one tie rod is made longer and the other one shorter, either the steering wheel or the front tires must turn. (Courtesy of Hunter Engineering)

quired the technician to convert one to the other depending on how the measuring scale on the equipment was made. Conversion of a dimension is fairly easy. Multiply the inch dimension by 25.4 to get millimeters; multiply millimeters by 0.04 to get inches. A chart is provided in the appendix to make this an even easier operation. (Fig. 18-79)

Several manufacturers are now printing their toe specifications in degrees per wheel; this is the angle that a toed wheel will take from straight ahead when viewed from above or below. Some of these manufacturers publish their specifications in both a toe angle and a distance dimension. Conversion from one dimension to the other is difficult because tire diameter affects the conversion. A 28 inch (71.12 cm) tire will have a smaller angle at 1/16 inch (1.58 mm) toe in than a 22 inch (55.88 mm) tire would. The chart in the appendix will help you make this conversion.

18.16 ADJUSTING TOE

Every car provides for adjusting toe; this is the only wheel alignment angle that can be ad-

STEERING WHEEL POSITION	TOE CORRECT (No change)	TO INCREASE TOE-IN	TO INCREASE TOE-OUT
	No adjustment needed	Make one-half of adjustment on each tie rod sleeve	Make one-half of adjustment on each tie rod sleeve
	Adjust tie rod sleeves equally in opposite directions	Adjust **right** tie rod sleeve	Adjust **left** tie rod sleeve
	Adjust tie rod sleeves equally in opposite directions	Adjust **left** tie rod sleeve	Adjust **right** tie rod sleeve

FIGURE 18-78.
When adjusting toe, follow these recommendations to end up with the correct toe and a centered steering wheel. When centering a steering wheel, one full revolution of each tie rod sleeve will move the steering wheel position about 2 inches. (Courtesy of Snap-On)

SPECIFICATIONS

			(1) SPECIFICATIONS FOR DIAGNOSIS FOR WARRANTY REPAIRS OR CUSTOMER PAID SERVICE	(2) SPECIFICATIONS FOR PERIODIC MOTOR VEHICLE INSPECTION	(3) SPECIFICATIONS FOR RESETTING ALIGNMENT
B SERIES	CASTER		+2° TO +4°	+1° TO +5°	+3° ± 0.5°
	CAMBER		0° TO +1.6°	-0.7° TO +2.3°	+0.8° ± 0.5°
	TOE-IN	INCHES (TOTAL)	1/16″ TO +1/4″	-3/16″ TO +9/16″	+1/8″ ± 1/16″
		DEG (PER WHEEL)	.05° TO +.25°	-.15° TO +.55°	+.15 ± .05°
G SERIES	CASTER	MAN. STR.	0 TO +2°	-1° TO +3°	+1° ± 0.5°
		POW. STR.	+2° TO +4°	+1° TO +5°	+3° ± 0.5°
	CAMBER		-0.3° TO +1.3°	-1.0° TO +2.0°	+0.5° ± 0.5°
	TOE-IN	INCHES (TOTAL)	1/16″ TO +1/4″	-3/16″ TO +9/16″	+1/8″ ± 1/16″
		DEG (PER WHEEL)	.05° TO +.25°	-.15° TO +.55°	+.15 ± .05°
ALL SERIES	CROSS CASTER		NO MORE THAN 1° SIDE TO SIDE VARIATION		NO MORE THAN 1/2° SIDE TO SIDE VARIATION
	CROSS CAMBER		NO MORE THAN 1° SIDE TO SIDE VARIATION		NO MORE THAN 1/2° SIDE TO SIDE VARIATION

101549

FIGURE 18–79.

Toe specifications are included in these wheel alignment specs.; note that they are given in both inches and degrees per wheel. When resetting toe, the ideal setting would be +1/8″ or +.15°. (Courtesy of Chevrolet)

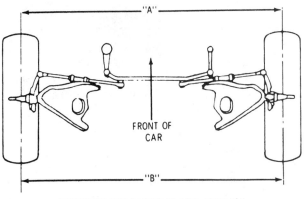

DIMENSION "A" SHOULD BE LESS THAN "B"

(A)

justed on all cars. Toe is adjusted by changing the length of the tie rods. Most cars with independent suspension will have two adjustable tie rods; vehicles with solid axles will have only one. The tie rods can be located either in front or in back of the tire center line, with the rear mounting being the most common. With the tie rods mounted at the rear, a longer tie rod will produce more toe in, and a shorter tie rod will produce more toe out. (Fig. 18–80)

Most tie rods are made in three parts: two ends and a center sleeve. One tie rod end and one end of the sleeve have right-hand threads while the other side has left-hand threads. The

FIGURE 18–80.

When the tie rods are mounted at the front, toe in is decreased if the tie rods are made longer; if they are at the rear, toe in is increased. (A is courtesy of Chevrolet; B is courtesy of Pontiac Division, GMC)

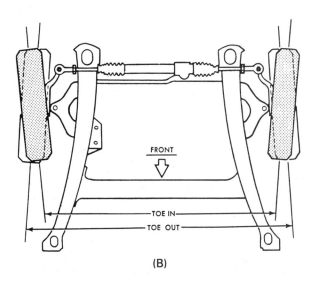

(B)

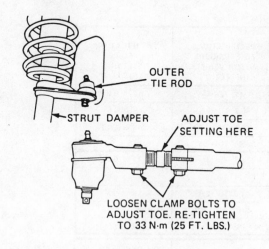

FIGURE 18-81.
Three-piece tie rods are adjusted by loosening the clamps and rotating the adjuster. (A is courtesy of Pontiac Division, GMC; B is courtesy of Moog Automotive)

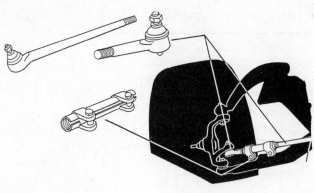

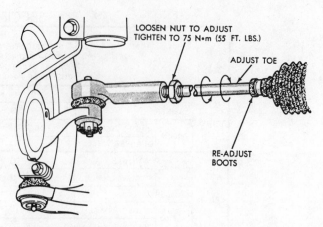

FIGURE 18-82.
A two-piece tie rod is adjusted by loosening the lock nut and boot clamp and then rotating the tie rod. (Courtesy of Chrysler Corporation)

sleeve is now like a turnbuckle; turning it one way makes the tie rod longer, turning it the other direction makes it shorter. Many cars with rack and pinion steering gears use a two-piece tie rod. The inner tie rod end is partially built onto the tie rod and partially onto the end of the rack, and the tie rod is threaded into the outer tie rod end. Toe adjustment is made by threading the tie rod into or out of the outer end. (Figs. 18-81 & 18-82)

Toe in should be the last adjustment made during a wheel alignment. Other adjustments will change the position of the steering knuckle and, therefore, change toe. Toe should always be checked and readjusted, if necessary, each time caster or camber is adjusted.

Many RWD cars use a toe specification of 1/16 to 3/16 inch (1.5 to 4.7 mm); some cars will have a range between the minimum and maximum setting as great as 1/4 inch (6.3 mm). Tire wear will often result if the toe is adjusted to the minimum or maximum specification. Normally the range is split, and toe is adjusted to the middle setting. A car with a toe specification of 1/16 to 3/16 inch would have a preferred setting of 1/8 inch. Check the tire for signs of toe wear when trying to determine the ideal setting for a particular car. For example, if a car with the above specification shows a slight toe out wear with 1/8 inch toe, the ideal setting for this particular car would be 1/16 inch.

Special wrenches are available for turning tie rod sleeves. Many technicians are tempted to use a pipe wrench or slip joint pliers for this, but this is a bad practice. The jaws of these tools not only mar the surface of the sleeve, they also tend to collapse the sleeve, making it harder to turn. A tie rod wrench grips the sleeve by the slot in the sleeve, and the pressure to turn the sleeve tends to open the sleeve, which makes it easier to turn. Tie rod wrenches are available in a large or small size and a variety of shapes to fit into tight locations. (Fig. 18-83)

Frozen, rusty tie rod sleeves can often be loosened by spraying penetrating oil on the

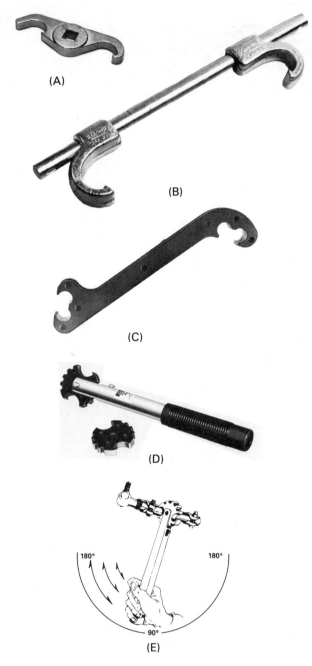

FIGURE 18-84.
If the sleeve is stuck or rusted in place, it can often be loosened using penetrating oil and by vibrating it using an air hammer with a special, blunt punch tool. (Courtesy of Superior Pneumatic)

FIGURE 18-83.
Wrenches to turn tie rod sleeves are designed to fit the two common sizes of sleeves, and they come in various shapes to permit turning the sleeves in tight locations. Each style is a type of spanner that grips the sleeve using the slot in the sleeve (E). (A, B, and C are courtesy of Specialized Products; D and E are courtesy of Branick Industries)

threads and in the slots in the sleeve and vibrating the sleeve using an air hammer and a special wide punch. (Fig. 18-84)

When completing a toe adjustment and it

comes time to retighten the clamps on the tie rod sleeves, two points are important: tie rod end bind and clamp location. The tie rod must be free to rotate within the limits of the ends; if the ends are incorrectly aligned, the amount of rotation will be limited. The easy way to align the tie rod ends is to rotate both of them toward the front, or rear, of the car as far as they will go, and then tighten the clamps. After both clamps are tightened, the tie rod ends can be rotated back to center. The clamps are usually positioned between two raised areas at the ends of the sleeves, and the open portion should not be in direct alignment with the slot in the sleeve. The clamps themselves should be positioned so they will not touch or strike any part of the frame or body during steering or suspension motions. After the clamps have been tightened to the correct torque, there must be a slight gap in the open portion of the clamps and the sleeve must be locked securely onto the tie rods. (Figs. 18-85 & 18-86)

18.16.1 Adjusting Toe, Standard Steering

Depending on the measuring system (total or individual measurements), toe is adjusted in one of two ways. Total toe is adjusted by turning each tie rod sleeve the same amount, unless a change of steering wheel position is desired. Individual toe is adjusted by turning the tie rod sleeve for one side until the tire on that side has the correct toe. Then, the tie rod sleeve for the

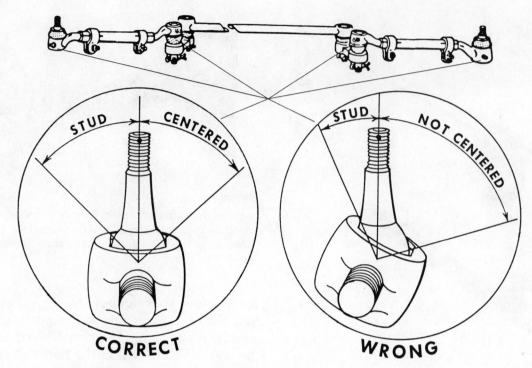

FIGURE 18-85.

When the tie rod sleeve clamps are tightened, both tie rod studs should be centered in their housings; misalignment of the tie rod ends will cause a binding during steering and/or suspension manuevers. (Courtesy of Moog Automotive)

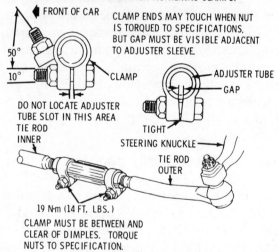

FIGURE 18-86.

As the clamps are tightened, these rules should be followed to ensure that the sleeves are locked securely to the tie rods. (Courtesy of Oldsmobile)

second side is turned enough to adjust the second side. Toe specifications are usually for the total amount of toe; individual tire settings are one-half of the total.

Steering wheel or centerlink holders are available to hold the steering linkage and gear steady while individual toe adjustments are being made. When adjusting toe on cars with power steering, it is a good practice to start the engine and rock the steering wheel slightly in a straight-ahead position to ensure that the steering gear is centered before installing the steering wheel holder; then shut off the engine. (Fig. 18-87)

To adjust toe on standard steering linkage, you should:

1. Park the car on a smooth level surface, preferably on an alignment rack, with the tires on turntables. Center and lock the steering if toe is to be adjusted individually.

2. Measure and record the amount of toe; compare the readings to the specifications, and determine what change is necessary.

FIGURE 18–88.
These wrenches combine a box wrench with a deep well socket to make loosening and tightening of the tie rod sleeve clamp bolt a one-hand operation.

FIGURE 18–87
A steering wheel holder has been installed to keep the steering wheel from turning during a toe adjustment; this is normally used if a toe gauge that gives a continuous read-out is being used. (Courtesy of Ammco)

3. Locate the tie rod adjusting sleeves, and determine whether they need to be lengthened or shortened. Clean off the threads at the ends of the sleeves and determine which direction the sleeves need to be rotated.

4. Loosen the tie rod sleeve clamps; special wrenches are available to make this a one-hand operation. (Fig. 18–88)

5. Engage the tie rod wrench in the slot in the adjuster sleeve and rotate the sleeve. When using a trammel bar, it is necessary to adjust both sleeves, remeasure toe, and then readjust both sleeves if needed. When using equipment that gives a continuous readout, observe the toe readings while turning the adjusting sleeves. Stop adjusting when toe is correct. (Fig. 18–89)

6. Rotate the adjusting sleeve clamps until they are in a good position, rotate the tie rod ends in the same direction to align them, and tighten the clamp bolts to the correct torque.

7. Road test the car to ensure that the steering wheel is centered; if necessary, center the steering wheel as described in Section 18.16.3.

TURN DOWNWARD TO DECREASE ROD LENGTH   TURN UPWARD TO INCREASE ROD LENGTH   TURN DOWNWARD TO INCREASE ROD LENGTH   TURN UPWARD TO DECREASE ROD LENGTH

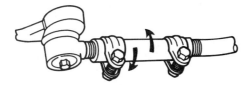

LEFT-HAND SLEEVE   VIEW A   RIGHT-HAND SLEEVE

FIGURE 18–89.
Rotating a tie rod sleeve one direction makes the tie rod longer; rotating it the opposite direction makes the tie rod shorter. (Courtesy of Ford Motor Company)

18.16.2 Adjusting Toe, Rack and Pinion Steering

If adjusting sleeves are used on the tie rods, toe adjustment on rack and pinion steering is no different than the procedure just described. When two-piece tie rods are used, a slightly different procedure is used; the tie rod is rotated to make the adjustment. When the tie rod is rotated, the bellows/boot clamp must be loosened, and care must be taken to ensure that the boot is not twisted. A jam nut is provided to lock the adjustment.

To adjust toe on rack and pinion steering, you should:

1. Follow Steps 1, 2, and 3 of Section 18.16.1.

2. Loosen the tie rod jam/lock nuts and the bellows/boot to tie rod clamps. (Fig. 18–90)

3. Rotate the tie rods until toe is correct. Make sure the tie rod turns inside of the boot and the boots do not get distorted.

4. When toe is correct, tighten the jam nuts and boot clamps, and road test the car.

18.16.3 Adjusting Toe to Center a Steering Wheel

After making an alignment adjustment, it is always a good practice to road test the car to ensure proper steering and a centered steering wheel. If the steering wheel is off center, one of the tie rods needs to be shortened and the other lengthened the same amount. Which tie rod to be shortened or lengthened depends on which side of the steering wheel is low and whether the tie rods are mounted in front or back of the tire center line. This is sometimes called a **clear vision** adjustment; the spokes of an off-center steering wheel can block the view of the dash instruments.

If the steering wheel is low on the left side, the tie rod sleeves should be adjusted to move the tires toward a left turn. If the tie rods are mounted behind the tire center line, this would mean shortening the left tie rod and lengthening the right one. The opposite would apply if the tie rods are in front of the tire center line or if the steering wheel was low on the right side. (Fig. 18–91)

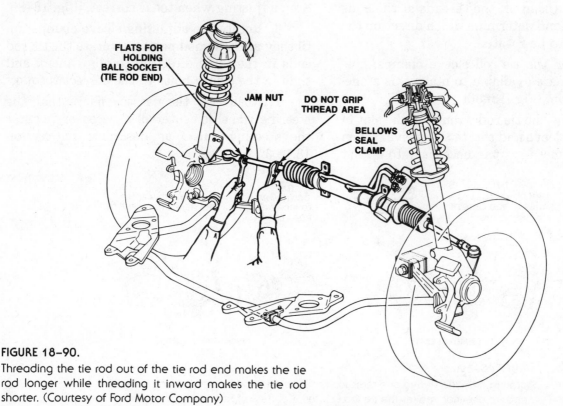

FLATS FOR HOLDING BALL SOCKET (TIE ROD END)

JAM NUT

DO NOT GRIP THREAD AREA

BELLOWS SEAL CLAMP

FIGURE 18–90.
Threading the tie rod out of the tie rod end makes the tie rod longer while threading it inward makes the tie rod shorter. (Courtesy of Ford Motor Company)

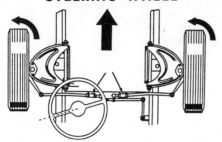

CENTERING STEERING WHEEL

When toe-in is correct: Turn both tie-rod adjusting sleeves equally so that front wheels move to the left.

When toe-in is not correct:
a. Lengthen R.H. Rod to increase toe-in.
b. Shorten L.H. Rod to decrease toe-in.

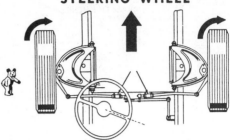

CENTERING STEERING WHEEL

When toe-in is correct: Turn both tie rod adjusting sleeves equally so that front wheels move to the right.

When toe-in is not correct:
a. Lengthen L.H. Rod to increase toe-in.
b. Shorten R.H. Rod to decrease toe-in.

FIGURE 18–91.
The tie rods can be readjusted to center the steering wheel. Note that the procedure is reversed if the tie rods are mounted at the front. (Courtesy of Bear Automotive)

How far each sleeve should be turned is a matter of trial and error and experience. For a starting point, one complete revolution of both tie rod sleeves (tie rods for rack and pinion steering) will move the rim of the steering wheel about 1 or 2 inches (25.4 or 50.8 mm). This depends on the pitch of the tie rod threads and the length of the steering arms.

18.17   REAR WHEEL ALIGNMENT

At one time, rear wheel alignments were seldom done. Occasionally, a car with IRS showed rear tire wear or had handling difficulties, which indicated a need for a rear wheel alignment. Today, rear wheel alignment is becoming common, and the newer alignment systems are designed for use on four wheels.

Portable camber and toe gauges and turntables can be used on rear wheels just like on the fronts; wheel rim mounting adapters are required to attach magnetic gauges to the wheels. Some two-wheel alignment systems and racks

require that the car be driven onto the rack backwards; others use long enough connecting leads for the wheel sensors to reach the rear wheels. A slip plate is required if the rear tires are raised off the rack; some racks provide rear

FIGURE 18–92.
The camber angle is being measured on this rear tire. Note that it is positioned on an alignment stand; the front tires are on turntables and alignment stands.

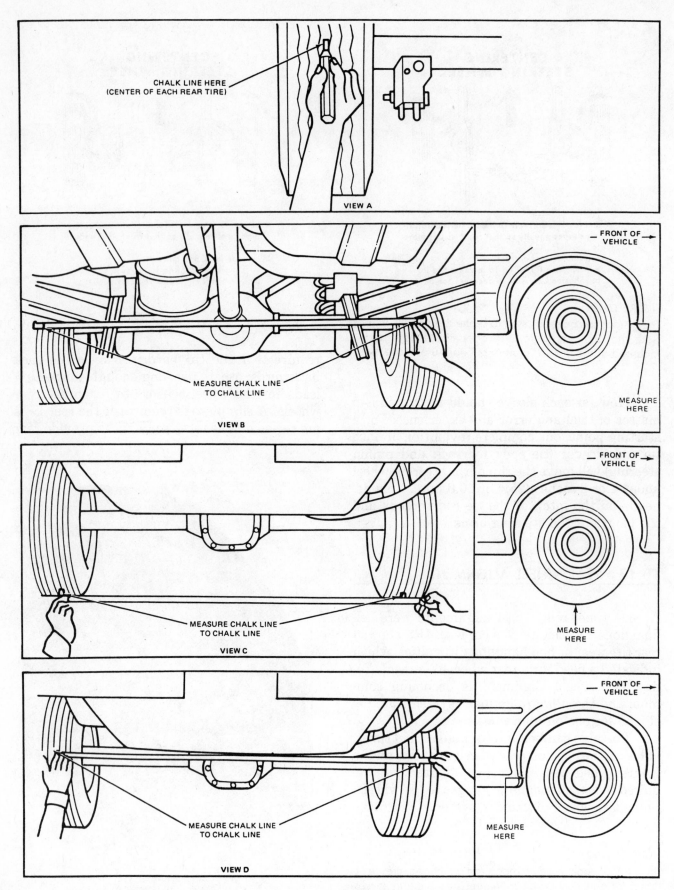

FIGURE 18-93.

This is an accurate procedure for checking for a bent rear axle housing; all three measurements should be the same. (Courtesy of Ford Motor Company)

tire slip plates. **Slip plates** allow the rear tires to slip sideways but not rotate like turntables. Side slippage is especially important to allow cars with IRS to return to normal position as the rear tires are lowered onto the rack. A four wheel system combines a rack with front turntables and rear slip plates with a measuring system using four wheel sensors and controls that allow measuring four wheels.

18.18 MEASURING REAR WHEEL CAMBER

Rear wheel camber is measured the same way as front wheel camber. The measuring procedure depends on the type of gauge and is the same as if the gauges were mounted on the front wheels. (Fig. 18-92)

Rear wheel camber specifications are published for most FWD cars and RWD cars with IRS. If no specifications are available for cars with solid rear axles, you can safely assume that the camber should be zero (0°).

Another method can be used to check for rear wheel camber on an RWD solid axle when you suspect a bent axle housing. Scribe a line around the two tires using a tire scribe or mark a spot on the inside of the sidewall on each tire. Measure the distance between the marks or the scribed lines at the very front, the very bottom, the very back, and if possible, the very top. If you are using the marked sidewall, it will be necessary to rotate the tires between measurements. Compare the measurements, and if they

are all the same, the housing is true. If the front and back are different, there is toe. If the bottom differs from the top or the front and back, there is camber. The axle housing should be straightened or replaced if there is camber or toe. (Fig. 18-93)

18.19 MEASURING REAR WHEEL TOE

Rear wheel toe is measured the same way as on the front wheels except that it is normally done from the rear of the car. This allows more working room for the gauges, but it also reverses the readings. Toe in readings become toe out and vice versa. Toe can be read using any of the different styles of toe gauges. (Fig. 18-94)

18.20 MEASURING REAR WHEEL TRACK/THRUST

Rear wheel track/thrust can be measured using different methods. **Track gauges** are relatively accurate and inexpensive, but they are becoming obsolete because they are slow and require two people. This gauge has three pointers that are adjusted so they just touch the front and rear of the rim, next to the wheel flange, on the rear wheel and the rear of the rim on the front wheel on one side of the car. The gauge is then moved to the other side of the car to compare the wheel positions; pointers that do not match indicate an error in wheel position. (Figs. 18-95 & 18-96)

REAR WHEEL ALIGNMENT		Acceptable Alignment Range	Preferred Setting
CAMBER			
L Body		−1.25° to −.25° (−1-1/4° to −1/4°)	−.75° ± .5° (1/2°)
K,E,G Bodies		−1.0° to 0° (−1° to 0°)	−.5° ± .5° (1/2°)
L/Z 28		−1.1° to −.1° (−1-1/8° to −1/8°)	−.6° ± .5° (1/2°)
TOE*			
L Body & L/Z 28	Specified in Inches	5/32″ **OUT** to 11/32″ **IN**	3/32″ **IN**
	Specified in Degrees	0.3° **OUT** to 0.7° **IN**	0.2° **IN**
K,E,G Bodies	Specified in Inches	3/16″ **OUT** to 3/16″ **IN**	0″ ± 1/8″
	Specified in Degrees	.38° **OUT** to .38° **IN**	0° ± .25°

*TOE OUT when backed on alignment rack is TOE IN when driving.

FIGURE 18-94.
Rear wheel alignment specifications, note the preferred (ideal) setting and the acceptable range. (Courtesy of Chrysler Corporation)

TRACKING MEASUREMENTS

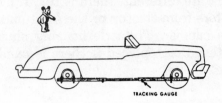

To check tracking: adjust gauge to the distance between the front and rear wheels on one side of the vehicle and compare with the other side. Mfg's Specs. on front pointer only:
  ⅛" tolerance for pass. cars
  ¼" tolerance for trucks
  ¼" tolerance for Pick ups (Providing it isn't short on right side)
No open pointers at the rear

FIGURE 18–95.
A track gauge has three pointers which are adjusted to fit the wheels at one side of the car; the gauge should fit the wheel positions at the other side of the car. (Courtesy of Bear Automotive)

TRACKING EXAMPLES

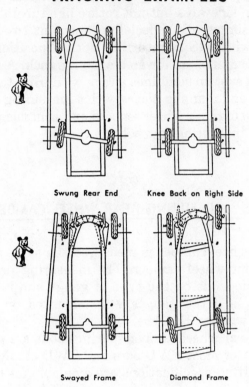

Swung Rear End   Knee Back on Right Side

Swayed Frame   Diamond Frame

A second method of checking thrust is done on the more sophisticated alignment systems. Measurements are made of the rear wheel positions, and these are compared with the front wheel positions by the machine.

As rear wheel toe is measured using toe gauges that give individual readings, a thrust error is indicated if the two readings are not the same. For example, 1/8 inch (3 mm) of toe in on the right rear tire, and 1/8 inch (3 mm) of toe out on the left will give zero total toe and a car that goes down the road with the rear wheels to the left of the fronts. If each tire has the same amount of toe in or toe out, the thrust line will be straight down the center of the car. (Fig. 18–97)

18.21 ADJUSTING REAR WHEEL CAMBER AND TOE

Most IRS cars have some provision for adjusting rear wheel camber and toe; the exact method will vary depending on the car make and model. Older Corvettes, for example, have an eccentric cam on the lower strut mount that

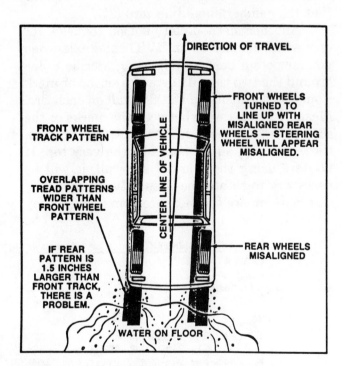

FIGURE 18–96.
A quick check for dog tracking is to drive the car through a large water spot and check the patterns made by the wet tires. If the pattern gets too much bigger after the rear tires pass, the alignment of the rear tires should be checked. (Courtesy of Ford Motor Company)

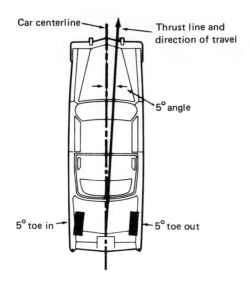

FIGURE 18–97.
If the left rear tire has toe in and the right rear tire has toe out, the rear wheel thrust line will be to the right; the car will dog track with the rear tires to the right; and the steering wheel will be low on the right side because a slight right turn will be necessary to go straight ahead.

will adjust camber and a set of shims at the front of the trailing arm to adjust toe. Newer Corvettes still use an eccentric cam for camber adjustment and a tie rod for the toe adjustment. (Fig. 18–98)

Many FWD cars use a rear wheel spindle that bolts to the rear axle or spindle support; shims are available to install between the spindle and the axle. Depending on the placement of the shim, camber or toe is changed. The shims are available in a tapered, full-sized style or a smaller design that fits under only two bolts. They come in various angles of taper or thickness to provide various amounts of adjustment. (Fig. 18–99)

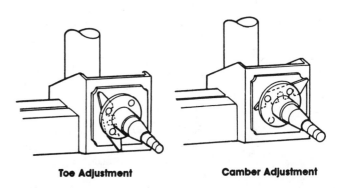

FIGURE 18–99.
Addition of these shims at the rear of an FWD car would adjust the tires toward toe out (left) and/or positive camber (right). (Courtesy of Moog Automotive)

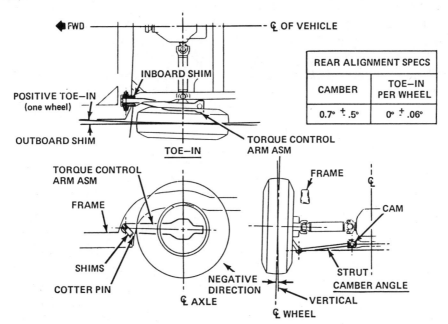

REAR ALIGNMENT SPECS	
CAMBER	TOE–IN PER WHEEL
0.7° ± .5°	0° ± .06°

FIGURE 18–98.
The rear wheel alignment specifications and adjustment locations for an older Corvette. Camber is adjusted using the cam at the inner end of the strut; toe is adjusted using shims at the front of the torque control (trailing) arm assembly. (Courtesy of Chevrolet)

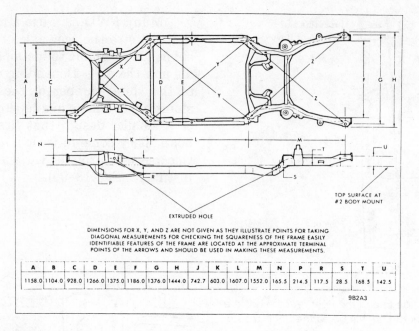

A	B	C	D	E	F	G	H	J	K	L	M	N	P	R	S	T	U
1158.0	1104.0	928.0	1266.0	1375.0	1186.0	1376.0	1444.0	742.7	603.0	1607.0	1552.0	165.5	214.5	117.5	28.5	168.5	142.5

FIGURE 18–100.
Frame/body dimensions are available for use in checking a frame/body for straightness; most measuring points are stamped or drilled holes. (Courtesy of Oldsmobile)

18.22 FRAME AND BODY ALIGNMENT

Occasionally there is a car with indications that the frame or body is out of alignment. Frame or body alignment is normally done by a body shop that is large enough to have the equipment and the skill needed for this operation. Frame and body alignment is checked in various ways.

Frame and body dimensions are available from most manufacturers; these dimensions provide the distances between various **datum** or **measuring points** under the car. These distances are measured for a particular car and compared with the specified dimensions; if they differ, the frame or body is out of alignment. Possible frame twists are checked by attaching a set of special **frame gauges** to the car and sighting along the gauges to locate any twists

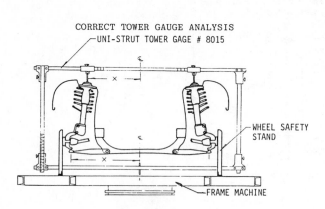

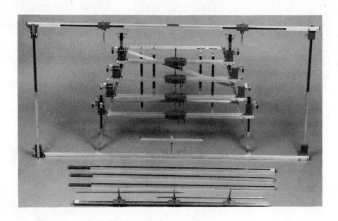

FIGURE 18–101.
A frame/body gauge set. The pieces are attached to the car at the various measuring points to check for bends or twists; note how the one gauge is used for checking the strut towers and lower control arm locations. (Courtesy of Arn-Wood)

or bends. Some of these gauges are large enough to fit the upper strut mounts on strut-equipped cars. Some shops use a special **jig**, which the car is compared to; if it does not fit the jig, the frame or body is bent. (Figs. 18–100 & 18–101)

18.23 ROAD TESTING AND TROUBLE SHOOTING

When trying to solve a driving or handling complaint or when completing a wheel alignment, the competent front end technician gives the car a road test. The car is generally driven in a particular pattern while the driver checks for various possible problems.

The road test often begins by driving the car over a **scuff tester**; this device gives a quick indication if the toe is incorrect or some other problems are causing tire scuff. On a road test, the car is driven at slow, medium, and high speeds over straight roads with some turning maneuvers so the technician can check for loose

FIGURE 18–102.
Driving a car over a scuff gauge. This gauge shows the amount of side scuff of the tires. Shops use this type of gauge as a quick check to determine if an alignment is needed and to check an alignment after completion. (Courtesy of Ammco)

or hard steering, pull, wander, poor returnability, shimmy, or vibrations. A guide such as Table 18–1 can be used to help diagnose the cause of any problems. You should be able to recognize most of these problems. (Fig. 18–102)

TABLE 18–1
A Suspension and Steering Trouble Shooting Chart.
(Courtesy of Moog Automotive)

SYMPTOM	PROBABLE CAUSE
Excessive tire wear on outside shoulder	Excessive positive camber.
Excessive tire wear on inside shoulder	Excessive negative camber.
Excessive tire wear on both shoulders	Rounding curves at high speeds. Under-inflated tires.
Saw-tooth tire wear	Too much toe-in or toe-out.
One tire wears more than the other	Improper camber. Defective brakes. Defective or worn strut cartridge.
Tire treads cupped or dished	Out-of-round tires. Out-of-balance condition. Loose wheel bearings. Defective or worn strut cartridge. Worn rack & pinion frame mount bushings.
Front wheels shimmy	Out-of-round tires. Out-of-balance condition.
Vehicle vibrates	Defective tires. One or more of all 4 tires out-of-round. One or more of all 4 tires out-of-balance. Drive shaft bent. Universal joints loose.
Car tends to wander either to the right or left	Improper toe setting. Looseness in steering assembly. Uneven caster.
Vehicle swerves or pulls to side when applying brakes	Uneven caster. Brakes need adjustment. Contaminated brake linings. Worn strut rod bushing.
Car tends to pull either to the right or left when taking hands off steering wheel	Improper camber. Unequal caster. Tires worn unevenly. Tire pressure unequal.
Car is hard to steer	Tires under-inflated. Power steering defective. Too much positive caster. Steering assembly too tight or binding.
Steering has excessive play or looseness	Loose wheel bearings. Loose tie rod ends. Loose bushings. Loose steering gear assembly; adjust. Worn rack & pinion frame mount bushings.

REVIEW QUESTIONS

1. Of the five wheel alignment angles,
   A. caster and camber are the tire-wearing angles.
   B. toe in and SAI are directional control angles.

   a. A only
   b. B only
   c. both A and B
   d. neither A nor B

2. Which of the following angles normally is not adjusted?

   a. caster
   b. camber
   c. SAI
   d. toe

3. Technician A says that the car must be placed on a level surface and be at the correct height when measuring alignment angles. Technician B says the front wheels must be straight ahead when measuring camber and toe. Who is right?

   a. A only
   b. B only
   c. both A and B
   d. neither A nor B

4. Caster-camber gauges are available that will magnetically attach to
   A. wheel rims.
   B. hub faces.

   a. A only
   b. B only
   c. both A and B
   d. neither A nor B

5. Technician A says that adapters are available to attach magnetic gauges to wheel rims, but they are not very accurate. Technician B says that caster is checked with the front wheels straight ahead. Who is right?

   a. A only
   b. B only
   c. both A and B
   d. neither A nor B

6. Depending on the car, camber is not adjusted by

   a. adding or removing shims at the control arm mounts.
   b. bending the spindle.

   c. turning one or two eccentrics at the control arm mount.
   d. sliding the control arm mount along a slotted hole.

7. Camber is made more positive by moving the upper ball joint or top of the strut
   A. inward.
   B. rearward.

   a. A only
   b. B only
   c. both A and B
   d. neither A nor B

8. Technician A says that it is a good practice to use a maximum of five shims at any single location. Technician B says that it is impossible to adjust camber on a solid or twin I-beam axle. Who is right?

   a. A only
   b. B only
   c. both A and B
   d. neither A nor B

9. Technician A says that the alignment cannot be adjusted on cars with strut suspension. Technician B says that the rear wheels on most FWD cars have alignment adjustments. Who is right?

   a. A only
   b. B only
   c. both A and B
   d. neither A nor B

10. Both front tires have a wear pattern that can be felt as your hand is slid outward over the tread but not inward. This tells us that the front tires have too much:

    a. negative camber
    b. positive camber
    c. toe out
    d. toe in

11. When a trammel bar is used to measure toe, the measurements are taken at

    a. the inner edge of the wheel rim.
    b. the inner edge of the tire tread.
    c. a scribed line on the tire tread.
    d. any of these.

12. Technician A says that toe in is increased by making the tie rods (front mounted) longer. Technician B says that toe is always the last angle to measure and adjust when doing a wheel alignment. Who is right?

    a. A only   **c.** both A and B
    b. B only   **d.** neither A nor B

13. Lengthening the left tie rod and shortening the right one an equal amount will
    A. increase toe in.
    B. change the steering wheel position.

    a. A only   **c.** both A and B
    b. B only   **d.** neither A nor B

14. Technician A says that both tie rod ends should be rotated in the same direction before tightening the clamps on the tie rod sleeve. Technician B says you need to make sure the rack and pinion gear boots do not get damaged while making toe adjustments. Who is right?

    a. A only   **c.** both A and B
    b. B only   **d.** neither A nor B

15. In order to measure caster, the tires
    A. must be turned to one position where the gauge is set, and then turned to a second position where the gauge is read.
    B. must be centered on a pair of turntables.

    a. A only   **c.** both A and B
    b. B only   **d.** neither A nor B

16. Technician A says that caster is made more positive by moving the front end of the upper control arm pivot shaft outward and the rear end inward. Technician B says that caster is made more positive by making a trailing strut rod shorter. Who is right?

    a. A only   **c.** both A and B
    b. B only   **d.** neither A nor B

17. Technician A says that road crown is compensated for by adjusting caster to a 1/4° to 1/2° spread. Technician B says that road crown is compensated for by adjusting the right wheel to a slightly more positive caster angle. Who is right?

    a. A only   **c.** both A and B
    b. B only   **d.** neither A nor B

18. SAI can
    a. not be measured.
    b. be adjusted similar to caster.
    c. not be adjusted.
    d. none of these.

19. The steering wheel is off center, down on the right side while driving straight down the road. Technician A says that there is set back, or the rear wheel thrust is off center and the tie rods need to be adjusted to center the wheel. Technician B says the right wheel caster is too positive. Who is right?

    a. A only   **c.** both A and B
    b. B only   **d.** neither A nor B

20. A car equipped with a manual rack and pinion steering gear has a shimmy. Technician A says that this can be caused by worn or loose tie rod ends. Technician B says that it could be caused by faulty rack mounting bushings. Who is right?

    a. A only   **c.** either A or B
    b. B only   **d.** neither A nor B

Chapter 19

SUSPENSION DYNAMICS

After completing this chapter, you should:

- Be familiar with the terms commonly used with vehicle dynamics.

- Be familiar with the physical forces which act on a vehicle during various handling maneuvers.

- Be familiar with the various modifications which can be made to improve a vehicle's handling abilities.

19.1   INTRODUCTION

The amount of grip or adhesion between the road and the tires is the major factor that limits how fast a vehicle can maneuver through a corner (or stop or accelerate). The greater the traction force, the faster or harder the car can corner, stop, or accelerate. The tire-to-road contact on a cornering vehicle is affected by several factors. **Suspension dynamics** is the study of these various forces and their effects. Our discussion will concentrate on these forces as they affect cornering and road handling, with some consideration given to how they affect acceleration and deceleration. Our discussion will also include road-racing or oval-track vehicles as well as high-performance street-driven cars. The forces described in this chapter affect almost every moving vehicle to some extent; most suspension system designers are familiar with them. Those of you that are interested in improving the handling of wheeled vehicles should be able to relate these forces to your particular vehicle.

If making any of the modifications described in this chapter, it must be remembered that:

- Most of these changes are trial-and-error; they must be tested to prove their effectiveness.
- Most of these changes interact with each other; they will usually, but not always, work better with each other.
- These changes are compromises; they will usually cause detrimental effects in some other aspect such as ride quality, ground clearance, or tire wear.
- Any changes must not violate any applicable vehicle code.
- Most of these changes are based on the performance of a good set of tires. Good handling cannot occur without good tires.

SAFETY NOTE

This chapter describes tests and modifications that are not normally done in the automotive repair industry and, in some cases, are not approved by the vehicle manufacturer. If these road tests are done, they should be carried out in a careful manner while observing these precautions:

- The vehicle must be in good mechanical condition.
- The tires must be in good, sound condition.
- The driver must wear lap and shoulder safety belts as well a safety helmet.
- The roadway used must have good, sound pavement and be clear of obstacles and other vehicles.
- Any tests conducted on public roads must comply with all of the applicable traffic regulations.

19.2   CORNERING FORCE

The ability of a car to corner, referred to as **cornering power** or **lateral acceleration**, is measured in g's; a g is also used to measure acceleration and deceleration. One g is equal to the force of gravity; one g of force will cause an object, dropped in a free fall, to accelerate at a rate of 32.2 feet per second (9.8 m/s) (21.95 MPH/35.3 KPH). Except for the direction, the sideways force we feel when turning a corner is basically the same as the rearward force we feel during acceleration or the forward force we feel when we brake hard. As a car goes faster and faster through a corner of a given radius, there will be a greater cornering force on the car. Cars that generate higher cornering forces can go through a particular corner at a higher speed or at the same speed in a more stable manner. (Fig. 19–1)

Most drivers use only a small portion of the car's ability to corner or brake fast; the average driver feels uncomfortable at cornering or braking forces above .2 or .3 g's. Many produc-

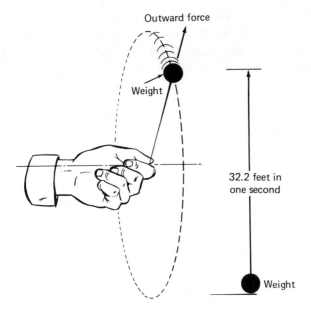

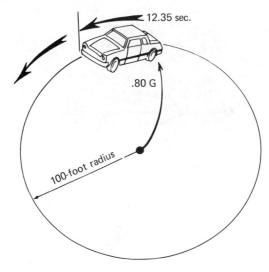

FIGURE 19–1.

The force of gravity will cause an object to accelerate at a speed of 32.2 feet per second during each second that it falls (assuming no aerodynamic drag). If an object (left) is spun fast enough, it can generate enough centrifugal force to keep the string tight even in a vertical position; this force would be 1 g or greater.

FIGURE 19–2.

Cornering forces can be measured on a skid pad, a large, flat circle. A car would generate .8 g's of force if it can drive around a 100 foot radius circle in 12.35 seconds.

tion cars have the ability to corner at a lateral force somewhat greater than this, between .7 and .8 g's. Recently, an American production sports car, in a slightly modified form, developed over 1 g of cornering power on a skid pad. This means that the sideways force on a 150 pound (68.2 Kg) driver would be greater than the weight of the driver. Racing vehicles, such as Formula 1 or Indianapolis cars, can generate cornering forces in the 1.2 to 1.7 g range. The ability to generate high cornering forces allows faster cornering speeds and, therefore, lower lap times.

Cornering forces are measured on a **skid pad**; this is merely a large, flat, paved area on which the car can be driven in a circle. (Fig. 19–2) The radius of the circle can vary, but too short a radius, under 100 feet (30.5 m), does not allow the car to reach a very high speed. One major car enthusiasts' magazine uses a 141 foot (42.9 m) radius skid pad for its test purposes. To measure cornering force, the car is driven around the skid pad at the highest possible speed without allowing the car to leave the designated circle. Skid pad tests begin at a fairly

low cornering speed, and the car is gradually accelerated until it reaches the maximum possible speed. The car would slide outward at any higher speed. The car is timed to determine how long it takes to make one revolution or lap, and the amount of cornering force is determined by using this formula:

$$g = \frac{R \times 1.225}{T \times T}$$

where g = centrifugal force in percentage of g

R = radius of skid pad in feet

T = time to complete one lap in seconds

Cornering power tests on a skid pad are normally made in both directions, clockwise and counter clockwise.

It can be argued that skid pad tests do not represent the real world, where right and left maneuvers are more common than driving around in a circle. Tests on a skid pad measure the car's **steady state** cornering ability; the car has stabilized or set in a cornering mode. Back-and-forth maneuvers are referred to as **transient state** conditions; **slalom type courses** are used to give an indication of the car's transient

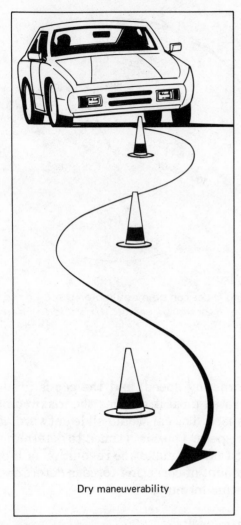

FIGURE 19–3.
Transient state manuevers can be checked as a car runs a slalom course. The car is driven on a zig-zag path through a series of pylons.

abilities. During a transient maneuver, the load on the car's tires and springs is changing. The tires, springs, and shock absorbers have a rather large effect on handling during transient conditions, or the time it takes the car to set up for a corner. (Fig. 19–3)

Skid pad tests do give a good indication of a car's cornering ability, and they are fairly easy to make. However, if you try one, you will find that skid pad tests are very hard on tires—the car's ability to corner is measured right up to the limit of the tires. While making a cornering power test on a skid pad, care should be taken to ensure that the engine does not lose its oil

pressure. Centrifugal force can move all of the oil to the side of the oil pan, away from the oil pump intake. This will cause starvation of the oil pump and the engine's bearings.

19.3   TRACTION/SLIP ANGLE

As mentioned earlier, cornering power is limited by the traction force of the tires. When the cornering power requirement for a particular manuever is less than the traction that can be provided by the tires, the car will go where it is pointed and steered. If the cornering force exceeds the available traction from the tires, the tires will slip across the road surface, and they will **skid**.

A tire will usually slip slightly on the road surface as it reaches its maximum limit; this is called **slip angle**. Slip angle is defined as the difference in angle between the center line of the tire tread and the center line of the wheel. In part, slip angle is caused by the flexibility of the tire's sidewall and tread; it is also caused by the tread rubber creeping into maximum contact with any imperfections in the road surface. (Fig. 19–4)

Most tires reach maximum traction at a slight slippage to the road surface. This varies with tread design and compound and the condition of the road surface; maximum traction with most tires occurs at about 10 to 25 percent slippage. Tire-to-road slippage greater than 25 percent usually results in a drastic loss in traction and a skid. (Fig. 19–5)

19.4   OVERSTEER/UNDERSTEER

A car is said to have **neutral steering** capabilities if the tire slip angles are the same at the front and the rear. A neutral steering car will require very little steering correction as it increases speed around a skid pad. When the cornering speed gets too great, the car will slide sideways off the skid pad, still pointed in the original direction of travel and with the same amount of steering input.

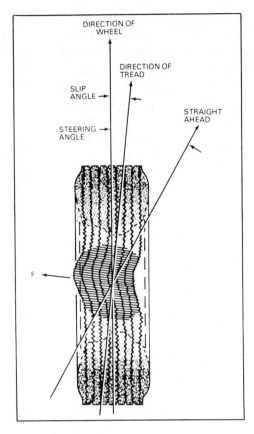

FIGURE 19–4.
During cornering manuevers, the tread portion of the tire distorts slightly as the tire develops a slip angle, the difference between the direction that the wheel is pointed and the direction the tire is pointed. (Courtesy of Chevrolet)

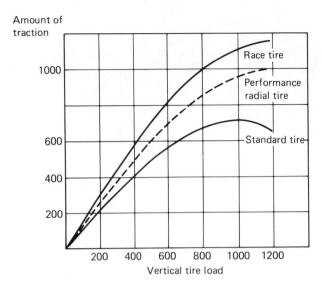

FIGURE 19–5.
The amount of tire traction usually increases as the load on the tire increases. Note that curves are not linear; doubling the tire load will not double the amount of traction. Also note that different styles of tire develop differing amounts of traction.

If the front tires have a greater slip angle than the rear tires, the car will **understeer.** The steering wheel will need to be turned sharper and sharper as the speed increases and the tires start to reach their traction limits. Eventually, the front tire traction will drop off drastically as the tires begin to skid, and the car will travel straight off the corner instead of turning. This is also called **pushing** or **plowing.** Many suspension engineers consider understeer to be the most safe condition for drivers without much experience in skids or fast cornering (the average driver). The normal reaction for a driver in a skid is to slow down, and this is usually the best thing to do when an understeering car gets into trouble on a corner. Let off the gas, but as we will see later, do not put on the brakes until the skid goes away. If you were to suddenly let off the gas on a neutral or oversteering car, the

change might cause a severe oversteer condition.

If the rear tires have a greater slip angle than the front tires, the car will need to be steered less, or more straight, as the speed increases. Eventually, if the speed gets too great, the rear tires will begin to skid and lose traction drastically. The back end of the car will swing outward, and the car will probably **spin out.** This is called **oversteer** or being **loose.** In certain conditions, an oversteering car can be fun to drive, but much driver skill is required to control an oversteering car. (Fig. 19–6)

19.5 TRACTION FORCE/CIRCLE OF TRACTION

The amount of traction or cornering power that the tire on a car can generate is affected by many factors; one vehicle-handling expert says that there are 31 different things that will affect traction. Some of these are:

1. Tire: construction (radial or bias ply)
   tread contact width
   side wall height (aspect ratio)

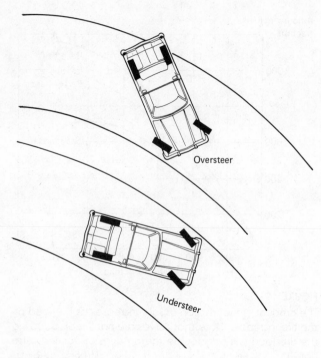

FIGURE 19-6.
An oversteering car appears to turn sharper than it should for a given steering wheel motion; an understeering car will turn less sharply.

        inflation pressure
        tread temperature
        tread pattern
        rubber compound

2. Road: surface material
    smoothness
    wetness

3. Alignment angle, camber and toe

4. Downward force/load on the tires

In this chapter, we will leave tire construction to the tire engineers and tire selection to the pocketbook of the car owner and will offer only the comment that some tires can generate more traction than others. The wider or softer the tread is, the more traction it will normally generate. However, tire selection is really a game of compromise—there is no perfect tire for all conditions. Modern performance tires have made tremendous progress in the areas of wet and dry pavement traction. Even racing tires come in various compounds and tread surfaces. (Fig. 19-7)

We will also treat road surface and condi-

tion as something we have no control over and hope that you realize that poor road surface conditions cause reduced traction and demand slower speeds. Those of you who have driven on glare ice have an idea of how small the circle of traction can become and how easy it can be to exceed it.

In tuning a chassis to increase cornering power, we do have control over a few of the factors related to traction; these are alignment angles and tire loading. As many people realize, tire traction increases as more load is put on the tire: the greater the load, the greater the traction. We describe this aspect more in the section dealing with weight transfer.

The amount of traction that a tire can generate is often referred to as the **circle of traction** or **friction circle**. A tire has almost equal traction in any direction relative to the road, sideways, forward, or backward. In actual practice, there is a slightly greater amount of traction in a direction parallel to the tread pattern, but the difference is very small. The tire also rolls best in a direction parallel to the tread, forward or backward, because it cannot roll to the side; it has to slip or skid to move sideways.

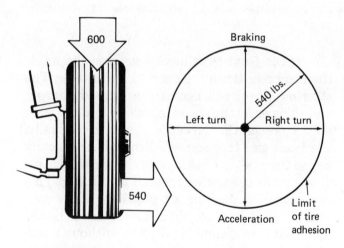

Weight x coefficient of friction = total force generated by tire

FIGURE 19-7.
The amount of traction that a tire can generate is basically determined by multiplying the coefficient of friction by the tire load; in this case, .9 x 600 = 540. We illustrate this amount of traction by a certain sized circle, "the circle of traction." This traction can be used to move the tire (and the car) forward, backward, or to the side.

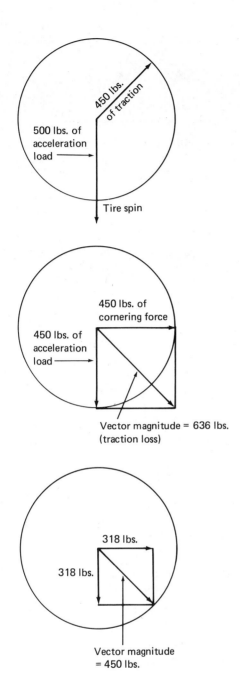

FIGURE 19–8.
Tire spin will occur if you apply a 500-pound acceleration load to a tire that has a 450-pound traction circle. If a 450-pound acceleration load and a 450-pound cornering load were applied to this tire, the tire would also slip and skid. This tire could handle a 318-pound acceleration and a 318-pound cornering load at the same time; the vector magnitude of these two loads is slightly less than 450 pounds.

When using the circle of traction, the diameter of the circle represents the amount of traction that a tire can generate; changes to in-

crease traction will increase the circle diameter. When the driver puts a forward, backward, or side load on the tire that is greater than the available traction, the tire will skid. Too much acceleration force will spin the tire, and too much braking or cornering force will skid the tire. When a tire does two things at once, such as the front tires braking and turning at the same time, these forces combine. The easiest way to describe the combination of forces is with vector analysis. This is illustrated in Figure 19–8. For example, let's say a given tire can generate 500 pounds (227 Kg) of traction. This amount of traction can be used for acceleration or cornering, but the car will slip or skid if it accelerates and corners at the same time at this rate.

The sizes of the four traction circles will change as a car is driven through a corner because of weight transfer across the car. Weight leaves the tires on the inside, reducing their traction circles, and moves to the outside tires, increasing their traction circles. The amount of traction also varies during acceleration and deceleration because of lengthwise weight transfer. This is described in more detail later in this chapter.

As the car is driven around a corner, the demands placed on the tires also vary. Braking, cornering, and acceleration loads can combine to come close to or exceed the size of the traction circle. Accelerating hard in a straight line can stress the limit of traction of the rear tires (RWD) almost equally. Accelerating hard while turning places a cornering load on all four tires, but the additional acceleration load on the rear tires often exceeds the limits of traction. The slip angle of the rear tires will increase drastically, and the car will probably go into severe oversteer and spin out. Or, a car can be cornering hard, near the traction limit, and the driver applies the brakes. The resulting weight transfer and increased braking load will probably exceed the traction limits of the front tires, and the car will plow (understeer) straight ahead. A skillful driver feels for signs of car slippage that indicate how hard the gas or brake pedal can be pushed or how hard the steering wheel can be turned. A slight amount of excessive tire slippage indicates a need to back off. (Fig. 19–9)

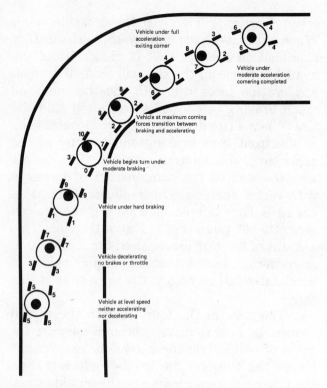

FIGURE 19-9.
Tire loading changes as a car accelerates, decelerates, and turns corners; the sizes of the four circles of traction at the tires is also changing along with the changes in tire load. (Courtesy of Quickor Engineering)

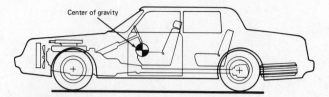

FIGURE 19-10.
The center of gravity is the balance point of the car; it is the exact center of all of the weight mass. (Courtesy of Quickor Engineering)

19.6   WEIGHT TRANSFER

Because of weight transfer, tire traction changes as a car goes around a corner (or accelerates or decelerates). Centrifugal force pushes outward on the car's **center of gravity, CG.** The CG is a point near the center of the car; it is the balance point of the car. Centrifugal force is resisted by the traction of the tires. The interaction of these two forces moves weight from the side of the car on the inside of the turn to the outside of the car, and the car leans and rolls. As this occurs, weight leaves the springs on the inside, and that side of the car raises. This weight goes to the springs on the outside, and that side of the car lowers. The total weight of the car will still be the same, but more of it is carried by the outside springs and tires. (Figs. 19-10 & 19-11)

The amount of weight that will transfer laterally (side to side) can be determined by using this formula:

$$WT = \frac{CF \times W \times CGH}{TW}$$

where WT = lateral weight transfer in pounds
   CF = centrifugal force in g's
   W = car weight in pounds
   CGH = center of gravity height in inches
   TW = track width in inches

Studying this formula shows us that the amount of weight that will transfer increases if the car corners faster, weighs more, or has a higher center of gravity or a narrower track width. Weight transfer and its effects can be reduced by cornering slower (no racer wants to do this), lowering the car's weight, lowering the car's height, or increasing the track width.

Lateral weight transfer moves weight from the inside tires, which lowers their traction, to the outside tires, which increases the traction there. The traction increase with the weight gain is always lower than the traction loss from an equal weight loss; this can be seen by studying a weight-to-traction chart such as Figure 19-5. Two tires with equal weight will always have more traction than two unequally loaded tires carrying the same amount of weight (all other things being equal). As weight is transferred from one tire to another, the total amount of traction for the two tires is reduced. Weight transfer should always be considered

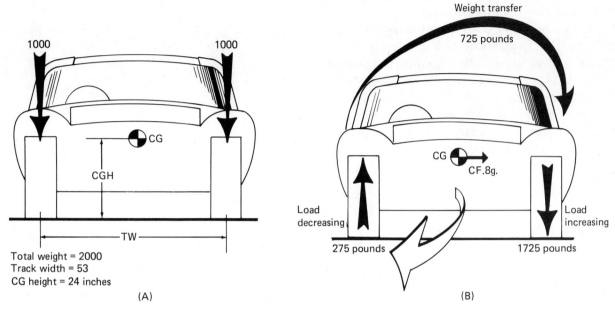

FIGURE 19-11.
As a car turns a corner (B), centrifugal force causes weight to transfer from the tires on the inside of the turn to the tires on the outside.

when building or modifying a car that is going to turn corners. (Fig. 19-12)

The formula for lengthwise or front-to-back or back-to-front weight transfer is essentially the same if we substitute acceleration or deceleration rate for centrifugal force and wheelbase for track width. The formula would look like this:

$$WT = \frac{R \times W \times CGH}{WB}$$

where WT = lengthwise weight transfer in pounds

R = rate of acceleration/deceleration in g's

W = car weight in pounds

CGH = center of gravity height in inches

WB = wheelbase in inches

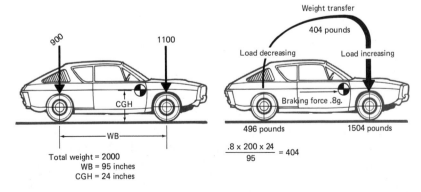

FIGURE 19-12.
As a car brakes (B), inertia will cause weight to transfer from the rear tires to the front tires; weight will transfer from the front to the back during acceleration.

Lengthwise weight transfer is very important on a drag race car. A high horsepower car will have the maximum amount of possible traction if all of the weight transfers to the rear tires. At this time, there will be no rolling friction from the front tires. A "wheelie" is evidence that a 100 percent weight transfer has occurred. When this occurs, there is an occasional problem in steering because there will be no traction at the front tires.

19.7   ROLL STEER

Weight transfer causes body lean and roll, and body roll will often cause changes in camber and toe. It can also cause a thrust angle change on rear axles resulting in **roll oversteer** or **roll understeer.**

If the front section of the rear leaf spring, or the control arms with coil springs, is inclined downward (at the rear), body roll will pull the axle forward on the inside (relative to the turn) end of the axle. Body lowering will push the outside end of the axle toward the rear. This action causes the axle thrust line to change so that it will steer the rear of the car toward the outside of the turn, causing roll oversteer. The rolling action of the car is causing an oversteer effect. Inclining the rear axle control arms the opposite way would produce the opposite effects; body roll would steer the rear of the car toward the inside of the turn. This second case would produce roll understeer. (Fig. 19–13)

Roll steer will also result from these camber or toe changes if they occur during body roll. Roll understeer will result if:

the inside front tire camber changes negative

the inside front tire toe changes inward

the outside front tire camber changes positive

the outside front tire toe changes outward

the inside rear tire camber changes positive

the inside rear tire toe changes outward

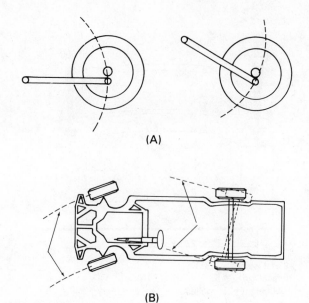

(A)

(B)

FIGURE 19–13.
Cornering maneuvers or any vertical body/frame motions will cause a trailing or leading axle control arm to change the wheelbase, the greater the starting angle the greater the change. If the wheelbase on each side of the car does not change an equal amount, roll under or oversteer can occur. Roll oversteer is shown at B.

the outside rear tire camber changes negative

the outside rear tire toe changes inward

Roll oversteer will result if these changes occur in the opposite direction.

The camber or toe change resulting from roll is designed into the suspension geometry of many cars. A properly designed S-L A suspension can handle the action quite well and sometimes produce an almost ideal camber change curve. Ideally, a tire should lean away a slight amount from the outside of a turn. A cornering bicycle or motorcycle provides an example of the ideal camber angle on corners. On a car, the outside tires should have negative camber while the inside tires should have positive camber. Because of the higher load and therefore greater traction, the outside tire is the most important while cornering.

This camber change is desired so the tire tread will have complete contact with the road

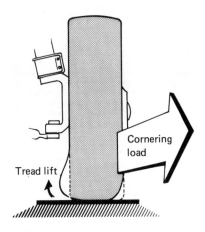

FIGURE 19-14.
A cornering load will cause a sideways distortion of the tire tread; when the force gets strong enough, it will begin lifting the edge of the tread from the road surface.

surface. The flexibility of the tire's tread and carcass causes the tread to roll under during a sideways stress; on a vertical tire, this action would lift part of the tread off the road and increase tire slippage. When the tire has negative camber, tire roll flattens the tread onto the road surface. A sensitive driver can notice the beginning of the slippage caused by tread lift and take it as a warning that the tires are approaching their traction limit. (Fig. 19-14)

Some suspension designs have bad camber change characteristics. The result of this is to reduce the cornering power of the tires while the car is entering a corner. These particular cars have very poor cornering abilities.

19.8 ROLL AXIS/ROLL CENTERS

Body roll is also affected by the location of the car's **roll axis**; this is an imaginary line that runs lengthwise through the car. The roll axis connects the front and rear roll centers. This is the pivot axis around which the body of the car rotates during cornering. (Fig. 19-15)

When a sideways force is placed on the car, this force acts on the mass of the car. For practical purposes, we can consider that it pushes sideways on the CG. As mentioned earlier, this force is resisted by the tires. The connection between the body of the car and the tires is

through the suspension members. The flexibility of the suspension is what allows body roll to take place. There will be some additional slight rolling action occurring at the tires, but this is usually minor. Each style of suspension system produces a **roll center**, the pivot point for body roll.

Normally, the car's CG is above the roll axis, so cornering forces will cause the car's body to roll outward on a turn. The CG is determined by the placement of the heavy parts (engine, transmission, etc.) of the car; good design usually dictates as low a CG as possible (remember that a lower CG produces less weight transfer). The ideal car would also have a CG in the middle of the car and halfway between the front and rear tires. The roll axis can be anywhere from below ground level to a foot or so above ground. The roll axis on a well-designed racing vehicle is within a few inches of ground level and an inch or so lower at the front than the rear. The vertical distance between the CG and the roll axis produces the leverage to produce body roll. If they were the same height, there would be no body roll; a greater distance between the CG and the roll axis would produce a greater amount of roll.

The location of the roll center at the front or rear depends on the type of suspension and

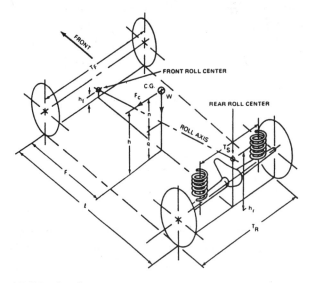

FIGURE 19-15.
The roll axis connects the front and rear roll centers; this is the line around which the car's body will roll during cornering manuevers. (Courtesy of Chevrolet)

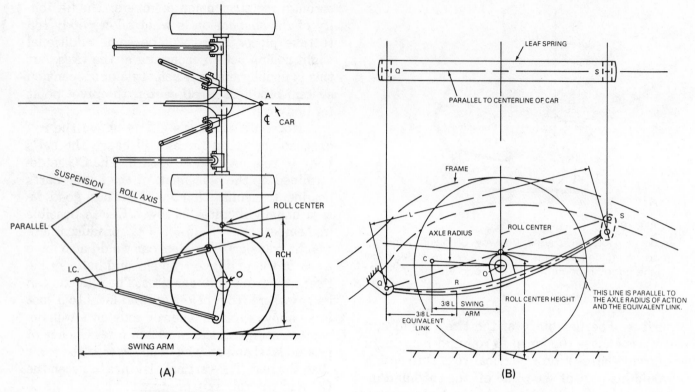

FIGURE 19–16.
The rear roll center is fairly easy to locate with solid axle suspensions; note the different procedure used depending on the spring/control arm type. (Courtesy of Chevrolet)

the placement of the springs or control members. The various suspension types are illustrated in Figures 19–16 and 19–17. In some cases, as with solid and swing axles, it is fairly easy to locate the roll center; in others, such as S-L A and strut suspensions, it requires some geometry and plotting of the suspension points. In some cars, it is possible for the suspension tuner to change the roll center(s) and roll axis; in others it is very difficult. Occasionally, someone will modify a car for one purpose, for example the "low rider" or "Baja off-road" look, and end up with a potentially dangerous CG and roll axis combination.

Plotting a roll center, in most cases, requires that you first plot the instant center, IC, of the suspension. In most cases you will be working with lever arm lengths or pivot points that are not easily seen. The IC of an S-L A suspension is that point where the two control arms would meet or converge if they were long enough. A strut suspension has only one control arm, but a line drawn at a right (90°) angle

from the upper pivot point provides the other portion of this imaginary lever arm.

On cars that use a Panhard rod, the roll center is located where the Panhard rod crosses the car's center line. During body roll, this point will change height; there can also be a slight side shifting of the body, especially if a short Panhard rod is used. Many race car constructors prefer to use a Watts link to reduce this shift of the body and roll center. (Fig. 19–18)

Chassis modifications to produce roll center and roll axis changes are beyond the scope of this text, but there are several well-written books that deal strictly with suspension dynamics, chassis tuning, or race car construction. My intent is to introduce you to this rather fascinating aspect of vehicle performance.

Many chassis tuners treat body roll as a necessary evil; you have to live with a certain amount of it in order to turn corners. The only way to eliminate body roll completely would be to stiffen the springs and antiroll bars to the point where the suspension is solid, and this

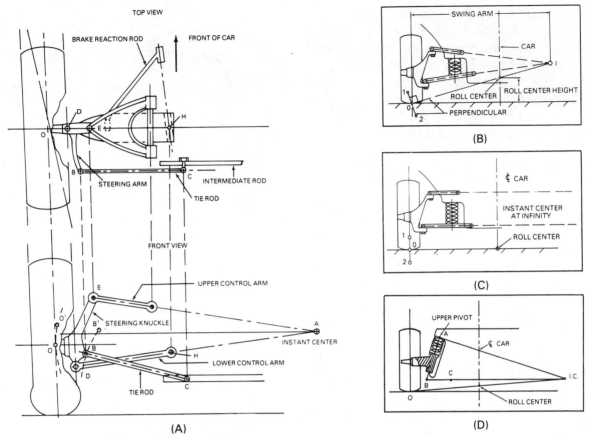

FIGURE 19–17.

The first step in locating the roll center on an independent suspension is to locate the instant center, IC; a line drawn between the IC and the tire tread at the road will cross the car's center line at the roll center. Note the different procedure used to plot the IC on the different types of suspensions. (Courtesy of Chevrolet)

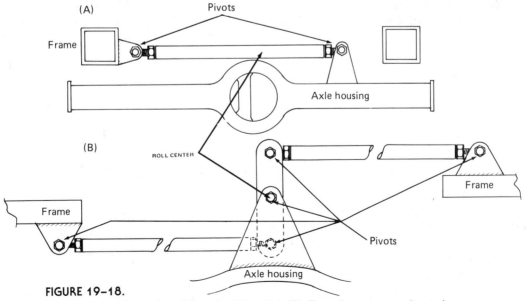

FIGURE 19–18.

A Panhard rod/sway bar (A) and a Watts link (B). Though more complicated, a Watts link will not cause body side motion or a roll center change during suspension travel.

would produce a horrible ride and drastic unsprung weight effects. Body roll is the result of having springs and suspension travel. Some body roll is desirable; it gives us an idea of how hard the car is cornering. Remember that a good suspension engineer or chassis tuner can sometimes let body roll help produce improved tire alignment angles. A bad suspension must be compensated for by using excessively stiff springs to reduce or eliminate motion.

19.9   UNSPRUNG WEIGHT

The tires must remain in contact with the road surface at all times if we want the car to corner, accelerate, or decelerate, and the springs must allow the tires and wheels to move up and down as the tires follow the road.

Most racing vehicles use lightweight, magnesium alloy wheels, thin, lightweight tires, special alloy brake components, or lightweight suspension members to reduce the unsprung weight. As mentioned earlier, in Chapter 10, that portion of the car supported by the springs is called sprung weight; everything that supports the springs is called unsprung weight. When an unsprung mass goes over a bump, it develops an upward traveling force that is proportionate to the weight and vertical speed of the unsprung mass. If the force is strong enough, the unsprung mass and the tires will leave the ground and fly for a distance, much like a motorcycle jumping a ramp in a thrill show. A lower-weight unsprung mass absorbs less energy from a bump and is more easily controlled by the springs and shock absorbers. Softer springs can be used, which should produce a better ride as well as better traction.

Because of the reduced bounce travel, lower unsprung weight allows the use of shorter springs, which in turn lowers the height of the car and the center of gravity. Remember that a lower center of gravity will produce less weight transfer and therefore less lean on corners. The springs must be long enough to allow sufficient tire travel without suspension bottoming. When the suspension bottoms against the bump stops, the whole vehicle becomes un-

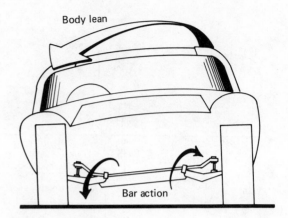

FIGURE 19-19.
The antiroll bar is forced to twist during body roll; the resistance of the bar tends to reduce the amount of roll.

sprung weight; tire loading increases drastically when the suspension bottoms.

19.10   ROLL RESISTANCE

Many chassis tuners use the antiroll bar as the major item to reduce body roll. If you are selecting an antiroll bar, remember that an antiroll bar resists single tire bumps and will produce a harsher ride on road surfaces, which affects tires individually. (Figs. 19-19 & 19-20)

An antiroll bar also has a definite effect on tire loading as a car goes around a corner. **Roll resistance** is the term used to describe those

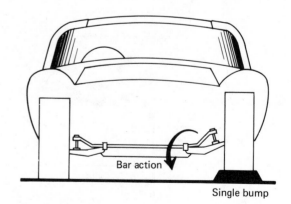

FIGURE 19-20.
A single tire bump will twist the antiroll bar; the rate of the antiroll bar will add to the rate of the spring during this condition making the bump more severe.

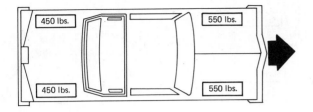

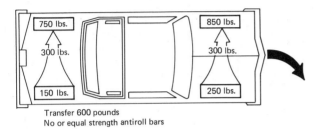

Transfer 600 pounds
No or equal strength antiroll bars

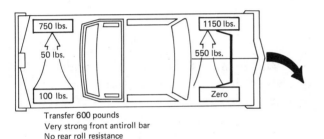

Transfer 600 pounds
Very strong front antiroll bar
No rear roll resistance

FIGURE 19-21.
If a car had no roll resistance other than the springs, weight
would transfer equally at each end of the car during turning
manuevers (center). If all the roll resistance was at one end
of the car, all of the weight transfer would occur at that end
(bottom).

suspension members that tend to keep the body
and frame from rolling on turns. Roll resistance
is generated by the antiroll bar, springs, shocks,
and, in most cases, the suspension bushings.
Roll forces, acting through roll resistance, will
increase the tire loading on the outside tires.

If a car has an equal amount of roll resistance at each end of the car, weight transfer increases the loading of the outside tires in an
amount equal to the load on the inside tires. If
the roll resistance is not equal, there is an unequal increase in tire loading. **Roll couple distribution** is the engineering term that refers to
relative amounts of weight that are transferred
to each outside tire. Imagine a car such as a

dragster with a solidly mounted rear axle and
a front axle with springs—there will be much
more roll resistance at the rear. If this vehicle
were to turn a corner, the outside rear tire would
accept all of the weight that was transferred;
the outside and inside front tires would have
about the same loading as when it was going
straight. (Fig. 19-21)

Next, we need to remember that tire traction changes with tire load; the greater the load,
the more traction. But also remember that the
traction increase is not proportionate to the
load; increasing the load 10 percent will produce a less than 10 percent increase in traction.
This principle allows us to easily change the understeer or oversteer characteristics of a car. We
can transfer less weight to the end of the car
with the greatest slip angle and more weight at
the other end. If we have a car that tends to
understeer, we install a stronger rear antiroll
bar. This will transfer more of the weight to the
outside rear tire and tend to equalize the front
and rear slip angles. (Fig. 19-22, Table 19-1)

19.11 WEDGE AND STAGGER

Wedge is a term used mostly with circle or oval
track racing; it refers to an unequal, static loading of the tires. Wedge works on the diagonal
of the car, the left front and the right rear or
the right front and the left rear. Wedge can be
put in a car by putting a stronger spring at one
or two corners (diagonal) of the car or by changing the length of one of the antiroll bar links.
A stronger spring or a longer antiroll bar link
at the right front will increase the load carried
by the right front tire. It will also increase the
load carried by the left rear tire and reduce the
load carried by the left front and right rear
tires. (Fig. 19-23)

The load carried by the two front springs
is determined by the mass above them; they
carry everything forward of the midpoint of the
car's wheelbase. The rear springs carry all the
load that is to the rear of the car's midpoint.
The two left springs carry all the weight on the
left half of the car, and the two right springs
carry all the weight to the right of the car's cen-

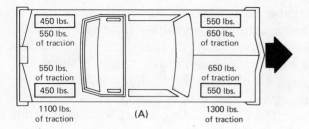

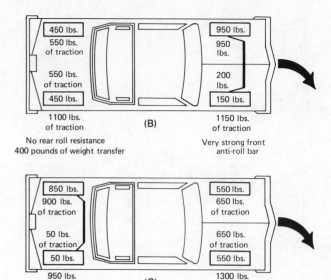

FIGURE 19–22.
The amount of traction at each end of the car during cornering manuevers can be changed by changing the strength of the antiroll bars; a stronger bar will cause a greater loss of traction. The car at B would probably understeer while the lower car (C) would probably oversteer.

ter line. If the two front springs are equal to each other, they transmit an equal load from the car to the tires, but if they are unequal, one tire is pushed harder into the ground. Pushing downward on one tire (or upward on the car at that corner) will tend to transfer one-half of the increase in load diagonally across the car to the other end.

Wedge is used to change the understeer or oversteer characteristics of a car. It should not be used on road cars because it can produce a car that will oversteer on right turns and understeer on left turns, or vice versa.

TABLE 19–1
Common Chassis Adjustments Used to Tune a Car to Neutral Steering*

Change	To Reduce Oversteer	To Reduce Understeer
Front tires	Decrease width	Increase width
	Lower pressure	Increase pressure
Rear tires	Increase width	Decrease width
	Increase pressure	Decrease pressure
Front anti-roll bar	Increase strength (thicker) (stronger)	Decrease strength (thinner) (weaker)
Rear anti-roll bar	Decrease strength (thinner) (weaker)	Increase strength (thicker) (stronger)
Front spring rate	Increase	Decrease
Rear spring rate	Decrease	Increase
Front wheel camber	Adjust more positive/ +	Adjust more negative/ −
Rear wheel camber	Adjust more negative/ −	Adjust more positive/ +
Front wheel toe	Adjust more inward	Adjust more outward
Front tire loading	Reduce	Increase
Rear tire loading	Increase	Reduce

*Note that a change to reduce oversteer will tend to increase understeer and vice versa.

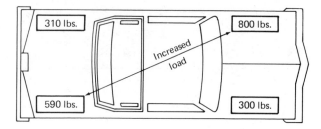

FIGURE 19-23.

A stronger spring at the left front or right rear has placed a "wedge" in the car; this car will corner much better to the left than the right.

Another feature that is only used on circle or oval track race cars is **tire stagger**; the outside, right rear tire will be selected that has a larger diameter and circumference than the inside, left rear tire. Under power, the effect of tire stagger is to cause the car to turn toward the inside of the turn, to the left. The correct amount of stagger will cause the car to turn itself to follow the race track. Tire stagger is used with cars having locked, no differential rear ends. Stagger can also be used on the front tires of a car that is racing on a circle or oval track with only left turns. (Fig. 19-24)

Many oval track cars are built to turn left naturally; they need to be turned right to go straight. Some of the more common changes are a shorter wheel base (as much as one inch) on one side of the car, a weight bias/CG shift to the left side, and a high amount of caster spread, more positive on the right.

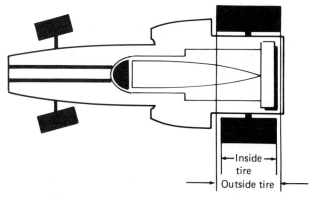

FIGURE 19-24.

A larger tire at the right rear has placed "stagger" into this car; it will now tend to turn toward the smaller tire, the left, by itself, especially under power.

19.12 CENTER OF GRAVITY LOCATION

Occasionally it is desirable to find the location of the CG of a particular car. This is a fairly easy but tedious operation that requires one or two pairs of fairly accurate scales and the use of a calculator. It requires two steps; the first determines the front-to-rear location of the CG and the second determines the height. To avoid the problem of calculating the CG height, some people estimate the CG height as being even with the camshaft of the engine; this is a fairly safe but not completely accurate assumption. Normally the CG is located on the car's center line; occasionally it will be slightly to the right or left if the engine or other heavy parts are mounted off-center. This can be determined by comparing the weight of the right and left sides; if they are the same, the CG is in the middle of the car.

To locate a car's CG, you should:

1. Either empty or fill the fuel tank, inflate the tires to an excessive pressure to reduce the tire bulge, and replace the shock absorbers with solid links to eliminate suspension travel.

2. Weigh the car by placing each of the tires on a scale. If only one or two scales are available, the tires can be weighed one or two at a time, but the other tires should be placed on blocks equal to the height of the scales. The car must remain level. Record the weight of each tire.

3. Add the weight of the two front tires together and the weight of the two rear tires together to determine the weight on each axle. Add the weight of the two axles together to determine the total weight of the car.

4. Measure the wheelbase of the car, the distance from the center of the front hub to the center of the rear axle.

5. Calculate the forward/backward location of the CG using these formulas (Fig. 19-25):

$$D = \frac{EW}{TW}$$

where D = Distance from rear axle to CG in percentage of wheelbase

       FW = Weight on front axle

       TW = Total weight of car

$$CGL = D \times WB$$

where CGL = CG location in front of rear axle

       WB = Wheelbase

6. Raise the rear tires, and place them on blocks 20 inches high (above the level of the front scales).

7. Remeasure the weight of the front tires.

8. Calculate the CG height using the following equation (Fig. 19-26):

$$CHG = \frac{{}^{\Delta}FW \times WB \times Cos\,\phi}{W \times Sin\,\phi}$$

where CGH = Center of gravity height in inches (above the center of the front tires)

    ${}^{\Delta}FW$ = Change in front tire weight

    WB = Wheelbase in inches

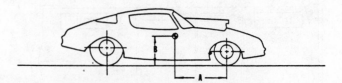

A = (% OF REAR WEIGHT) × WHEELBASE
     EXAMPLE: A = 45% × 108 INCHES
     A = 48.6 INCHES
B = HEIGHT OF CAMSHAFT CENTERLINE

FIGURE 19-26.

Step one to determine A is as illustrated in Figure 19-25; CG height is estimated as B. (Courtesy of Alston Engineering, Ph. 916-381-3291)

     $Sin\,\phi$ = Sine of the angle ϕ, the tilt angle in degrees, calculated from $20 \div WB$

     W = Total weight of car

     $Cos\,\phi$ = The cosine of angle ϕ

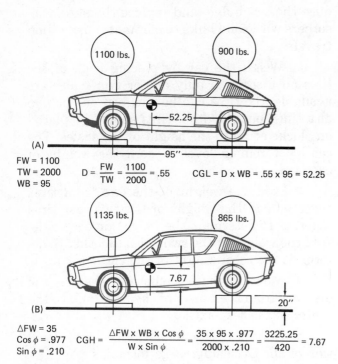

(A)

FW = 1100
TW = 2000 $D = \dfrac{FW}{TW} = \dfrac{1100}{2000} = .55$ $CGL = D \times WB = .55 \times 95 = 52.25$
WB = 95

(B)

$\Delta FW = 35$
$Cos\,\phi = .977$ $CGH = \dfrac{\Delta FW \times WB \times Cos\,\phi}{W \times Sin\,\phi} = \dfrac{35 \times 95 \times .977}{2000 \times .210} = \dfrac{3225.25}{420} = 7.67$
$Sin\,\phi = .210$

FIGURE 19-25.

The exact location of the center of gravity of a car can be plotted by following this two-step process. If the math is too complex, refer to Figure 19-26.

Special Notes on Trig functions

Many students and mechanics share a fear of anything mathematical and a special fear of algebra or trigonometry. Engineers commonly use skills in these areas to calculate forces to determine the effects that changes can be expected to produce.

Even though math formulas look quite intimidating, most of us have discovered that a math formula is a helpful guide while working our way through a math problem. Pocket calculators have made most of the calculations needed by a mechanic very easy, but some of the calculations needed for chassis modifications involve some trig functions. Most of these functions are concerned with sine and cosine, functions related to a right triangle. Knowing some of the rules relating to a right triangle helps us solve some unknowns and calculate some difficult-to-measure distances. Trigonometry is commonly used to solve the size of an angle or the length of a line.

Remember that a right triangle has one 90° angle plus two acute angles, and the sum of these three angles is 180°. The side opposite to the 90° angle is the hypotenuse, and the hypotenuse squared equals the sum of the squares of the two sides (Pythagorean Theorem).

The trig functions we need for Figure 19–25 are:

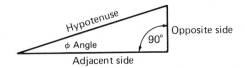

Sine ϕ = opposite side ÷ hypotenuse
Cosine ϕ = adjacent side ÷ hypotenuse

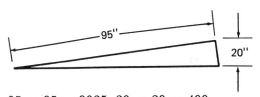

95 × 95 = 9025, 20 × 20 = 400,
9025 − 400 = 8625
the square root of 8625 is 92.86
Sine ϕ = 20 ÷ 95 = .210
Cosine ϕ = 92.87 ÷ 95 = .977

19.13   ANTIDIVE SUSPENSION GEOMETRY

Antidive is a suspension design characteristic that resists the forward pitch and front end dive during braking. Since most of the brake dive occurs at the front (there is some lifting of the rear), the front suspension geometry is the primary location of antidive.

Antidive is accomplished by mounting the control arms at an angle; the angle is such that an imaginary line drawn through the inner control arm pivots will meet or converge somewhere under the CG. This convergence point, called the **instant center,** becomes the end of a force arm. The downward motion of the front suspension during braking is converted into a lifting force at the instant center. (Fig. 19–27)

Enough antidive force can be created to give the vehicle a completely flat stop. In ac-

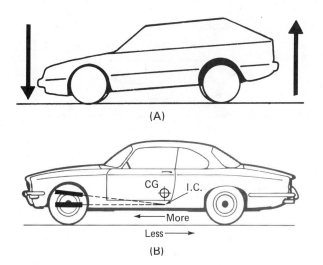

FIGURE 19–27.
Brake dive causes the front of the car to lower (A) during braking; antidive is achieved by placing the mounting points of the front control arms at an angle. (A is courtesy of Quickor Engineering)

tual practice, production cars have somewhere around 25 to 50 percent antidive. Too much antidive has drawbacks; the greatest is probably that the angle of the control arms to give antidive creates a situation where caster will increase during a bump, making steering more difficult. Also, the antidive forces tend to inhibit or remove normal suspension action and generate vibration problems. A certain amount of brake dive is desirable for feedback to the driver; we often judge how hard we are stopping by the amount of dive that is occurring.

Normally antidive is not changed or tuned after the car is built. The average person setting up a car for street operation or moderate competition does not change antidive.

19.14   ANTISQUAT

Antisquat, also called **power squat** or **percentage of rise,** is a suspension design characteristic that resists the rearward pitch, the squat and lowering of the back of the car during acceleration. The forces involved are similar to the reverse of antidive. Since most of the power squat occurs at the rear of the car (there is some lifting of the front), the rear axle suspension is

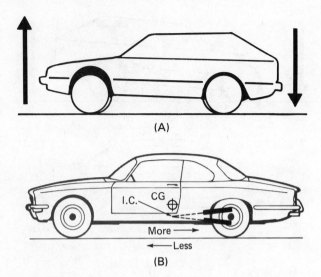

FIGURE 19-28.
Power squat will cause the back end of a car to lower (A) during acceleration; antisquat is achieved by mounting the rear axle control arms at an angle. (A is courtesy of Quickor Engineering)

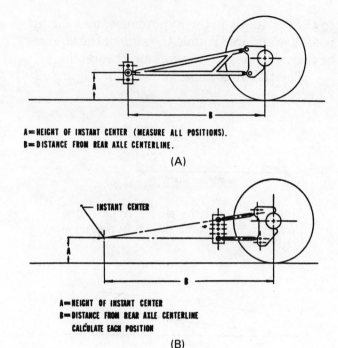

A = HEIGHT OF INSTANT CENTER (MEASURE ALL POSITIONS).
B = DISTANCE FROM REAR AXLE CENTERLINE.

(A)

A = HEIGHT OF INSTANT CENTER
B = DISTANCE FROM REAR AXLE CENTERLINE
CALCULATE EACH POSITION

(B)

FIGURE 19-29.
A three-link (A) and a four-link (B) rear suspensions as used in a drag race car. (Courtesy of Alston Engineering, Ph 916-381-3291)

the primary location of antisquat suspension geometry.

Antisquat is used on RWD cars with solid rear axles; cars with FWD or IRS do not lend themselves to antisquat. Antisquat is accomplished by mounting the locating members for the rear axle at an angle; the angles are such that an imaginary line drawn through the locating members will converge at the instant center, somewhere under the CG. This instant center becomes the end of a force arm. Torque reaction of the rear axle during acceleration becomes a lifting force at the instant center. This force can eliminate power squat. The drawback with antisquat geometry is that the control arm angles can create roll oversteer during cornering and too-short control arms can induce **power hop**, a rapid bouncing of the rear tires during very hard acceleration. (Fig. 19-28)

Drag race cars work with antisquat linkages in an attempt to gain rear tire traction. The lifting force at the forward end of the suspension linkage is also a downward force at the axle and tires. **Three** and **four-link/bar** suspensions have been developed to take advantage of this force. The links serve the same purpose as stock, coil spring, solid axle suspension links. They are usually constructed to offer a great

deal of adjustment. Antisquat force is changed by moving the instant center of the control links around a line drawn between the CG and the road contact point of the rear tire. (Figs. 19-29 & 19-30)

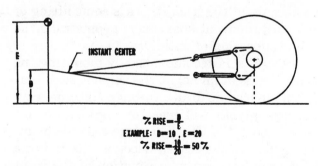

INSTANT CENTER

% RISE = $\frac{D}{E}$
EXAMPLE: D = 10 , E = 20
% RISE = $\frac{10}{20}$ = 50 %

FIGURE 19-30.
The location of the IC of a three- or four-link suspension relative to the CG is used to determine the expected percentage of rise/power squat during acceleration. (Courtesy of Alston Engineering, Ph. 916-381-3291)

19.15   POLAR MOMENT OF INERTIA

The distribution of a car's weight in reference to the yaw axis is referred to as the **polar moment of inertia**; the yaw axis is that point around which the body of a car would pivot if the car were to spin. If the heavy portions of the car are close to the yaw axis, like in a mid-engine car, the car would have a low polar moment of inertia. When the heavy portions of the car are farther away from the yaw axis, such as when the engine hangs over the front tire center line, the car has a high polar moment of inertia. (Fig. 19-31)

A higher polar moment of inertia will develop a high degree of inertial resistance to spin but also a high amount of inertia to maintain a spin if it should begin. The polar moment of inertia has a large effect on the stability of the car. Cars with a high polar moment of inertia tend to be very stable in a straight line; a lot of weight or mass has to move a fairly long distance around the yaw axis in order to turn. However, once a turn begins and this mass is moving, it tends to keep moving; once this car starts to spin, it tends to keep spinning. A car with a center-mounted engine and a low polar moment of inertia pivots on its yaw axis and turns much easier; the weight mass moves a much shorter distance so it has less inertial effect. This car is much more maneuverable and also less stable; it will deviate from a straight line much easier.

The polar moment of inertia is difficult to change on a production car, but it is a definite design factor for racing vehicles. Cars built for road circuits, where maneuverability is important, tend to be mid-engine (if the rules allow) with as many of the heavy parts of the car as possible close to the car's CG. A car built for very high speeds, where stability is important, would benefit from a higher polar moment of inertia; the weight should be moved toward the ends of the car.

19.16   AERODYNAMICS

Aerodynamics is important to car builders for several reasons. Improving the air flow past the car, called **streamlining**, reduces the horsepower required to push the car through the air and therefore can increase the speed. Improved air flow in the engine's intake system can increase horsepower, and in the radiator and braking system it can improve efficiency. Also, aerodynamics can generate a downward force at the tires and can increase the tires' circle of traction.

The aerodynamic downforce generated by inverted wings mounted on the car or ground effects between the road and the bottom of the car can more than double the load on the car's springs and tires. Much of the credit for some race cars being able to generate cornering and braking forces greater than one or two g's should go to this increased downforce. The traction power of the tire is increased without an increase in car weight, which would cause an increase in inertial or centrifugal load on the tires. (Fig. 19-32)

A drawback with aerodynamic downforce is that the devices generating downforce usu-

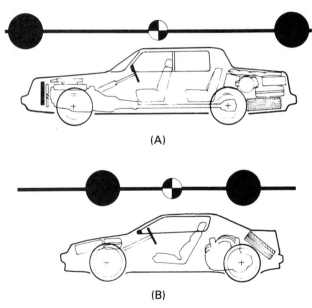

(A)

(B)

FIGURE 19-31.
How far the heavy parts of a car are from the CG determine a car's polar moment of inertia. A front engined car (A) has a high polar moment of inertia and should be quite stable. A mid-engine car (B) has a much lower polar moment of inertia and should be quite manueverable. (Courtesy of Quickor Engineering)

FIGURE 19-32.
Tire load and traction can be increased by using a wing to create an aerodynamic downforce.

ally increase the aerodynamic drag; cornering speed is increased at the expense of straightaway speed. Another problem is that downforce and traction are lost if the air flow to the car becomes disrupted by another vehicle or wind gust.

19.17   SUSPENSION TUNING/ MODIFICATIONS

Depending on the intended use of the vehicle, several modifications can be made to production vehicles to improve their handling capabilities. When planning to modify or tune a car's suspension, it should be understood that each aspect requires a compromise; there are pros and cons with each step. Many race cars handle fantastically, but they would make horrible commuting vehicles. Also, regardless of what is said by many "bench racers," most major automobile manufacturers realize what handling is and how to get good handling. In several cases, an optional production car "handling package" is very close to the ideal compromise between good handling and driveability. (Table 19-2)

Normally the first step in improving the handling capabilities of a road car is the purchase of a set of good tires designed for road performance. In some cases the improvement provided by the tires is enough to satisfy the desires of the car owner. It is often a real surprise how much improvement a good set of tires can make without the expense and some of the disadvantages of other modifications. Be sure to follow the recommendations given in the early chapters of this book when choosing replacement tires or wheels. Also, it is a good practice to start with a tire change, because the other changes will not really work well without good tires.

When tuning or modifying a vehicle that is to be driven on public roads, don't forget that the car must be kept legal as well as practical. Some states and countries have laws dealing with suspension modifications. Some states allow no modifications whatsoever, while others will allow some changes as long as the vehicle

TABLE 19-2
Comparison of Production and Racing Car Suspensions*

Production sedan	Grand touring sports car	Slalom car	Race car
Relatively soft ride comfort	**Bushing compliance**		Metal/no compliance
Fairly soft	**Roll stiffness**		Fairly high/adjustable
Fairly high for ground clearance	**CG height**		As low as track conditions allow
Fairly soft	**Spring and shock rates**		Relatively firm
Relatively flexible, #1 criteria is tire life	**Alignment angles**		Very precise, #1 criteria is traction
Not too critical	**Unsprung weight**		As low as possible

*The suspension of a production car is very similar to that of a race car, but the demands of the race track require a high degree of tire-to-road contact precision while ride comfort and tire life are high priorities with the production car.

remains at a certain height or complies with other vehicle code regulations. The Department of Motor Vehicles, the Vehicle Code, or the law enforcement department should be checked to ensure future legality.

Also don't forget that vehicle dynamics involve a multitude of interacting physical principles and variables. Some changes—to alignment, springs, shocks, antiroll bar, tires, etc.—will not always produce the desired result; in fact, some changes will seem to produce a step backward as one change interacts with another. Many people follow a rule that states: "Make only one change at a time so you can determine the effect of that change." Most race teams spend countless hours in track testing and computer time making tire tests and chassis tests to achieve maximum performance from a car.

Suspension tuning falls into several categories that can be combined or treated separately: reducing suspension bushing compliance, changing roll stiffness or roll couple distribution, lowering or reducing CG height, changing spring or shock absorber rate, changing the alignment angles, and reducing unsprung weight. The advantages and disadvantages of these changes are compared in Table 19-3.

TABLE 19-3
Changes Commonly Made to Improve Handling Characteristics and Probable Results of These Changes

Change	Positive effect	Negative effect
Increase tire width	Increased dry traction	Hydroplaning in wet conditions Requires wider wheels Fender well clearance Often uses lower air pressure
Increase wheel offset, positive	Increased track width Reduced weight transfer	Aerodynamic drag Increased wheel bearing load Reduced tire clearance
Lower CG	Reduced weight transfer Reduced body roll	Reduced ground clearance Reduced wheel travel
Increase spring rate	Reduced body roll	Increased bump shock
Increased shock rate	Reduced body roll (transient)	Harsher ride
Increased antiroll bar rate	Reduced body roll	Increased single bump harshness
Increased negative/− camber	Increased tire cornering power	Tire wear Steering/braking pull Reduced straight line traction
Increased positive/+ caster	Straight line stability Improved camber on turns	Harder steering Possible shimmy
Lower roll center	Less camber change on roll	Increased tire scrub on roll
Replace suspension bushings, more solid	Reduced alignment deflection Faster antiroll bar response	Increased vibration Harsher ride Replacement bushings usually require lubrication

19.17.1   Reducing Suspension Bushing Compliance/Deflection

The rubber bushings used in most suspension systems are designed to have a certain amount of compliance. As mentioned earlier, bushing compliance allows the control arms and antiroll bar to move or pivot through one or more planes of travel without binding. These bushings also dampen the vibrations and noises generated as the tire follows the road surface and keeps these annoyances from entering the passenger compartment. (Fig. 19-33)

The drawback with bushing compliance is that the softness of the bushing allows some control arm deflection under cornering and braking loads. This deflection, in turn, will change the position of the control arm and therefore the alignment angles. Camber, caster, and toe can change during cornering or braking maneuvers; the result can be a poor- or "mushy"-handling car.

Bushings of materials other than rubber can be used at the pivot points. The degree of bushing rigidity desired and the amount of noise, vibration, and deflection that can be tolerated are considered when a new bushing material is selected. The closest to stock replacement is a bushing of **high durometer** rubber; this bushing appears stock but uses a harder rubber material. It is relatively low cost, easy to install, relatively quiet, and reduces deflection somewhat from stock. (Fig. 19-34)

FIGURE 19-34.
Although they appear stock, these bushings use a high durometer/hard rubber that will allow less deflection during cornering loads. (Courtesy of Guldstrand Engineering)

Still less compliance is allowed by a slightly different bushing, one made from a plastic, polyurethane material. These bushings are low cost and allow very little deflection, but they usually require a control arm modification, can be noisy when fitted loosely enough to allow free suspension movement, and can cold-flow under prolonged load. Cold-flow can change bushing shape, alignment angles, or operating clearances. Delrin (a brand name) is readily available and easily machined; many racers use it to make their own bushings. (Fig. 19-35)

Steel suspension bushings are also available. They can be purchased with or without a nylon liner; the nylon liner provides a better bearing surface. These bushings offer zero deflection, but they require control arm modification, will transmit many of the road vibra-

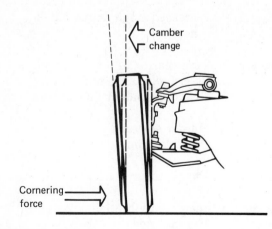

FIGURE 19-33.
Cornering loads can cause suspension bushing distortion, which will cause alignment changes.

FIGURE 19-35.
These control arm bushings are made from polyurethane, which is firmer than hard rubber. (Courtesy of Vette Products)

FIGURE 19–36.
These are metal control arm bushings. They will not allow any deflection, but they will transmit vibration and must be lubricated regularly. (Courtesy of Guldstrand Engineering)

tions and noises, and must be lubricated. (Fig. 19–36)

Many competition vehicles use self-aligning, spherical bearings for pivot bushings; these are often called **Heim joints**. This bushing is very effective where a twist of the control arm is required in addition to the normal pivoting motion. Threaded, spherical-bearing rod ends are often used to provide a means of adjusting the alignment angle as well as providing a pivot point. Like metal bushings, spherical bearings require lubrication and a large degree of control

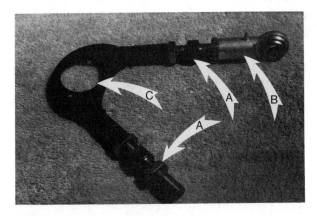

FIGURE 19–37.
This tubular upper control arm has two double-bolts (A), which are used to adjust caster and camber. The inner bushings are a pair of spherical rod ends (B) (the lower one is missing from this photograph). The ball joint, which will connect to the steering knuckle, will be bolted to the arm at "C".

arm modification and will transmit road vibrations and noise. (Fig. 19–37)

19.17.2 Changing Roll Stiffness / Roll Couple Distribution

Usually roll stiffness, the amount of resistance to body roll, is changed by installing a different antiroll bar or by changing the length of the bar ends. Bars of various diameters are available from some vehicle manufacturers and various aftermarket sources; these different bars will provide varying amounts of roll stiffness. Some bars have provision for mounting the end links at various points on the bar ends. The bar becomes softer and has less resistance the farther out that the end link is mounted on the bar end. (Fig. 19–38)

When changing bars, care should be taken to ensure that the roll couple distribution is not disrupted. Increasing the roll stiffness at the front or rear will change the relative resistance between the front and rear. Increasing the diameter of the rear bar tends to induce oversteer, while an increase in the front bar will induce understeer. It should be noted when changing antiroll bars that the rate of the bar is a function of the fourth power of the diameter; a small change in bar diameter can cause a large change in bar stiffness. Refer to Section 10.12, Figure 10–47 for the formula to compute antiroll bar rate.

The response or lag time of an antiroll bar can be improved by replacing the end links and bushings. The rubber material of these bushings allows some vertical suspension travel or compliance before the bar begins to twist. This lag is reduced by changing the bushings in which the bar rotates to bushings made from high durometer rubber, polyurethane, or metal. The link bushings can also be replaced with polyurethane bushings, or the links themselves can be replaced with ones using spherical rod ends. (Fig. 19–39)

19.17.3 Lowering the CG Height

Lowering the car to lower the CG height results in less weight transfer and less body roll; also,

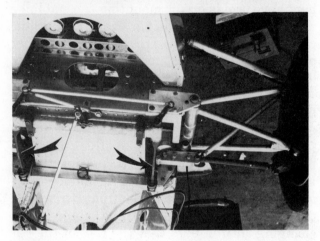

FIGURE 19-38.
The adjustable anti-roll bar at the rear of the upper race car is adjusted firmer by sliding the link connection toward the pivot bushing. The driver-adjustable anti-roll bar at the front of the bottom race car is in the stiffest position; it can be made softer by rotating the lever arms (arrows) to a more horizontal position.

FIGURE 19-39.
Either of these two different antiroll bar end links will allow less bushing deflection during cornering. (Courtesy of Vette Products)

the reduced car height often improves gas mileage because of reduced road drag. A rule with some race car designers and suspension tuners is that "the car is too high if the body does not touch the ground occasionally." Formula 1 and Indianapolis cars are now using "rub plates" so the body of the car can occasionally drag on the road surface.

Lowering a car's height is normally done by installing shorter springs, cutting or reshaping the springs, or installing spacer or lowering blocks between the axle and spring. If shorter springs are used, don't forget that increased tire scrub might occur on cars with S-L A suspension. Dropped spindles and axles are available for some cars to provide a lowered car with normal suspension action. A dropped spindle has the spindle positioned higher up on the steering knuckle. Lower springs usually must be of a higher rate. Lower cars have less room for suspension travel; a shorter spring having the same rate would let the suspension strike the bump stops.

Cutting one-half or one coil from a coil spring shortens the spring and also increases the spring rate; removing one coil from a spring having ten coils will increase the spring rate by 1/10, or 10 percent. This, of course, will produce a harsher ride at a higher frequency. If a coil spring is shortened, the cut end of the coil should be reshaped to conform to the original shape and the shape of the spring pocket in the control arm.

An axle suspended on leaf springs can often be lowered by the addition of lowering blocks or spacers with longer "U" bolts, by de-arcing the spring, or by reversing the arc of the main spring leaf so the spring eyes are downward instead of upward.

Changing the arc of a leaf spring or cutting a coil should be done by someone with an understanding of spring metallurgy. Most springs are made from tempered steel, and heating the spring can easily alter it's temper. It is possible to perform these operations with very little damage to the spring; it is also possible to weaken a spring in such a way that a potentially dangerous situation is created.

19.17.4 Changing the Spring/Shock Absorber Rate

Many softly sprung cars wallow, lean, and bounce excessively on turns and braking maneuvers. On these cars, an increase in spring or shock absorber rate often makes a definite improvement on ride and handling qualities. Replacement springs with various spring rates and free lengths are often available from car manufacturers and aftermarket suppliers.

When selecting a replacement spring, divide the load on the spring by the spring rate to determine how far the spring will compress;

subtract this distance from the free length to determine the installed length of the spring (how long it is when installed). This calculated, installed length can be compared with the existing spring to determine whether a change in car height will occur as the spring is replaced. (Fig. 19–40)

If we know the wheel rate of the spring, we can determine the "undampened suspension frequency," and this will give us a good indication of the ride quality that will result from suspension or spring changes. The wheel rate for a particular suspension can be calculated using this formula (Fig. 19–41):

$$WSR = SR \times \frac{(SAL)^2}{(CAL)} \times \frac{(TIC)^2}{(TBJ)}$$

where WSR = Wheel spring rate in pounds per inch
SR = Spring rate in pounds per inch
SAL = Spring arm length, distance from the center of the spring to the inner control arm pivot in inches
CAL = Control arm length, from inner pivot to ball joint in inches
TIC = Distance from tire center line to instant center in inches
TBJ = Distance from tire center line to ball joint in inches

	A	B	C	D	E	F
Inside diameter	5.57	5.57	5.57	5.57	5.57	5.57
Wire diameter	.610	.580	.580	.580	.590	.590
Free height	16.13	14.09	17.53	19.37	19.29	17.84
Spring rate	157.2	166.9	124.3	110.9	119.5	131.8
Load	1200	1200	1200	1200	1200	1200
Amount of compression during installation	7.63	7.19	9.65	10.82	10.04	9.1
Installed height	8.5	6.9	7.88	8.55	9.25	8.74

Amount of compression = $\dfrac{\text{load}}{\text{spring rate}}$

Installed height = free length − amount of compression

Note: all lengths are in inches and spring rates and load are in pounds.

FIGURE 19–40

Because of a similar ID and wire diameter, all of these springs will probably fit the same car. Spring B would give the firmest ride and also the lowest car. Spring D would give the softest ride. Spring E would give the highest car.

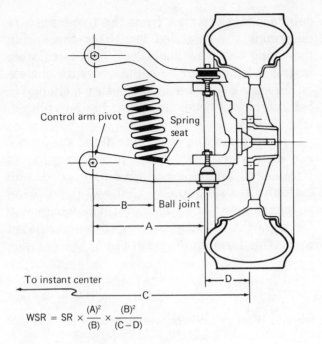

$$WSR = SR \times \frac{(A)^2}{(B)} \times \frac{(B)^2}{(C-D)}$$

WSR = Wheel spring rate
SR = Spring rate in pounds per inch
A, B, C, & D are dimensions in inches

FIGURE 19–41.
The wheel rate of a spring can be determined using this formula.

It should be noted that the wheel rate of the spring is affected by the lever ratio of the spring on the control arm and the amount of wheel offset.

Wheel rate can also be measured by lowering the tire of the car onto a scale, measuring the height of the body when the tire contacts the scale, and noting the weight increase relative to the height change as the car is lowered. Also, weight can be added to the body of the car, and the change in height can be compared to the change in weight. Don't forget that weight that is added might be divided by more than one tire, depending on where the weight is placed. It should be placed directly over the tire being weighed to have a 100 percent effect on that tire. If a car dropped three inches as the scale showed a weight increase of 1,500 pounds, the wheel rate would be 500 pounds per inch.

Undampened suspension frequencies between 1 and 2 cycles per second (CPS) are comfortable to the human body; higher frequencies are not only uncomfortable, with a vibrating nature, they can be harmful. Most passenger car suspensions are designed with a suspension frequency between 1.2 and 1.9 CPS with the rear suspension usually lower than the front (flat-ride tuning). High performance production cars work best with a suspension frequency about 1.8 to 1.95 at the front and 1.5 and 1.65 at the rear. The suspension frequency of racing vehicles is about 2 to 2.1 at the front and 1.4 to 1.6 at the rear. A formula that can be used to compute the suspension frequency is:

$$SF = .159 \times \frac{WR}{(SW \div 384)}$$

where SF = Suspension frequency in CPS
         WR = Wheel rate of spring in pounds per inch
         SW = Sprung weight on spring in pounds

Selection of the right shock absorber for a particular use on a certain vehicle is more difficult because performance diagrams are not generally available for most production shock absorbers. Also, most of us have no way of telling how much compression and extension dampening is necessary or desirable. Some manufacturers of high-performance or racing shock absorbers do make their performance curves available and will provide help in making a selection. Most racing shock absorbers are easily adjustable. It is a common tendency to get more dampening resistance than needed when selecting shock absorbers. This results in an unnecessarily harsh ride and the possibility of tire patter on washboard surfaces. Adjustable shock absorbers are very convenient during suspension tuning because they allow the comparison of different settings.

Several aftermarket suppliers have assembled handling/suspension kits for various popular makes and models of production cars. These kits usually include springs, shock absorbers, and antiroll bars; in addition, some of them include suspension bushings and links. These kits are usually designed to provide a lower car with the correct spring and shock absorber rates for improved handling. They are often available in various levels or stages from

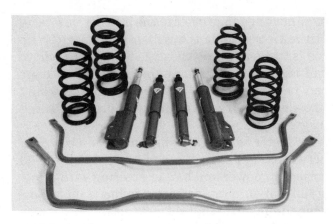

FIGURE 19–42.

A suspension kit for a car with a modified strut front and a coil spring rear suspension. The rates of the springs, shocks, and anti-roll bars and the length of the springs has been designed to produce a lower car having a sportier, more responsive ride quality. (Courtesy of Koni)

high-performance street use to autocross or gymkhana competition. (Fig. 19–42)

19.17.5 Changing Alignment Angles

Alignment angles for street-driven cars are generally optimized to give maximum tire life, which also provides maximum fuel mileage and fairly good handling and stopping qualities. Race cars use alignment angles that are optimized to give maximum cornering power or straight-line speed, depending on the type of racing. A NASCAR sedan, Formula 1 car, and top-fuel dragster will use different alignment angles. A car on an oval track is aligned to turn left; a car on a road course needs cornering power in both directions with maximum traction in turns; a drag race car needs maximum forward traction power of the rear tires and minimum rolling drag of the front tires. The alignment angles that can be changed to improve handling are camber, caster, and toe at the front and camber and toe at the rear.

To obtain maximum tire life and reduce sideways pull, a tire should have zero camber as it rolls down the road. Remember that a static setting of 1/4 to 1/2° positive usually produces this. Cornering power will increase if the tire has negative camber; tire roll during cor-

nering flattens the tire tread out to produce total tire-to-road contact. This applies to both front and rear tires. A good starting point for alignment settings to improve cornering on a street-driven car is 0 to 1/2° negative, and for competition cars it is 1 to 2° negative.

Some race teams use a tire pyrometer to help optimize camber settings. A tire pyrometer is a unit, much like a thermometer, which is used to measure tire temperatures at three locations across the tread. A tire that is aligned correctly with the correct inflation pressure will generate an even temperature across the tread. If the tire is running the hottest at the outside, coolest at the inside, and somewhere in between at the center, the camber setting is too positive; it should be adjusted to a more negative setting. When readjusting a suspension, don't forget that 1 or 2° of negative camber will definitely cause tire wear, reduce traction, and cause stability or pull problems while traveling in a straight line. (Fig. 19–43)

To obtain maximum tire life, the tire should have zero toe as it travels down the road; remember that a static setting of 1/16 to 1/8 inch is usually used for RWD cars. Any toe in or out can produce tire scrub, which increases tire drag and wear. Most street and competition vehicles should optimize their front and rear toe settings to be zero toe when running

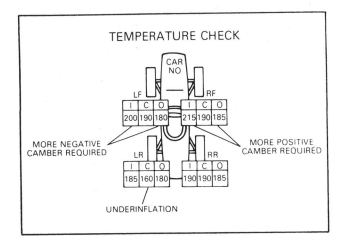

FIGURE 19–43.

The tire temperatures on this car show that the front tire camber angles and the right rear tire pressure should be changed. (Courtesy of Chevrolet)

straight. Where handling and increased tire response is important, a slight toe in setting is used to preload the tire's tread and sidewall; this reduces the lag time needed to set the tire when entering a corner.

Toe change becomes a concern as a suspension is being modified; this term refers to a change in the toe setting as the suspension rises and falls. It is also called **bump steer**. To check toe change, toe is measured at various heights from full compression to full extension, usually in 1-inch increments. This can be done using some conventional toe gauges; or a toe change gauge can be fabricated rather easily. This gauge is shown in Figure 19-44. A version of this gauge is positioned so both of its pointers touch the tire at ride height. The car is then lifted by the frame, 1 inch at a time, and the distance between the pointers and the tires is checked to determine whether toe is changing. After the car is checked at its maximum height (jounce), it is checked at various compression (bounce) heights by adding weight to lower the car. Ideally, the springs are removed, and the height of the vehicle is controlled by jacks. The ideal toe change is zero—no change. If a

significant change occurs, the inner or outer end of the tie rod is at the wrong height or the tie rod is too long or too short; the toe change curve can be used to determine the cause. Toe change can be corrected by someone with a working knowledge of race car fabrication and metallurgy. (Fig. 19-45)

Caster is also a compromise; in this case, between high speed stability with improved cornering power and harder steering with a possibility of shimmy as caster is made more positive. Remember that positive caster will cause the camber on the inside tire to become more positive and the outside tire more negative as the steering is turned into a corner. Many road cars will use positive caster to produce this more ideal camber change during cornering.

As alignment angles become more important, frame rigidity also becomes important. Any twisting or flexing of the frame can allow the suspension mounting points to change position and therefore change the alignment angles. A well-designed competition vehicle will include a well-designed and constructed roll cage. This cage serves two purposes: the first is to protect the driver from injury in case something goes wrong; the second is to reinforce the suspension mounting points in the frame to provide a rigid platform for the suspension and steering systems. (Fig. 19-46)

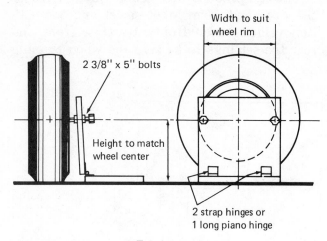

Toe change gauge

FIGURE 19-44.
A toe change gauge can be made from a few simple parts as shown here. Weight is placed on the base to keep the gauge from moving while the suspension is raised and lowered to various heights (ride height plus up and down travel). The amount of toe change is indicated by the size of the gap between the wheel and one of the pointers. One of the pointers can be replaced with a dial indicator to produce quicker measurements.

19.17.6 Reducing Unsprung Weight

For most street-driven cars, unsprung weight reduction is usually accomplished by using wheels made from a lightweight alloy, aluminum or magnesium. Lighter racing tires reduce unsprung weight, but they are not approved by DOT and can and should not be used on public roads. Usually, when alloy wheels are purchased, they are wider to allow for a larger, wider tire. It is doubtful that an improvement in unsprung weight actually occurs; the added weight of the wider tire usually offsets much of the weight reduction from the wheels. However, there should be a handling benefit from the wider replacement tires. Competition vehicles also use lightweight braking and suspen-

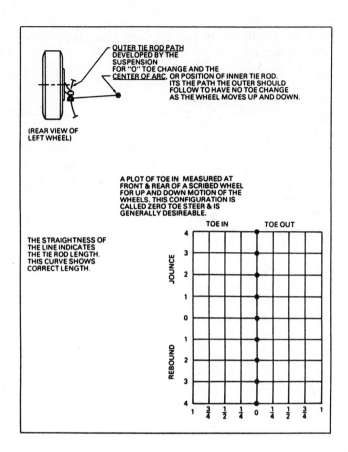

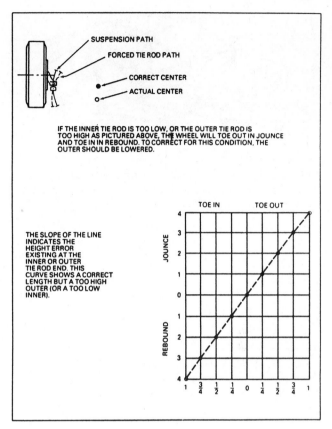

FIGURE 19-45.

If the tie rod correctly matches the geometry of the lower control arm, there will be no toe change. If excessive toe change occurs, the length or height of the tie rod should be changed. (Courtesy of Chevrolet)

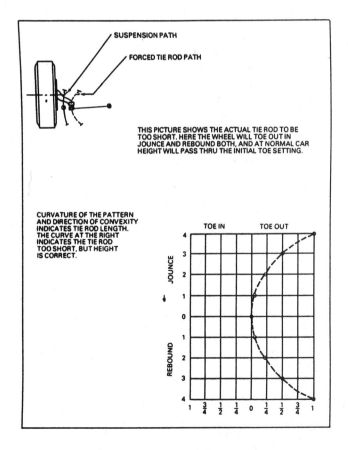

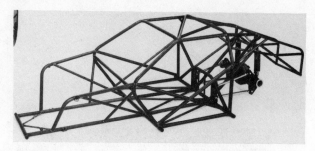

FIGURE 19-46.
A competition car should have a strong and rigid roll cage to protect the driver and to provide for rigid suspension mounting points. (Courtesy of Alston Engineering, Ph. 916-383-3291)

sion components to reduce unsprung weight to a minimum.

19.18 CONCLUSION

This chapter was written to introduce you to the aspects, possibilities, and maybe some of the excitement of suspension tuning. It is impossible to cover all of the aspects needed to develop a suspension system for all of the various types of cars. At this time there are more than a few excellent books available that cover suspension tuning and race car construction. These books describe suspension development and modification or tuning in greater detail.

Chassis/suspension tuning or modifying is not normally done in alignment or front end shops because of the time consumed and the skill required. This type of work is usually left to highly specialized shops, race car constructors, or highly knowledgeable hobbyists.

GLOSSARY

A arm: See Control Arm

Ackerman Angle: The angle of the two steering arms which produce toe out on turns.

Aftermarket: A replacement part that is produced and sold by a company other than the original manufacturer.

Air Suspension: A system in which air-filled, elastic springs are used in place of metallic springs.

Alignment: The process of adjusting the position of the tires and steering axis to bring them to a specified, predetermined position.

ASE, Automotive Service Excellence: A group, the National Institute for Automotive Service Excellence, which promotes excellence in the automotive repair field through the voluntary testing and certification of competent technicians.

Aspect Ratio: The relationship between the width and height of a tire.

Axial: A direction that is parallel to the rotating axis.

Axis: A line or point marking the center of rotation of an object or thing.

Axle: A cross support for a vehicle which is designed to carry the weight of the car.

Balance, Chassis: A ride condition that gives a level, flat front-to-rear flowing sensation without pitch.

Balance, Tire: A condition in which the tire can spin without causing a vibration of the suspension or car.

Ball Joint Angle, BJA: See Steering Axis Inclination.

Bearing: A device that allows rotation or linear motion with a minimum of friction. It usually uses a series of balls or rollers so there is a rotating motion of the internal parts.

Bellows: A flexible, accordian-like seal used where angular or lateral motions require a large degree of movement.

Boot: See Bellows.

Boss: An area, usually enlarged, where a bolt or fastener is to be installed.

Bounce: Straight-line motions of the sprung mass of a car in a vertical direction.

Bottoming: A noise and jolt created when the compression cycle of the suspension ends at the bump stops.

Bubble Balancer: A device used to statically balance tires which uses a bubble level as an indicator.

Bump Steer: A steering action caused by bounce motion of the suspension.

Bump Stop: An elastic member which increases the spring rate near the end of the compression and extension travel to reduce the effects of bottoming and/or topping.

Bushing: A device that allows rotation or linear motion. It usually uses a sliding motion of the internal parts.

Camber: A vertical angle of the tire seen when viewing the tire from the front or rear; used primarily to reduce tire wear.

Caster: An angle of the steering axis in which the top of the steering axis leans forward or rearward; used primarily for directional control.

Centrifugal Force: A force acting on a turning body which pushes the body outward.

CG, Center of Gravity: The balance point of a car.

Chassis: The portions of a car which remain after the body has been removed. It includes suspension and steering systems.

Coefficient of Friction: The amount of friction there is between two items that is dependent on the material composition of the two materials.

Compliance: The ability to yield elastically to change position.

Computer Balancer: A tire balancer which uses a computer to indicate the location and amount of weight needed to kinetically and dynamically balance a tire.

Computer Controlled Suspension: A suspension system which uses a computer to change the shock absorber settings and/or air spring pressure to suit various driving conditions.

Control Arm: A suspension member used to determine the position of a steering knuckle or axle, usually in a lateral direction.

Cornering Force: The tractional force in a lateral direction that is generated by a tire.

Curb Height: The distance from the ground to a specified point of the car to determine if the car is at the correct height. It is usually measured with the car at the correct curb weight.

Curb Weight: The weight of an empty vehicle that has a full supply of fuel, oil, and water.

CV Joint, Constant Velocity Joint: A universal joint that will deliver power at an angle with no change in velocity between the input and output.

Damped, Dampened: A force or action opposing a vibrating motion to reduce the amount of vibration.

Deflection: A movement which changes a shape or position, reacting to an outside force.

Directional Stability: Ability of a car to travel in a straight line with a minimum of correction from the driver.

Dive: A pitching motion of the sprung mass of a car downward at the front which usually occurs during braking.

Drift: See Wander.

Dynamic Balance: A balancing of the lateral, centrifugal forces of a spinning tire and wheel.

Feather Edge: An abnormal tire wear condition in which each tire rib wears in a tapered, angled fashion.

Float: A slow, low-frequency movement of the car which produces a sensation of continuous front-to-rear, vertical movement of the suspension.

Flutter: A forced oscillation of the steerable wheels about their axes occurring at speeds of 50 to 88 MPH.

Force: Physical power which will cause a movement.

Frequency: The speed at which an action occurs.

Friction Loaded Joint: A flexible joint in which internal springs provide a friction to remove any free play from the joint.

Front End Geometry: The angular relationship of the front end and steering parts in their various positions of operation.

G Force, g: A measurement of the amount of acceleration, braking, or cornering force that a car can generate.

Handling: The relative ability of a vehicle to maneuver through turns and go where the driver wants it to go.

Harmonic Vibration: An increase in vibration amplitude that occurs when the speed of the vibrating force matches the natural frequency of a vibrating object such as a spring or suspension system.

Harshness: Vibrations that can be felt and/or heard that are caused by interaction between the tire and the road surface. It can be caused by tire or road irregularities.

Hop: The vertical oscillations of a tire which can be caused by radial runout and/or static or kinetic unbalance.

Included Angle: The angle between the camber angle and SAI.

Jounce: A bounce motion during which the tire travels upward, relative to the car, compressing the spring and shock absorber.

Kick: An undampened reaction of the suspension that causes a jolting recoil sensation.

Kinetic Balance: A balancing of the radial, centrifugal forces of a spinning tire and wheel.

King Bolt, King Pin: A sturdy steel shaft used to connect the steering knuckle to an axle. It provides the pivot axis.

King Pin Inclination, KPI: See Steering Axis Inclination.

King Pin Offset: See Scrub Radius.

Lateral: A direction that is to the side.

Lateral Runout: A side-to-side motion or wobble of a rotating tire or wheel.

Lead: See Pull.

Linkage: A system of levers and rods used to transmit motion or force.

Load Range: A system of measuring and labeling the carrying capacity of the tires.

Loose: See oversteer.

Lurch: Straight-line motions of the sprung mass of a car in a horizontal direction.

MacPherson Strut: See Strut.

Neutral Steer: A cornering situation where the slip angles of the front and rear tires are equal. No steering correction is required.

NIASE, National Institute of Automotive Service Excellence: See ASE.

OEM, Original Equipment Manufacturer: The company that made the parts that were originally used on a car.

Oscillation: A back-and-forth, repeating motion.

Oversteer: A tendency of the car to turn more sharply than it should for the amount of steering input. A steering correction toward straight ahead or the opposite direction is required.

Panhard Rod: A device which connects the axle to the frame that controls sideways motions of the frame/body.

Play: Free movement of an item allowed by internal clearances.

Plow: See Understeer.

Pitch: Rotary motions of the sprung mass of a car around the transverse axis. The front end will rise while the rear lowers and vice versa.

Preload: A slight thrust load adjusted into a bearing set to eliminate all axial motion.

Pressure: A unit of force applied on a given area.

Pull: A tendency for a car to steer toward one side.

Radial: An outward direction that is at a right angle to the center.

Radial Runout: A variation of the tire's radius which causes a vertical oscillation or bounce of the tire.

Radius Arm: A type of control arm which attaches to an axle at one end and pivots at the other end. It is often mounted in a lengthwise direction.

R.M.A., Rubber Manufacturers Association: A group which sets standards in the tire industry.

Rebound: A bounce motion during which the tire travels downward relative to the car and the spring and shock absorber extend.

Revolutions per Mile: The number of revolutions a tire must make to travel one mile.

Ride Height: See Curb Height.

Road Crown: The raised portion in the center of a road to promote water runoff.

Road Shock: A harsh force transmitted from the tires through the suspension or steering linkage.

Roll: A rotary motion of the sprung mass of a car around the longitudinal (lengthwise) axis that results in body roll.

Roll Rate: The amount of resistance generated by the suspension components that resist roll.

Roll Steer: A steering motion of the front or rear tires that occurs as a result of the car's body rotating on its roll axis.

Roughness: A heard or felt vibration generated by a rolling tire on a smooth road surface that produces the sensation of rolling on a coarse or irregular road surface.

S.A.E., Society of Automotive Engineers: A group which establishes standards in the automotive and petroleum industries.

Scrub Radius: The distance between the center of the tire and the steering axis when measured at the road surface.

Shake: A resonance set up in the suspension which produces a nervous ride quality.

Shim: A spacer used to adjust the distance or angle of an item.

Shimmy: A series of rotary (turning) oscillations of the front wheels around the steering axis.

Shock Absorber: A device, usually hydraulic, used to dampen or reduce the amount of spring oscillations after a bump.

S-L A, Short-Long Arm Suspension: A suspension system which uses a relatively short upper control arm and a longer lower control arm.

Skid: A sliding rather than rolling action of the tire across the road.

Slip Angle: The angular difference between the wheel's centerline and the actual direction of travel.

Spring: A flexible suspension member which allows bounce travel of the suspension.

Spring Rate: The change of load on a spring per unit of deflection.

Sprung Weight: The total weight of the portions of the car which are carried by the springs.

Squat: A pitching motion of the sprung mass downward at the rear which often occurs during acceleration.

Stabilizer Bar: A suspension member used to reduce body lean during cornering.

Standing Wave: A deformation of the tire sidewall and tread which results when the tire velocity exceeds a critical point.

Stabilizer Bar: A "U"-shaped bar of steel used to reduce body roll.

Static: Stationary, not moving.

Static Balance: A balancing of a tire and wheel so it will remain stationary and not rotate or lean because of unequal weights.

Steady State: The conditions of a car's body and suspension that exist while the body and suspension are not changing roll, pitch, or yaw attitude.

Steering Axis: The line around which the front tires turn when a car turns a corner.

Steering Axis Inclination, SAI: An angle of the steering axis in which the top of the steering axis leans inward.

Steering Geometry: The angular relationship of the various steering and suspension components during their different positions of travel.

Steering Knuckle: The front suspension component which attaches the front tire and wheel to the steering axis and steering linkage.

Strut: A suspension system type which utilizes the shock absorber as the upper tire position locating member.

Strut Rod: A suspension member which is used to brace the control arm to keep it from moving forward or backward.

Sway Bar: See Stabilizer Bar.

Thump: A periodic vibration and/or sound generated by a tire producing a pounding sensation which occurs in time with the tire rotation.

Tire Contact Area, Tire Print: The amount of tire tread that is in contact with the road surface. Also called the "footprint."

Toe: An angle of a tire, relative to straightahead, if viewed from above.

Toe Angle: The actual amount that the tire differs from pointing straight ahead.

Toe In: A condition where both tires of an axle are positioned so they are closer together at the front than the rear.

Toe Out: A condition where both tires of an axle are positioned so they are closer together at the rear than the front.

Toe Out On Turns: A condition where the front tires toe outward during a turning action of the car.

Topping: A noise and jolt when the extension cycle of the suspension travel ends at the travel stop.

Torque: A twisting or turning force.

Torsilastic Bushing: A bushing which allows motion through the elastic nature of rubber.

Torsion: A rotating motion which causes a twisting action.

Torsion Bar: A spring which allows suspension motion by twisting.

Track: The center-to-center distance between the two tires on an axle.

Tracking: The degree in which the rear tires follow behind the front tires.

Traction: The ability of a tire to grip the road surface.

Tramp: A rotary motion of a solid axle upward at one end and downward at the other end which reverses in a series of oscillations.

Transient State: The conditions which result in a car while the body and suspension are changing position.

Transverse: A direction that goes across the car.

Tread Width: The outside-edge-to-outside-edge width of the two tires on an axle.

Trim Height: See Curb Height.

Undamped, Undampened: A system where there are no forces or actions which oppose vibrating motions.

Understeer: A tendency of the car that requires a greater steering motion to make a given turn; the car turns less than it should.

Unequal-Length Control Arms: See Short-Long Arm.

Unsprung Weight: The total weight of the portions of the car which support the springs.

Vibration: A periodic motion or oscillation of an item which often causes an annoying motion or sound.

Wander, Weave: The tendency of a car not to follow a straight line; it requires continuous correction from the driver.

Watts Link: A suspension member consisting of two rods and a pivoting bell crank that is used to keep the body from moving sideways relative to the axle.

Weight Transfer: The amount of weight which moves laterally because of cornering forces or lengthwise because of acceleration or braking forces.

Wheelbase: The center-to-center distance between the front and rear tires.

Wheel Hop: A rapid vertical oscillation of the tires resulting from a loss of traction control.

Wheel Offset: The lateral distance between the centerline of a wheel and the inner side of the wheel mounting flange.

Yaw: The rotary motion of the sprung mass of a car around a vertical axis that is encountered in a spin.

Zerk Fitting: A brand of a grease fitting.

Appendix 1
ENGLISH-METRIC CONVERSION

The following conversion factors can help you convert a dimension from one measuring system to another. Simply multiply the dimension you have by the factor to get the dimension you want.

Unit	Multiply	by	to get
LENGTH	inch	25.4	millimeter (mm)
	foot	0.305	meter (m)
	yard	0.914	meter
	mile	1.609	kilometer (km)
	millimeter	0.04	inch
	centimeter	0.4	inch
	meter	3.28	feet
	kilometer	0.62	mile
AREA	inch^2	645.2	millimeter^2 (mm^2)
	foot^2	0.093	meter^2 (m^2)
	millimeter^2	0.0016	inch^2
	centimeter^2	0.16	inch^2
	meter^2	10.76	foot^2
VOLUME	inch^3	16,387	millimeter^3 (mm^3)
	quart	0.164	liter (l)
	gallon	3.785	liter
	millimeter^3	0.000061	inch^3
	liter	1.06	quart
	liter	0.26	gallon
WEIGHT	ounce	28.4	gram (g)
	pound	0.45	kilogram (kg)
	ton	907.18	kilogram
	gram	0.035	ounce
	kilogram	2.2	pound
FORCE	kilogram	9.807	newton (N)
	ounce	0.278	newton
	pound	4.448	newton
PRESSURE	in. of water (H_2O)	0.2488	kilopascals (kPa)
	pounds/inch^2	6.895	kilopascals
	kilopascals	0.145	pounds/in.^{2}
	kilopascals	0.296	in. of mercury (Hg)

(continued)

Unit	Multiply	by	to get
POWER	horsepower	0.746	kilowatt (kw)
	kilowatts	1.34	horsepower
TORQUE	inch-pound	0.113	newton-meter (N-m)
	foot-pound	1.356	newton-meter
	newton-meter	8.857	inch-pound
	newton-meter	0.737	foot-pound
SPEED	miles per hour	1.609	kilometer per hour (km/h)
	km per hour	0.621	miles/hour
ACCELERATION/ DECELERATION	feet/sec.^{2}	0.345	meter/sec.^{2}
	inch/sec.^{2}	0.025	meter/sec.^{2}
	meter/sec.^{2}	3.28	feet/sec.^{2}
FUEL ECONOMY	miles/gallon	0.425	km/liter (km/l)
	km/liter	2.35	miles per gal.
TEMPERATURE	Fahrenheit, degree	0.556 ($^\circ$F $-$ 32)	Celsius, deg. ($^\circ$C)
	Celsius, degree	1.8 ($^\circ$C $+$ 32)	Fahrenheit, deg. ($^\circ$F)

Appendix 2
DISTANCE AND ANGULAR EQUIVALENTS

The following table provides you with the length and angular equivalencies you might need to change dimensions from one measuring system to another.

Inch (Fractional)	Inch (Decimal)	Metric (mm)	Degrees (Decimal)	Degrees (Fractional)	Degrees (Minutes)
1/32	.0312	.793	.0625	1/16	3.75
1/16	.0625	1.587	.125	1/8	7.5
3/32	.0937	2.381	.1875	3/16	11.25
1/8	.125	3.175	.25	1/4	15
5/32	.1562	3.968	.3125	5/16	18.75
3/16	.1875	4.762	.375	3/8	22.5
7/32	.2187	5.556	.4375	7/16	26.25
1/4	.250	6.35	.5	1/8	30
9/32	.2812	7.143	.56625	9/16	33.75
5/16	.3125	7.937	.625	5/8	37.5
11/32	.343	8.7317	.6875	11/16	41.25
3/8	.375	9.525	.75	3/4	45
13/32	.4062	10.318	.8125	13/16	48.75
7/16	.4375	11.112	.875	7/8	52.5
15/32	.4687	11.906	.9375	15/16	56.25
1/2	.500	12.7	1.0	1	60

BOLT TORQUE TIGHTENING CHART

Torque tightening values for a bolt will generally vary depending on the diameter of the bolt, the grade of the bolt material, the pitch of the bolt thread, whether the threads are lubricated and the type of lubricant used, and the material into which the bolt is threaded. Tightening a bolt too tightly might stretch the bolt to the yield point where the bolt might break or it might cause stripping of the threads of the bolt or nut. Tightening a bolt to too low a torque value might allow the bolt to come loose prematurely.

If the tightening torque for a particular bolt cannot be located, the values in the following table can be used as a guide.

Grade:

| S.A.E. | 1 & 2 | | 5 | | 8 | | |
| Metric | | 5 | | 8 | | 10 | 12 |

Size/Diameter U.S.	Metric	S.A.E. 1 & 2	Metric 5	S.A.E. 5	Metric 8	S.A.E. 8	Metric 10	Metric 12
	6		5		9		11	13
1/4		5		7		10		
5/16		9		14		22		
	8		12		21		26	32
3/8		15		25		37		
	10		23		40		50	60
7/16		24		40		60		
	12		40		70		87	105
1/2		37		60		90		
	14		65		110		135	160

Note: All torque values are given in foot-pounds and for clean, lubricated bolts. The values given are for steel-to-steel threads using motor oil for a lubricant.

To convert these values to inch-pounds, multiply them by 12. To convert them to newton-meters, multiply them by 1.356.

INDEX

A

A-arm, 185, 209
Ackerman linkage, 469–70
Aerodynamics, 555–56
Air shock absorbers (*see* Shock absorbers)
Air suspension, 18–19
  aftermarket, 239–40
  electronic, 227–28, 237–39
Alignment (*see* Wheel alignment)
Antidive, 185–86, 261–63, 553
Antiroll, 263
Antiroll bar (*see* Stabilizer)
Antisquat, 263, 553–54
Aspect ratio, 36, 59

B

Back spacing, 57–58
Balance, static and kinetic, 95
Balancing, tire, 97–111
  bubble, 100–2
  computer, 107–11
  mechanical, 102–4
  spin, 98–100
  strobe light, 104–7
Ball joint, 15–16, 187–88
  checks, 342–54
  clearance specifications, 345
  free play, checking, 351
  I-Beam/solid axle, 352–54
  inclination (*see* Steering axis inclination)
  load carrying, 347–52
  lubricating, 342
  preload, checking, 352
  replacement, 377–78, 380
    pressed-in, 381–83
    riveted/bolted-in, 383–84
    threaded, 384–85
  stud, replacing, 377
  taper, breaking, 373–77
  wear indicator, 345–47
Bandag Corporation, 48

Bead wires, 29
Beam axle (*see* Solid axle)
Bearing-retained axle, 123
  replacement of bearing, 145–48
Bearing:
  adjusting, 141–44
  axle, 116, 145–48
  ball, 117
  cleaning, 137–40
  clearance, 119, 347
  diagnosis, 130–32
  disassembling, 135–37
  drive axle, independent suspension, 124–25
  drive axle, nonserviceable, 125–26
  drive axle, solid axle, 121–24
  end play, 119, 120, 125, 130, 141
  frictionless, 116–17
  full-floating axle, 122
  inspecting, 137–40
  maintenance and lubrication, 130
  needle, 117
  nondrive axle, nonserviceable, 121
  nondrive axle, serviceable, 120–21
  packing, 141–44
  parts, 116–18
  preload, 119, 120, 125, 313, 352, 420, 424, 426–27
  repacking, 134–44
  repairing:
    drive axle, solid axle, 145–52
    nonserviceable, 144–45
    nonserviceable FWD front, 154–55
    serviceable FWD front, 152–54
  seals, 119–120
  semi-floating axle, 122
  tapered roller, 117
  types, 116
  worm shaft, 420, 424, 426–27
Bell crank steering linkage, 22
Belt, power steering (*see* Power steering)
Belts, tire, 28, 32–33
Body alignment, 530–31
Bolts (*see* Fasteners; Lug bolts)
Boot, CV joint (*see* Constant velocity universal joint)
Bounce, 222
Bounce test, 265, 271

Brakes:
    hydro-boost, 327
    and wheel bearing service, 134
Bubble level gauge, 479
Bump steer, 329–30, 334, 514, 564
Bump stop, 228, 248, 271
Burping, CV boot, 176
Bushing, 13–15, 230–31
    compliance, 558–59
    control arm:
        checking, 355–56
        installing, 392–94
        replacing, 385–92
    idler arm, checking, 361
    strut rod:
        checking, 356–57
        removing and replacing, 394–95

C

"C" lock axle, 123, 145
    bearing replacement, 148–52
Cadillac, 498
Camaro, 496
Camber, 5, 8, 10, 77–78, 184, 454–59
    adjusting, 489–501, 563
    change, 459
    and change in caster, 464–65
    and included angle, 467–69
    measuring, 483–88
    positive/negative, 455–56
    rear wheel, 471–72, 527
    and roll steer, 544–45
    and scrub radius, 456–57
    specifications, 488–89
    and tire wear, 457–58
    spread, and road crown, 458–59
Cap base, 31
Caster, 88, 184, 454, 462–66
    adjusting, 505–10, 564
    effects, 464–65, 466
    and load, 502
    measuring, 501–5
    and road crown, 466
    spread, 466, 509, 551
    things affecting, 465
Center of gravity, determining, 542, 551–53
Center link, 332–34
    removing and replacing, 418
Chapman strut, 10, 209
Chassis, components of, 4
Chassis height (see Ride height)
Chevrolet, 209
Chrysler Corporation, 292, 322
Clear vision adjustment, 524
Coefficient of friction, 41
Coil over-shock, 240
Coil spring suspension, 206–7 (see also Springs)
Constant velocity universal joint, 124, 152, 154
    assembling, inner type, 174
    assembling, outer fixed-type, 172–73
    boot, 162, 165, 170, 174–76
    checking, 164–65
    disassembling, inner type, 173–74

    disassembling, outer fixed-type, 171–72
    operation, 160–64
    servicing, 169
Control arm, 8, 9, 11, 12–13
    bushings (see Bushings, control arm)
    in rear suspensions, 207
    in short-long arm suspensions, 184–87
    removing, 378–80
Control valve, steering gear, 320–23
Cornering force/power, 536–38 (see also Traction force)
Corvette, 209, 227, 473, 528–29
Cradle, 6
Cupping, 78
Curb height (see Ride height)
Custom capping, 49
CV joint (see Constant velocity universal joint)

D

Damper:
    shock absorber as, 248
    steering, 314
    strut, 190–91, 357–58
de Doin axle, 210–11
Department of Transportation, U.S., 34–35
Dessicant, 238
Directional control angles, 454, 478
Dogtracking, 204, 472, 528–29
DOT number, 34–35
Double wishbone, 213
Drag link, 22–23, 330, 334–35
    socket, removing and replacing, 419–20
Driver, bushing, 388–89
Drop center rim, 54–55
Dry park test, 358–59
Dual compounding, 31
Dynamic Tracking Suspension System, 213–17
Dynamometer, shock absorber, 251

E

Eccentric cam
    adjusting camber with, 494–95
    adjusting caster with, 506–7
Elliott king pin, 16
End play:
    bearing, 119, 120, 125, 130, 141
    sector gear/shaft, 424–26

F

Fasteners, 132–34
Firebird, 496
Fixed joint, 161
Flanges, 54
Flat ride tuning, 226
Flexible coupler, 310
Floor stand, 479–80
Fluid, power steering (see Steering, power)
Ford Motor Company, 213, 291, 498
Frame, 6
    alignment, 530–31
    rigidity, 564

Frame gauge, 530–31
Friction plunger, 430
Front wheel drive suspension:
  axles, 197–99
  drive shafts, 159–81
    construction, 160–64
    disassembly, 168
    installing, 176–80
    removal, 165–68
    servicing CV joints, 169–76
  rear, 211–13
Full-floating axle, 122

G

Gear lash, 313, 420, 424–25, 427
General Motors, 227, 291, 332, 380, 387, 414, 496

H

Half shaft, 124, 160
Haltenberger, 22
Height (see Ride height)
Height sensor, 237–39
Heim joint, 559
Hoist rack, 482–83
Hoses, power steering (see Power steering)
Hotchkiss drive shaft, 205
Hydraulic actuator, 314
Hydraulics, principles of, 315–17
Hydro-boost brakes, 327
Hydroplaning, 29, 33–34

I

Idler arm, 332–34
  checking, 360–61
  removing and replacing, 415–18
Ignition switch, 308
Included angle, 467–69, 490–91
Independent suspension, 6–7
  rear, 208–10
Instant center, 546, 553
Interleaf friction, 229

J

Jacking, 210
Jounce, 222

K

Kickback, 314
King pin, 16
  and camber adjustment, 499
  clearance, 354
  inclination/angle (see Steering axis inclination)
  offset, 58
  removing and replacing, 395–98
Knock-off wheels, 64

L

Lateral acceleration (see Cornering force)
Lead (see Pull)
Leading strut rod, 185
Leaf spring suspension, 206 (see also Springs)
Light beam, 513
Light gauge, 515–16
Limited-slip differential, 46
  and balancing, 99, 100
Load ratings, of tires, 37–40
Loads, wheel, types of, 54
Lock washer (see Fasteners)
Low pivot axle, 210
Lug bolts, 55, 62–63
  replacing, 71–73
  runout, measuring, 92–93
Lug nut, 62–63
  torque, 71

M

MacPherson strut, 10, 184, 189 (see also Strut
  suspension)
Magnetic gauge, 479–80
  measuring camber with, 484–86
  measuring caster with, 502–4
  measuring SAI with, 510–11
Memory steer, 292, 406

N

National Highway Safety Administration (NHSA), 40
Neutral steering, 538
  tuning to, 550
No-flat tire, 46
Nonindependent suspension, 6–7
Nut breaker, 275
Nuts (see Fasteners)

O

Offset, 56–58
Original Equipment Manufacturer (OEM), 44
Over center adjustment, 313, 425
Oversteer, 45, 184, 538–39, 541, 549
  and wedge, 550

P

Panhard rod, 207, 211, 546
Percentage of rise (see Antisquat)
Pickle fork, 373
Pinch-bolt, 377
Pitch, 226–27
Pit rack, 482–83
Pitman arm, 20
  removing and replacing, 418–19
Pitman shaft, 20, 310
Plies, tire, 29–32

Plowing (*see* Understeer)
Plus 1/2/3 conversion, 58–61
P-metric system, 36–37
  and load ratings, 38–39
Polar moment of inertia, 555
Power hop, 554
Power squat (*see* Antisquat)
Power steering, 314–28
  belt, 364–67, 435
  and caster, 465
  checking, 364–67, 408
  diagnosis, 434–37
  fluid change, 437–40
  gear service, 446–49
  hoses, 326–29, 364–67, 436
  pressure testing, 440–41
  pump:
    operation, 317–20
    service, 441–46
  servicing, 404, 434–39, 446–49
Precured recapping, 47–49
Preload, bearing, 119, 120, 125, 313, 352, 420, 424,
    426–27
Pressure pad, 430
Pull, tire, 88–89, 458, 531
Puller, 373, 375–76, 388
Pump, power steering, 317–20 (*see also* Steering, power)
  checking, 364–67
  overhaul, 444–47
  service, 441–44
Pushing (*see* Understeer)

Q

Quick-change wheels, 64

R

R & R, 116
Rack and pinion steering, 20–21, 306, 314
  adjustments, 430–33, 524
  linkage check, 361–63
  power, 323–24
  replacement, 404, 422–24
  tie rods, 330–32
Rack support/yoke/bearing, 430
Racks, wheel alignment, 482–83
Radius rod, 8
Rear spacing (*see* Back spacing)
Rear suspension, 203–17
  service, 398–99
  three-/four-link, 554
Rebound, 222
Recapping (*see* Retreading)
Retreading, 47–49
Reversed Elliott king pin, 16
Ride height:
  and alignment angles, 478
  checking, 271–74
  lowering, 559–61
  and steering linkage, 330
  and toe measurement, 462, 514
Rim mount adapter, 485

Rims, wheel, 54–55
Rising rate, 228
Road crown:
  and camber spread, 458–59
  and caster, 466
Road testing, after alignment, 531
Roll axis, 545
Roll cage, 564, 566
Roll center, 545–48
Roll couple distribution, 549–559
Roll resistance, 548–49
Roll steer, 544–45
Rubber Manufacturers' Association (RMA), 40, 55,
    68, 79
Run-flat tire, 47
Runout:
  checking for, 89
  defined, 88
  lateral, 89–90, 93–94
  lug bolt, measuring, 92–93
  radial, 90–91
  tolerances, 91
  unloaded, 91
  and wheel alignment, 481
  wheel, measuring, 91
Rzeppa CV joint, 160–61, 171, 173

S

Safety bead, 54
Sag, 18, 227, 232–34
  torsion bar, 235
Schrader valve, 43
Scrub, defined, 10
Scrub radius, 58
  and camber, 456–57
  and toe, 460
Scuff tester, 531
Seals, bearing, 119–20
SEMA Foundation, 62
Semi-floating axle, 122
Semi-trailing arm suspension, 12, 208–9
Separator tool, 373, 375
Set back, 471, 513
Shims:
  adjusting camber with, 491–94
  adjusting caster with, 506
  tapered, 499–500, 508–9
  adjusting toe out on turns with, 512
Shimmy, 466, 531
Shock absorbers, 17
  adjustable, 562
  aftermarket, 263, 562
  air, 239–40, 263
  coil-over, 240, 263
  computer-controlled, 260
  damping force, 251–52
  damping ratios, 250–51
  de Carbon (*see* single-tube)
  double-tube, 252–58
  dynamometer, 251
  failures, 265
  friction, 249
  gas-charged, 257

gas pressure (*see* single-tube)
inspection, 270–74
lever, 250
load-carrying, 263
monotube (*see* single-tube)
operation, 248–50
quality, 264–65
removing and replacing, 275–79
replacement, 264–65
selecting, 562
single tube 258–59
strut, 259–60
Short-long arm suspension, 9–10, 184–89
  rear, 213
  springs and ball joints, 187–88
  wear factors, 188–89
Side shake, 195
Skid pad tests, 537–38
Sliding adjustment:
  adjusting camber with, 495–96
  adjusting caster with, 507–8
Sliding pillar suspension, 197
Sliding valve, 322–23
Slip angle, 538
Slip plate, 479, 527
Solid axle, 7
  front, 193–95
  rear, 204–7, 211
Spare tire (*see* Tires, spare)
Specialty Equipment Manufacturers Association (SEMA), 62
Speed ratings, tire, 41–42
Speed reset, 263
Speedometer error, 59
Spindle:
  checking, 140
  dropped, 560
Split rim, 55
Spoke wheels, and tire tubes, 44
Spreader, 373
Springs, 16–18, 222–45
  air suspension, 18–19, 237–40
  coil, 18, 232–34
    removing and replacing, 279–85
  inspection, 270–74
  leaf, 17, 229–32
    service, 285–86
  materials, 227–29
  operation, 222–23
  overload, 240
  spring weight and frequency, 223–25, 561–63
  strut, removal and replacement, 294–97
  torsion bars, 18, 234–36
    service, 286–87
  and wedge, 549–50
  wheel rate and frequency, 225–27, 561–63
Sprung weight, 222–23, 548
Stabilizer bar, 240–43, 548–50, 559
  service, 298–301
Staking, 179
Standing wave, 42
Steering arm, 19
Steering axis, 19, 58
Steering axis inclination (SAI), 184, 454, 466–69, 478
  adjusting, 478, 511

and caster, 463–65
  measuring, 510–11
Steering column, 307–11
  collapsible, 309
Steering damper, 314
Steering gear, 19–21
  checking, 363–64
  power, 446–49
  rack and pinion, 314
    adjustments, 430–31
    overhaul, 431–33
    removal and replacement, 422–24
  service, 420–33, 446–49
  standard, 310–313
    adjustments, 424–27
    overhaul, 427–30
    removal and replacement, 421–22
Steering linkage, 21–23, 329–34
  checking, 358–63
  parallelism adjustment, 417
  parallelogram, 22, 332–34, 358, 360–61
  replacement, 404–20
Steering, power (*see* Power steering)
Steering, rack and pinion (*see* Rack and pinion steering)
Steering ratio, 306
Steering, rear-wheel, 213–17
Steering shaft, 20, 310
  preload, 424–25
Steering system:
  components of, 5, 19
  inspection points, 340–43
  service, 404–49
Steering wheel:
  centering, 517–18, 524–25
  removing and replacing, 433–34
Stowaway spare, 46
Streamlining (*see* Aerodynamics)
Stress raiser, 233, 235
Strike out bumper, 228
Strut suspension, 10–13, 184, 189–93
  advantages, 193
  ball joint stud, replacing, 377
  bent strut, diagnosing, 490–91
  cartridge, installation of, 297–98
  checking, 356–58
  modified, 191–93
  rear, 209–10, 213
  service, 287–98
  spring, removal and replacement, 294–97
  strut mount, upper, 190–91
  strut rod, 13, 356, 394–95
  wear factors, 193
Support yoke, 430
Suspension system (*see also* Air suspension; Rear suspension; Semi-trailing arm suspension; Short-long arm suspension; Strut suspension; Trailing arm suspension)
  components of, 4
  dynamics of, 536–66
  frequency, 226, 561–63
  inspection points, 340–43
  modifying (*see* tuning)
  racing, 554, 556–66
  service, 372–99
  stages of operation, 251–52

Suspension system (*cont.*)
  tuning, 556–66
  wheel rate, calculating, 561–62
Swing axle suspension, 7–9, 195–97, 210

T

Taper, breaking, 373–77
  using pressure, 375–77
  using separator tool, 375
  using shock, 374–75
  and steering linkage replacement, 404–5
Temperature grades, tire, 41
Temporary spare, 46
Thread pitch, 132
Thrust/thrust line, 204, 455, 471, 472–73, 527–28
Tie rods, 21–22, 23, 330–32
  checking, 359–60, 362–63
  inner, removing and replacing, 407–15
  outer, removing and replacing, 405–7
Tires:
  balancing, 94–110
  belted, 28, 77
  conicity, 88
  construction, 29–34
  driving problems related to, 87–94
  grade labeling, 40–41
  inflating, 82–83
  installing, 68–74
  load ratings, 37–40
  matching, 91–92
  no flat/run flat, 46–47
  and power steering diagnosis, 434
  pressure, 76–77
  pull, 88–89, 458
  radius, 458
  rear, 211
  repairing, 85–87
  replacement, 44–45, 556
  replacing, 80–84
  retread, 47–49
  rotation, 79–80, 211
  scribe, 513
  sidewall information, 34–35
  sizing, 35–37
  slip angle, 538
  spare, 46, 80
  speed ratings, 41–42
  and spring rate, 240
  stagger, 551
  traction force, 539–42
  tread, 33–34, 75–76
  truing, 94
  tube and tubeless, 43–44
  valve, replacing, 84
  wear:
    and camber, 457–58
    inspection, 74–79, 340
    and rear wheel alignment, 471–72
    and toe, 461, 471, 513
    and torque steer, 471
  and wheel alignment, 478
  and wheel bearing diagnosis, 130
Toe, 5, 78, 184, 454, 459–62

  adjusting, 519–25, 563–64
  angles, converting, 459
  change, 461–62, 564–65
  measuring, 513–17
  rear wheel, 471–72, 527
  and roll steer, 544–45
  specifications, 517–19, 520
  and steering wheel position, 517–18
  things affecting, 460–61
  and tire wear, 461, 471, 513
Toe out on turns, 459–71
  adjusting, 512
  measuring, 511–12
Torque arm, 205
Torque steer, 88–89, 471
Torque tube drive shaft, 205
Torsion bars, 18, 234–36
  service, 286–87
Track, 4, 8, 9, 455, 471, 472, 527–28
Track bar, 207
Track gauge, 527–28
Traction, defined, 28
Traction force, 539–42
Traction grades, tire, 41
Trailing arm suspension, 12, 197
  rear, 209, 211–13
Trailing strut rod, 185
Trammel bar, 514–15
Tramp, 195, 466
Transaxle, 160
Transverse engine, 12, 160
Trim height (*see* Ride height)
Tripod joint, 161, 173–76 (*see also* Constant velocity
    universal joint)
Tripot (*see* Tripod joint)
Troubleshooting, steering and suspension, 531
Tubes, tire, 43–44
Tulip joint (*see* Tripod joint)
Turning angle/radius, 454, 469
Turning torque, measuring, 352–54
Turntable, 479–80
Twin I-beam axle (*see* Swing axle; Suspension system)
Two-plane balancer, 95–96

U

Understeer, 538–39, 541, 549
  and wedge, 550
Under tread, 31
Unequal arm suspension (*see* Short-long arm
    suspension)
Unibody construction, 6
Uniform Tire Quality Grade Labeling, 40
Universal joints, 160
Unsprung weight, 222–23, 548
  reducing, 564

V

Valve stem, 43–44
Vibration, tire-caused, 89
Vulcanizing, 33

W

Waddle, tire, 89
Wander, 466, 517, 531
Watt's link, 207, 546
Wedge, 549–50
Weight transfer, 542–44
Wheel alignment
  angles, 184, 454–55
    changing, 563–64
    measuring, 478–83
  camber, 455–59
    adjusting, 489–501, 563
    measuring, 483–87
    specifications, 488–89
  caster, 462–66
    measuring, 501–5
    adjusting, 505–10, 564
  four-wheel, 472
  and frame and body alignment, 530–31
  machines (*see* Systems)
  racks, 481–83
  rear, 204, 205, 471–73, 525–29
  road testing, 531
  set back, 471, 513
  sequence, 483
  steering axis inclination, 466–69
    adjusting, 511
    measuring, 510–11
  systems, 481–82
    measuring camber with, 486–88
    measuring caster with, 504–5
    measuring toe with, 513, 516–17
  toe, 459–62
    adjusting, 519–25, 563–65
    measuring, 513–17
    specifications, 517–19
    and steering wheel position, 517
    toe out on turns, 469–71
      adjusting, 512
      measuring, 511–12
    track/thrust line, 472–73
    troubleshooting, 531
Wheel base:
  defined, 4
  and caster, 465
Wheel bearings (*see* Bearings, wheel)
Wheel hop, 206
Wheel offset, 56–58
Wheel rate:
  defined, 225–27
  measuring, 561–62
Wheels:
  aftermarket, 58–62, 564
  attachment, 62–64
  backspacing, 57–58
  balancing, 94–111
  cast, 63
  composite, 55
  construction, 54–55
  knock-off, 64
  loads, types of, 54
  removing and replacing, 68–74
  sizes, 56
  spoke, 44
  weights, 96–97
Wishbone, 185
  double, 213
Windup, 195, 204, 206
Worm and roller, 310
Worm shaft, 310, 420, 424, 426–27

Y

Yaw, 195, 206
Yaw axis, 555